Lecture Notes in Mathematics

Edited by A. Dold and B. Eckmann

1130

Methods in Mathematical Logic

Proceedings of the 6th Latin American Symposium on Mathematical Logic held in Caracas, Venezuela August 1–6, 1983

Edited by C. A. Di Prisco

Springer-Verlag
Berlin Heidelberg New York Tokyo

Editor

Carlos Augusto Di Prisco
Instituto Venezolano de Investigaciones Científicas
Departamento de Matemáticas
Apartado 1827, Caracas 1010-A, Venezuela

ISBN 3-540-15236-9 Springer-Verlag Berlin Heidelberg New York Tokyo
ISBN 0-387-15236-9 Springer-Verlag New York Heidelberg Berlin Tokyo

This work is subject to copyright. All rights are reserved, whether the whole or part of the material is concerned, specifically those of translation, reprinting, re-use of illustrations, broadcasting, reproduction by photocopying machine or similar means, and storage in data banks. Under § 54 of the German Copyright Law where copies are made for other than private use, a fee is payable to "Verwertungsgesellschaft Wort", Munich.

© by Springer-Verlag Berlin Heidelberg 1985
Printed in Germany

Printing and binding: Beltz Offsetdruck, Hemsbach/Bergstr.
2146/3140-543210

PREFACE

The VI Latin American Symposium on Mathematical Logic was held in Caracas, Venezuela from the 1st to the 6th of August 1983. The meeting was sponsored by Asociación Venezolana para el Avance de la Ciencia (AsoVAC), Consejo Nacional de Investigaciones Científicas y Tecnológicas (CONICIT), Fundación Polar, IBM de Venezuela, Instituto Venezolano de Investigaciones Científicas (IVIC), Universidad Central de Venezuela, The National Science Foundation of the United States of America, The British Council, The French Government, Division of Logic, Methodology and Philosophy of Science of the International Union for History and Philosophy of Science, Organization of American States.

The Program Committee was formed by Xavier Caicedo (Universidad de los Andes, Bogotá, Colombia), Rolando Chuaqui (Pontificia Universidad Católica de Chile, Santiago, Chile), Newton C. A. da Costa (Universidade de Sao Paulo, Brasil) and Carlos Augusto Di Prisco (Instituto Venezolano de Investigaciones Científicas y Universidad Central de Venezuela).

IVIC's Centro de Estudios Avanzados sponsored a seminar consisting in five short courses of ten hours each. The courses were: Non-Classical Logics by Newton C. A. da Costa, Some Aspects of the Theory of Large Cardinals by Wiktor Marek (Warsaw University), Mathematical Practice and Subsystems of Second Order Arithmetic by Stephen Simpson (Pennsylvania State University), The ω-rule by E. G. K. Lopez-Escobar (University of Maryland), and Logic, Real Algebra and Real Geometry by Máximo Dickmann (Université de Paris VII-CNRS). The courses were attended by participants fron Argentina, Chile, Mexico, Italy and Venezuela.

The Ateneo de Caracas and the Consejo Nacional para el Desarrollo de la Energía Nuclear hosted a panel discusssion on the Philosophy of Mathematics with the participation of Newton C. A. da Costa, Rolando Chuaqui, George Wilmers (Manchester University) and Vincenzo P. Lo Monaco, Pedro Luberes and Juan Nuño of the Institue of Philosophy of the Universidad Central de Venezuela.

The local Organizing Committee included Arturo Rodriguez Lemoine, (Universidad Central de Venezuela), Pedro Lluberes who organized and chaired the panel discussion on Philosophy of Mathematics and Jorge Baralt (Universidad Simón Bolívar). Professor Wiktor Marek provided invaluable help organizing the symposium.

The final version of these proceedings was prepared by Mrs. Magally Arvelo-Osorio. My sincere appreciation for her skilled work. Carlos Uzcátegui proofread large portions of the book.

This volume was edited with the collaboration of Xavier Caicedo, Rolando Chuaqui and Newton C. A. da Costa.

Carlos Augusto Di Prisco

LIST OF PARTICIPANTS

José A. Amor
Ayda Arruda
Jorge Baralt
Susana Berestovoy
Lenore Blum
Everett Bull
Xavier Caicedo
Ana Cavalli
Rolando Chuaqui
Roberto Cignoli
Peter Clote
Manuel Corrada
Luis Jaime Corredor
Newton C.A. Da Costa
Máximo Dickmann
Carlos A. Di Prisco
Itala M. D'Ottaviano
Sergio Fajardo
James Henle
Jorge Herrera
Jaime Ihoda
Thomas Jech
E.G.K. López Escobar
Alain Louveau
María Jimena Llopis
Pedro Lluberes
Menachem Magidor
Jerome I. Malitz
Wiktor Marek
María V. Marshall
Adrian R.D. Mathias
Gisela Méndez
Irene Mikenberg
Julia Muller
Kenneth McAloon
Anil Nerode
Juan Nuño
Jeffrey Paris
Ramón Pino
Rubén Preiss
Alexander Prestel
Cecylia Rauszer
William Reinhardt
Arturo Rodríguez Lemoine
Rafael Rojas Barbachano
Gerald Sacks
María G. Schwartze
Antonio M. Sette
Stephen Simpson
Roger Soler
Jacques Stern
Elías Tahhan
Carlos Uzcátegui
Carlos Vasco
George Wilmers

Professor Michael Makkai was unable to attend the meeting. His paper is, nevertheless, included in the proceedings.

TABLE OF CONTENTS

FAILURE OF INTERPOLATION FOR QUANTIFIERS OF MONADIC TYPE

Xavier Caicedo
Universidad de los Andes - Departamento de Matemáticas
Apartado Aéreo 4976
Botogtá D.E., Colombia

ABSTRACT

It is shown that no proper extension of first order logic by Lindström-Mostowski quantifiers of monadic type, that is quantifiers of the form $Qx_1 \ldots x_n(\phi_1(x_1),\ldots,\phi_n(x_n))$, satisfies the many sorted Craig's interpolation lemma or even the one sorted, if closed under relativizations. For example $L_{\omega_1\omega}$ or any of its admissible fragments can not be generated by any number of these quantifiers. This generalizes previous results of the same type shown under stronger hypothesis. In contrast, all monadic logics generated by cardinal quantifiers satisfy interpolation.

§0. INTRODUCTION. In the context of abstract model theory few "natural" logics seem to satisfy Craig's interpolation lemma. No proper compact extension of first order logic satisfying this property is known; and besides $L_{\omega_1\omega}$ and its admissible fragments there are not too many uncompact examples.

On the other hand, there are several general non-interpolation results, starting with Lindström [9], who shows that if a logic extends $L_{\omega\omega}$ in infinite models, is generated by finitely many quantifiers, and satisfies the downward Löwenheim-Skolem Theorem, then it does not satisfy even Beth's definability theorem.

In [7], Friedman shows that no logic between $L_{\omega\omega}(\mathrm{Ch})$ or $L_{\omega\omega}(Q_\alpha)$ with $\alpha \geq 1$ and the large logic $L_{\infty\omega}(\mathrm{Ch},Q_\alpha|\alpha \in \mathrm{Ord})$ satisfies Beth's theorem, where Ch is Chang's quantifier and Q_α is the quantifier "there are at least ω_α...". In [3], we observe that this generalizes to sublogics of $L_{\omega\omega}(Q|Q \in \mathrm{Mon})$ where Mon denotes the class of all quantifiers of monadic type.

Makowsky and Shelah prove in [11] that no logic of the form $L_{\omega\omega}(Q^i\ i \in I)$ with $Q^i \in \mathrm{Mon}$ satisfies many sorted interpolation, provided it satisfies Robinson's joint consistency lemma, the Feferman-Vaught property for sums of structures, and also $|I| \leq \aleph_\omega$ where $\aleph_\omega$ is assumed to be a strong limit cardinal. See also Makowsky [10] and Mundici [13] for related results.

In this paper we show that for quantifiers of monadic type, the above hypothesis, as well as the assumption that the logic contains one of the quantifiers Ch or Q_α, $\alpha \geq 1$, are unnecessary. No logic of the form $L_{\omega\omega}(Q^i|i \in I)$, $Q^i \in \mathrm{Mon}$, satisfies many sorted interpolation, and it does not satisfy even single sorted interpolation in case it admits relativizations of sentences.

After considering some simple applications we show an analogous result for extensions of $L_{\omega_1\omega}$. Then, we observe that in contrast, the result does not hold for monadic logics, since all extensions of first order monadic logic by cardinal quantifiers satisfy interpolation.

§1. PRELIMINARIES. We assume that the reader is acquainted with the basic notions of abstract model theory as presented, for example, in [1], [8] or [11]. Universes of structures $\mathfrak{A}, \mathfrak{B}, \mathfrak{C}, \ldots$ will be denoted by A,B,C,..., respectively, $|A|$ denotes the cardinal number of A. For a formula $\phi(\vec{x})$ in any logic, $\phi^{\mathfrak{A}} = \{(a_1,\ldots,a_n) \in A^n | \mathfrak{A} \models \phi[a_1,\ldots,a_n]\}$. Elementary equivalence between two structures with respect to a logic L will be denoted "$\mathfrak{A} \equiv \mathfrak{B}$ mod L", elementary equivalence in $L_{\omega\omega}$ will be simply denoted "$\mathfrak{A} \equiv \mathfrak{B}$".

We will consider quantifiers in the sense of Mostowski [12] as generalized by Lindstrom [8]. A quantifier Q is a class of structures of some finite type $\langle n_1,\ldots,n_k \rangle$ closed under isomorphism, to which it is associated a syntactical rule which allows to form, from formulae $\phi_1(\vec{x}_1),\ldots\phi_n(\vec{x}_n)$, the new formula $\sigma = Q\vec{x}_1,\ldots\vec{x}_n(\phi_1(\vec{x}_1),\ldots,\phi_n(\vec{x}_n))$, with the meaning: $\mathfrak{A} \models \sigma \Leftrightarrow (A,\phi_1^{\mathfrak{A}},\ldots,\phi_n^{\mathfrak{A}}) \in Q$. Note that we use the same symbol for the quantifier itself and for its syntactical expression. Given a family of quantifiers Q^i, $i \in I$, the smallest logic closed under the first order logical operations and the quantifiers Q^i will be denoted $L_{\omega\omega}(Q^i|i \in I)$ and we will say that it is generated by the quantifiers Q^i. For cardinals $\kappa \geq \omega$, $L_{\kappa\omega}(Q^i|i \in I)$ is constructed allowing conjunctions of size less than κ; the subscript $\kappa = \infty$ will denote closure under conjuctions of arbitrary sets of sentences. Finally, $L^\gamma_{\infty\omega}(Q^i|i \in I)$ represents the sublogic obtained by restricting to sentences of quantifier rank less than the ordinal γ (cf.[3]).

Note that the infinitary character of a logic is a relative matter, and definitively it is not a model theoretical property, since under very weak conditions <u>any logic has the form</u> $L_{\omega\omega}(Q^i | i \in I)$ for adequate families (could be proper classes) of quantifiers. This includes the infinitary logics construed above (for γ limit). This is also the case of any logic satisfying the interpolation lemma.

This paper studies the case of logics generated by quantifiers of monadic type <1> or <1,...,1>. Well known examples of these are the cardinal quantifiers $Q_\alpha = \{(A,B) \mid |B| \geq \omega_\alpha\}$, as well as Chang's $Ch = \{(A,B) \mid |B| = |A|\}$ and Hartig's $H = \{(A,B,C) \mid |B| = |C|\}$. One could consider many other possibilities such as

$$\mathfrak{A} \models Qx_1, \dots x_n(\phi_1(x_1), \dots, \phi_n(x_n)) \Leftrightarrow p(|\phi_1^{\mathfrak{A}}|, \dots, |\phi_n^{\mathfrak{A}}|) = 0$$

where $P(x_1, \dots, x_n)$ is a given diofantine polynomial, or

$$Qxyz(\phi(x), \psi(y), \sigma(z)) \Leftrightarrow |\phi^{\mathfrak{A}}| \to (|\psi^{\mathfrak{A}}|)^2_{|\sigma^{\mathfrak{A}}|}$$

in the sense of Erdös-Rado partition calculus. Obviously these do not include the quantifiers of Henkin, Magidor and Malitz, and all sort of order quantifiers (cf.[11]).

Without misgivings about classes of classes, let Mon be the class of all quantifiers of monadic type (cf. [8]). We will abbreviate $L_{\infty\omega}(\mathrm{Mon})$ for $L_{\infty\omega}(Q | Q \in \mathrm{Mon})$.

Given formulae $\phi_1(x), \dots, \phi_n(x)$, which may allow extra free variables, we introduce for each function $\delta: \{0,1,\dots,n-1\} \to \{0,1\}$ the formula

$$\phi_\delta := \bigwedge_{i<n} \phi_{i+1}(x)^{\delta(i)}$$

where $\phi_i^{\,0}$ is ϕ_i and $\phi_i^{\,1}$ is $\neg\phi_i$. Given a monadic language $\{P_1, \dots, P_n\}$, a corresponding monadic structure $\mathfrak{C} = (C, P_1^{\mathfrak{C}}, \dots, P_n^{\mathfrak{C}})$ is completely determined up to isomorphism, by the cardinals

$$K_\delta = |P_\delta(x)^{\mathfrak{C}}| \qquad \delta \in 2^n.$$

Hence, the associated quantifier $Q^{\mathfrak{C}} = \{\mathfrak{C}' | \mathfrak{C}' \cong \mathfrak{C}\}$ is expressible, for $|C| \leq \omega_\alpha$, in terms of the cardinal quantifiers Q_β, $\beta \leq \alpha+1$, and $\exists^{=m}$ (there are exactly m), $m < \omega$:

$$Q^{\mathfrak{C}} x_1, \dots, x_n(P_1, \dots P_n) \Leftrightarrow$$

$$\bigwedge_{K_\delta = \omega_\beta} (Q_\beta x P_\delta(x) \wedge \neg\, Q_{\beta+1} x P_\delta(x)) \wedge \bigwedge_{K_\delta = m<\omega} (\exists^{=m} x\, P_\delta(x)).$$

We have taken the quantifiers $\exists^{=m}$ as primitives in order that the two equivalent formulae above have the same quantifier rank. Therefore, by a simple induction, for every ordinal (even finite)

$$L^{\gamma}_{\infty\omega}\ (Q^{\mathfrak{C}} \mid |C| \leq \omega_{\alpha}) \prec L^{\gamma}_{\infty\omega}(Q_{\beta},\ \exists^{=m} \mid \beta \leq \alpha+1,\ n<\omega) \qquad (1)$$

where $\mathfrak{C}$ runs through all monadic structures of size at most ω_{α}. Then we have the following lemma, shown in [3] for $\alpha = 1$ and $\gamma = \infty$.

LEMMA 1.1. Let $\mathfrak{A}$ and $\mathfrak{B}$ be structures of power at most ω_{α}, then for each ordinal γ, $\mathfrak{A} \equiv \mathfrak{B}$ mod $L^{\gamma}_{\infty\omega}$ $(Q_{\beta},\ \exists^{=m} \mid \beta \leq \alpha,\ m<\omega)$ implies $\mathfrak{A} \equiv \mathfrak{B}$ mod $L^{\gamma}_{\infty\omega}$ (Mon).

PROOF. We can dispense the quantifier $Q_{\alpha+1}$ since it acts trivially in $\mathfrak{A}$ and $\mathfrak{B}$. Hence, if we have the first elementary equivalence, by (1) it is enough to show that $\mathfrak{A} \equiv \mathfrak{B}$ mod $L^{\gamma}_{\infty\omega}$ $(Q^{\mathfrak{C}} \mid |C| \leq \omega_{\alpha})$ implies $\mathfrak{A} \equiv \mathfrak{B}$ mod $L^{\gamma}_{\infty\omega}$(Mon). This is done by a back-and-forth argument as in [3], using the characterization of elementary equivalence for arbitrary quantifiers given by the author there (cf. theorems 3.1, 3.3 and lemmas 4.1, 4.2). □

Note that if $\mathrm{cof}(\gamma) \geq \omega$, we may ignore the quantifiers $\exists^{=m}$ in the hypothesis of the lemma, because they are definable (from $\exists$) without raising the quantifier rank up to γ.

Now, we exploit a well known counterexample for interpolation in $L_{\omega\omega}(Q_1)$. Let $K \geq M$ be arbitrary cardinals, then define $\mathcal{D}(K,M)$ to be a structure (D,E) such that $|D| = K$ and E is an equivalence relation in D having M equivalence classes, each one of power K. The following lemma appears in [2] for $K = M = \omega_1$, $M' = \omega$, and quantifiers of type <1>. As Makowsky observes in [7], its generalization to all cardinals and all quantifiers of monadic type is straightforward:

LEMMA 1.2. If $K \geq M \geq M' \geq \omega$ then $\mathcal{D}(K,M) \equiv \mathcal{D}(K,M')$ mod $L_{\infty\omega}$(Mon)

Obviously, this lemma does not hold for M or M' finite. However, we have the following finite version.

LEMMA 1.3. If $K \geq M \geq M' \geq n$ with $n \in \omega$ then $\mathcal{D}(K,M) \equiv \mathcal{D}(K,M')$ mod $L^{n}_{\infty\omega}$(Mon).

PROOF. By lemma 1.1, it suffices to show elementary equivalence in the logic $L^{n}_{\infty\omega}(Q_{\beta},\ \exists^{=m} \mid \beta \in \mathrm{Ord},\ m \in \omega)$.

For this purpose we introduce the following back-and-forth relation between the two structures (cf. definitions 2.1, 2.2 and theorem 3.3 in [3]) with parameters in $(n,<)$. Let $\mathcal{D}(K,M) = (A,E)$, $\mathcal{D}(K,M') = (B,E')$; then for $(x_1,\ldots,x_r),\ (y_1,\ldots,y_r) \in A^r \cup B^r$ and $p = 0, 1, \ldots, n-1$: $(x_1,\ldots,x_r) \overset{p}{\frown} (y_1,\ldots y_r)$ if and only if the assignment $x_i \mapsto y_i$ is a partial isomorphism, and $r \leq n-p$.

Let us recall that given $\vec{a} = (a_1,\ldots,a_r) \in A^r$ and $a \in A$, then

$$[a]^p_{\vec{a}} = \{x \in A \mid \vec{a}x \overset{p}{\sim} \vec{a}a\},$$

here, we write $\vec{a}x$ for $(a_1,\ldots,a_r,x)$. Analogously, for $\vec{b} \in B^r$ and and $b \in B$:

$$[b]^p_{\vec{b}} = \{x \in B \mid \vec{b}x \overset{p}{\sim} \vec{b}b\}.$$

To conclude that we have elementary equivalence we must verify:

(i) $\emptyset \overset{p}{\sim} \emptyset$, where $\emptyset$ is the empty sequence.

(ii) Given $\vec{a} \overset{p+1}{\sim} \vec{b}$ with $\vec{a} \in A^r$ and $\vec{b} \in B^r$, there exists $f: A \to B$ such that
(A) $\vec{a}a \overset{p}{\sim} \vec{b}f(a)$ for any $a \in A$,
(B) for any $X \subseteq A$ and any quantifier Q generating the logic,

$$\bigcup_{a \in X} [a]^p_{\vec{a}} \in Q \quad \text{implies} \quad \bigcup_{a \in X} [f(a)]^p_{\vec{p}} \in Q.$$

(iii) As (ii), interchanging the role of A and B.

(iv) If $(x_1,\ldots,x_r) \overset{p}{\sim} (y_1,\ldots,y_r)$ then the assignment $x_i \mapsto y_i$ is a partial isomorphism between the structures to which $\vec{x}$ and $\vec{y}$ belong, respectively.

We verify only the extension property (ii), the rest is trivial. If $\vec{a} \overset{p+1}{\sim} \vec{b}$, $\vec{a} \in A^r$, $b \in B^r$ then $r \le n-(p+1) < n$; hence, there will be equivalence classes in A and B (with respect to the relations E and E', respectively) not occupied by any a_i or or b_i, and so we can define $f: A \to B$ such that

(i) f establishes a bijection between the equivalence class of a_i and that of b_i, with $f(a_i) = b_i$, for $i = 1,\ldots,r$.

(ii) f establishes a bijection between the union of the equivalence classes of A not occupied by any a_i and the union of the equivalence classes of B not occupied by any b_i.

It is easy to see that $\vec{a}\,a \overset{p}{\sim} \vec{b}\,f(a)$, because we still have a partial isomorphism and length of $(\vec{a}\,a) \le n-(p+1)+1=n-p$. Also, for any $a \in A$, $|[a]^p_{\vec{a}}| = |[f(a)]^p_{\vec{b}}|$ and disjoint sets of the form $[a]^p_{\vec{a}}$ are sent to disjoint sets $[f(a)]^p_{\vec{b}}$ by f.

Hence, for any $X \subseteq A$ and any quantifier $Q = Q_\alpha$ or $\exists^{=m}$ we have the

the extension property:

$$\bigcup_{a\in X}[a]^{P}_{\vec{a}}\in Q \quad \text{implies} \quad \bigcup_{a\in X}[f(a)]^{P}_{\vec{a}}\in Q.$$

Therefore, in the notation of [3]:

$$(\mathcal{D}(K,M),\ Q_\alpha,\ \exists^{=m})_{\alpha,m} \underset{\sim}{(n,<)} (\mathcal{D}(K,M'),\ Q_\alpha,\ \exists^{=m})_{\alpha,m}$$

which shows elementary equivalence. □

§2. FAILURE OF INTERPOLATION. For simplicity, we consider first the case of single sorted logics closed under relativizations. The many sorted case will follow readily. A logic is closed under relativizations if for every sentence of type τ and monadic predicate V, there is ϕ^V of type $\tau\cup\{V\}$ such that for all $\mathfrak{A}$

$$\mathfrak{A}\models\phi^V \iff \mathfrak{A}\restriction V^{\mathfrak{A}}\models\phi.$$

Most known logics have this property.

Recall that a class K of structures of type τ is PC (projective) in a logic L if there is a sentence θ in L of type $\tau'\supseteq\tau$ such that $K=\{\mathfrak{A}\restriction\tau \mid \mathfrak{A}\models\theta\}$. The logic has the interpolation property if any two disjoint PC classes may be separated by an elementary class.

In the next lemma we only need closure under negation and conjunctions.

2.1 LEMMA. Assume that $L \prec L_{\infty\omega}$ (Mon) and L is closed under relativizations. If L has the interpolation property, then for any pair of monadic structures (of finite type) $\mathfrak{A}\equiv\mathfrak{B}$ implies $\mathfrak{A}\equiv\mathfrak{B}$ mod L.

PROOF. Suppose that the conclusion does not hold then there are structures $\mathfrak{A}=(A,P_i^{\mathfrak{A}})_i$ and $\mathfrak{B}=(B,P_i^{\mathfrak{B}})_i$ such that

(1) $\mathfrak{A}\equiv\mathfrak{B}$ mod $L_{\omega\omega}$ (2) $\mathfrak{A}\not\equiv\mathfrak{B}$ mod L.

By the downward Löwenheim-Skolem Theorem in $L_{\omega\omega}$ we may assume without loss of generality that $|B|\le\omega$. Then by (1) we have the following possibilities for each $\delta\in 2^n$:

$$|P_\delta^{\mathfrak{A}}| = |P_\delta^{\mathfrak{B}}| = m<\omega$$

$$|P_\delta^{\mathfrak{A}}| \ge |P_\delta^{\mathfrak{B}}| = |B| = \omega.$$

By (2), $\mathfrak{A}\not\equiv\mathfrak{B}$, and so $|P_\delta^{\mathfrak{A}}| > |P_\delta^{\mathfrak{B}}|=\omega$ for some δ. Hence $|A|>|B|=\omega$. Choose $\phi\in L$ such that $\mathfrak{A}\models\phi$ and $\mathfrak{B}\not\models\phi$. For each monadic predicate P_i introduce a new predicate $\bar{P}_i$, let V be another new monadic predicate. With the help of additional (binary) predicates to express the cardinality relations indicated symbolically below, it is possible to say in the logic L the following sentences:

i) $|V| \geq \omega$

ii) $\forall x\ (\overline{P}_i(x) \to V(x))$

iii) $\exists^{=m} x\ P_\delta(x) \wedge \exists^{=m} x\ (\overline{P}_\delta(x) \wedge V(x))$ for $|P^{\mathfrak{B}}_\delta| = m < \omega$

iv) $(|\overline{P}_\delta \cap V| = |V|) \wedge (|\overline{P}_\delta \cap V| \leq |P|)$ for $|P^{\mathfrak{B}}_\delta| = \omega$.

v) $\phi \wedge \neg\, \phi^V$

Then $(\mathfrak{A}, \mathfrak{B}, \ldots)$, with V interpreted as B and $\overline{P}_i$ interpreted as $P^{\mathfrak{B}}_i$, satisfies the sentences. Moreover, any $(\mathfrak{A}', \mathfrak{B}', \ldots)$ satisfying the sentences must have $|A'| > |B'| \geq \omega$, because if $|A'| = |B'|$ we would have by (iii), (iv), $|\overline{P}_\delta \cap V| = |P_\delta|$ for all $\delta \in 2^n$, and so by (ii) $\mathfrak{A}' \approx \mathfrak{B}'$, contradicting (v).

Now, consider sentences with an additional predicate E and as many additional predicates as necessary to say :

vi) "E is an equivalence relation of the universe"

vii) "Each equivalence class is equipotent to the universe"

viii) "The number of equivalence classes is equipotent to V".

Then, if Θ is the conjunction of (i) to (viii), the class

$$K_1 = \{(A',E') \mid (\mathfrak{A}', \mathfrak{B}', \ldots, E' \ldots) \models \Theta \text{ for some } \mathfrak{A}', \mathfrak{B}', \ldots\}$$

is PC in L and contains $\mathcal{D}(|A|, |B|)$ with $\mathfrak{A}$ and $\mathfrak{B}$ as above. Moreover, it is disjoint of the class

$$K_2 = \{(A',E') \mid (A',E') \approx \mathcal{D}(K,K),\ K \in \text{Card}\}$$

which is trivially PC in L. Since we have $\mathcal{D}(|A|, |B|) \equiv \mathcal{D}(|A|, |A|)$ mod $L_{\infty\omega}(\text{Mon})$ by lemma 1.2, interpolation fails in L. □

<u>REMARK.</u> Clearly the conclusion of the lemma follows from the weaker hypothesis: Interpolation $(L, L_{\infty\omega}(\text{Mon}))$.

2.2 <u>THEOREM.</u> If $L = L_{\omega\omega}(Q^i \mid i \in I)$ with $Q^i \in \text{Mon}$ is closed under relativization and extends properly $L_{\omega\omega}$ then it does not satisfy interpolation.

<u>PROOF.</u> If the logic has the form $L = L_{\omega\omega}(Q^i \mid i \in I)$ with each Q^i of monadic type, then for any monadic structures $\mathfrak{A}, \mathfrak{B}$ such that $\mathfrak{A} \equiv \mathfrak{B}$, we have by the lemma: $\mathfrak{A} \in Q^i$ if and only if $\mathfrak{B} \in Q^i$. This means that Q^i does not distinguish infinite cardinals in the sense of Mostowski [12] and Lindström [8]. Applying Corollary 3.2 in Lindström's paper, we have that L must satisfy the downward Löwenheim-Skolem Theorem. By Lindström's corollary I in [9], there is a PC class K in L with the following properties:

i) $(A,B) \in K$ implies $|B| < \omega$.

ii) For all $n \in \omega$ there is $(A,B) \in K$ such that $|B| = n + 1$.

Let $\phi(P)$ be the sentence which characterizes K (with the help of additional predicates) when P is interpreted as B. Add new predicate symbols $<$, (binary), and V (monadic). Then the conjunction of the following sentences

"$<$ is a linear order without last element"

"E is an equivalence relation"

$\forall x(P(x) \to V(x)) \wedge \phi(P)^V \wedge \forall x \exists ! y(P(y) \wedge yEx)$

shows that the class of equivalence relations:

$K = \{(A,E) \mid \ |A| \geq \omega,\ E$ has finitely many equivalence classes$\}$

is PC in L, because $\mathfrak{A} \models \phi(P)^V$ if and only if $|P^{\mathfrak{A}}| < \omega$. Since this class is disjoint of the PC class

$K' = \{(A,E) \mid (A,E) \approx \mathcal{D}(K,K),\ K \subset \mathrm{Card}\}$,

and by lemma 1.3 we have

$\mathcal{D}(\omega,n) \equiv \mathcal{D}(\omega,\omega) \mod L^n_{\infty\omega}$ (Mon),

these two classes are inseparable by any finite quantifier rank formulae and so by L. □

Up to now we have considered single sorted types only. If we consider many sorted types then any sentence may be stated about each one of the sorts (or universes) of the type. Hence, wherever we used relativized sentences ϕ^V, we may use instead the sentence ϕ, construed as speaking about a second sort. Thus, we have:

2.3 THEOREM. Let $L = L_{\omega\omega}(Q^i \mid i \in I)$, $Q^i \in \mathrm{Mon}$, be a logic satisfying many (two) sorted interpolation then $L \equiv L_{\omega\omega}$.

This improves considerably theorem 6.15 in Makowsky and Shelah [11], shown under stronger additional hypothesis (see introduction) which imply full compactness.

§3. AN APPLICATION. Note that theorems 2.2 and 2.3 can not be extended to arbitrary sublogics of $L_{\infty\omega}$(Mon) since $L_{\omega_1\omega}$ is a such sublogic satisfying interpolation and having relativizations. The same can be said of any admissible fragment of $L_{\omega_1\omega}$, for example the smallest one, $\Delta L_{\omega\omega}(Q_o)$, where Δ denotes closure under Δ-interpolation (cf. [11]). However, these logics are generated by adequately chosen families of quantifiers. But $\Delta L_{\omega\omega}(Q_o)$ may not be generated by finitely many quantifiers because it satisfies the downward Löwenheim-Skolem

Theorem (Lindström [9]). For quantifiers of monadic type we have more generally:

3.1 <u>COROLLARY.</u> $L_{\omega_1\omega}$ and its admissible fragments are not generated by any number of quantifiers of monadic type.

As an interesting example, introduce for each set of natural numbers S the quantifier Q^S, where

$$\mathfrak{A} \models Q^S x\phi(x) \Leftrightarrow |\phi^{\mathfrak{A}}| \in S.$$

Obviously $L_{\omega\omega}(Q^S|S\subseteq\omega) \prec L_{\omega_1\omega}$ and by the corollary the extension must be proper. However, this logic is strong enough to have $\Delta L_{\omega\omega}(Q^S|S\subseteq\omega) = L_{\omega_1\omega}$. In the same way, if A is a countable admissible set, $L_{\omega\omega}(Q^S|S\in A)$ is a <u>proper</u> sublogic of L_A whose Δ-closure is L_A.

Now, we turn our attention to the extensions of $L_{\omega_1\omega}$, showing an analogue of theorems 2.2 and 2.3.

3.2 <u>THEOREM</u>. A proper extension of $L_{\omega_1\omega}$ of the form $L_{\omega_1\omega}(Q^i|i\in I)$ with $Q^i\in \mathrm{Mon}$ does not satisfy the many sorted interpolation lemma. If it is closed under relativization it does not satisfy single sorted interpolation either.

<u>PROOF.</u> If the logic (call it L) satisfies interpolation then lemma 2.1 applies and each class of monadic structures which is elementary in L is characterized by its countable models. Hence, for each $i\in I$,

$$Q^i x_1,\dots,x_n(P_1(x_1),\dots,P_n(x_n)) \equiv \bigvee\{\phi_{\mathfrak{A}}|\,\mathfrak{A}\in Q^i\ \ |A|\le\omega\} \qquad (1)$$

where $\phi_{\mathfrak{A}}$ is a "Scott sentence" for $\mathfrak{A}$ in $L_{\omega_1\omega}$. Since there are only countably many non-isomorphism structures of a given (finite) monadic type, then the disjunction in the right hand side of (1) may be taken in $L_{\omega_1\omega}$. By an obvious inductive argument, $L \prec L_{\omega_1\omega}$. □

Actually, by the remark after lemma 2.1, we have under the hypothesis of theorem 3.2 that interpolation $(L, L_{\infty\omega}(\mathrm{Mon}))$ fails also.

§4. <u>MONADIC LOGICS.</u> The non-interpolation result does not hold in monadic logics. In [4] it is shown, using compactness, that $M_{\omega\omega}(Q_1)$ satisfies interpolation (we use M to denote monadic logics in general). Since the logic $M_{\omega\omega}(Q_1,Q_2,\dots,Q_m)$ is also countably compact, as shown in [5], the same method yields that this logic satisfies (many sorted) interpolation (cf. [6] theorem 1.3.4, for generalizations). However, we may give a more general result.

4.1 <u>THEOREM.</u> For any class of ordinals J, the logic $M_{\omega\omega}(Q_\alpha|\alpha\in J)$ has the interpolation property.

Since interpolation is a validity matter and any sentence uses finitely many quantifiers, by the remarks above it is enough to show the following lemma.

4.2 LEMMA. Let ϕ be a sentence of $M_{\omega\omega}(Q_{\alpha_1},\ldots,Q_{\alpha_m})$, $\alpha_1<\ldots<\alpha_m$, and let ϕ' be the sentence of $M_{\omega\omega}(Q_1,\ldots,Q_m)$ which results of substituting the occurrences of Q_{α_i} in ϕ for Q_i, then: $\models\phi \Leftrightarrow \models\phi'$.

PROOF. Associate to each monadic structure $\mathfrak{A} = (A,P_1^{\mathfrak{A}},\ldots,P_n^{\mathfrak{A}})$ with $|P_\delta^{\mathfrak{A}}| = K_\alpha$, $\delta\in 2^n$, another structure $\mathfrak{A}' = (A',P_1^{\mathfrak{A}'},\ldots, P_n^{\mathfrak{A}'})$ by

$$|P_\delta^{\mathfrak{A}'}| = \begin{cases} K_\delta & \text{if} \quad K_\delta < \omega_{\alpha_1} \\ \omega_i & \text{if} \quad \omega_{\alpha_i} \le K_\delta < \omega_{\alpha_{i+1}} \\ \omega_m & \text{if} \quad \omega_{\alpha_m} \le K_\delta . \end{cases}$$

A straightforward back-and-forth argument shows that $(\mathfrak{A},Q_{\alpha_1},\ldots,Q_{\alpha_m}) \equiv (\mathfrak{A}',Q_1,\ldots,Q_m)$, in the notation of [3] . Hence, $\mathfrak{A}'\models\phi' \Leftrightarrow \mathfrak{A}\models\phi$ and so

$$\models \phi' \Longrightarrow \models \phi \tag{1}$$

The converse implication is more delicate and we give the details. Let $N > qr(\phi)$, $N\in\omega$. Given $\mathfrak{A}$ as before, associate $\mathfrak{B}$ as follows:

$$|P_\delta^{\mathfrak{B}}| = \begin{cases} K_\delta & \text{if} \quad K_\delta < N \\ N & \text{if} \quad N \le K_\delta < \omega_1 \\ \omega_{\alpha_i} & \text{if} \quad \omega_i \le K_\delta < \omega_{i+1}, \ 1\le i<m \\ \omega_{\alpha_m} & \text{if} \quad \omega_m \le K_\delta \end{cases} \tag{2}$$

Now, define a back-and-forth relation between $\mathfrak{A}$ and $\mathfrak{B}$, with parameters in $(N,<)$, by : $\bar a \overset{p}{\sim} \bar b \Leftrightarrow a_i \to b_i$ is a partial isomorphism and length$(\bar a)$ = length$(\bar b) \le N-p$, $p = 0,1,\ldots,N-1$. Assume now that $\bar a \overset{p+1}{\sim} \bar b$, then K = length$(\bar a)$ = length$(b) < N$ and so by (2), for each δ:

$$P_\delta^{\mathfrak{A}} - \{a_i \mid i<k\} \neq \emptyset \Leftrightarrow P_\delta^{\mathfrak{B}} - \{b_i \mid i<k\} \neq \emptyset .$$

This permits to define $f:A\to B$ such that $f(a_i) = b_i$ and for $a\in P_\delta^{\mathfrak{A}} - \{a_i|i<k\}$, $f(a)\in P_\delta^{\mathfrak{B}} - \{b_i|\ i<k\}$. Obviously $\vec{a}a \overset{p}{\sim} \vec{b}f(a)$. Since $[a]^p_{\vec a}$ and $[f(a)]^p_{\vec b}$ have simultaneously the possible forms $\{a_i\}$ and $\{b_i\}$, or $P_\delta^{\mathfrak{A}} - \{a_i|i<k\}$ and $P_\delta^{\mathfrak{B}} - \{b_i|i<k\}$, we have by (2)

$$|\bigcup_{a\in X}[a]^{P}_{\vec{a}}| \geq \omega_i \Rightarrow |\bigcup_{a\in X}[f(a)]^{P}_{\vec{a}}| \geq \omega_{\alpha_i}, \ (1 \leq i \leq m),$$

the extension property in one direction. The other direction is similar Hence (cf. theorem 3.3, [3]),

$$(\mathfrak{A}, Q_1, \ldots, Q_m) \underset{\sim}{\overset{(N,<)}{}} (\mathfrak{B}, Q_{\alpha_1}, \ldots, Q_{\alpha_m})$$

and so $\mathfrak{A} \models \phi' \Longleftrightarrow \mathfrak{B} = \phi$. Thus,

$$\models\phi \Rightarrow \models\phi' \tag{3}$$

The claim follows from (1) and (3). □

As an application, $M_{\omega\omega}(Q_o)$ satisfies interpolation. This logic is also equivalent to $M_{\omega\omega}(Q_1)$ in validities and so it is recursively axiomatizable.

REFERENCES

[1] Barwise, K.J. Axioms for abstract model theory. Ann.Math.Logic 7(1974), pp.221-265.

[2] Caicedo, X. Maximality and interpolation in abstract logics. Thesis, University of Maryland (1978)

[3] ________. Back-and-forth systems for arbitrary quantifiers, IN: "Mathematical Logic in Latin America", Proc. IV Latin American Symposium on Math.Logic, North Holland (1980).

[4] ________ On extensions of $L_{\omega\omega}(Q_1)$, Notre Dame J. of Formal Logic 22 (1981), pp. 85-93.

[5] Fajardo, S. Compacidad y decidibilidad en lógicas monádicas con cuantificadores cardinales, Rev.Colombiana de Mat. 14(1980), pp. 173-196.

[6] Flum, J. Characterizing logics. Preprint (1982).

[7] Friedman, H. Beth's theorem in cardinality logics. Israel J. Math. 14(1973), pp.205-212.

[8] Lindstrom, P. First order predicate calculus with generalized quantifiers. Theoria 32(1966), pp. 187-195.

[9] ________ On extensions of elementary logic, Theoria 35 (1969), pp. 1-11.

[10] Makowsky, J.A. Characterizing monadic and equivalence quantifiers, Preprint (1978)

[11] Makowsky, J.A. and Shelah, S. The theorems of Beth and Craig in abstract model theory I, Trans.AMS 256(1979) pp.215-239.

[12] Mostowski, A. On a generalization of quantifiers. Fund.Math 44(1957), pp.12-36.

[13] Mundici, D. Quantifiers, an owerview. Preprint (1981).

APPROXIMATION TO TRUTH AND THEORY OF ERRORS[1]

Rolando Chuaqui[2] Leopoldo Bertossi[3]
Universidad Católica de Chile
Casilla 114-D
Santiago, Chile

1. Introduction.

The purpose of this paper is twofold. On the one hand, we shall give in Section 2, a model for the theory of errors based on the probability structures of Chuaqui 1983 and 1984. The construction of this model uses recent result in non-standard analysis. On the other hand, a formalization of the notion of "approximation to truth", which includes the possibility of random errors in measurement, will be presented in Section 3. This last section is mainly the work of the first author, who takes full responsability for it.

A formalization of the notion of approximation to truth was presented in a paper inspired by ideas of N.da Costa(Mikenberg, da Costa, and Chuaqui 198+, MDC, for short), but there, the possibility of random errors in measurement was not taken into account. As an introduction to our ideas in the present paper, we shall briefly summarize the formalization of MDC.

A relational structure $\mathfrak{A} = \langle A, R_i \rangle_{i \in I}$ is thought of as a theoretical physical structure about the objects in A. We think of this total structure (i.e. for each n-tuples of elements of A, it is determined whether it belongs to the n-ary relation R_i or not) as what the theory gives us for the objects in A. A natural way to formalize this point of view, is to consider scientific theories as set-theoretical predicates, as in Suppes 1957, Chapter 12. What is actually known about the objects in A, is in the partial structure $\mathfrak{A}' = \langle A, R_i' \rangle_{i \in I}$, where each R_i' is only a partially defined relation over elements of A. That is, only for some tuples it is determined whether they belong R_i or not. For the rest of the tuples, it is undetermined. A theoretical structure $\mathfrak{A}$ is adequate for (or compatible with) $\mathfrak{A}'$, if $\mathfrak{A}$ is an extension of $\mathfrak{A}'$, i.e. R_i coincides with R_i', where the latter is defined. We define a sentence to be true in $\mathfrak{A}'$, if it is true in all extensions with the same universe A. A theoretical structure adequate for $\mathfrak{A}'$ is one of these extensions, $\mathfrak{A}$, i.e. what is true in $\mathfrak{A}'$ is also true in $\mathfrak{A}$, and nothing that is false in $\mathfrak{A}'$ is true in $\mathfrak{A}$. Thus, there may be several theoretical structures which are adequate for $\mathfrak{A}'$, each one determined (possibly) by different theories. When our knowledge about the objects in A increases we obtain another partial structure $\mathfrak{A}''$, which is an extension of $\mathfrak{A}'$. If $\mathfrak{A}$ were compatible with $\mathfrak{A}'$, but not with $\mathfrak{A}''$, then it must be changed and the corresponding theory also.

Another possibility for the rejection of the theory is the following. Suppose that $\mathfrak{A}'$ is what we know for the objects in A and that a theory T, say classical mechanics, determines a total structure $\mathfrak{A}$, with universe (or domain) A, that is compatible with $\mathfrak{A}'$. Assume that if we take a different domain B (which may include A), then $\mathfrak{B}'$ is what we know about the objects in B, and that the theory T determines, for B, a total structure $\mathfrak{B}$ that is incompatible with $\mathfrak{B}'$. Then T should be rejected, but we would say that it is still approximately true for the objects in A. For instance, if we take A as the medium sized objects at slow velocities and T as classical mechanics, then T is approximately true for A.

This picture, however, is not completely accurate. The theoretical structure $\mathfrak{A}$, given by T, in general does not extend the partial structure $\mathfrak{A}'$ exactly. We usually say that $\mathfrak{A}$ coincides with $\mathfrak{A}'$, except for possible errors in measurement. It is this last factor that we want to formalize in the present paper.

In the measurement of a certain quantity, we usually assume that there is a theoretical value, given in the total structure $\mathfrak{A}$, but that the actual value, obtained,i.e. occurring in $\mathfrak{A}'$, may differ from it because of errors in the procedure of measurement. There are three main sources for this error. Sometimes, there is a systematic error derived from the procedure itself. Second, there may be an error produced by the limit of precision of the measurement method. Lastly there may be random errors. We shall disregard the first two types of error and consider just random errors, which, in general are the most important. The account presented in Section 2 could be modified so as to take into consideration the other types of errors.

Thus, if we include random errors in measurement, then the theoretical structure $\mathfrak{A}$ might not be an extension of the partial structure $\mathfrak{A}'$, but, anyway, be considered compatible with it. That is, there might be a sentence true in $\mathfrak{A}'$, but false, strictly speaking, in $\mathfrak{A}$, without this fact being enough ground to reject $\mathfrak{A}$ (and hence T). In order to formalize this situation, we do the following. In the first place, we associate with $\mathfrak{A}$ another structure that we call an error-structure $\mathfrak{A}_e$. Now, a sentence ϕ, instead of being true or false in $\mathfrak{A}_e$, has a probability $P_{\mathfrak{A}_e}(\phi)$, which depends on $\mathfrak{A}_e$. If there is no sentence true in $\mathfrak{A}'$ that has low probability according to $\mathfrak{A}_e$, and high probability according to $\mathfrak{B}_e$, where $\mathfrak{B}$ is an alternative theoretical structure of $\mathfrak{A}$, then $\mathfrak{A}$ is compatible with $\mathfrak{A}'$.

Probability is thought of as "degree of partial truth" (see Chuaqui 1977, for a justification for this view). Because of the possibility of error, we cannot get to truth; thus, we should strive to get as close as possible to it, i.e. to high probability. In a similar way, a false consequence means rejection, but we cannot get to falsehood, but just to low probability, which is approximate falsehood. In fact, we should try to approximate falsehood as much as possible, i.e. given any $\varepsilon > 0$, to try to get a sentence ϕ true in $\mathfrak{A}'$ but with $P_{\mathfrak{A}_e}(\phi) < \varepsilon$. In Chuaqui 198+, there is a discussion of how to approximate falsehood with a sequence of probabilities decreasing to zero. In order to do this, we need sequences of trials of the same experiment. Thus, our structures $\mathfrak{A}_e$, have to be complicated somewhat for this purpose: we construct from $\mathfrak{A}_e$, the structure $\mathfrak{A}_e^{\omega}$ that formalizes an unlimited number of repetitions of the experiments and define probabilities for sentences according to $\mathfrak{A}_e^{\omega}$, derived from $P_{\mathfrak{A}_e}$. These probabilities are defined using the methods of Chuaqui 198+'

This presentation is offered, not as a program for practical implementation, but only as a way to illuminate the relations between theory and evidence, and between truth and probability, in science. It is clear, that the models presented here are a preliminary version that is oversimplified. In particular, we just consider deterministic theories. We hope to improve these models in the future, and include nondeterministic theories.

2. A theoretical model for the theory of errors.

In this section, we present a model for the theory of random errors. In this theory, a quantity is supposed to have a theoretical value, but the procedure of measurement introduces a random error that comes from a combination of a large number of independent causes, each one producing a very small error in the positive or negative direction. The total error for a particular measurement is obtained by adding up the errors produced by the different causes.

In the compound probability structures of Chuaqui 1983, the causes are represented by a causal tree T, with a partial ordering relation. Since here the causes are independent, two different elements of T are never related, i.e. one has no influence upon the other. The fact that there is a large number of causes, can be represented, in non standard analytic terms, by taking T to be the internal set $T = \{t_o, t_1, \ldots, t_\eta\}$ where η is a non-standard infinite natural number, i.e. $\eta \in {}^*\mathbb{N} \sim \mathbb{N}$. That is, T is an infinite, but hyperfinite internal set. With each cause t_k we associate a positive infinitesimal number ε_k and t_k may cause the error ε_k or $-\varepsilon_k$ with equal probability. In order to describe the action of t_k, we introduce a simple probability structure K_k, as in Chuaqui 1984. K_k consists of two relational structures, namely,

$$K_k = \{ \langle\{-\varepsilon_k, \varepsilon_k\}, \{-\varepsilon_k\}\rangle, \langle\{-\varepsilon_k, \varepsilon_k\}, \{\varepsilon_k\}\rangle\}$$

The universe of K_k is $\{-\varepsilon_k, \varepsilon_k\}$. The first structure obtains, if at t_k, $-\varepsilon_k$ is produced, and the second, if ε_k is. The algebra of events naturally consists of all subsets of K_k. In order to obtain the probability measure μ_k, we need a group of permutations of $\{-\varepsilon_k, \varepsilon_k\}$, G_{K_k}. In this case, it clearly contains all permutations of this set and μ_k, the G_{K_k} - invariant measure, assigns 1/2 to each of the models of K_k.

The probability structure for the action of all causes is a compound structure with causal tree $\langle T, =\rangle$. In order to keep the total error with bounds, we must assume that $\sum_{k=0}^{\eta} \varepsilon_k^2 = \varepsilon^2$ where ε is a finite positive number.

The set of compounds outcomes is

$$H = \Pi \langle K_k : t_k \in T\rangle,$$

i.e., H consists of the functions ξ with domain T and such that $\xi(t_k) \in K_k$ for each $k \leq \eta$. Each $\xi \in H$, represents a possible measurement (we assume that the theoretical value to be measured is 0). As in Chuaqui 1983 the probability measure μ defined on subsets of H is the product measure of the μ_k for $k < \eta$.

The result of the measurement represented by an outcome $\xi \in H$ is given as follows. For each $k \leq \eta$, we first define a random variable $X_k \to {}^*\mathbb{R}$ (the non-standard reals or hyperreal numbers) by,

$$X_k(\xi) = \delta_k, \text{ if } \xi(t_k) = \langle\{\varepsilon_k, -\varepsilon_k\}, \{\delta_k\}\rangle$$

(i.e., $\delta_k = \varepsilon_k$ or $\delta_k = -\varepsilon_k$).

Then. the result of the measurement in outcome $\xi \in H$ is

$$f(\xi) = \sum_{k=0}^{\eta} \ _k(\xi)$$

In order to study the distribution of f, we need the central limit theorem with the Lindeberg condition in non-standard form as given in Stoll 1982:

If $\eta \in {}^*\mathbb{N} - \mathbb{N}$ and $< Y_k : k \leq \eta >$ is an internal sequence of *independent random variables in an internal probability space (Ω, A, μ) such that $E(Y_k) = 0$ and $E(Y_k{}^2) = 1$, and $< \alpha_k : k \leq \eta >$ is a sequence of infinitesimal weights $\alpha_k \in {}^*\mathbb{R}$ such that $\sum_{k=0}^{\eta} \alpha_k{}^2 = \alpha^2$ with $0 < {}^\circ\alpha \in \mathbb{R}^+$, then

$$\mu([\sum_{k=0}^{\eta} \alpha_k Y_k \leq \lambda]) \simeq {}^* \Phi(\frac{\lambda}{\alpha}) \text{ for all } \lambda \in {}^*\mathbb{R}.$$

Here $^*\Phi$ is the non-standard normal distribution with mean 0 and standard deviation 1.

Taking standard parts, one can show that if γ is Loeb's measure generated by $^\circ\mu$, then

$$\gamma([{}^\circ \sum_{k=0}^{\eta} \alpha_k Y_k \leq \lambda]) = \Phi({}^\circ\frac{\lambda}{\alpha}) \text{, for all } \lambda \in \mathbb{R},$$ where ϕ is the standard normal distribution.

This work of Stoll is based on Loeb 1975 and Anderson 1976.

Thus, we have for our random variable $f: H \to {}^*\mathbb{R}$, the following distribution:

$$\mu([f \leq \lambda]) = \mu([\sum_{k=0}^{\eta} \varepsilon_k \left[\frac{X_k}{\varepsilon_k}\right] \leq \lambda])$$

$$\simeq {}^*\phi({}^\circ\frac{\lambda}{\varepsilon}),$$

for each $\lambda \in {}^*\mathbb{R}$.

Then, its standard distribution is,

$$\gamma([{}^\circ f \leq \lambda]) = \phi({}^\circ\frac{\lambda}{\varepsilon}),$$

for each $\lambda \in \mathbb{R}$

Thus, the result of the measurement is normally distributed with mean 0 and standard deviation $^\circ\varepsilon$. It is easy to modify the construction so that the mean (i.e. the theoretical measurement) is any number $r \in \mathbb{R}$. Thus, in order to obtain a model for the measurement of a certain quantity, we must be given two parameters: the theoretical measurement r and the standard error $°\varepsilon$. With these two numbers, we construct a compound probability structure as above (that will be called an error probability structures). Then the random variable f, which gives the actual value of the measurement, will be normally distributed with mean r and standard deviation $°\varepsilon$.

The mean r represents the theoretical measurement and may be obtained by calculations from the theory or be estimated from the data. For instance, a measurement of length is usually obtained directly from the data, but other quantities may be consequences or the theory.

The standard deviation $°\varepsilon$, on the other hand, depends on the method of measurement. If the method is more accurate $\varepsilon°$ will be smaller, i.e. there may be less causes of error of the error produced by each cause may be smaller. There may be several procedures of measurement for the same quantity all should give the same mean, but possibly have different standard deviations. This standard error is usually estimated from the distribution of actual measurements, but occasionally it is roughly estimated from theoretical considerations concerning the supposed precision of the procedure.

These values of the mean and standard error are, then, compared with a series of actual values using the usual statistical techniques. If the distribution of the actual values is very improbable according to the theoretical distribution then this last one is rejected as a model of the real state of affairs.

Although the mean has a theoretical significance, the standard error has not, since we are just considering deterministic phenomena. Thus, it is usually important just to test the appropriateness of the theoretical value of the measurement as compared with the actual values obtained. Which is the theoretical standard error (i.e. the standard deviation of the error probability structure) is not important. A statistical test for testing the mean with unknown standard deviation is Student's Test. For using this test, we calculate the quantity

$$t = \frac{M - r}{S_M} ,$$

when r is the theoretical value of the measurement, M the mean value of the values actually obtained, and S_M the standard deviation of these values. We can then compute the probability of $|t| < a$ for an $a \in \mathbb{R}$, for any error probability structure with mean r. We can obtain a value of a for which these probabilities (one for each error probability structure) are less than a certain α. Thus, whatever may the standard deviation of the error probability structure be, the probability of

$|t|<a$, is less than α. If we take α sufficiently small, then the probability of the event $|t|<a$, will be small for r being the theoretical value of the measurement. Here, the hypothesis that r is this value, is to be rejected.

3. Structures with errors in measurement.

We are now ready to introduce a theoretical structure that includes measurement with random errors, and the corresponding partial structure representing what we actually know. In fact, we shall introduce two new types of theoretical structures : a pure measurement structure or $\mathbb{R}$-structure, and the error structures associated with it. We shall discuss later the structures that represent our knowledge and which correspond to the partial structures of the old setting sketched in Section 1.

An $\mathbb{R}$-structure is a system of the form

$$\mathfrak{A} = \langle A, f_i^{\mathfrak{A}}, R_j^{\mathfrak{A}} \rangle_{i \in I, j \in J},$$

where each $R_j^{\mathfrak{A}}$, for $j \in J$, is an n_j-ary relation between elements of A, and each $f_i^{\mathfrak{A}}$, for $i \in I$, is an n_i-ary operation from A into $\mathbb{R}$. We could also have operations from A into A, or distinguished elements of A, but, for simplicity, we shall not include them, since they can be replaced by relations. We think of $\mathfrak{A}$ as what we accept theoretically to be true of the elements of A. The $R_j^{\mathfrak{A}}$'s represent possible relations between these elements, and the $f_i^{\mathfrak{A}}$'s, measurements performed on them. Thus, $f_i^{\mathfrak{A}}(a_o,\ldots,a_{n_i-1})$ is a real number that measures some property of the system $(a_o,\ldots,a_{n-1})$ of elements of A. There could be two measurements $f_i^{\mathfrak{A}}$ and $f_k^{\mathfrak{A}}$, with $i \neq k$, of the same quantity. In the $\mathbb{R}$-structure $\mathfrak{A}$ they could coincide. However, in the error structures and the partial structures to be introduced below, they may differ. As in Section 1, $\mathfrak{A}$ is what the scientific theory prescribes for the elements of A. In this paper, we only consider deterministic theories.

The language for $\mathbb{R}$-structures is a one-sorted language, with variables $x,y,z,\ldots$, that constains the following types of atomic formulas:

$$x = y$$

$$R_j\, x_o,\ldots,x_{n_j-1} \quad \text{for each } j \in J,$$

and

$$[f_i(x_o,\ldots,x_{n_i-1}) \geq r] \text{ , for each } r \in \mathbb{Q} \text{ (the rational numbers), and each } i \in I.$$

These formulas are combined in an $L_{\omega_1\omega}$-language with negation, countable conjunctions and disjunctions, and finitely many quantifiers.

The variables are assigned elements of A. For each formula ϕ and each assigment s of the variables in A, we define when s satisfies ϕ in $\mathfrak{A}$., in symbols $\mathfrak{A} \models \phi[s]$. Most of the clauses are the usual ones, plus

$$\models [f_i(x_o,\ldots,x_{n_i-1}) \geq r][s] \quad \text{iff} \quad f_i^{\mathfrak{A}}(s(x_o),\ldots,x(x_{n_i-1})) \geq r.$$

We could also have a two-sorted language with variables and operations for the real numbers, but we shall not need this in this paper.

Most of the mathematical results obtained in MDC could easily be extended to $\mathbb{R}$-structures, with the natural notion of partial $\mathbb{R}$-structure. We shall not pursue this line here. Instead, we shall introduce another type of theoretical structure: the error-structures (or briefly, E-structures) associated with the $\mathbb{R}$-structure $\mathfrak{A}$. While $\mathfrak{A}$ determines whether a formula is satisfied by an assignment or not, an E-structure $\mathfrak{A}_e$, determines only the probability that is assigned to the formula.

An <u>E-structure associated with</u> $\mathfrak{A}$ is a system of the form

$$\mathfrak{A}_e = \langle A, f_i^{\mathfrak{A}_e}, R_j^{\mathfrak{A}_e} \rangle_{i\in I, j\in J},$$

where $R_j^{\mathfrak{A}_e} = R_j^{\mathfrak{A}}$, for $j\epsilon J$, and, if $f_i^{\mathfrak{A}}$ is an n_i-ary operation, then for each $a_o,\ldots,a_{n_i} \epsilon A$, $f_i^{\mathfrak{A}e}(a_o,\ldots,a_{n_i-1})$ is a random variables whose distribution is given by an error probability structure (see Section 2); $f_i{}^{e}(a_o,\ldots,a_{n_i-1})$ is a random variable having a normal distribution with mean $f_i^{\mathfrak{A}}(a_o,\ldots,a_{n_i-1})$, for each $i\epsilon I$. More precisely, $f_i^{\mathfrak{A}e}(a_o,\ldots,a_{n_i-1}) = f_i^{\mathfrak{A}}(a_o,\ldots,a_{n_i-1}) + \varepsilon_i(a_o,\ldots,a_{n_i-1})$, where $\varepsilon_i(a_o,\ldots,a_{n_i-1})$ is random variable with mean 0, which represents the error in the measurement. The distribution of $\varepsilon_i(a_o,\ldots,a_{n_i-1})$ is determined by an error probability structure (of Section 2) whose universe H is the domain of $\varepsilon_i(a_o,\ldots,a_{n_i-1})$.

Each $f_i^{\mathfrak{A}}$, for $i\epsilon I$, represents one method of measurement for a quantity. There may be several methods for the same quantity indexed by different elements of I. Thus, $f_i^{\mathfrak{A}}$ and $f_k^{\mathfrak{A}}$ may be measurements of length, say by a ruler and by wavelengths. In this case, $f_i^{\mathfrak{A}}(a_o,\ldots,a_{n-1}) = f_k^{\mathfrak{A}}(a_o,\ldots,a_{n-1})$ for every $a_o,\ldots,a_{n-1}, \epsilon A$. Hence, $f_i^{\mathfrak{A}e}(a_o,\ldots,a_{n-1})$ and $f_k^{\mathfrak{A}e}(a_o,\ldots,a_{n-1})$ are random variables with the same mean. But, their standard deviations may be different. This deviation depends on the procedure of measurement, and may be determined by the theory of the method or the empirical data. as was explained in Section 2.

It may be more reasonable to assume that for each $i \in I$, $f_i^{\mathfrak{A}_e}$ is defined only for a subset of A, namely, for those objects that are possible to measure with the procedure involved. This would introduce additional inessential complications to our models, so that we shall assume that for each tuple $a_o,\ldots,a_{n-1}$ of elements of A, $f_i^{\mathfrak{A}_e}(a_o,\ldots,a_{n-1})$ is defined (i.e. it is a random variable with values in $\mathbb{R}$).

The language for the E-structures $\mathfrak{A}_e$ associated with $\mathfrak{A}$, is the same as that for $\mathfrak{A}$. However, instead of satisfaction, $\mathfrak{A}_e$ determines a probability. For each formula ϕ and each assignments of the variables in A, we define the probability that s assigns to ϕ in $\mathfrak{A}_e$, in symbols $P_{\mathfrak{A}_e}(\phi,s)$. The definition that will be given below is based in that of Scott and Krauss 1966. The main differences with Scott and Krauss are that we use assignments instead of constants, and that we give the definition jointly for all relations and operations instead of doing it separately, and then joining them by their method of independent unions. In any case, this means that we assume the different $f_i^{\mathfrak{A}_e}$ and R_j, for $i \in I$ and $j \in J$, to be stochastically independent.

We now proceed to state the definition of $P_{\mathfrak{A}_e}$ in several stages. Let $B_i^{\mathfrak{A}_e}(a_o,\ldots,a_{n-1})$ be the measure algebra of the error probability structure where $f_i^{\mathfrak{A}_e}(a_o,\ldots,a_{n-1})$ is defined, and $\mu_i^{\mathfrak{A}_e}(a_o,\ldots,a_{n-1})$, its measure. We shall always consider, now and in what follows, strictly positive measures, i.e. measures that vanish only on the zero of the algebra, and their corresponding measure algebras. This is needed because in our definitions we must have complete algebras, i.e. algebras where the suprema and infima are always defined. If necessary, to achieve a strictly positive measure, we take the algebra (and the measure) modulo its null sets. In what follows, we shall suppose that this is done, without mentioning it.

For each $i \in I$, $B_i^{\mathfrak{A}_e}$ is defined to be the product algebra.

$$B_i^{\mathfrak{A}_e} = \Pi < B_i^{\mathfrak{A}_e}(a_o,,,.a_{n-1}) : a_o,\ldots,a_{n-1} \in A >.$$

and $\mu_i^{\mathfrak{A}_e}$ its corresponding (strictly positive) measure.

For each $j \in J$, $B_j^{\mathfrak{A}_e}$ is the two element measure algebra $\{0_j, 1_j\}$, and $\mu_j^{\mathfrak{A}_e}$ the measure that assigns 1 to 1_j and 0 to 0_j.

We consider, now, the product $B^{\mathfrak{A}_e}$ of all these algebras:

$$\mathcal{B}^{\mathfrak{A}_e} = \Pi < \mathcal{B}_k^{\mathfrak{A}_e} : k \in I \cup J >,$$

and its corresponding product measure, $\mu^{\mathfrak{A}_e}$.

An element in $\mathcal{B}^{\mathfrak{A}_e}$ is a system

$$\mu = < \mu_k : k \in I \cup J > ;$$

on its turn, if $i \in I$, then μ_i is a system

$$\mu_i = < \mu_i (a_o, \ldots, a_{n_i-1}) : a_o, \ldots, a_{n_i-1} \in A >$$

We call the unit of $\mathcal{B}^{\mathfrak{A}_e}$, $\mathbb{1}$, and its zero $\mathbb{0}$. Similarly, $1_k, 0_k$ will be the corresponding elements of $\mathcal{B}_k^{\mathfrak{A}_e}$, for $k \in I \cup J$, where, if $i \in I$, $1_i = < 1_i(a_o, \ldots, a_{n_i-1}) : a_o, \ldots, a_{n_i-1} \in A >$, and $0_i = < 0_i(a_o, \ldots, a_{n_i-1}) : a_o, \ldots, a_{n_i-1} \in A >$.

For each formula ϕ and assignment s of the variables in A, we define a valuation $h(\phi;s) \in \mathcal{B}^{\mathfrak{A}_e}$, by recursion:

(i) $h(x=y;s) = \begin{cases} \mathbb{1}, & \text{if } s(x) = s(y) \\ \mathbb{0}, & \text{otherwise} \end{cases}$

(ii) $h(R_j x_o, \ldots, x_{n-1}, s) = \mu$, where its components μ_j are given by

$$\mu_j = \begin{cases} 1_j, & \text{if } < s(x_o), \ldots, s(x_{n-1}0 > \in R_j^{\mathfrak{A}} \\ 0_j, & \text{otherwise,} \end{cases}$$

and $\mu_k = 1_k$, for all $k \in I \cup J$ with $j \neq k$.

(iii) $h([f_i(x_o, \ldots, x_{n-1}) \geq r] ; s) = \mu$, where

$$\mu_i(s(x_o), \ldots, s(x_{n-1})) = [f_i^{\mathfrak{A}_e}(s(x_o), \ldots, s(x_{n-1})) \geq r]$$

(i.e. the corresponding element of $\mathcal{B}_i^{\mathfrak{A}_e}(s(x_o), \ldots, s(x_{n-1}))$,

$$\mu_i(a_o, \ldots, a_{n-1}) = 1_i(a_o, \ldots, a_{n-1}) \text{ for } (a_o, \ldots, a_{n-1}) \neq$$

$(s(x_o), \ldots, s(x_{n-1}))$, and

$$\mu_k = 1_k, \text{ for } k \in I \cup J, \ k \neq i.$$

(iv) $h(\neg \phi;s) = \mathbb{1} - h(\phi,s)$

(v) $h(\bigvee_{n \in \mathbb{N}} \phi_n;s) = \bigvee_{n \in \mathbb{N}} h(\phi_n,s)$

(vi) $h(\bigwedge_{n \in \mathbb{N}} \phi_n;s) = \bigwedge_{n \in \mathbb{N}} h(\phi_n,s)$

(vii) $h(\exists x\phi;s) = \bigvee_{a\in A} h(\phi,s^x_a)$

(viii) $h(\forall x\phi;s) = \bigwedge_{a\in A} h(\phi,s^x_a)$

Here, s^x_a is the assigment that coincides with s everywhere, except, possibly, on x where it assigns a.

Now we are ready to define $P_{\mathfrak{A}_e}(\phi,s)$, the probability that s assigns $\mathfrak{A}_e$. This is simply given by

$$P_{\mathfrak{A}_e}(\phi,s) = \mu^{\mathfrak{A}_e}(h(\phi;s)).$$

We shall now proceed to the discussion of the structures that represent what we actually know and their relation to the theoretical structures. In order to study this relationship, it is not enough to consider one $\mathbb{R}$-structure $\mathfrak{A}$, but need to consider all of its alternatives, as well. An <u>alternative to</u> the $\mathbb{R}$-structure $\mathfrak{A}$ is an $\mathbb{R}$-structure $\mathfrak{B}$, with the same universe A and the same similarity type. That is, if $\mathfrak{A} = \langle A, f_i^{\mathfrak{A}}, R_j^{\mathfrak{A}} \rangle_{i\in I, j\in J}$, then $\mathfrak{B} = \langle A, f_i^{\mathfrak{B}}, R_j^{\mathfrak{B}} \rangle_{i\in I, j\in J}$ where $f_i^{\mathfrak{B}}$ and $R_j^{\mathfrak{B}}$ are of the same arity as $f_i^{\mathfrak{A}}$ and $R_j^{\mathfrak{A}}$, respectively (The similarity type τ determines for each $k\in I\cup J$, where the symbol indexed by k is an operation or a relation, and its arity). The set of alternatives with universe A and similarity type τ, we call the A,τ-alternatives.

For each alternative $\mathfrak{B}$ to $\mathfrak{A}$, we construct the corresponding E-structure. If $\mathfrak{A}_e$ is an E-structure associated with $\mathfrak{A}$, we shall designate by $\mathfrak{B}_e$, the E-structure associated to $\mathfrak{B}$ in which the distribution of $f_i^{\mathfrak{B}_e}(a_0,\dots,a_{n-1})$ has the same standard deviation as that of $f_i^{\mathfrak{A}_e}(a_0,\dots,a_{n-1})$.

Now, we define an A,τ-partial $\mathbb{R}$-structure $\mathfrak{A}'$, where A is a universe (i.e. a nonempty set) and τ a similarity type. $\mathfrak{A}'$ is a system of the form:

$$\mathfrak{A}' = \langle A, f_i^{\mathfrak{A}'}, R_j^{\mathfrak{A}'} \rangle_{i\in I,\ j\in J}$$

where $f_i^{\mathfrak{A}'}$, for $i\in I$, is an n_i-ary partial function from A into $\mathbb{R}$, and $R_j^{\mathfrak{A}'}$, for $j\in J$, is an n_j partial relation. For describing partial structures, it is better to replace relations by their characteristic functions, i.e. we write

$$R_j^{\mathfrak{A}'}(a_o,\ldots,a_{n_j-1}) = 1, \text{ if } \langle a_o,\ldots,a_{n_j-1}\rangle \in R_j^{\mathfrak{A}'}$$

$$= 0, \text{ otherwise}$$

Then, a partial relation is a partial function from A into {0,1}.

A complete extension $\mathfrak{B}$ of the A,τ-partial ℝ-structure $\mathfrak{A}'$ is an A,τ-alternative (i.e. an ℝ-structure with universe A and similarity type τ) such that the operations and relation of $\mathfrak{B}$ are extension of those in $\mathfrak{A}'$. We already have defined satisfaction for ℝ-structures. We can now define satisfaction for partial ℝ-structures $\mathfrak{A}'$ as in MDC, namely, for any formula ϕ, and assignment s in A:

$\mathfrak{A}' \models_T \phi[s]$ iff for every complete extension $\mathfrak{B}$ of $\mathfrak{A}'$, we have

$$\mathfrak{B} \models \phi[s]$$

$\mathfrak{A}' \models_F \phi[s]$ iff $\mathfrak{A}' \models_T \neg\phi[s]$

$\mathfrak{A}' \models_U \phi[s]$, otherwise

Thus, a formula may be satisfied, not satisfied, or left undetermined by an assignment s in $\mathfrak{A}'$.

Notice that for atomic formulas, the definition of satisfaction given above can be translated to :

$\mathfrak{A}' \models_T [f_i(x_o,\ldots,x_{n-1}) \geq r][s]$ iff $f_i^{\mathfrak{A}'}(s(x_o),\ldots,s(x_{n-1})$, is defined and $\geq r$;

$\mathfrak{A} \models_T R_j\, x_o,\ldots,x_{n-1}[s]$ iff $R_j^{\mathfrak{A}'}(s(x_o),\ldots,s(x_{n-1}))$ is defined and equal to 1.

A partial ℝ-structure $\mathfrak{A}'$ represents what we actually know, or, at least, accept and are not willing to change. In MDC, the theoretical structures $\mathfrak{B}$ compatible with $\mathfrak{A}'$ (i.e. that are possible given $\mathfrak{A}'$) are the complete extensions of $\mathfrak{A}'$. Here, the situation will be different. There may be compatible theoretical structures which are not extensions of $\mathfrak{A}'$.

Now we are ready to relate $\mathfrak{A}$ to $\mathfrak{A}'$. We say that $\mathfrak{A}'$ is incompatible with the total ℝ-structure $\mathfrak{A}$ (given $\mathfrak{A}_e$), iff there is a formula ϕ and an assignment s in $\mathfrak{A}$ such that,

(i) $\mathfrak{A}' \models_T \phi[s]$,

(ii) $P_{\mathfrak{A}_e}(\phi, s)$ is low,

and

(iii) $P_{\mathfrak{B}_e}(\phi, s)$ is high, for some alternative to $\mathfrak{A}$, $\mathfrak{B}$.

In the account without considering random errors of MDC, $\mathfrak{A}$ was to be rejected, if it was not an extension of $\mathfrak{A}$. That is, if a sentence true in $\mathfrak{A}$, was false in $\mathfrak{A}'$ (or, more precisely, if there is a formula ϕ and an assignment s such that $\mathfrak{A} \models \phi[s]$, but $\mathfrak{A}' \models_T \neg \phi[s]$). In our present account, $\mathfrak{A}$ might not be an extension of $\mathfrak{A}'$, but anyway compatible with it, if $P_{\mathfrak{A}_e}(\phi, s)$ is high for all ϕ and s with $\mathfrak{A}' \models_T \phi[s]$. That is, everything that is approximately true in $\mathfrak{A}$ (i.e. has high probability in $\mathfrak{A}_e$) is true in $\mathfrak{A}'$, and there is nothing true in $\mathfrak{A}'$ that is approximately false in $\mathfrak{A}$ (i.e. has low probability in $\mathfrak{A}_e$).

How low the probabilities should be to reject $\mathfrak{A}$, depends fundamentally on the alternatives available. If there is a "reasonable alternative that assigns high probabilities to all sentences true in $\mathfrak{A}'$, then we might reject $\mathfrak{A}$, even though the probabilities in $\mathfrak{A}_e$ might not be very low. With no reasonable alternative, we would need very low probabilities, in order to reject $\mathfrak{A}$. The following is a possible explanation of what a reasonable alternative is. First, a definition. We say that the theory τ (in the similarity type τ) is confirmed by the B-τ-partial structure $\mathfrak{B}'$ (given $\mathfrak{A}_e$) if the total $\mathbb{R}$-structure $\mathfrak{B}$ determined by T for the objects in B, has the property that for all formulas ϕ and assignments s in B, if $\mathfrak{B}' \models_T \phi[s]$ then $P_{\mathfrak{B}_e}(\phi, s)$ is high. Suppose that if B is a set of objects that has been studied in a science, then $\mathfrak{B}_B$ is the B-τ-partial structure that is accepted as true, and assume that T is confirmed by all such $\mathfrak{B}_B$. Then, if Σ is the total $\mathbb{R}$-structure determined by such a T for the objects in A, it is a <u>reasonable alternative</u> to $\mathfrak{A}$.

The account given up to now is unrealistic in that it assumes that we measure each object just once. We could solve this problem by having several measurements, but assign one value to $f_i^{\mathfrak{A}'}(a_o, \ldots, a_{n-1})$, namely, their average. However, by using this procedure we lose some of the statistical power that may be available. In particular, with just one value assigned, we have no real hope of getting rid of $\mathfrak{A}_e$ in the definition of incompatibility. As given, we defined $\mathfrak{A}$ incompatible with $\mathfrak{A}'$ (given $\mathfrak{A}_e$). The standard deviations included in $\mathfrak{A}_e$ are not, usually, important for scientific theories.

In order to include repetitions of measurements, we introduce, for each $\mathfrak{A}$ and $\mathfrak{A}_e$, the structure $\mathfrak{A}_e^\omega$, called an ω-E-structure, with a language for this structure and a definition of probability for its formulas.

To all operations and relations in $\mathfrak{A}$, we add one more place to range over ω, the natural numbers; $f_i^{\mathfrak{A}_e^\omega}(a_o,\ldots,a_{n-1},t)$ is a random variable for each $a_o,\ldots,a_{n-1} \in A$ and $t \in \omega$, with the same distribution as $f_i^{\mathfrak{A}_e}(a_o,\ldots,a_{n-1})$; similarly, with relations, $R_j^{\mathfrak{A}_e^\omega}(a_o,\ldots,a_{n-1},t) = R_j^{\mathfrak{A}}(a_o,\ldots a_{n-1})$ for all $a_o,\ldots,a_{n-1}, \in A$, $t \in \omega$.

The language is now a two-sorted language with variables $x,y,z\ldots$ for elements of A, and m,n for elements of ω. The atomic formulas are:

$x = y$

$m = n$

$R_j\, x_o,\ldots,x_{n_j-1},m$, for each $j \in J$

$[f_i(x_o,\ldots x_{n_i-1},m) \geq r]$, for each $r \in \mathbb{Q}$, $i \in I$.

This language will be a two-sorted $L_{\omega_1\omega}$-language with finitely many quantifiers for both types of variables. The assignments s, now, adscribe elements of A for the variables $x,y,z\ldots$, and elements of ω for the the other sort. Just as for $\mathfrak{A}_e$, $\mathfrak{A}_e^\omega$ assigns probabilities to formulas.

Let $\mathcal{B}_i^{\mathfrak{A}_e^\omega}(a_o,\ldots,a_{n-1})$ be the product algebra of $\mathcal{B}_i^{\mathfrak{A}_e}(a_o,\ldots a_{n-1})$ ω-times, and $\mu_i^{\mathfrak{A}_e^\omega}(a_o,\ldots,a_{n-1})$ its product measure. Then,

$\mathcal{B}_i^{\mathfrak{A}_e^\omega} = \Pi < \mathcal{B}_i^{\mathfrak{A}_e^\omega}(a_o,\ldots,a_{n-1}) : a_o,\ldots,a_{n-1} \in A >$

and $\mu_i^{\mathfrak{A}_e^\omega}$ is its corresponding product measure. For $j \in J$,

$\mathcal{B}_j^{\mathfrak{A}_e^\omega}$ and $\mu_j^{\mathfrak{A}_e^\omega}$ are defined analogously. Finally, let

$\mathcal{B}^{\mathfrak{A}_e^\omega} = \Pi < \mathcal{B}_k^{\mathfrak{A}_e^\omega} : k \in I \cup J >$,

and let $\mu^{\mathfrak{A}_e^\omega}$ be its corresponding product measure.

An element $\mu \in \mathcal{B}^{\mathfrak{A}_e^\omega}$ is a system

$\mu = < \mu_k : k \in I \cup J >$.

If $k \in J$, then $\mu_k = < \mu_k(t) : t \in \omega >$ where $\mu_k(t) \in \mathcal{B}_k^{\mathfrak{A}_e}$. If $i \in I$, then

$\mu_i = \langle \mu_i(a_o,\ldots,a_{n-1},t) : a_o,\ldots,a_{n-1}\in A, t\in\omega\rangle$, where $\mu_i(a_o,\ldots,a_{n-1},t)\in B_i^{\mathfrak{A}_e}(a_o\ldots,a_{n-1})$ for each $t\in\omega$.

h, now, assigns to each formula ϕ and assignment s of the new language an element of $B^{\mathfrak{A}_e^\omega}$, as follows:

(i) $$h(x=y;s) = \begin{cases} \mathbb{1}, & \text{if } s(x) = s(y) \\ \mathbb{0}, & \text{otherwise.}\end{cases}$$

$$h(n = m;s) = \begin{cases} \mathbb{1}, & \text{if } s(n) = s(m) \\ \mathbb{0}, & \text{otherwise.}\end{cases}$$

(ii) $h(R_j\, x_o,\ldots,x_{n-1},m;s) = \mu$ where

$$\mu_j(s(m)) = \begin{cases} 1_j, & \text{if } R_j^{\mathfrak{A}}(s(x_o),\ldots,s(x_{n-1})) = 1 \\ \mathbb{0}_j, & \text{otherwise,}\end{cases}$$

and $\mu_k(t) = 1_k$, for all $k\in I\cup J$, $t\in\omega$, with $k\neq j$ or $t\neq s(m)$.

(iii) $h([f_i(x_o,\ldots,x_{n-1},m)\geq r];s) = \mu$ where

$$\mu_i(s(x_o),\ldots,s(x_{n-1}),s(m)) = [f_i(s(x_o),\ldots,s(x_{n-1}))\geq r]$$

and $\mu_k(a_o,\ldots,a_{n-1}t) = 1_k(a_o,\ldots,a_{n-1})$, for all $k\neq i$ or

$$(a_o,\ldots,a_{n-1},t)\neq(s(x_o),\ldots,s(x_{n-1}),s(m))$$

(iv), (v), (vii), and (viii) are the same as before.

We need two more clauses:

(ix) $$h(\exists n\phi;s) = \bigvee_{t\in\omega} h(\phi,s_t^n).$$

(x) $$h(\forall n\phi;s) = \bigwedge_{t\in\omega} h(\phi,s_t^n).$$

Just as before, the probability in $\mathfrak{A}_e^\omega$ is given by:

$$P_{\mathfrak{A}_e^\omega}(\phi,s) = \mu^{\mathfrak{A}_e^\omega}(h(\phi;s)).$$

Now, the A,t- partial ω-structures (or partial structures with repetition), $\bar{\mathfrak{A}}$ are of the form

$$\bar{\mathfrak{A}} = \langle A, f_i^{\bar{\mathfrak{A}}}, R_j^{\bar{\mathfrak{A}}} \rangle_{i \in I,\ j \in J},$$

where $f_i^{\bar{\mathfrak{A}}}$ is a partial operation defined on ${}^{n_i}A \times \omega$ into $\mathbb{R}$ and $R_j^{\bar{\mathfrak{A}}}$ is a partial relation on ${}^{n_j}A \times \omega$. (Here, ${}^{n}A$ is the set of n-tuples of A). These functions may be partially defined on A, ω or both; e.g. $f_i^{\bar{\mathfrak{A}}}(a_o,\ldots,a_{n-1},t)$ may be defined only for some $a_o,\ldots,a_{n-1} \in A$ and $t \in \omega$. In general, if we assume that $\bar{\mathfrak{A}}$ represents our actual knowledge, then, for each $a_o,\ldots,a_{n-1} \in A$ there will only be finitely many $t \in \omega$ with $f_i^{\bar{\mathfrak{A}}}(a_o,\ldots,a_{n-1}t)$ defined.

A complete extension $\bar{\mathfrak{B}}$ of $\bar{\mathfrak{A}}$ will have these functions defined everywhere in A and ω, and extend those of $\bar{\mathfrak{A}}$. Observe that in $\bar{\mathfrak{A}}$, or in any of its extensions $\bar{\mathfrak{B}}$, we may have $f_i^{\bar{\mathfrak{A}}}(a_o,\ldots,a_{n-1},t) \neq f_i^{\bar{\mathfrak{A}}}(a_o,\ldots,a_{n-1},v)$, for $t, v \in \omega$ with $t \neq v$.

Satisfaction for $\bar{\mathfrak{A}}$ is defined just as for the partial structures without repetitions $\mathfrak{A}'$.

In the language that we have introduced there is a formula ϕ and an assignment s such that

$$\bar{\mathfrak{A}} \models_T \phi[s] \quad \text{iff} \quad |t_i(a_o,\ldots,a_{n-1})| < a$$

where $t_i(a_o,\ldots,a_{n-1})$ is Student's t for the measurement $f_i^{\bar{\mathfrak{A}}}(a_o,\ldots,a_{n-1},v)$ with $v \in \omega$ that are defined in $\bar{\mathfrak{A}}$, and $a \in \mathbb{R}$.

That is

$$t_i(a_o,\ldots,a_{n-1}) = \frac{M - f_i^{\bar{\mathfrak{A}}}(a_o,\ldots,a_{n-1})}{S_M},$$

where M is the average of the sequence

$\langle f_i^{\bar{\mathfrak{A}}}(a_o,\ldots,a_{n-1},v) : v \in \omega$ and $f_i^{\bar{\mathfrak{A}}}(a_o,\ldots,a_{n-1}v)$ is defined in $\bar{\mathfrak{A}} \rangle$ and S_M is its sample standard deviation.

As we mentioned in Section 2, there is an $a \in \mathbb{R}$, such that $P_{\bar{\mathfrak{A}}_e}\omega(\phi,s) \cdot$

is low for all ω-E-structures $\mathfrak{A}_e^{\omega}$, associated with $\mathfrak{A}$. Thus, the following definition makes sense.

We say that the partial ω-structure $\bar{\mathfrak{A}}$ is incompatible with $\mathfrak{A}$ iff there is a formula ϕ and an assignment s such that,

(i) $\bar{\mathfrak{A}} \models_T \phi[s]$,

(ii) $P_{\mathfrak{A}_e^{\omega}}(\phi,s)$ is low, for every ω-E-structure $\mathfrak{A}_e^{\omega}$ associated with $\mathfrak{A}$.

(iii) $P_{\mathfrak{B}_e^{\omega}}(\phi,s)$ is high, for a certain alternative to $\mathfrak{A}$, $\mathfrak{B}$, and a certain ω-E- structure $\mathfrak{B}_e^{\omega}$ associated with $\mathfrak{B}$.

If a certain ω-E-structure $\mathfrak{A}_e^{\omega}$ is preferred, because of theoretical reasons, over all other ω-E-structures associated with $\mathfrak{A}$, then we might relativize the definition of compatibility to this $\mathfrak{A}_e^{\omega}$, by changing (ii) to (ii)' : $P_{\mathfrak{A}_e^{\omega}}(\phi,s)$ is low.

However, the definition given (with (ii) instead of (ii)') is preferable, because it is independent of inessential theoretical features, such as standard deviations.

It can be shown, by arguments similar to those presented in Chuaqui 198+, that the statistical tests for hypothesis are a special case of these definitions for the situation of this paper. In particular, we can explain, in this fashion the approximation to falsehood by a sequence of probabilities decreasing to zero.

Two possible extensions of the models discussed here may be mentioned. In the first place, $f_i^{\mathfrak{A}_e}(a_o,\ldots,a_{n-1})$ may have a different distribution than the normal one. This may happen with some methods of measurement. The second possible extension is to non-deterministic theories. In this case, the theoretical structure $\mathfrak{A}$ itself may have random variables, i.e. $f_i^{\mathfrak{A}}(a_o,\ldots,a_{n-1})$ may itself be a random variable. This is a possible line of inquiring that we have not yet pursued.

REFERENCES

Anderson, R.M. [1976] A non-standard representation for Brownian Motion and Ito integration, Israel J.Math. vol. 25, pp.15-46.

Chuaqui, R. [1977] A semantical definition of probability, in Non-Classical Logics, Model Theory, and Computability, Arruda, da Costa, and Chuaqui (editors), North Holland Public.Co. Amsterdam pp. 135-167.

[1983] Factual and cognitive probability, to appear in the Proceedings of the V Latin American Logic Symposium, Caicedo (editor), Marcel Dekker Inc., New York.

[1984] Models for probability, to appear in the Proceedings of the First Chilean Symposium on Analysis, Geometry, and Probability, Chuaqui (editor), Marcel Dekker, Inc., New York.

[1985] How to decide between statistical methods. To appear in Mathematical Logic and Formal Systems. (volume in honor of N.C.A.da Costa), de Alcantara (editor), Marcel Dekker Inc., New York.

Loeb, P.A. [1975] Conversion from non-standard to standard measure spaces and applications in probability theory, Trans.Am.Math.Soc.vol. 211, pp. 113-122

Mikenberg, I., N.C.A.da Costa and R.Chuaqui [198+] Pragmatic truth and approximation to truth. To appear

Scott, D. and P. Krauss [1966] Assigning probabilities to logical formulas, in Aspects of Inductive Logic, Hintikka and Suppes (editors), North-Holland Pubblic.Co., Amsterdam, pp. 219-264.

Stoll, A. [1982] A non-standard construction of Lévy Brownian motion with applications to invariance principles, Diplomarbeit (Mathematik), Universitat, Freiburg, BRD.

Suppes, P. [1957] Introduction to Logic, D. Van Nostrand Co., Inc., Princeton.

(1) This paper was partially supported by a grant of the Scientific and Technological Development Program of the Organization of American States and the Dirección de Investigación (DIUC) of the Pontificia Universidad Católica de Chile.

(2) The paper was partially written when the first author was at the Institute for Mathematical Studies in the Social Sciences at Stanford University, financed in part by a John Simon Guggenheim Memorial Foundation Fellowship.

(3) The authors would like to thank N.C.A.da Costa for many useful comments.

PARTITION RELATIONS IN ARITHMETIC

P. Clote[1]
UNIVERSITE PARIS VII
U.E.R. de Mathématiques et Informatique
Tour 45-55 5ème étage - 2 Place Jussieu
75230 Paris Cedex 05, France

§0. Introduction.

Recall the folklore result that Ramsey's Theorem , as a definable scheme, is provable in first order Peano arithmetic. In this paper, we give (infinite) combinatorial equivalents for certain subsystems of Peano arithmetic. Part of the original motivation for this work was to produce a certain amount of machinery to allow one to formalize combinatorial arguments in certain subsystems of arithmetic.

The results presented herein contribute to the proof theoretic study of fragments of Peano arithmetic - specifically that of Σ_n induction ($I\Sigma_n$) and Σ_n collection or bounding principle ($B\Sigma_n$). To see combinatorial significance of these subsystems, recall the well-known result of J. Paris [20] that a recursive function f is primitive recursive (or even Kalmar elementary) in the function $g_{n,m}$ for some positive integer m

$$g_{n,m}(x) = \text{least } y \text{ such that } [x,y] \underset{*}{\dashrightarrow} (n+2)^{n+1}_m$$

if and only if

$$I\Sigma_n \vdash \forall x\, \exists y\, \text{"}f(x) = y\text{"}$$

if and only if

$$B\Sigma_{n+1} \vdash \forall x \exists y\, \text{"}f(x) = y\text{"}$$

---

[1] These results were obtained and presented while giving a course in Models of Arithmetic in the fall semester of 1982-83 at the Université Paris VII. Author's present address: Department of Computer Science, Boston College, Chesnut Hill, MA 02167 USA.

This result links essentially the growth rate of a fast growing recursive function with the proof-theoretic notion of a function being provably recursive in a system of arithmetic. Our original motivation in this paper was to find infinite Ramsey Theorem-type characterizations of subsystems of Peano arithmetic with the intent to simplify some of the combinatorial wizardry of [20]. The main result, Theorem 11, turns out (in one direction) to be essentially a proof-theoretic formalization of the recursion theoretic analysis of Ramsey's Theorem done by C.G. Jockusch, Jr. in [6]. Our result is that a model M satisfies Σ_{n+1} collection if and only if for any recursive partition of n-element subsets of M into (possibly non-standard) boundedly many pieces, there is an unbounded homogeneous set. Schematically

$$M \models B\Sigma_{n+1} \quad \text{iff} \quad M \xrightarrow[\Delta_1]{} (M)^n_{<M}.$$

This characterization of the collection scheme then answers a question about initial segments posed by G. Mills and J. Paris in [9] : an initial segment I of M is n-Ramsey if and only if I is a model of $B\Sigma^*_{n+1}$. Along the way, we give an easy combinatorial equivalent of Σ_n induction. A model M satisfies Σ_n induction if and only if M satisfies a certain infinitary Δ_n pigeon hole principle : when any unbounded Δ_n definable subset of M is partitioned into (possibly non-standard) boundedly many pieces, then at least two elements go to the same piece (Theorem 4). This yield the useful corollary (Corollary 6) that Σ_n induction implies induction for Boolean combinations of Σ_n formulas, even prefixed by bounded quantifiers. In Theorem 8 we give an alternative formulation and proof of a result due to G. Mills and J. Paris [9] : a model M satisfies Σ_{2n} collection if and only if the filter product of the Fréchet filter with itself n-times

$$\underbrace{F \times F \times F \times \ldots \times F}_{n\text{-times}}$$

is Δ_0-complete.

The techniques used are model theoretic (involving end extensions and ultraproducts) and more especially recursion theoretic (involving formalized versions of the limit lemma and the Jockusch-Soare low-basis theorem, Lemma 14). Forthcoming work by the author has further exploited the low-basis theorem and other recursion theoretic techniques to answer some of the questions posed in this article.

1. Preliminaries.

A formula in the language of Peano arithmetic $\{+,\cdot,0,1,<\}$ is said to be bounded if it is built up from atomic formulas by the Boolean operations and by bounded quantification : $\exists x<y$ and $\forall x<y$.

$$\Delta_o = \Sigma_o = \Pi_o = \{\phi : \phi \text{ is a bounded formula in the language of arithmetic}\}.$$

Let Σ_{n+1} be the class of formulas on the form $\exists x_1, \ldots \exists x_m \theta$ where $m \in N$ and θ is in Π_n; let Π_{n+1} be the class of formulas of the form $\forall x_1, \ldots, \forall x_m \theta$ where $m \in N$ and θ is in Σ_n. If T is a theory in the language of arithmetic, then $\Sigma_n(T)$ and $\Pi_n(T)$ are the classes of formulas which are provably equivalent in T to Σ_n respectively Π_n formulas; also $\Delta_n(T) = \Sigma_n(T) \cap \Pi_n(T)$. As shown in [11,p.201] it requires a certain amount of induction or collection (see below) in a theory T to allow one to pass bounded quantifiers to the right of unbounded quantifiers thus obtaining T-equivalent formulas.

The theory P^- is the usual finitely axiomatized theory of a discretely ordered semi-ring (see [11], p. 200).

$I\Sigma_n$ (Σ_n-induction with parameters) is the scheme

$$\forall \vec{u}[(\phi(0,\vec{u}) \quad \& \quad \forall x(\phi(x,\vec{u}) \to \phi(x+1,\vec{u}))) \to \forall x\phi(x,\vec{u})]$$

where ϕ is a Σ_n formula.

$B\Sigma_n$(Σ_n-collection or bounding principle with parameters) is the scheme

$$\forall \vec{u}\ [\forall x < a\ \exists y\phi(x,y,\vec{u}) \to \exists b\ \forall x < a\ \exists y < b\phi(x,y,\vec{u})]$$

where ϕ is a Σ_n formula.

We remark that A. Wilkie has shown that $I\Sigma_o$ is equivalent over P^- to the scheme of Σ_o-induction without parameters, although of course for $n \geq 1$ the scheme with parameters is stronger. (This latter follows from Proposition 3). Peano arithmetic, denoted P, is the theory P^- together with all of the induction schemes. Throughout this paper, we adopt the convention that M is a countable model of the weak base theory $P^- + I\Sigma_o$.

A set X or function f will be said to be Σ_n(respectively Δ_n) in a model M if it is definable by a Σ_n(resp. both a Σ_n and a Π_n) formula with parameters in M (a function is definable if its graph is definable). This is denoted as $X \in \Sigma_n(M)$, etc...

DEFINITION.- $M \xrightarrow[\Delta_n]{} (M)^1_{<M}$ means that for any Δ_n definable unbounded set X of M and any Δ_n partition $F:X \to a$ (for partitions, we identify numbers with the set of their predecessors so that $F : X \to \{0,1,\dots,a-1\}$) where $a \in M$, there exists $i < a$ such that $F^{-1}(i)$ is unbounded in M. This is the infinitary Δ_n pigeonhole scheme : any Δ_n partition of a Δ_n unbounded subset into boundedly many pieces has one piece with unboundedly many elements.

Recall the well-known

FACT 1.- If $M \models P^- + I\Sigma_n$ for $n \geq 1$, then one can define functions by Σ_n recursion in M. More precisely, if g,h are total Σ_n functions in M, then so is f, where f is defined by

$$f(0,\vec{x}) = g(\vec{x}) \qquad f(a+1,\vec{x}) = h(a,\vec{x},f(a,\vec{x})).$$

PROOF. Since $n \geq 1$ and all primitive recursive functions are provably total in $I\Sigma_1$, we have

$$f(a,\vec{x})=y \underset{df}{\longleftrightarrow} \exists s\in Seq[lh(s)=a+1\ \&\ (s)_o=g(x)\ \&\ \forall\ i<a\ \ (s)_{i+1}=h(i,\vec{x},(s)_i)$$

$$\&\ \ (s)_a = y]$$

$\longleftrightarrow \forall s \in Seq\ [lh(s)=a+1\ \&\ (s)_o=g(x)\ \&\ \forall i<a\ \ (s)_{i+1}=h(i,\vec{x},(s)_i) \rightarrow (s)_a=y]$

Since $I\Sigma_n$ implies $B\Sigma_n$, the above formulas are equivalent to a Σ_n resp. Π_n formula. □

Recall that M' is an n-elementary end extension of M (denoted by $M \preccurlyeq_{n,e} M'$) iff M' is an extension of M satisfying

(i) for $\vec{u} \in M$ and θ a Σ_n or Π_n formula, $M \models \theta(\vec{u})$ iff $M' \models \theta(\vec{u})$ and

(ii) M' is an end extension of M (equivalently, M is an initial segment of M' - denoted by $M \subset_e M'$): $\forall m \in M\ \forall a \in M'\ (M' \models a \leq m \rightarrow \exists b \in M\ (M' \models a = b))$.

Our point of departure is from

THEOREM 2. (Kirby-Paris [11] Theorems A and B) Modulo $P^- + I\Sigma_o$, for $n \geq 0$

$$I\Sigma_{n+1} \Longrightarrow B\Sigma_{n+1} \Longleftrightarrow B\Pi_n \Longrightarrow I\Sigma_n$$

and these implications are strict. Furthermore, for M a countable model of $P^- + I\Sigma_o$ and for $n \geq 1$,

$$M \models B\Sigma_{n+1} \quad \text{iff} \quad M \underset{\Delta_n}{\rightarrow} (M)^1_{<M}$$

iff M admits a proper n + 1-elementary end extension which is a model of $P^- + I\Sigma_o$.

REMARK. By induction on $n \geq 1$ using Fact 1 and Theorem 2, it is easy to show that

$$M \xrightarrow[\Delta_n]{} (M)^1_{<M} \quad \text{iff} \quad M \xrightarrow[\Delta_n]{} (M)^1_{<M}$$

where the latter means : given any Δ_n definable partition $F : M \rightarrow a$ where $a \in M$, there is an $i < a$ for which $F^{-1}(i)$ is unbounded in M (i.e. the partition relation on the left concerns a partition of an unbounded subset into boundedly many pieces, whereas that on the right concerns a partition of the entire universe). The reason that we have defined the partition relations given before Fact 1 in terms of a partition on an unbounded subset will be clear when considering other partition relations (Theorems 5 and 9).

We remark here that the equivalence of $B\Sigma_{n+1}$ with $M \xrightarrow[\Delta_n]{} (M)^1_{<M}$ is more or less implicit in [11]; however, definable partition relations were first introduced by E. Kranakis, who proved the above equivalence in the context of set theory (Theorem 4.3 of [12]).

For instance, suppose that $M \xrightarrow[\Delta_1]{} (M)^1_{<M}$. We show first that $M \models I\Sigma_1$ by proving that every non-empty Σ_1 definable subset of M has a least element. Suppose that $M \models \exists y\theta(x,y)$ where θ is a bounded formula. Let $F : M \to a + 2$ be defined by

$$f(s) = \begin{cases} \text{least } x \le a \text{ such that } \exists\, y < s\theta(x,y) \text{ if such exists} \\ a + 1 \text{ otherwise.} \end{cases}$$

If $i \le a + 1$ is such that $F^{-1}(i)$ is unbounded in M, then clearly i is the least element in the given Σ_1 set. Now if X is a Δ_1 definable unbounded subset of M then by Fact 1, X is order isomorphic with M via a Δ_1 function and so

$$M \xrightarrow[\Delta_1]{} (M)^1_{<M} \text{ implies } M \xrightarrow[\Delta_1]{} (M)^1_{<M}.$$

For $n > 1$ argue in the same fashion, but using the induction hypothesis and Theorem 2 to have that $M \models B\Sigma_n$ and hence ([11]Lemma 3) a formula of the form $\forall\, y < s\, \theta$ where θ is Σ_n(respectively $\exists\, y < s\, \psi$ where ψ is Π_n)is equivalent in M to a Σ_n(respectively Π_n)formula.

Before going on, several things are worth noting. First is that in the proof that for $n \ge 1$

$M \models B\Sigma_{n+1}$ implies the existence of a proper $n + 1$-elementary end extension K of M which is itself a model of $P^- + I\Sigma_0$,

Kirby and Paris constructed a "Σ_n ultrafilter" $\mathcal{U}$ on $\Sigma_n(M)$, the Σ_n definable subsets of M by defining an ω-sequence $X_0 \supseteq X_1 \supseteq X_2 \supseteq \ldots$ such that

(1) each X_i is an unbounded Δ_n definable subset of M,

(2) given the ith Σ_n definable <u>partial</u> function f_i with bounded

range in an ω-listing with infinite repetitions of all such functions (recall that Σ_n definable means definable with parameters - it is at this point that countability of the model M is used), either X_{i+1} is disjoint from the domain of f_i or f_i is constant on X_{i+1}.

The "ultrafilter" $\mathcal{U}$ is then defined to be $\mathcal{U} = \{X \in \Sigma_n(M) : \exists\, i \in \omega(X \supseteq X_i)\}$. $\mathcal{U}$ is a "Σ_n ultrafilter" in the sense that for a Σ_n definable subset X of M, either $X \in \mathcal{U}$ or there is a disjoint Σ_n definable set Y with $Y \in \mathcal{U}$. Defining the restricted ultraproduct

$$K = \{f : f \text{ is a } \Sigma_n \text{ partial function with } \mathrm{dom}(f) \in \mathcal{U} \text{ and } \mathrm{rng}(f) \subset M\}/\mathcal{U},$$

then by using the fact that $M \models I\Sigma_n$, the following form of Łoś' Theorem holds : if ϕ is a Σ_n formula with m free variables then

$$K \models \phi([f_1],\ldots,[f_m]) \text{ iff } \{a \in M : a \text{ is in domain of } f_1,\ldots,f_m \text{ and } M \models \phi(f_1(a),\ldots,f_m(a))\} \in \mathcal{U}.$$

One then easily verifies that K is a proper n + 1-elementary end extension of M which satisfies $P^- + I\Sigma_0$. Each instance

$$\phi(0,\vec{u}) \;\&\; \forall\, x(\phi(x,\vec{u}) \to \phi(x+1,\vec{u})) \to \forall\, x\, \phi(x,\vec{u})$$

of a Σ_n or Π_n induction axiom with parameters restricted to M is of the form $\Sigma_n \vee \Sigma_{n+1} \vee \Pi_{n+1}$ (resp. $\Sigma_n \vee \Sigma_{n+1} \vee \Pi_n$) and so true in K. Similarly every instance of a Σ_{n-1} or Π_{n-1} induction axiom with parameters in K is true in K.

Thus it is a natural question whether given a model of $B\Sigma_{n+1}$ for $n \geq 1$, one can construct a proper n + 1-elementary end extension K satisfying $P^- + I\Sigma_n$. A negative answer follows from

<u>PROPOSITION 3</u>. If K is a proper n + 1-elementary end extension of M satisfying $I\Sigma_n$ with $n \geq 0$, then M is a model of $B\Sigma_{n+2}$.

<u>PROOF.</u> (The case for n = 0 is part of Theorem B in [11]. The proof for $n \geq 1$ arose in a conversation with Z. Adamowicz and A. Wilkie). We show that $M \models B\Pi_{n+1}$.

Suppose that $M \models \forall i<a \ \exists t \ \forall x \ \Phi$, where Φ is Σ_n. Then for every $d \in K - M$ we have

$$\forall i<a \ \exists t<d \quad M \models \forall x \ \Phi(i,t,x).$$

Fix $e_0 \in K - M$. Then for every $d \in K - M$,

$$\forall i<a \ \exists t<d \quad K \models \forall x<e_o \Phi(i,t,x).$$

$$K \models \forall i<a \ \exists t<d \ \forall x<e_o \Phi(i,t,x).$$

Since $K \models I\Sigma_n$, let d_o be the least d satisfying the above formula. As every $d \in K - M$ satisfies this formula, $d_o \in M$. It follows that

$$M \models \exists d \ \forall i<a \ \exists t<d \ \forall x \ \Phi.$$

QUESTIONS. 1) The version in arithmetic of a question posed by M. Kaufmann (see [8] p. 102) is whether, for $n \geq 1$ and any countable model M for $P^- + B\Sigma_{n+1}$ there is necessarily a proper $n+1$-elementary end extension of K of M which is a model of $P^- + B\Sigma_n$.

2) Does Theorem 2 hold for uncountable models? We mention that the natural attempts to solve this problem affirmatively proceed by trying to formalize a tree argument (as in Lemma 14) or by the arithmetized completeness theorem ([14], p. 252). However, these attempts fail, since the arguments appear to require more than the desired amount of collection.

3) To our knowledge, it is an open problem to characterize those models M with a proper n-elementary end extension K for which $K \models P^- + \Sigma_m$ or even $K \models P$. This is related to a problem due to J-P Ressayre : characterize those models M having a k-tower of n-elementary end extensions, i.e. $M \preccurlyeq_{n,e} M_1 \preccurlyeq_{n,e} M_2 \preccurlyeq_{n,e} \cdots \preccurlyeq_{n,e} M_k$. In this connection we also mention an open problem (with unpublished partial results) due to J. Paris and A. Wilkie : if M is a countable model of $P^- + I\Sigma_o + B\Sigma_1$, then does there exist a proper end extension K of M which is a model of $P^- + I\Sigma_o$ (i.e. an arithmetic analogue of a corollary to the Barwise compactness theorem)?

2. INDUCTION SCHEMES.

In [5] H. Friedman defined a weak subsystem of second order arithmetic "arithmetical comprehension axiom with restricted induction" ACA_o

with language $\{+,.,0,1,<,\epsilon\}$ and as axioms : P^-, extensionality, the induction axiom (and not the second order scheme)

$$\forall x[(0 \in X \ \& \ \forall x(x \in X \rightarrow x+1 \in X)) \rightarrow \forall x \ (x \in X)]$$

and arithmetical comprehension

$$\exists X \forall x \ (x \in X \longleftrightarrow \phi(x))$$

where ϕ is an arithmetical formula (i.e. quantification only over individuals and not sets) possibly with free second order variables other than X.

If M is a model of P then (M, Def M) is a model of ACA_o, where Def M is the class of subsets of M definable with parameters. Thus ACA_o is a conservative extension of P. Since ACA_o is equivalent to Σ_1-CA_o (where arithmetical comprehension is replaced by Σ_1 comprehension), one can show that ACA_o is finitely axiomatizable and that its essential axiom is equivalent to certain second order combinatorial theorems (König's Lemma, $\omega \rightarrow (\omega)^3_2$). By the Paris-Kirby miniaturization technique, this produces indicators for strong initial segments I of M (i.e. $(I, R_M I) \models ACA_o$ where $R_M I = \{A \cap I : A \in \text{Def } M\}$) and hence independence results for Peano arithmetic.

The motivation for the following easily established result was to solve the equation $\frac{P}{ACA_o} = \frac{I\Sigma_n}{?}$ and to possibly simplify independence results for the subsystems $I\Sigma_n$. See also the discussion at the end of [2].[1]

THEOREM 4. For $n \geq 1$ and M a model of $P^- + I\Sigma_o$, the following are equivalent.

1) $M \xrightarrow[\Delta_n]{} (2)^1_{<M}$

2) $M = \forall a \exists b \forall x \leq a(\exists y\Phi \longleftrightarrow \exists y \leq b\Phi)$, where Φ is any Π_{n-1} formula.

3) $M \models \forall a \exists b_1 \exists b_2 \ldots \exists b_n \forall x \leq a \ (Q_1x_1 \ldots Q_nx_n\Phi \longleftrightarrow Q_1x_1 \leq b_1 \ldots Q_nx_n \leq b_n\Phi)$ where Φ is any bounded formula and the Q_i quantifiers.

4) $(M, \text{Def}_o M) \models w - \Sigma_n CA_o$.

5) $M \models I\Sigma_n$.

---

[1] The referee has pointed out that $I\Sigma^o_n$ (second order Σ^o_n induction) seems to be the natural theory for solving this equation.

Here M $\xrightarrow[\Delta_n]{} (2)^1_{<M}$ means that for any Δ_n definable unbounded subset X of M and any Δ_n definable partition $F : X \to a$, where $a \in M$, there is an $i < a$ with card $F^{-1}(i) \geq 2$. This is a seemingly trivial infinitary pigeonhole principle asserting that any Δ_n partition of a Δ_n unbounded set into boundedly many pieces has a piece with at least two elements. $\mathrm{Def}_o M$ designates the collection of Δ_o definable (with parameters) subsets of M. The subsystem "weak Σ_n comprehension" $w - \Sigma_n CA_o$ is the theory with language $\{+,.,0,1,<,\epsilon\}$ and as axioms P^-, extensionality, induction axiom and the "localization" of Σ_n comprehension,

$$\forall m \exists X \forall x \leq m (x \in X \longleftrightarrow \phi(x))$$

where Φ is an Σ_n formula possibly with free second order parameters other than X.

PROOF. 1) → 2). Let $X = \{<x,b> : M \models x < a \,\&\, \Phi(x,b) \,\&\, \forall b'<b \neg \Phi(x,b')\}$. By the induction hypothesis, M is a model of $I\Sigma_{n-1}$ hence of $B\Sigma_{n-1}$, so X is Δ_n definable. If X is bounded, we obtain our conclusion. If X is unbounded, then let $F : X \to a$ be defined by $F(<x,b>) = x$. But then F violates the hypothesis 1).

2) → 3). By induction on the number of quantifiers.

3) → 4). Since M is a model of $I\Sigma_o$, $(M, \mathrm{Def}_o M) \models$ axiom of induction. Weak Σ_n comprehension follows from the hypothesis 3).

4) → 5). One shows the least element principle for Σ_n formulas: if $M \models \exists x \Phi(x,\vec{u})$ then $M \models \phi(a,\vec{u})$ for some $a \in M$; obtain $X \in \mathrm{Def}_o M$ with

$$(M, \mathrm{Def}_o M) \models \forall x \leq a (x \in X \longleftrightarrow \phi(x,\vec{u}))$$

and apply the induction axiom.

5) → 1). First show that

(*) $\quad P^- + I\Sigma_1 \vdash$ "for all a and all finite functions $f : a \to a - 1$ there exist two elements with the same image".

Now since M is a model of $I\Sigma_n$, for $F : X \to a$ a given Δ_n partition,

----

(*) We remark that the equivalence of 2) and 5) was noticed independently and much earlier by H. Friedman in [4].

$M \models \forall x \exists s \in Seq(lh(s) = x \,\&\, \forall i < x((s)_i$ is the $(i+1)^{st}$ element of $X))$.

Let g be a finite function in M with domain $\{0,\dots,a\}$ such that g(i) is the $(i+1)^{st}$ element of K. The $f = F \circ g$ satisfies (*), so that 1) holds.

REMARKS. 1) Notice that

$$M \underset{\Delta_n}{\longrightarrow} (2)^1_{<M} \text{ iff } M \underset{\Delta_n}{\longrightarrow} (m)^1_{<M} \text{ for any } m \in M.$$

This holds as well for $n = 0$.

2) In this notation, for an initial segment I of M, I is semi-regular iff $I \underset{\text{coded}}{\longrightarrow} (2)^1_{<I}$ and I is regular iff $I \underset{\text{coded}}{\longrightarrow} (I)^1_{<I}$.

3) $B\Sigma_n$ asserts that "the range of a Σ_n function with bounded domain is bounded" and $I\Sigma_n$ asserts that "the range of a Σ_n partial function with bounded domain is bounded".

COROLLARY 5. The Σ_n definable elements of a model of $P^- + I\Sigma_n$ are not cofinal in the model.

PROOF. Recall that an element $a \in M$ is Σ_n-definable if there is a Σ_n formula Φ such that

$$M \models \exists!\, x\Phi(x) \;\;\&\;\; \Phi(a).$$

There is a Σ_n satisfaction predicate Sat_n for Σ_n formulas. Let $a \in M \setminus N$ and apply 2) to obtain

$$M \models \exists b \,\forall x < a \,(\text{"}\exists y \text{ satisfying the } \Sigma_n \text{ formula coded by } x \longleftrightarrow \exists y < b \text{ satisfying the } \Sigma_n \text{ formula coded by } x\text{"}).$$

COROLLARY 6. $I\Sigma_n$ is equivalent over P^- to IB^0_n, where the latter is the induction scheme for the closure under bounded quantification of the collection of Boolean combinations of Σ_n formulas.

SKETCH PROOF. To obtain a least element for a non-empty set defined by $\exists y \phi \vee \forall y \psi$ where ϕ(resp. ψ) is Π_{n-1}(resp. Σ_{n-1}), suppose that

$$M \models \exists y\phi(a,y) \vee \forall y\psi(a,y).$$

Apply 2) to obtain b such that

$$M \models \forall x \le a((\exists y\phi(x,y) \longleftrightarrow \exists y < b\phi(x,y)) \ \& \ (\forall y\psi(x,y) \longleftrightarrow \forall y < b\psi(x,y))).$$

Now apply $I\Sigma_n$ to obtain a least element for the set

$$X = \{x \le a : M \models \exists y < b\phi(x,y) \vee \forall y < b\psi(x,y)\}.$$

We remark that the above corollary has been independently observed by R. Kossak, J. Paris and doubtless many others.

If $M \subseteq_e K$ then $K \models M - B\Sigma_n$ means that for any $a \in M$ and any Σ_n formula $\Phi(x,y,\vec{m})$ where $\vec{m} \in K$

$$K \models \forall x < a \exists y\Phi(x,y,\vec{m}) \rightarrow \exists b \forall x < a \exists y < b\Phi(x,y,\vec{m}).$$

A Σ_n-ultrafilter U on M is said to be Σ_n-complete if for any Σ_n-definable set $X \subseteq M^2$ and $a \in M$,

$$\forall i < a \ (X)_i \in U \Longrightarrow \bigcap_{i<a} (X)_i \in U.$$

where $(X)_i = \{z : (i,z) \in X\}$.

If U is a Σ_n-complete Σ_n-ultrafilter on M, then letting

$$K = \{f : f \text{ is an } \Sigma_n \text{ partial function with } \mathrm{dom}(f) \in U \text{ and } \mathrm{rng}(f) \subset M\}/U,$$

it is easy to see that K is an n + 1-elementary end extension of M and that $K \models M - B\Sigma_n$.

The following yields some insight about question 1) after proposition 3.

PROPOSITION 7. If $n \ge 2$ and M is a countable model of $P^- + I\Sigma_{n+1}$, then M has a proper n-elementary end extension $\bar{K}$ which is a model of $P^- + B\Sigma_n$.

PROOF. Using $I\Sigma_{n+1}$, we first construct a Σ_n-complete Σ_n-ultrafilter U on M.

Let $\{X_m : m \in \omega\}$ be an infinitely repetitive ω-list of all Σ_n-definable (with parameters) subsets of M^2 and let $\{a_m : m \in \omega\}$ be an infinitely repetitive ω-list of all elements of M. Let $<,>$ be a bijective pairing function and $(\ )_0$ and $(\ )_1$ be the first and second projection function.

We construct an ω-sequence

$$Y_o \supset Y_1 \supset \ldots \supset Y_k \supset \ldots$$

of Σ_n-definable sets such that

(i) each Y_k is unbounded in M

(ii) $Y_o = M$

(iii) given Y_k, if $Y_k \cap \bigcap_{i<(k)_o} (X_{(k)_1})_i$

is unbounded in M then set Y_{k+1} to be this set; otherwise by $I\Sigma_{n+1}$ we can find the largest $j < (k)_o$ such that

$$Y_k \cap \bigcap_{i<j} (X_{(k)_1})_i$$

is unbounded in M and we can find $\ell \in M$ such that

$$Y_k \cap \bigcap_{i<j+1} (X_{(k)_1})_i$$

is bounded by ℓ. In this case set

$$Y_{k+1} = \{x \in M : x > \ell \text{ and } x \in Y_k \cap \bigcap_{i<j} (X_{(k)_1})_i\}.$$

Now let

$$\mathcal{U} = \{Z \in \Sigma_n (M) : \text{there exists } k \in \omega \text{ such that } Z \supseteq Y_k\}.$$

Clearly $\mathcal{U}$ is a Σ_n-complete Σ_n-ultrafilter on M. As previously discussed, let K be an n + 1-elementary end extension of M satisfying M - $B\Sigma_n$.

Now form a "Σ_{n-1} ultrafilter" $\mathcal{U}'$ on Σ_{n-1} definable subsets (with parameters) of K which additionally satisfies

1) every element of $\mathcal{U}'$ is an unbounded Σ_{n-1} subset of K

2) if f is a $\Sigma_{n-1}(K)$ partial function with $\mathrm{dom}(f) \in \mathcal{U}'$ and $\mathrm{rng}(f) \subseteq m$ for some $m \in M$, then there exists $B \in \mathcal{U}'$ with $B \subseteq \mathrm{dom}(f)$ and f constant on B.

Now let

$L = \{f : f$ is $\Sigma_{n-1}(K)$ partial function with $\mathrm{dom}(f) \in \mathcal{U}'$ and $\mathrm{rng}(f) \subseteq K\}/\mathcal{U}'$.

Then Łoś' Theorem holds in the sense that

$L \models \theta([f_1],\ldots,[f_r])$ iff $\{i \in K : K \models \theta(f_1(i),\ldots,f_r(i))\} \in \mathcal{U}'$,

for all Σ_{n-1} formulas θ. It is now straightforward to check that L is a proper non-cofinal n-elementary extension of K, which furthermore is an end extension of M. (L is said to be a cofinal extension of M if $\forall a \in L \, \exists m \in M \, (L \models a \le m)$. Also we write $K \preccurlyeq_{n,M} L$ to mean that L is an n-elementary extension of K where both K and L are end extensions of M). Let $\bar{K}$ be the cofinal closure of K in L:

$$\bar{K} = \{a \in L : \exists m \in K (L \models a \le m)\}.$$

Now since $M \preccurlyeq_{n+1} K \preccurlyeq_n L$, by the discussion before Proposition 3, we have $K \models I\Sigma_{n-1}$ and $L \models I\Sigma_{n-2}$.

<u>CLAIM.</u> $\bar{K} \preccurlyeq_{n-1,e} L$.

<u>PROOF OF CLAIM.</u> Clearly L is an end extension of $\bar{K}$. We show by induction on $k \le n-1$ that $\bar{K} \preccurlyeq_{k,e} L$. This is clear for $k = 0$ since $\bar{K} \subseteq_e L$. Consider the case $k+1$: suppose that $L \models \exists x \phi(x,[g])$ where ϕ is Π_k and $[g] \in \bar{K}$ and $m \in K$ and $\bar{K} \models [g] \le m$. Then $A = \{i \in K : K \models g(i) \le m \;\&\; \exists x \phi(x,g(i))\} \in \mathcal{U}'$. Since $K \models I\Sigma_{n-2}$, by Theorem 4, there is a $b \in K$ such that

$$K \models \forall z \le m \, (\exists x \phi(x,z) \longleftrightarrow \exists x \le b \phi(x,z)).$$

Now let $\mathrm{dom}(h) = A$ and

$$h : i \longmapsto \text{least } x \in K \text{ such that } K \models \phi(x,z).$$

Then h is a $\Sigma_{n-1}(K)$ partial function and $L \models [k] \le b$, thus $[h] \in \bar{K}$ and $\bar{K} \models \phi([h],[g])$. Thus $\bar{K} \models \exists x \phi(x,[g])$.

<u>CLAIM.</u> $M \preccurlyeq_{n,e} \bar{K}$.

<u>PROOF OF CLAIM.</u> Clearly $\bar{K}$ is an end extension of M and so $M \preccurlyeq_0 \bar{K}$. We show that $M \preccurlyeq_n \bar{K}$. Suppose that $\bar{K} \models \exists x \phi(x,m)$ where $m \in M$ and ϕ is a Π_{n-1} formula. Then $\bar{K} \models \phi([f],m)$ for some $[f] \in \bar{K}$ and so by the previous claim, $L \models \phi([f],m)$. Thus

$$A = \{i \in K : K \models \neg \phi(f(i),m)\} \notin \mathcal{U}' \quad \text{so there is a } B \in \mathcal{U}' \text{ with } B \cap A = \emptyset$$

and $B \subseteq \{i \in K : K \models \phi(f(i),m)\}$.

(We write this in terms of a double negation since Los' Theorem holds only for Σ_{n-1} formulas). Let i_o be such that $K = \phi(f(i_o),m)$. Then $\{i \in M : M \models \phi(f(i_o)(i),m)\} \in \mathcal{U}$, so that $M \models \exists x\, \phi(x,m)$.

Now since M has a proper n-elementary end extension $\bar{K}$ which itself has a proper n-1-elementary end extension $L \models P^- + I\Sigma_{n-2}$, so by Proposition 3, $\bar{K} \models P^- + B\Sigma_n$.

REMARKS. 1) Can one show that $M \models P^- + B\Sigma_{n+1}$ iff there exists an n-elementary end extension $\bar{K}$ which is a model of $P^- + B\Sigma_n$? (The direction from right to left follows from Proposition 3).

2) Notice that in the proof of Theorem 4, we actually show that

$$M \xrightarrow[\Delta_o]{} (2)^1_{<M} \quad \text{iff} \quad M \models I\Sigma_1.$$

One can also show that

$$M \xrightarrow[\Delta_o]{} (M)^1_{<M} \quad \text{iff} \quad M \models B\Sigma_2.$$

Recall that the primitive recursive functions are exactly those which are provably total in $P^- + I\Sigma_1$ (theorem due to Minc) and hence in the above schemes. Also recall the results in complexity theory:

Wrathall's Theorem [18]. The linear time hierarchy = Rudimentary.

Bennett's Theorem [1]. Rudimentary = $Def_o N$.

Since it is still an open question whether the linear time hierarchy collapses or not, G. Wilmers raised the question whether

$$M \xrightarrow[E_n]{} (2)^1_{<M} \text{ is strictly weaker than } M \xrightarrow[E_{n+1}]{} (2)^1_{<M}$$

where an E_n formula is one with n alternating blocks of bounded quantifiers beginning with bounded existential quantifiers.

A. Wilkie showed that the answer is "no", the exact level where $M \xrightarrow[E_n]{} (2)^1_{<M}$ implies that M is a model of $I\Sigma_1$ depending on the logical complexity of the Δ_o formula expressing the graph of the exponential function (in the proof of Bennett's Theorem). Whether the graph can be E_1 is an open question. We present A. Wilkie's proof.

PROOF. Let n_o be such that $\phi(x,y,z)$ is an E_{n_o} formula representing $x^y = z$. If $M \models \forall a,b \exists c(a^b = c)$ for some $a,b \in M$ then let

$$X = \{\langle y,z\rangle : M \models a^y = z \ \& \ y \le b\}.$$

If X is bounded then by $I\Sigma_o$, X has a maximum element - say $\langle y_o,z_o\rangle$. But then $\langle y_o + 1, a\circ z\rangle \in X$ thus contradicting maximality of $\langle y_o,z_o\rangle$. If X is unbounded in M, then let $F : X \to \{0,\dots,b\}$ by $F(\langle y,z\rangle) = y$. But this contradicts our assumption that $M \xrightarrow[E_{n_o}]{} (2)^1_{<M}$. (This part of the argument is due to G. Wilmers). Hence

$$M \xrightarrow[E_{n_o}]{} (2)^1_{<M} \text{ implies that } M \models \mathrm{Exp},$$

where Exp is the statement $\forall x \forall y \exists z \phi(x,y,z)$. By Dimitracopoulos ([3],p.31), $P^- + I\Sigma_o + \mathrm{Exp} \vdash$ Matijasevic' Theorem. We now show that M satisfies the least element principle for Σ_1 formulas. Suppose that $M \models \exists x\theta(x,\vec{u})$ & $\theta(a,\vec{u})$ where θ is Σ_1. By Matijasevic' Theorem,

$$M \models \forall \vec{u} \ (\theta(x,\vec{u}) \longleftrightarrow \exists \vec{w}(p(x,\vec{u},\vec{w}) = q(x,\vec{u},\vec{w}))$$

where p and q are polynomials with non-negative integer coefficients. Coding $\vec{w}$ by a single element w, let

$$X = \{\langle x,w\rangle : M \models x \le a \ \& \ p(x,\vec{u},w) = q(x,\vec{u},w) \ \& \ \forall w' < w \ p(x,\vec{u},w') \ne q(x,\vec{u},w')\}.$$

If X is bounded above by d then

$$M \models \forall x \le a(\theta(x,u) \longleftrightarrow \exists \vec{w} \le d(p(x,\vec{u},\vec{w}) = q(x,\vec{u},\vec{w})))$$

and as M is a model of $I\Sigma_o$, we are finished. If X is bounded in M, then define $F : X \to \{0,\dots,a\}$ by $F(\langle x,w\rangle) = x$ and we have a violation of $M \xrightarrow[E_{n_o}]{} (2)^1_{<M}$. □

Notice also that the infinitary Δ_o pigeonhole scheme has little to do with the finitary Δ_o pigeonhole scheme (due to A. Macintyre). A. Macintyre's question

$$P^- + I\Sigma_o \vdash \text{"finitary } \Delta_o \text{ pigeonhole scheme"}$$

is still an open question, where the latter scheme is

$$\forall \vec{u}\ (\forall x(\forall i \le x\ \exists !\ y < x\theta(i,y,\vec{u}) \rightarrow \exists\ i \ne i' \le x\ \exists\ y < x(\theta(i,y,\vec{u})\ \&\ \theta(i',y,\vec{u})))$$

where θ is a Δ_o formula. In this connection, by an ingenious argument, A. Woods [17] has shown that

$$P^- + I\Sigma_o \vdash \text{"finitary } \Delta_o \text{ pigeonhole"} \rightarrow \forall\ x\ \exists\ y(x \le y \le 2x\ \&\ y \text{ is prime})$$

3. COLLECTION SCHEMES.

Let us first introduce the

DEFINITION. For M a model of $P^- + I\Sigma_o$, the Fréchet filter F_M (we drop the subscript M if clear from the context) is the collection of all co-bounded subsets of M (sets A whose complement is bounded in M). If $A \subseteq M^{k+1}$ and $a \in M$ then the ath section of A is $A_a = \{(b_1,\ldots,b_k) : (a,b_1,\ldots,b_k) \in A\}$. Now let F^k be the usual k-fold filter product of F; that is,

$$F^1 = F;\ F^{k+1} = F \times F^k = \{A \subseteq M^{k+1} : \{a \in M : A_a \in F^k\} \in F\}.$$

Similarly a set $A \subseteq M^{k+1}$ is unbounded if $\{a \in M : A_a \text{ unbounded in } M^k\}$ is unbounded.

A filter F on M^k is Δ_n-complete if for all Δ_n definable sets $X \subseteq M^{k+1}$ and $a \in M$, if $\forall\ i < a\ X_i \in F$ then $\bigcap_{i<a} X_i \in F$.

In this terminology, we can rephrase and give a different proof of a theorem due to G. Mills and J. Paris [9] .

The following lemma simplifies a small step in the proof we give of the Mills-Paris result and will be used later in the proof of Theorem 11.

LIMIT LEMMA. If $M \models P^- + B\Sigma_{n+1}$ where $n \ge 0$ and if $f \in \Sigma_{n+1}(M)$ is a total function from M into M, then there exists an "approximating" total function $g \in \Sigma_n(M)$ from M^2 into M such that for all $a \in M$

$$f(a) = \lim_{s \in M} g(a,s) .$$

Here, in the obvious sense,

$$\lim_{s\in M} g(a,s) = b \quad \text{iff} \quad M \models \exists\, s\, \forall\, t \geq s\; g(a,t) = b$$

PROOF. This is trivial if $n = 0$ - one just bounds the existential quantifier to obtain an approximation; the general result is just an arithmetical version of the well-known limit lemma in recursion theory - that a Δ_2 function can be approximated by a Δ_1 function. We adapt the proof due to Putnam ([16],p. 51). We show first that if M is a model of $P^- + I\Sigma_n$ and if $X \in \Delta_{n+1}(M)$, then there exists a Δ_n definable function $h: M^2 \to \{0,1\}$ which approximates the characteristic function of X. Suppose that

$$a \in X \longleftrightarrow M \models \exists\, y\, \forall\, z \phi(a,y,z)$$

$$a \notin X \longleftrightarrow M \models \exists\, y\, \forall\, z \psi(a,y,z), \phi \text{ and } \psi \text{ both } \Sigma_{n-1} \text{ formulas}$$

Define $\theta(a,y,s) \longleftrightarrow$ y is the least element in M satisfying

$$(\forall\, z \leq s\phi(a,y,z)\ \&\ \neg \forall\, z \leq s\psi(a,y,z)) \text{ or } (\forall\, z \leq s\psi(a,y,z)$$
$$\&\ \neg \forall\, z \leq s\phi(a,y,z)).$$

Now define

$$h(a,s) = \begin{cases} 0 & \text{if } \exists\, y \leq s(\theta(a,y,s)\ \&\ \forall\, z \leq s\phi(a,y,z)\ \&\ \neg \forall\, z \leq s\psi(a,y,z)) \\ 1 & \text{otherwise.} \end{cases}$$

Since $M \models B\Sigma_n$ the above is a Δ_n definition in M. One than shows that

$$\lim_s h(a,s) = 0 \longleftrightarrow a \in X$$

$$\lim_s h(a,s) = 1 \longleftrightarrow a \notin X.$$

IDEA OF PROOF. Suppose that $a \in X$. Let y_o = least y such that $\forall\, z\phi(a,y,z)$ - this exists by $I\Sigma_n$. Thus

$$\forall\, y \leq y_o \exists\, z \neg \psi(a,y,z) \text{ and } \forall\, y < y_o \exists\, z \neg \phi(a,y,z)$$

and by $B\Sigma_n$ we can bound all these z by an s larger than y_o. Then for any t larger than s, we have $h(a,t) = 0$. One argues similarly if $a \notin X$.

Now if $f : M \to M$ is Σ_{n+1} definable, the graph of f

$$X = \{\langle a,b\rangle : f(a) = b\}$$

is Δ_{n+1} definable. By the above discussion, let $h \in \Sigma_n(M)$ be an approximation of the characteristic function of X. Define

$$g(a,s) \begin{cases} \text{least } y \le s(h(<a,y>) = 0) \text{ if such exists} \\ 0 \quad \text{otherwise} \end{cases}$$

If f(a) = b then

$$M \models \forall y < b \exists s \forall t \ge s(h(<a,y>) = 1 \ \& \ \exists s \forall t \ge s \ h(<a,t>) = 0.$$

So by $B\Sigma_{n+1}$, g eventually takes on the value b. □

THEOREM 8. (Mills-Paris). For M a model of $P^- + I\Sigma_o$ and $k \ge 1$

$$M \models B\Sigma_{2k} \text{ iff } F^k \text{ is } \Delta_1\text{-complete (or even } \Delta_o\text{-complete).}$$

PROOF. We first recall the underlying idea in the proof of Mills-Paris. The difficult direction is from right to left. Suppose that M satisfies the hypothesis but not the conclusion of an instance of $B\Pi_{2k-1}$: say ϕ is Π_{2k-1} and so is of the form $\forall x_2 \ldots \forall x_{2k}\theta$ where θ is bounded. Suppose that

$$M \models \forall i < a \exists x_1 \forall x_2 \ldots \exists x_{2k-1} \forall x_{2k}\theta.$$

Mills-Paris then show that modulo $B\Sigma_{2k-2}$ any Π_{2k} formula is equivalent to a U_k formula, where, as in recursion theory, the quantifier U is interpreted as "there exist unboundedly many". Hence by the induction hypothesis,

$$M \models \forall i < a \neg Uy_1 \ldots Uy_k\psi \text{ where } \psi \text{ is bounded.}$$

But this is equivalent to saying that

$$\forall i < a \ X_i \in F^k \quad \text{where } X = \{(y_2,\ldots,y_k) : M = \neg \psi(y_1,\ldots,y_k)\}.$$

By the hypothesis that F^k is Δ_o-complete, $\bigcap_{i<a} X_i \in F^k$ and so (essentially) $M \models \neg Uy_1 \ldots Uy_k \forall i < a\psi$ and we then obtain a bound b such that

$$M \models \forall i < a \exists x < b\phi.$$

The upshot of the argument is that since $\forall \exists$ is equivalent to U, in order to obtain the Δ_1-completeness of F^k, we need collection for Σ_{2k} formulas.

Using what has gone before, we proceed slightly differently to prove that for $k \geq 1$ and $n \geq 0$

$$M \models B\Sigma_{2k+n} \text{ iff } F^k \text{ is } \Delta_{n+1}\text{-complete.}$$

First we introduce the following definition due to J-P Ressayre:

$$M^k \xrightarrow[\Delta_n]{} (M^k)^1_{<M}$$

means that for any Δ_n definable unbounded set $X \subseteq M^k$ and any Δ_n definable partition $F : X \to a$ where $a \in M$, there is an $i < a$ with $F^{-1}(i)$ unbounded.

<u>FACT.</u> $M^k \xrightarrow[\Delta_n]{} (M^k)^1_{<M}$ iff F^k is Δ_n-complete.

This fact follows easily from the definitions and is left to the reader. Now it is easy to verify that

$$M \models B\Sigma_{2k+n} \text{ implies } M^k \xrightarrow[\Delta_{n+1}]{} (M^k)^1_{<M} .$$

For the converse, we proceed by double induction : for k fixed, we induct on n.

If $k = 1$, then this is just Theorem 2. Now suppose that $k = 2$ and $n = 0$ and that

$$M^2 \xrightarrow[\Delta_1]{} (M^2)^1_{<M}.$$

The idea is to "step down" the exponent 2 to 1 and add 2 to the complexity of the partition (removing one occurence of "unbounded" adds a factor of 2 to the complexity of the partition). Such an idea of a "step down" is originally due to E. Kranakis [13] (see also the proof of Theorem 11). Thus we propose to prove that

$$M \xrightarrow[\Delta_3]{} (M)^1_{<M}$$

which by Theorem 3 is equivalent with $B\Sigma_4$.

<u>LEMMA A.</u> For $n \geq 0$ and every Σ_{n+1} formula of the form $\exists x \Phi$ where Φ is Π_n, there exists a formula $\phi \in \Pi_n (P^- + I\Sigma_n)$ such that

$$P^- + I\Sigma_n \vdash \exists x\Phi \longleftrightarrow \exists ! x\phi.$$

<u>PROOF.</u> This fact was observed in recursion theory by [19] - notice that the formula ϕ is Π_n and not simply of the form $\Phi(x)$ & $\forall i < x \neg \Phi(i)$which would be equivalent to a "Π_n & Σ_n" formula. This is proved by induction on n: for n = 0, let $\phi(x)$ be $\Phi(x)$ & $\forall i < x \neg \Phi(i)$ - this uses $I\Sigma_o$. For n = 1,

$$P^- + I\Sigma_1 \vdash \exists x\Phi \longleftrightarrow \exists x(\Phi(x) \ \& \ \forall i < x \neg \Phi(i)).$$

Let $\neg \Phi(x)$ be equivalent to $\exists y\theta(x,y)$ with θ bounded. By the induction hypothesis, $\forall i < x \neg \Phi(i)$ is equivalent to $\forall i < x \exists ! y\theta^*(i,y)$ and so $B\Sigma_1 + I\Sigma_1$ to

$\exists ! x \forall i < x\theta^*(i,(z)_i)$, using the usual projection functions for the ith element of a sequence coded by z.

Thus $\exists x\Phi$ is equivalent to $\exists ! u(\Phi((u)_o) \ \& \ \forall i < (u)_o\theta^*(i,((u)_1)_i))$.

Proceed in a similar manner for $n \geq 2$. □

We leave it to the reader to define $\exists ! x_1 \forall x_2 \exists ! x_3 \ldots$ $\ldots \forall x_{2m} \exists ! x_{2m+1}\theta$.

For instance, $\exists ! x \forall y \exists ! z\theta$ is an abbreviation for

$$\exists x(\forall y \exists ! z\theta'(x,y,z) \ \& \ \forall i \neq x \neg (\forall y \exists z\theta(i,y,z))).$$

<u>LEMMA B.</u> $M^2 \xrightarrow[\Delta_1]{} (M^2)^1_{<M}$ implies that M is a model of $I\Sigma_2$.

<u>PROOF.</u> The hypothesis implies that $M \xrightarrow[\Delta_1]{} (M)^1_{<M}$ and so M is a model of $B\Sigma_2$. To see that $M \xrightarrow[\Delta_2]{} (M)^1_{<M}$ and hence that M satisfies $B\Sigma_3$ and so $I\Sigma_2$, suppose that $F : M \to \{0,\ldots,a-1\}$ is a Δ_2 partition and $a \in M$; by the limit lemma let g be a Δ_1 total approximating function with

$$\text{"}\lim_s\text{"}\ g(x,s) = i\text{"} \longleftrightarrow F(x) = i.$$

Then $g : M^2 \to \{0,\ldots,a-1\}$ and if (by the hypothesis) $g^{-1}(i) \in F^2$, then it is clear that $F^{-1}(i)$ is unbounded in M. (A variant of this argument works if we had begun with $F : X \to \{0,\ldots,a-1\}$ where X is a Δ_2 definable unbounded subset of M; but by the remark after Theorem 2, this argument suffices). □

LEMMA C. $M^2 \underset{\Delta_1}{\longrightarrow} (M^2)^1_{<M}$ implies that $M \underset{\Delta_3}{\longrightarrow} (M)^1_{<M}$.

<u>PROOF.</u> Suppose that X is a Σ_3 definable unbounded subset of M and that $F : X \to a$ is a Σ_3 partition and $a \in M$. Then by $I\Sigma_2$,

$$x \in X \longleftrightarrow M \models \exists\, !\, y\ \forall z\ \exists\, !\, w \phi(x,y,z,w)$$

$$F(x) = i \longleftrightarrow M \models \exists\, !\, y\ \forall z\ \exists\, !\, w \psi(x,i,y,z,w)$$

where ϕ and ψ are bounded.

Define

$$Z = \{(\langle x,i,y_1,y_2\rangle, \langle z,u_z,v_z\rangle) : M \models i < a \ \& \ \phi(x,y_1,z,u_z) \ \& \ \psi(x,i,y_2,z,v_z)\}.$$

Consider the picture:

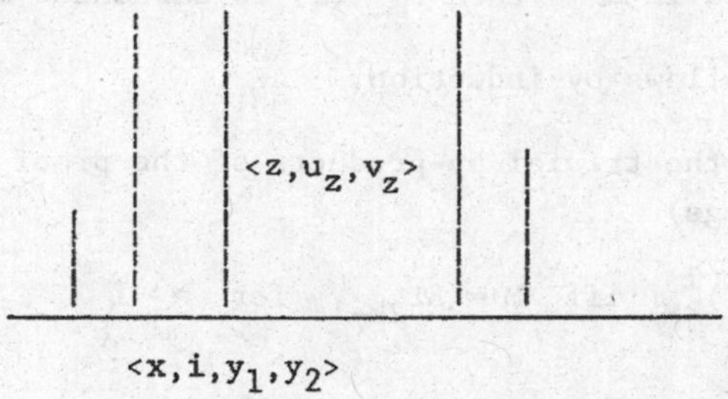

Then it is easy to see that $x \in X$ and $F(x) = i$ iff the section $Z_{\langle x,i,y_1,y_2\rangle}$ is unbounded for some unique y_1, y_2. Let $G(\langle x,i,y_1,y_2\rangle, \langle z,u_z,v_z\rangle) = i$. Then Z is a Δ_o definable "unbounded" subset of M^2 and G is a Δ_o partition of Z into $\{0,\ldots,a-1\}$.

Clearly if $G^{-1}(i)$ is unbounded then $F^{-1}(i)$ is an unbounded subset of X.

In a similar manner, we have

$$M^2 \underset{\Delta_n}{\longrightarrow} (M^2)^1_{<M} \text{ implies that } M \underset{\Delta_{n+2}}{\longrightarrow} (M)^1_{<M}. \qquad \square$$

To obtain $M^{k+1} \underset{\Delta_1}{\longrightarrow} (M^{k+1})^1_{<M}$ implies $M^k \underset{\Delta_3}{\longrightarrow} (M^k)^1_{<M}$, suppose that X is a Σ_3 unbounded subset of M^k and $F : M \to \{0,\ldots,a-1\}$ is a Σ_3 partition of X into a-many pieces where $a \in M$ By $I\Sigma_2$ let

$$(x_1,\ldots x_k) \in X \longleftrightarrow M \models \exists\, !y \forall z \exists\, ! w\, \phi$$

$$F(x_1,\ldots,x_k) = i \longleftrightarrow M \models \exists\, ! y \forall z \exists\, ! w\, \psi$$

where ϕ and ψ are bounded

Define $Z = \{(x_1,x_2,\ldots,x_{k-1},\langle x_k,i,y_1,y_2\rangle, \langle z,u_z v_z\rangle) :$

$$M \models i < a \ \&\ \phi(x_1,\ldots,x_k,y_1,z,u_z) \ \&\ \psi(x_1,\ldots,x_k,i,y_2,z,v_z)\}$$

and $G : Z \to \{0,\ldots,a-1\}$ by $G((x_1,\ldots,x_{k-1},\langle x_k,i,y_1,y_2\rangle, \langle z,u_z,v_z\rangle) = i.$

Then Z and G are Δ_o definable and Z is unbounded in M^{k+1}. Clearly if $G^{-1}(i)$ is unbounded in M^{k+1} then F^{-1} (i) is unbounded in M^k.

The result now follows by induction.

We state some of the trivial by-products of the proof : (with the obvious intended meanings)

$$M^k \underset{\Delta_n}{\longrightarrow} (M^k)^1_{<M} \quad \text{iff} \quad M \models B\Sigma_{2k-1} \quad \text{for} \quad n \geq 1$$

$$M^k \underset{\Delta_o}{\longrightarrow} (M^k)^1_{<M} \quad \text{iff} \quad M \models B\Sigma_{2k}.$$

We define a Δ_n-based ultrafilter $\mathcal{U}$ on M^k to be a collection of subsets of M^k (not necessarily Δ_n-definable) such that

(i) $A \in \mathcal{U} \to$ there exists $B \in \Delta_n(M)$ with $B \subseteq A$ and $B \in \mathcal{U}$ and B unbounded in M^k.

(ii) $A,B \in \mathcal{U} \to A \cap B \in \mathcal{U}$

(iii) $A \in \mathcal{U}$ and $A \subseteq B \subseteq M \to B \in \mathcal{U}$

(iv) $A \in \Delta_n(M) \to$ either $A \in \mathcal{U}$ or $\bar{A} = M^k - A \in \mathcal{U}$.

With this notation, we mention

<u>PROPOSITION 9</u>. (Pino [15]). For $n,k \geq 1$, if M is a countable model of $B\Sigma_{2k+n-1}$, then there is a Δ_n-complete Δ_n-based ultrafilter $\mathcal{U}$ on M^k.

<u>THEOREM 10.</u> (Ressayre-Pino [15]). For $n,k \geq 1$, if there is a Δ_n-complete

Δ_n-based ultrafilter $\mathcal{U}$ on M^k, then there is a k-tower of proper n+1-elementary end extensions

$$M \prec_{n+1,e} M_1 \prec_{n+1,e} M_2 \prec_{n+1,e} \cdots \prec_{n+1,e} M_k \models P^- + I\Sigma_o.$$

In view of Theorem 9, it is natural question (due to Ressayre-Pino [15]) whether the converse or Proposition 9 and Theorem 10 holds. (By the limit lemma, it is easy to show that if there exists a Δ_n-complete Δ_n-based ultrafilter $\mathcal{U}$ on M^k then $M \models B\Sigma_{n+k}$).

We come now to a final combinatorial equivalence of certain collection schemes. The result turns out to be an arithmetical version of C.G Jockusch's recursion theoretic analysis of Ramsey's Theorem [6]. The proof requires much of the machinery thus far introduced. Our original motivation was to consider the "2-3 question" [10] from a proof theoretical angle. Another motivation was to find combinatorial schemes equivalent to certain collection schemes so as to facilitate the formalization of combinatorial type arguments in fragments of Peano arithmetic.

<u>DEFINITION</u>. $M \xrightarrow[\Delta_m]{} (M)^n_{<M}$ means that given any Δ_m unbounded subset X of M and any Δ_m partition $F : [X]^n \to a$, where $a \in M$, of the n-element subsets of X into a-many pieces, there is an unbounded <u>homogeneous</u> set $Y \subseteq X$ (i.e. there exists an $i < a$ with $F''[Y]^n = \{i\}$).

A word of warning: it turns out that Y will be definable in M, but the above scheme is written with the second order variable Y. Also, by the remark after Theorem 2, it suffices to consider partitions $F : [M]^n \to a$.

<u>THEOREM 11</u>. For $n, m \geq 1$

$$M \xrightarrow[\Delta_m]{} (M)^n_{<M} \quad \text{iff} \quad M \models B\Sigma_{n+m}.$$

<u>PROOF.</u> ($\to$) This part is easy and uses the limit lemma to step down the exponent of a partition while stepping up the complexity of the partition. By induction on n : if n = 1, this is Theorem 2. Suppose that n = k + 1 and

$$M \xrightarrow[\Delta_m]{} (M)^{k+1}_{<M}$$

We show that

$$M \xrightarrow[\Delta_{m+1}]{} (M)^k_{<M}.$$

Using $B\Sigma_{m+1}$, apply the limit lemma to obtain a Δ_m approximation

$$G : [M]^{k+1} \to a \text{ with } \lim_s G(x_1 m, \ldots, x_k, s) = F(x_1, \ldots, x_k).$$

If Y is homogeneous for G and unbounded in M, then Y is also homogeneous for F.

($\leftarrow$) This direction is more difficult and proceeds by a series of lemmas.

First, let $I\Delta_n$ be the scheme

$$\forall \vec{u}(\forall x(\phi(x,\vec{u}) \longleftrightarrow \psi(x,\vec{u})) \to ((\phi(0,\vec{u}) \,\&\, \forall x(\phi(x,\vec{u}) \to \phi(x+1,\vec{u}))) \to \forall x\phi(x,\vec{u})))$$

where ϕ is Σ_n and ψ is Π_n.

<u>LEMMA 12.</u> For $n \geq 0$, $P^- + I\Sigma_o + B\Sigma_{n+1} \vdash I\Delta_{n+1}$.

<u>PROOF.</u> This is well-known and trivial for $n = 0$. For $n \geq 1$, we show that the least element principle holds for non-empty Δ_{n+1} definable sets. Suppose that $a \in X \in \Delta_{n+1}(M)$. By the limit lemma, let g be a Δ_n approximation for the characteristic function of X; let

$$X_s = \{x \leq a : M \models g(x,s) = 0\}.$$

Then $M \models \forall x \leq a \,\exists s \,\forall t \geq s \; g(x,s) = g(x,t)$, so by $B\Sigma_{n+1}$, let s_o be sufficiently large to bound all these values of s. Now apply $I\Sigma_n$ to the set X_s to obtain a least element - this is also the least element of X.

<u>REMARK</u>. Independently and much earlier, H. Friedman noticed this lemma. In fact, in [4] it is stated that $B\Sigma_{n+1}$ and $I\Delta_{n+1}$ are equivalent over the base theory $P^- + I\Sigma_o + \forall x \,\exists y(2^x = y)$.

Before giving details of the proof of Theorem 11, we give a sketch of the idea behind the proof that

$$M \models B\Sigma_3 \quad \text{implies} \quad M \xrightarrow[\Delta_1]{} (M)^2_{<M}$$

First, in the usual manner, associate a Δ_1 definable "Erdos-Rado" tree $T \subseteq M^{<M}$ with the partition $F : [M]^2 \to a$ such that an unbounded branch B of the tree is "pseudo-homogeneous" in the sense that if $b_i, b_j, b_k \in B$ and $b_i < b_j$ & $b_i < b_k$ then $F(b_i, b_j) = F(b_i, b_k)$. Suppose, in order to obtain a contradiction, that there is no unbounded monochromatic subset of B. Then

$$M \models \forall i < a \, \exists x \, \forall y,z \geq x (y,z \in B \;\&\; y < z \to F(y,z) \neq i).$$

A trivial basis theorem produces an unbounded branch B which is Δ_3- thus if M is a model of $B\Sigma_4$ then

$$M \models \exists t \, \forall i < a \, \exists x \leq t \, \forall y,z \geq x (y,z \in B \;\&\; y < z \to F(y,z) \neq i)$$

which is absurd since B is unbounded and F is defined on $[M]^2$.

To push through this argument when M is only a model of $B\Sigma_3$, we formalize the Jockusch-Soare "low basis" theorem [7]: if $T \subseteq \omega^{<\omega}$ is a recursive infinite finite branching tree, then there exists an infinite branch B whose jump B' is Δ_3 definable. To formalize this, we modify the Σ_1 satisfaction predicate for Σ_1 formulas (to obtain an analogue of the jump operator) and obtain an unbounded Δ_3 definable "1-generic" branch B, so that via the satisfaction predicate all Σ_1 and Π_1 formulas in the extended language $\{+, \cdot, 0, 1, <, B\}$ are equivalent in M to a Δ_3 formula : then

$$\forall y,z \geq x (y,z \in B \;\&\; y < z \to F(y,z) \neq i$$

is equivalent to a Δ_3 formula and so $B\Sigma_3$ suffices to obtain a contradiction.

For the following discussion, fix a (not necessarily countable) model M of $P^- + I\Sigma_0$. Recall that Seq is a Δ_1 definition of sequence numbers. We say that T is a Σ_n-<u>tree in M</u> if T is a Σ_n definable (with parameters) subset of M, $T \subseteq Seq^M = \{a \in M : M \models Seq(a)\}$ and

$$M \models \forall s,t(\text{"}t \in T\text{"} \;\&\; lh(s) \leq lh(t) \;\&\; \forall i < lh(s)((s)_i = (t)_i \to \text{"}s \in T\text{"}).$$

To make things more readable, we work in M and adopt usual set theoretic conventions : "$t \in T$" abbreviates $\Phi(t,\vec{u})$ where Φ is the Σ_n definition of T with parameters $\vec{u}$, $s \lessdot t$ means $Seq(s)$ & $Seq(t)$ & $lh(s) < lh(t)$ & $\forall i < lh(s)((s)_i = (t)_i)$, etc...

T is said to be <u>finite branching</u> (in M) if

$$M \models \forall s \in T\ \exists a\ \forall b \geq a\ (s^\frown\langle b\rangle \notin T).$$

A definable function $f : M \to M$ is said to be piecewise coded in M if for any $a \in M$ there is an $s \in Seq^M$ such that

$$\forall i < a\ f(i) = (s)_i\ \&\ lh(s) = a.$$

(This notion is due to R. Kossak).

A branch f of T is a definable piecewise coded total function $f:M \to M$ such that for all $a \in M$ $\langle f(0),\ldots,f(a)\rangle \in T$. Let $L(f) = \{+,\cdot,0,1,<,f\}$ where f is a new function symbol of one argument. In an obvious way, there is an effective manner to associate to every bounded formula $\phi(\vec{z})$ in $L(f)$ a bounded formula $\phi^*(z_o,\vec{z})$ in $L = \{+,\cdot,0,1,<\}$ such that if the function symbol f is interpreted by the function

$$f_s = \{(i,(s)_i) : i < lh(s)\} \quad \text{where } s \in Seq^M,$$

then for any $\vec{a} \in M$,

$$(M,f_s) \models \phi(\vec{a}) \quad \text{iff} \quad M \models \phi^*(s,\vec{a}).$$

(One replaces occurrance of $f(x) = y$ in ϕ by $(v)_x = y$ to obtain ϕ': ϕ^* is then $Seq(v)\ \&\ \phi'$).

Furthermore, if $M \models P^- + I\Sigma_o + Exp$, then the usual Δ_1 formalization of bounded satisfaction allows

$$(M,f_s) \models \phi(\vec{a}) \quad \text{iff} \quad M \models Sat_o\ (\ulcorner \phi^* \urcorner,\ \langle s,\vec{a}\rangle),$$

where ϕ is bounded.

FACT. If $M \models P^- + I\Sigma_o + B\Sigma_n$ for $n \geq 2$, then any total Σ_n definable function f is piecewise coded in M.

PROOF. By hypothesis, the graph of f is Δ_n definable. Let $a \in M$ given. By $B\Sigma_n$ let b be sufficiently large so that

$$M \models \forall i < a\ \exists y < b\ f(i) = y.$$

Now

$$M \models \exists s \leq p_b^{b+1}\ (s \in Seq\ \&\ lh(s) = a + 1\ \&\ \forall i \leq a((s)_i = f(i)))$$

by induction on $i \leq a$ using $I\Delta_n$. Here $p_b = b + 1$ - st prime; we work in $B\Sigma_n$ for $n \geq 2$ to be able to use the function p_b by $I\Sigma_1$).

A formula in the language $L(f)$ is bounded if it is built up from atomic formulas (including $f(x) = y$) by Boolean operations and bounded quantifiers $\exists x<y$ and $\forall x<y$.

LEMMA 13. Suppose that $f : M \to M$ is a definable piecewise coded total function in a model M of $P^- + I\Sigma_0$ and that $\phi(x,\vec{y})$ is a bounded formula in $L(f)$ with those free variables indicated. Suppose furthermore that for each $a \in M$, $s_a \in \mathrm{Seq}^M$ and

$$s_a = \langle f(0),\dots,f(a)\rangle .$$

Then for $b \in M$ the following are equivalent

(1) $(M,f) \models \exists x\, \phi(x,\vec{b})$

(2) $\exists a_0 \in M \forall a \geq a_0\ M \models \exists x < a+1\ \phi^*(s_a,x,b)$

(3) $\forall a_0 \in M\ \exists a \geq a_0\ M \models \exists x < a+1\phi^*(s_a,x,b)$.

Given the formula $\exists x\phi(x,\vec{y})$, let $\tilde{\phi}$ be the formula $\exists x < lh(u)\ \phi^*(u,x,\vec{y})$. If additionally $M \models \mathrm{Exp}$, then (1) is equivalent with the following

(2') $M \models \exists z \forall u \geq z(u \in \mathrm{Seq}\ \&\ lh(u) \geq z \& \forall i < lh(u)((u)_i = f(i)) \to \mathrm{Sat}_0(\ulcorner\tilde{\phi}\urcorner, u^\frown\langle x,\vec{b}\rangle)$

(3') $M \models \forall z\ \exists u \geq z(u \in \mathrm{Seq} \& lh(u) \geq z \& \forall i < lh(u)((u)_i = f(i)) \to \mathrm{Sat}_0(\ulcorner\tilde{\phi}\urcorner, u^\frown\langle x,\vec{b}\rangle)$.

PROOF. We can suppose that ϕ is in disjunctive normal form:

$\phi(x,\vec{y})$ is $\exists x_1 < r_1 \dots Qx_n < r_n \bigvee \bigwedge$ (atomic or negation of atomic formula).

First, we show by induction on n = number of bounded quantifiers in $\phi(x,\vec{y})$ that for $c,\vec{b} \in M$

$$(+)\ (M,f) \models \phi(c,\vec{b}) \to \exists a_0 \in M \forall a \geq a_0\ M \models \phi^*(s_a,c,\vec{b}).$$

If ϕ is an atomic formula or negation of atomic formula, then the only cases where there is anything to prove are when ϕ is of the form $f(t_1) = t_2$, $f(t_1) \neq t_2$, $f(t_1) < t_2$, $f(t_1) \not< t_2$, $t_2 < f(t_1)$, $t_2 \not< f(t_1)$, with $t_1 \in \{c,b_1,\dots,b_k,\bar{0},\bar{1},\bar{2},\dots\}$ (here $\vec{b}$ is $b_1,\dots,b_k$). Then for a_0 sufficiently large (so that t_1 is in the domain of s_{a_0}) (+) holds.

Disjunction and conjunction pose no problem and by choosing a_0 sufficiently large, we obtain the induction step for showing (+) when $n = j + 1$. (It is important at this point that we define bounded quantifiers by $\exists x<y$ and $\forall x<y$ rather than $\exists x<t$ and $\forall x<t$ where t is a term in $L(f)$).

It now follows that for $\phi(x,\vec{y})$ a bounded formula in $L(f)$,

$$(M,f) \models \phi(c,\vec{b}) \text{ iff } \exists a_o \in M \ \forall a \geq a_o \ M \models \phi(s_a,c,\vec{b})$$

$$\text{iff } \forall a_o \ M \ \exists a \geq a_o \ M \models \phi(s_a,c,\vec{b}).$$

Notice that for $\phi(x,\vec{y})$ a bounded formula in $L(\dot{f})$,

$$(++) \text{if } M \models \phi^*(s_s,c,b) \text{ then for all } a' \geq a \ M \models \phi^*(s_a,c,b).$$

From this fact and the previous discussion, (1) and (2) easily follow.

If the exponential function is total in M, then the usual satisfaction relation for bounded formulas is Δ_1 definable and yields (2') and (3') □

A Σ_n definable total function f in M is said to be 1-generic if every formula of the form $\exists x\phi(x,\vec{b})$ or $\forall x\phi(x,\vec{b})$, where ϕ is a bounded formula in $L(\dot{f})$, is equivalent in M to both a Σ_n and a Π_n formula in L; i.e. there are Σ_n formulas θ, ψ and Π_n formulas θ', ψ' in L such that

$$(M,f) \models \exists x\phi(x,\vec{b}) \text{ iff } M \models \theta(\vec{b}) \text{ iff } M \models \theta'(\vec{b})$$

and $$(M,f) \models \forall x\phi(x,\vec{b}) \text{ iff } M \models \psi(\vec{b}) \text{ iff } M \models \psi'(\vec{b})$$

LEMMA 14. If $M \models P^- + B\Sigma_{n+2}$ where $n \geq 1$ and T is an unbounded finite branching Σ_n tree in M, then there is a Δ_{n+2} definable 1-generic branch f of T. (This is the arithmetized version of the "low basis" theorem).

PROOF. We work in M. First define a function k bounding the sequence numbers of the levels:

$k(0) = 0 =$ sequence number for <>

$k(a+1) =$ least x such that $\forall y \geq x \ \forall s \leq \bar{k}(a+1)(lh(s)=a \rightarrow s^\frown\langle y\rangle \notin T)$.

Here $\bar{k}(a+1) = \langle k(0),\dots,k(a)\rangle$. Using $I\Sigma_{n+1}$, it follows that $M \models \forall a \ \exists_1 y \ k(a) = y$.

Now define in M

$g(0) = 0 =$ sequence number for < >

$$g(1)=\begin{cases} 0 & \text{if } \forall m \ \exists t \in T(lh(t) \geq m \ \& \ \neg \exists y < lh(t) \ Sat_o((0)_o,t,\langle y\rangle^\frown(i)_1)) \\ 1 & \text{otherwise} \end{cases}$$

and

$$g(2a+2) = \text{least } x(\forall m \, \exists t \in T(lh(t) \geq m \,\&\, \langle g(0), g(2), \ldots, g(2a)\rangle^\frown\langle x\rangle \preccurlyeq t \,\&$$

$$\forall i \leq a(g(2i+1)=0 \rightarrow \neg \exists y < lh(t) \; Sat_o((i)_o, t, \langle y\rangle^\frown(i)_1))))$$

$$g(2a+3) = \begin{cases} 0 \text{ if } \forall m \, \exists t \in T(lh(t) \geq m \,\&\, \langle g(0), g(2), \ldots, g(2a+2)\rangle \preccurlyeq t \,\& \\ \quad \forall i \leq a(g(2i+1)=0 \rightarrow \neg \exists y < lh(t) \; Sat_o((i)_o, t, y(i)_1))) \,\& \\ \quad \neg \exists y < lh(t) \; Sat_o((a+1)_o, t, \langle y\rangle^\frown(a+1)_1) \\ 1 \text{ otherwise.} \end{cases}$$

Recall that $\neg \exists y < lh(t) \; Sat_o((i)_o, t, \langle y\rangle^\frown(i)_1)$ is the formalization of "there is no $y <$ length of t such that $(i)_o$ is the Gödel number of a bounded formula ϕ, $t \in Seq$ and $\phi(y,\vec{u})$ is true, where $\vec{u}$ is the sequence coded by $(i)_1$". Thus when i runs through M, we consider all possible bounded formulas with parameters in M.

<u>CLAIM.</u> The function g is total in M.

<u>PROOF.</u> Notice that for $m \in M$, if $g(m)$ is defined, then $g(m) \leq k_o(m)$, where

$$k_o(2m) = k(m) \quad \text{and} \quad k_o(2m+1) = 1.$$

We wish to show that

$$\underbrace{M \models \forall m \, \exists y \leq k_o(m) \text{"} g(m) = y \text{"}}_{\theta(m)}$$

where the defining clause for $g(m)$ can be written in the form

$$\underbrace{\exists s \leq p_{k(m+1)+1}^{k(m+1)+1} (s \in Seq \,\&\, \Pi_{n+1} \,\&\, \Sigma_{n+1})}_{\psi}$$

Then

$$M \models \theta(m) \longleftrightarrow \exists z(\underbrace{k_o(m) = z}_{\Delta_{n+1}} \,\&\, \underbrace{\exists y \leq z\psi(m)}_{\Delta_{n+2}(B\Sigma_{n+2})})$$

$$M \models \theta(m) \longleftrightarrow \forall z(\underbrace{k_o(m) = z}_{\Delta_{m+1}} \rightarrow \underbrace{\exists y \leq z\psi(m)}_{\Delta_{n+2}(B\Sigma_{n+2})})$$

and hence θ is $\Delta_{n+2}(P^- + B\Sigma_{n+2})$.

CLAIM. $M \models \theta(0) \ \& \ \forall x(\theta(x) \to \theta(x+1))$.

PROOF. Suppose that $M \models \theta(a)$ where $a = 2b + 2$, so $g(0),\ldots,g(2b+1)$ are defined. Then

$$M \models \forall m \ \exists t \in T(lh(t) \geq m \ \& \ \langle g(0),\ldots,g(2b)\rangle \preccurlyeq t \ \&$$
$$\forall i \leq b(g(2i+1) = 0 \to \neg \exists y < lh(t) Sat_o((i)_o, t\langle y\rangle^\frown (i)_1))).$$

Let $X = \{t \in T : \langle g(0),g(2),\ldots,g(2b)\rangle \preccurlyeq t \ \&$
$$\forall i \leq b(g(2i+1) = 0 \to \neg \exists y < lh(t) Sat_o((i)_o, t\langle y\rangle^\frown (i)_1)))\}.$$

Then X is an unbounded Δ_1 definable set. Let $F : X \to \{0,\ldots,k_o(2b+2)\}$ be defined by

$$F(t) = x \quad \text{if} \quad \langle g(0),g(2),\ldots,g(2b)\rangle^\frown \langle x\rangle \preccurlyeq t.$$

Since $B\Sigma_{n+2}$ implies

$$M \xrightarrow[\Delta_{n+1}]{} (M)^1_{<M},$$

there is an $x \leq k_o(2b+2)$ such that

$$X_x = \{t \in T: \langle g(0),g(2),\ldots,g(2b),x\rangle \preccurlyeq t \ \&$$
$$\forall i \leq b(g(2i+1) = 0 \to \neg \exists y < lh(t) Sat_o((i)_o, \langle y\rangle^\frown (i)_1))\}$$

is unbounded. Define $g(2b+2)$ to be the least $x \leq k_o(2b+2)$ such that X_x is unbounded.

To show that $g(2b+3)$ is defined is similar and even easier. QED claim.

Now M is a model of $B\Sigma_{n+2}$ and hence of $I\Delta_{n+2}$ and $\theta \in \Delta_{n+2}(P^- + B\Sigma_{n+2})$

$$M \models \theta(0) \ \& \ \forall x(\theta(x) \to \theta(x+1))$$

hence

$$M \models \forall m \ \exists y \ \text{"}g(m) = y\text{"}$$

By lemma 13 it is clear that g is a 1-generic Δ_{n+2} definable branch in the tree T. □

LEMMA 15. If M is a model of $P^- + B\Sigma_3$, then $M \xrightarrow[\Delta_1]{} (M)^2_{<M}$.

PROOF. Since $M \models I\Sigma_1$, any unbounded Δ_1 definable subset of M is order-isomorphic with M via a Δ_1 isomorphism, so we can suppose that we are given a Δ_1 partition

$$F : [M]^2 \to a, \quad \text{where } a \in M.$$

Define the Erdos-Rado tree

$$T = \{t \in \mathrm{Seq} : \forall i < j < lh(t)((t)_i < (t)_j \;\&\; (lh(t) > 0 \to (t)_o = 0 \;\&$$

$$\forall i < lh(t) - 1[(t)_{i+1} \text{ is a minimal compatible extension of } \langle (t)_o, \ldots, (t)_i \rangle]$$

where the expression in square brackets is

+ "compatibility" $\forall j < k \le k' \le i(F((t)_j,(t)_k) = F((t)_j,(t)_{k'}) = F((t)_j,(t)_{i+1})))$

and

++ "minimality of comportment"

$$\forall y((t)_i < y \le (t)_{i+1} \& \forall j < k \le k' \le i(F((t)_j,(t)_k) = F((t)_j,(t)_{k'}) =$$

$$F((t)_j, y)) \;\&\; F((t)_j, y) = F((t)_i,(t)_{i+1}) \to y = (t)_{i+1}).$$

Clearly, T is a Δ_1 tree.

CLAIM. T is finite branching and unbounded in M.

PROOF. We must show

$$M \models \forall m \forall s \in T(lh(s) = m \to \exists x \forall y \ge z(s^\frown\langle y\rangle \notin T)).$$

This can be done using $I\Sigma_2$ noting the fact that if $F : [M]^2 \to a$ was the original partition, then there are at most a-many new minimal comportments possible, so at most a-many immediate successors of each node $s \in T$.

To see that T is unbounded in M, let

$X = \{s \in T$: there are unboundedly many $x > (s)_{lh(s)-1}$ such that

$$\forall i < lh(s) - 1(F((s)_i, x) = F((s)_i,(s)_{i+1}))\}.$$

CLAIM. X is unbounded in M.

PROOF. If not, then since X has a Π_2 definition, by $I\Sigma_2$ the set X has a maximum element t. Then

$$Y = \{x : x > (t)_{lh(t)-1} \;\&\; \forall i < lh(t)-1(F((t)_i, x) = F((t)_i,(t)_{i+1}))\}$$

is unbounded in M. Let $H : Y \to a$ be a Δ_1 partition defined by

$$H(x) = i \quad \text{iff} \quad F((t)_{lh(t)-1}, x) = i.$$

Then by $M \xrightarrow[\Delta_1]{} (M)^1_{<M}$, there is an $i_o < a$ such that $H^{-1}(i_o)$ is unbounded in M. Let x_o be the least element of $H^{-1}(i_o)$. Then

$$t' = \langle (t)_o, \ldots, (t)_{lh(t)-1}, x_o \rangle$$

is in X and is larger than t, contradicting maximality of t. Thus X is unbounded and hence T is unbounded in M.

By Lemma 14, let f be a Δ_3 definable 1-generic branch of T. By construction of T, for $i < J < k \in M$

$$F(f(i),f(j)) = F(f(i),f(k)) \underset{df}{=} c_i < a.$$

Now define $G : \{f(x) : x \in M\} = \{m : \exists\, x \le m(m = f(x))\} \to a$

by $\quad G(f(x)) = F(f(x), f(x + 1))$.

<u>CLAIM.</u> There exists an $i < a$ such that $G^{-1}(i)$ is unbounded in M.

<u>PROOF</u>. If not, then

$$M \models \forall\, i < a\ \exists\, x\ [\forall\, y \ge x(F(f(y),f(y + 1)) \ne i)].$$

By 1-genericity, the expression in square brackets is Δ_3, so a final application of $B\Sigma_3$ insures a contradiction. This final claim completes the proof of Lemma 15. □

<u>LEMMA 16.</u> If M is a model of $P^- + B\Sigma_{n+2}$, then $M \xrightarrow[\Delta_1]{} (M)^n_{<M}$.

<u>PROOF.</u> Suppose this has been shown for n and consider n + 1 : given a partition $F : [M]^{n+1} \to a$, let

$$T = \{s\ Seq : \forall\, i < j < lh(s)((s)_i < (s)_j)\ \&\ \forall j \le n(lh(s) \ge j \to \forall i < j((s)_i = i))\ \&$$
$$\forall\, i \le lh(s) - 1[(i \ge n \to (s)_{i+1} \text{ is a minimal compatible extension of } \langle (s)_o, \ldots, (s)_i \rangle)]\},$$

where the expression in square brackets is

+ "compatibility"

$$\forall\, i < \ldots < i_n < k < k' \le i(F((s)_{i_1}, \ldots, (s)_{i_n}, (s)_k) = F((s)_{i_1}, \ldots, (s)_{i_n}, (s)_{k'}) = F((s)_{i_1}, \ldots, (s)_{i+1}))$$

++ "minimality of comportment"

$\forall y((s)_i < y \le (s)_{i+1}$ & $\forall i_1 < \ldots < i_n < k < k' \le i(F((s)_{i_1}, \ldots, (s)_{i_n}, (s)_k) =$

$$F((s)_{i_1}, \ldots, (s)_{i_n}, (s)_{k'}) = F((s)_{i_1}, \ldots, (s)_{i_n}, y)) \quad \&$$

$\forall i_1 < \ldots < i_{n-1} < i(F((s)_{i_1}, \ldots, (s)_{i_{n-1}}, (s)_i, y) = F((s)_{i_1}, \ldots, (s)_{i_{n-1}},$

$$(s)_i, (s)_{i+1})) \rightarrow y = (s)_{i+1}).$$

Using $I\Sigma_2$, one can show that T is a finite branching unbounded tree in M (a node $s \in T$ has at most $\binom{lh(s)}{n-1}$- many immediate successors). By Lemma 14, let f be a 1-generic Δ_3 definable branch of T. Then by construction of T,

$$\{m \in M : \exists x \le m\, f(x) = m\} = \{f(x) : x \in M\}$$

is pseudo-homogeneous for F; i.e. for $x_1 < x_2 < \ldots < x_{n+2}$

$$F(f(x_1), \ldots, f(x_n), f(x_{n+1})) = F(f(x_1), \ldots, f(x_n), f(x_{n+2})).$$

An examination of the proof of Lemma 15 reveals that if M is a model of $B\Sigma_3$ then

$$M \xrightarrow[\text{low } \Delta_2]{} (M)^2_{<M}$$

by which is meant that there exists an unbounded homogeneous set for appropriate partitions whose graph is low Δ_2. Similarly by relativization, one obtains that if M is a model of $B\Sigma_{n+2}$ then

$$M \xrightarrow[\text{low } \Delta_{n+1}]{} (M)^2_{<M}$$

This observation together with the induction hypothesis yields that if $M \models B\Sigma_{n+2}$ then

$$(*) \qquad M \xrightarrow[\text{low } \Delta_3]{} (M)^{n-1}_{<M} .$$

As usual, let

$$G : [f''\, M]^{n-1} \longrightarrow a \text{ be the induced partition defined}$$

by $G(f(x_1), \ldots, f(x_{n-1})) = F(f(x_1), \ldots, f(x_{n-1}), f(x_{n-1} + 1))$.

By the property (*) there is an unbounded set homogeneous for G and by pseudo-homogeneity thus also homogeneous for F.

This completes the proof of Lemma 16 and hence of Theorem 11. □

The following corollary was pointed out by J. Paris. We first introduce some notation. If I is a proper initial segment of M, then the collection $R_M(I)$ of reals of I or coded subsets of I is the collection of traces on I of M-definable sets:

$$R_M(I) = \{A \cap I : A \text{ definable with parameters in } M\}.$$

Similarly, we can speak of functions being coded, etc... An initial segment I of M is n-Ramsey if for any $a \in I$ and any coded $F : [I]^n \to a$ there exists an $i < a$ such that $F^{-1}(i)$ is unbounded in I^n. Initial segments I of M which are 1-Ramsey are also called regular. We write $I \models B\Sigma_n^*$ to mean that I satisfies the usual Σ_n-collection scheme, where coded parameters are allowed (see [14], p. 251). In [20] it is shown that for $I \subset_e M$, I is 1-Ramsey iff $I \models B\Sigma_2^*$ and in Proposition 3 of [20] that if I is n-Ramsey then $I \models B\Sigma_{n+1}^*$.

In a remark on p. 11 of [9], it is suggested that for $n > 2$, the notions of n-Ramsey and $B\Sigma_{n+1}^*$ do not necessarily coincide. As a corollary of Theorem 11, we have the surprising fact that these notions do in fact coincide.

COROLLARY 17. If $I \subset_e M$ with $I \models P^- + I\Sigma_0$ then I is n-Ramsey iff $I \models B\Sigma_{n+1}^*$.

PROOF. ($\to$) By Proposition 3 of [20].

($\leftarrow$) Suppose that $F : [I]^n \to a$ where $a \in I$ and F is coded in M. Repeat the proof of Theorem 11 but with a single coded parameter. Thus there is an $i < a$ and $X \subseteq I$ unbounded in I with $F''[X]^n = \{i\}$. So then $F^{-1}(i)$ is unbounded in I^n. □

ACKNOWLEDGEMENTS. I would like to thank J.P. Ressayre, A. Wilkie and especially J. Paris for several conversations on the subject matter of this paper. Special thanks to J. Paris for kindly pointing out Corollary 17.

REFERENCES

[1] BENNETT, J.H. On spectra, Ph.D.dissertation, Princeton University (1962).

[2] CLOTE, P. Anti-basis theorems and their relation to independence results in Peano Arithmetic, in Model Theory and Arithmetic, Springer Lecture Notes in Mathematics 890(1980) pp. 115-133.

[3] DIMITRACOPOULOS, C. Matijasevic's theorem and fragments of arithmetic, Ph.D. Dissertation, University of Manchester, (1980).

[4] FRIEDMAN, H. On fragments of Peano arithmetic, preprint.

[5] ________ Systems of second order arithmetic with restricted induction. (Abstract) J.Symbolic Logic 41(1976) pp. 557-559.

[6] JOCKUSCH, C.G. Jr. Ramsey's theorem and recursion theory. J.Symbolic Logic 37, (1972), pp. 268-280.

[7] JOCKUSCH, G.G. Jr. & SOARE, R.I. Π_1^0 classes and degrees of theories. Transactions of Am.Math.Soc. 173(1972) pp. 33-56.

[8] KAUFMANN, M. On existence of Σ_n end extensions. Logic Year 1979-80, Springer Lecture Notes in Mathematics, 859(1980) pp. 92-103.

[9] MILLS, G. & PARIS, J. Regularity in models of arithmetic. J. Symbolic Logic, 49, Nº 1 (1984), pp. 272-280.

[10] KIRBY, L. Ultrafilters and types on models of arithmetic. Preprint.

[11] KIRBY, L. & PARIS, J. Σ_n-collection schemas in arithmetic. Logic Colloquium 77, pp. 199-209, North-Holland Publishing Co., (1978).

[12] KRANAKIS, E. Partition and reflection properties of admissible ordinals. Annals of Mathematical Logic 22(1982), pp. 213-242.

[13] ________ Stepping up lemmas in definable partitions. J. Symbolic Logic 49, Nº1, (1984), pp. 22-31.

[14] PARIS, J. Some conservation results for fragments of arithmetic, in: Model Theory and Arithmetic, Springer Lecture Notes in Mathematics, № 890 (1980) pp. 251-262.

[15] PINO, R. Π_n-collection, indicatrices et ultrafiltres définissables. Thèse de 3°cycle, Université Paris VII (1983).

[16] PUTNAM, H. Trial and error predicates and the solution to a problem of Mostowski, J. Symbolic Logic, 30 (1965), pp. 49-57.

[17] WOODS, A.R. Some problems in logic and number theory, Ph.D. dissertation, University of Manchester (1981).

[18] WRATHALL, C. Rudimentary predicates and relative computation, SIAM J. Comput. 7(1976), pp. 194-209.

[19] HAY, L. MANASTER, A. & ROSENSTEIN, J. Small recursive ordinals, many-one degrees, and the arithmetical difference hierarchy, Annals of Math.Logic, 8(1975), pp. 297-343.

[20] PARIS, J. A hierarchy of cuts in models of arithmetic, in Model Theory of Algebra and Arithmetic, Springer Lecture Notes in Mahtematics, № 834, pp. 312-337.

ON THE AXIOMATIZABILITY OF SETS IN A CLASS THEORY

Manuel Corrada C.
Universidad Católica de Chile
Santiago, Chile.

Among the systems of set theory frequently used in mathematical practice we find Zermelo-Fraenkel's system ZF, Von-Neumann-Bernays-Gödel theory VNBG, or the Kelley-Morse system, symbolically denoted KM. The difference between these systems is that some of them admit the existence of very comprehensive objects-classes or, in Cantor's terminology, "inconsistent multiplicities"-and others do not. The reason to introduce such kind of objects appears at the very beginning of the development of set theory as a formalized discipline in order to avoid the paradoxes; some authors call this the "limitation of size" doctrine. However this type of question is far from the working mathematician's interest. As Kreisel[1971] says, "day-to-day foundational research does not affect day-to-day practice. This conviction may well be behind the mathematician's reluctance to get involved in foundations...".

During a long time it has been maintained that classes are at least for the mathematicians, but a façon de parler. This point of view seems justified; if we call $VNBG^m$ the set of formulae referring to sets that are provable in VNBG, then Novak [1950] proves $VNBG^m = ZF$. That is, classes do not add new information about sets to that obtained by means of set-theoretic axioms.

But in the case of the system KM things are different. Here we can prove the existence of a universal class, call it V, and also $V \models ZF$. But then we have proved in KM the consistency of ZF. Unless ZF is inconsistent, Godel's second theorem gives $ZF \nvdash Con_{ZF}$ (Con_{ZF} is the formal (number-theoretic)statement expressing the consistency of ZF). Hence we obtain $Con_{ZF} \in KM^m - ZF$, from which $ZF \subsetneq KM^m$. It has been argued that the formula Con_{ZF} is not of mathematical character, and neither are the infinitely many formulae $\phi \in KM^m - ZF$ which are obtained by results in Kreisel-Levy[1968]. Recent results by Friedman show, however that this is not the case; KM proves predicative formulae of mathematical interest which cannot be proved in ZF (see Friedman [1981]).

$VNBG^m$ is axiomatized by the axioms of ZF. KM^m is a recursively enumerable set of sentences, then using the device of Craig [1953], KM^m is recursively axiomatizable; The question arises, then, how to find axioms of set-theoretical character for KM^m?.

The natural formalization of a theory with two kinds of objects, sets and classes, is in a two-sorted language. The set of all formulae in this language will be denoted by Φ. Therefore the set of set formulae of KM will be $\Phi^m = \{\phi : \text{there exists } \Psi \in \Phi \ (\phi \equiv_{KM} \Psi^V)\}$.

Due to the presence in the set of axioms of KM, Ax_{KM}, of the impredicative axiom schema of comprehension we have the characteristic structural property of possible models of KM, that is every model of KM can be represented as $\langle S, C \cup S, E \rangle$ where $C \subseteq P(S)$ ($P(S)$: power class of S) and E is the binary relation which interpretates the binary predicate ϵ. We are interested in structures in which S is a transitive set and E is standard. In these cases we will write $\langle S, C \rangle$ instead of the corresponding structure. For each ordinal number γ we will denote by N_γ the class which consists of all structures of the form $\langle R_\gamma, T, \epsilon \rangle$ where $T \subseteq R_{\gamma+1}$ and R_ξ is the usual cumulative hierarchy. $L_\kappa(R_\beta)$ is relative constructibility over R_β constructing the subsets of R_β until level κ so $L_\kappa(R_\beta) \subseteq R_{\beta+1}$.

Dg T will denote the interpretability degree of the theory T, i.e. the class of all theories mutually interpretable with T.

The following theorem due to W, Reinhardt (private communication) is important in the subsequent discussion.

<u>THEOREM 0</u>. For each n there exists a closed unbounded class of ordinals Ω_n such that for each $\alpha \in \Omega_n$ there exists an ordinal δ_α such that $\langle R_\alpha, L_{\delta_\alpha}(R_\alpha) \rangle \models \Delta^1_n$ comprehension + replacement.

<u>PROOF</u>: By reflection principle (see Chuaqui [1980]) there is a well order type S such that $L_S[V] <_{\Delta^1_n} L[V]$. Take the first. Then $L_S[V] \models \Delta^1_n$-comprehension + replacement.
But there is a closed unbounded class of ordinals Ω_n such that if $\alpha \in \Omega_n$ then $L_{S \upharpoonright R_\alpha}(R_\alpha) < L_S[V]$. Let δ^n_α be the least β such that $L_\beta(R_\alpha) \models \Delta^1_n$-comprehension + replacement (it always exists). But $S \upharpoonright R_\alpha \cong \gamma$ for a certain ordinal γ. Then $S \upharpoonright R_\alpha = \delta^n_\alpha$. □

As a consequence we have

THEOREM 1. For each finite subset of the axioms for KM,Σ, there exists a class of structures K, such that KM$\vdash$ ($K\models\Sigma$) and a closed unbounded class of ordinals Ω such that for every $\gamma\in\Omega, N_\gamma\cap K\neq 0$.

The following immediate consequence is obtained by means of Godel's second theorem:

THEOREM 2. (a) KM is not finitely axiomatizable

(b) KM^m is not finitely axiomatizable

This can be extended to

THEOREM 3. KM is not finitely axiomatizable over ZF.

PROOF: Assume the contrary, i.e. KM=ZF+Σ for some finite $\Sigma\subseteq\Phi$. Let Γ be a closed unbounded class of ordinals such that for every $\delta\in\Gamma$, $\langle R_\delta, R_\delta, \epsilon\rangle\models$ ZF.
By theorem 1 let Ω be a closed unbounded class of ordinals such that for every $\gamma\in\Omega$, $\langle R_\gamma, T, \epsilon\rangle\models\Sigma$ where $T\subseteq R_{\gamma+1}$. Because $\Omega\cap\Gamma\neq 0$ take $\alpha\in\Omega\cap\Gamma$. Hence $\langle R_\alpha, T, \epsilon\rangle\models$ ZF+Σ=KM where $T\subseteq R_{\alpha+1}$. Therefore KM$\models$ Con_{KM} which is impossible unless KM will be inconsistent. □

As a trivial consequence we obtain the

THEOREM 4. It cannot hold that KM^m will be finitely axiomatizable over ZF and KM will be finitely axiomatizable over KM^m.

By an argument similar to those in the proof of theorem 3 it can be proved that KM^m is not finitely axiomatizable over ZF. Moreover, using technics in Kreisel and Levy [1968] this can be extended to (see Chuaqui [1980]):

THEOREM 5. (a) KM^m is not finitely axiomatizable by formulae of bounded depth over ZF.

(b) KM is not finitely axiomatizable by formulae of bounded depth over KM^m.

In spite of this unboundness results a certain kind of finitization may be obtained,

THEOREM 6. (a) There exists a finite set of formulae Σ, $\Sigma\subseteq\Phi$ such that $KM^m = \{\phi : \Sigma\vdash\phi\}\cap\Phi^m$

(b) There exists a sentence $\Theta\in\Phi^m$ such that $DgKM^m$ = Dg(ZF+Θ).

PROOF: (a) See Krajewski [1974] ; (b) Similar to Lindström [1979] □

The following theorem will be useful for our purposes

THEOREM 7. If $\phi\in\Phi^m$ and $n\in\omega$ then
$KM\vdash((\exists\alpha)(\exists F)(Ord(\alpha)\wedge F\subseteq R_{\alpha+1}\wedge\langle R_\alpha,F,\epsilon\rangle\models KM_n)\wedge(\phi\leftrightarrow\phi^{R_\alpha}))$ (KM_n is KM with comprehension restricted to Σ^1_n formulae)

PROOF: By Reflection Theorem there exists a closed unbounded class of ordinals Γ such that

$$\phi \leftrightarrow \phi^{R_\delta}, \quad \delta\in\Gamma$$

Because KM_n is finitely axiomatizable, using Theorem 1 get a class of structures K such that $K\models KM_n$.

Again by Theorem 1 let Ω be a class of ordinals such that $K\cap N_\gamma\neq 0$ for each $\gamma\in\Omega$. Take $\alpha\in\Omega\cap\Gamma$. □

The above theorem is closely connected with the notion of extendability of models of set-theories (see Marek & Mostowski [1974]). In fact $(\exists F)$ $(F\subseteq R_{\alpha+1}\wedge\langle R_\alpha, F,\epsilon\rangle\models KM_n)$ means that R_α is KM_n-extendable. So we can rewrite the theorem as:

THEOREM 7'. If $\phi\in\Phi^m$ and $n\in\omega$ then,

$KM\vdash(\exists\alpha)(Ord(\alpha)\wedge R_\alpha$ is KM_n extendable$)\wedge(\phi\leftrightarrow\phi^{R_\alpha})$.

An improvement of Vaught [1967] as presented in Ratajczyk [1979] gives the following

THEOREM 8. If j is an interpretation of Peano's arithmetic P in KM and $(\forall\phi\in\Phi^m)(KM\vdash\phi\rightarrow Con_j(\Delta_\phi+\Delta_S)$ for each finite fragment S of KM then the set of sentences $\vec{j}(P)+\{\phi\rightarrow Con_j(\Delta_\phi+\Delta_S):\phi\in\Phi^m, S\subseteq_{finite}Ax_{KM}\}$ is an axiomatization of KM^m.

PROOF: Ratajczyk [1979], Theorem 0. □

By results in Montague [1960] it is known that $(\forall\phi\in\Phi^m)(KM\vdash\phi\rightarrow Con_j(\Delta_\phi+\Delta_S))$ so we obtain

THEOREM 8'. If j is an interpretation of P in KM, then $\vec{j}(P)+\{\phi\rightarrow Con_j(\Delta_\phi+\Delta_S):\phi\in\Phi^m, S\subseteq_{finite}Ax_{KM}\}$ as an axiomatization of KM^m.

Using different methods the following theorem has been proved by Ratajczyk [1979].

THEOREM 9. The set
ZF+$\{(\exists_\alpha)(R_\alpha$ is KM_n extendable$\wedge\phi\longleftrightarrow\phi^{R_\alpha}:\phi\in\Phi^m\wedge n\in\omega\}$ is an axiomatization of KM^m

PROOF: Let $\Gamma=\{(\exists_\alpha)(R_\alpha$ is KM_n-extendable$\wedge\phi^{R_\alpha}\leftrightarrow\phi):\phi\in\Phi^m\wedge n\in\omega\}$
Enumerate Γ, i.e. $\Gamma=\{X_{\phi_n}:\phi\in\Phi^m\wedge n\in\omega\}$. Then for each $\phi\in\Phi^m$ and each $n\in\omega$, by theorem 7', $KM\vdash X_{\phi_n}$ from which $\Gamma\subseteq KM^m$, hence $KM^m\vdash\Gamma$. Call $\Lambda=\{\phi\to Con_j(\Delta_\phi+\Delta_S):\phi\in\Phi^m\wedge S\subseteq_{finite}Ax_{KM}\}$. We will prove $ZF+\Gamma\vdash\vec{\exists}(P)+\Lambda$.
Let $\langle M,E\rangle\models ZF$ and suppose for each n, and each ϕ, $\langle M,E\rangle\models X_{\phi,n}$. Let $\phi\in\Phi^m$ and let $S\subseteq_{finite}Ax_{KM}$ Assume $\langle M,E\rangle\models\phi$. Because $S\subseteq_{finite}Ax_{KM}$ there exist $\alpha\in OR^M$ and $F\in M$ such that $\langle M,E\rangle\models F\subseteq R_{\alpha+1}\wedge(\langle F,R_\alpha,\epsilon\rangle\models S)\wedge$ $(\phi^{R_\alpha}\longleftrightarrow\phi)$. As we have assumed $\langle M,E\rangle\models\phi$ then $\langle M,E\rangle\models F\subseteq R_{\alpha+1}\wedge(\langle F,R_\alpha\epsilon\rangle\models S)\wedge\phi^{R_\alpha}$. Therefore $\langle M,E\rangle\models Con(\Delta_\phi+\Delta_S)$ from which $\langle M,E\rangle\models Con_j(\Delta_\phi+\Delta_S)$.

Although this theorem solves the problem of finding a set-theoretic set theoretic set of axioms for KM^m we will give a simpler axiomatization which is based on reflection principles.
For each $n\in\omega$ and each $\Psi\in\Phi^m$ let Γ be the following set schemata:

$$\exists\alpha\,\exists\beta\,(\lim_{>\omega}(\alpha)\wedge\lim_{>\alpha}(\beta)\wedge\langle R_\alpha,L_{\delta^n_\alpha}(R_\alpha)\rangle\prec\langle R_\beta,L_{\delta^n_\beta}(R_\beta)\rangle\wedge\psi^{R_\alpha}\leftrightarrow\psi)$$

were δ^n_κ is the first β such that $\langle R_\kappa,L_\beta(R_\kappa)\rangle\models\Delta^1_n$-comprehension, and $\langle R_\alpha,L_{\delta^n_\alpha}(R_\alpha)\rangle\prec\langle R_\beta,L_{\delta^n_\beta}(R_\beta)\rangle$ is the formal statement that $\langle R_\alpha,L_{\delta^n_\alpha}(R_\alpha)\rangle$ is an elementary submodel of $\langle R_\beta,L_{\delta^n_\beta}(R_\beta)\rangle$ and the relation occurs only with set parameters.

THEOREM 10. (W. Reinhardt) Let Θ be an instance of Γ. Then $KM\vdash\Theta$, and therefore $KM^m\vdash\Theta$.

PROOF: By Theorem 0 take $\beta\in\Omega_n$ and get a structure $\langle R_\beta,L_{\delta^n_\beta}(R_\beta)\rangle$. Let $\mathcal{D}$ be the set of definable elements of R_β in $L_{\delta^n_\beta}(R_\beta)$ using parameters from R_β, i.e. $t\in\mathcal{D}\longleftrightarrow$ there exists a formula θ and parameters $b_1,b_2,\ldots,b_n$ in R_β such that $t=\{x\in R_\beta:\langle R_\beta,L_{\delta^n_\beta}(R_\beta)\rangle\models\theta(x,b_1,\ldots,b_n)\}$. It is obvious that $\mathcal{D}\subset L_{\delta^n_\beta}(R_\beta)$ and moreover we have $\mathcal{D}\prec L_{\delta^n_\beta}(R_\beta)$. Collapse $\mathcal{D}$ to get $\langle R_\alpha,L_{\delta^n_\alpha}(R_\alpha)\rangle\cong\mathcal{D}$. This is the desired structure. The reflection $\psi^{R_\alpha}\longleftrightarrow\psi$ is inmediate. □

THEOREM 11: If $\psi\in\Phi^m$ and Σ is a finite subset of the axioms for KM then, $ZF+\Gamma \vdash (\exists\alpha)(\exists F)((Ord(\alpha)\wedge F\subseteq R_{\alpha+1}\wedge \langle R_\alpha, F\rangle \models \Sigma)\wedge \psi \longleftrightarrow \psi^{R_\alpha})$.

PROOF: Let ψ and Σ be given. Assume the more complex instance of comprehension which occurs in Σ is Δ^1_n for a certain n. Then clearly

$$\langle R_\alpha, L_{\delta_\alpha^n}(R_\alpha)\rangle \models \Sigma \text{ and } \psi \longleftrightarrow \psi^{R_\alpha}.$$ □

Using these results we can give the following axiomatization for KM^m using the "reflection principle" Γ,

THEOREM 12. The set $ZF+\Gamma$ is an axiomatization for KM^m.

PROOF: Let Λ $\{\phi \rightarrow Con_j(\Delta_\phi+\Delta_\Sigma) : \phi\in\Phi^m$, Σa finite subset of the axioms for KM$\}$.
By Theorem 10 it is sufficient to prove $ZF+\Gamma \vdash \vec{j}(P)+\Lambda$
Let $\langle M,E\rangle \models ZF$ and assume $\langle M,E\rangle$ satisfies all the instances of Γ, i.e. $\langle M,E\rangle \models ZF+\Gamma$.
Let $\phi\in\Phi^m$ and let Σ be a finite subset of the axioms for KM. Assume $\langle M,E\rangle \models \Gamma$. By Theorem 11, since $\langle M,E\rangle \models \Gamma$ there exists $\alpha\in OR^M$ and $F\in M$ such that $\langle M,E\rangle \models F\subseteq R_{\alpha+1} \wedge ((R_\alpha, F)\models\Sigma) \wedge \phi^{R_\alpha} \longleftrightarrow \phi$.
Since we have assumed that $\langle M,E\rangle \models \phi$, we have $\langle M,E\rangle \models F\subseteq R_{\alpha+1} \wedge (\langle R_\alpha, F\rangle \models \Sigma) \wedge \phi^{R_\alpha}$, then $\langle M,E\rangle \models Con(\Delta_\phi+\Delta_\Sigma)$ from which follows $\langle M,E\rangle \models Con_j(\Delta_\phi+\Delta_\Sigma)$. □

NOTE: This paper was partially written while the author was at the Instituto Venezolano de Investigaciones Científicas supported by a grant from PNUD/UNESCO (Matemática). The author thanks Dr. Carlos A. Di Prisco for some very helpful conversations.

REFERENCES

Chuaqui, R. [1980]. Internal and forcing models for the impedicative theory of classes, Dissertationes Mathematicae CLXXVI, Warzawa.

Craig, W. [1953] On axiomatizability within a system, J.Symbolic Logic 18,pp. 30-32.

Friedman, H.[1981] On the necessary use of abstract set theory, Advances in Mathematics 41, pp. 209-280.

Krajewski, S[1974] Predicative expansions of axiomatic theories, Z.Math.Logik, vol. 20, pp.435-452.

Keisel,G. [1971] Observations on popular discussions of foundations, Axiomatic Set Theory, ed. by Dana Scott, American Mathematical Society, pp. 189-198.

Kreisel, G. and Levy, A. [1968] Reflection principles and their use for establishing the complexity of axiomatic systems, Z.Math. Logik, vol. 14, pp. 97-112.

Lindström, P [1979] Some results in interpretability, Proceedings from 5th Scandinavian Logic Symposium, Aalborg University Press, pp. 329-361.

Marek, W and Mostowski, A [1974] On extendability of models of ZF set theory to the models of KM theory of classes, Lecture Notes in Mathematics, vol. 499.

Montague, R. [1960] Semantical closure and non finite axiomatizability, Infinitistic Methods, P.W.N. pp 45-69.

Novak, I.L. [1950] A construction for models of consistent systems, Fundamenta Mathematicae, vol. 37, pp. 310-317.

Ratajczyk, Z.[1979] On sentences provable in impredicative extensions of theories, Dissertationes Mathematicae CLXXVIII, pp 1-44.

Vaught, R.L. [1967] Axiomatizability by a schema, J. Symbolic Logic 32,pp. 473-479.

APPLICATIONS OF MODEL THEORY
TO REAL ALGEBRAIC GEOMETRY
A SURVEY.

M.A. Dickmann
CNRS - University of Paris VII
Paris, France

CONTENTS.

§1. INTRODUCTION.

This survey deals with real algebraic geometry, a subject still in its infancy, and with the application of model-theoretic techniques in this area.

Let V be an algebraic variety defined over a field K. Deviating slightly from standard usage, we will think of V as being given by a specific finite set $P_1,\ldots,P_\ell$ of polynomials in $K(X_1,\ldots,X_n)$ for some n - or equivalently, by the formal expression "$\bigwedge_{i=1}^{\ell} P_i(X_1,\ldots,X_n) = 0$". For a field F containing K we let V(F) denote the set of F-rational points (which we will call F-points) of V:

$$V(F) = \{\bar{a}\in F^n \mid P_i(\bar{a}) = 0 \quad \text{for} \quad i = 1,\ldots,\ell\}.$$

Classical algebraic geometry was primarily the study of the complex points $V(\mathbb{C})$, and in spite of the attention payed to rationality questions in the development of the subject, even the general theory initiated by Grotendieck has been set up to deliver results corresponding to the complex case. Thus, some of the most natural questions connected with the study of the real points $V(\mathbb{R})$, like the number of connected components in the Euclidean topology, or even the existence of non-singular real points, simply do not arise in that context.

Indeed, such questions -which form the proper subject matter of real algebraic geometry have received detailed consideration only relatively recently. In the last ten years especially, a body of theory has emerged in which real closed fields assume their proper role, the real spectrum (with a suitable topology) has been constructed, and the geometry of semi-algebraic sets and rings of Nash functions have been explored (subjects with no meaningful parallel in the classical theory). The necessary definitions and background will be given below as the occasion arises .

Model theoretic techniques - notably a systematic use of quantifier elimination for real closed fields and related ideas- have a role to play in all of this. The precise purpose of this article is to survey this interaction of model theory with various kinds of geometrical investigations.

As we intend to give only an overview of the subject, we omit many proofs, in order to concentrate on those which illustrate the use of an important technique combined with some aspect of mathematical logic. At the same time we have tried to supply many examples and references for further study. Some familiarity with the elementary algebra of real closed fields is assumed.

We will begin our survey by reminding the reader of some "pathology" associated with the geometry of the sets $V(\mathbb{R})$ (see §2). The general analysis of the behavior of such sets depends on a body of theory based on the following four ingredients:

(1) Quantifier elimination for the theory of real closed fields (see §3). This provides a substitute for the techniques of classical elimination theory. Quantifier elimination in algebraically closed fields is a rather trivial affair which was replaced by other devices; in the real closed case it constitutes a powerful tool, at least for the present. In particular, the presence of an ordering introduces features having no parallel in the classical case.

(2) The "separation theorem" and related tools (see §5). These yield considerable information concerning the topology of real algebraic varieties and, more generally, of "semi-algebraic" sets (arbitrary definable subsets of $\mathbb{R}^n$) with respect to the ordinary Euclidean topology, where for example the connected components, the interior or the closure of a set are of interest. They are also used in transfering such notions to other real closed fields.

(3) The real spectrum of a ring (see §6). The use of this instrument in real algebraic geometry is similar to that of the prime spectrum in the classical theory. It turns out, however, that its topological properties are better, yielding information about the sets $V(\mathbb{R})$ of real points with respect to the Euclidean, rather than the Zariski topology.

(4) Nash functions (see §10, §11). This class of functions (analytic functions with a definable graph) is the tool which makes possible an explicit description of the "branches" and the connected components of real varieties, in spite of the complexity of the situations arising in this case, without a common measure with those occurring in classical algebraic geometry.

Applications of quantifier elimination to real algebra are surveyed in

§4, while the study of real algebraic varieties properly is carried out in §7, §8. We have also included a section (§9) on continuous semi-algebraic functions, the analysis of which leads naturally to the theory of Nash functions.

This article (and the author's interest on real algebraic geometry) owes much to Coste/Coste-Roy [62], a clear, deep and inspiring exposition of the core of the subject. Thanks are due to the referee, whose suggestions helped to improve substantially the original manuscript.

§2. REAL VERSUS COMPLEX GEOMETRY.

We illustrate some of the comments made above by recalling results of complex algebraic geometry which rule out phenomena occurring quite commonly among real algebraic varieties. First we must fix some notation.

2.1 NOTATION. Let k be a field and F a field extension of k (when $F = k$ we will drop it from our notation).

(a) To each subset $X \subseteq k^n$ we associate the ideal

$$I_F(X) = \{P \in F[X_1,\ldots,X_n] \mid P(\overline{x}) = 0 \text{ for all } \overline{x} \in X\}.$$

(b) Given a set $S \subseteq k[X_1,\ldots,X_n]$ we define

$$V_F(S) = \{\overline{x} \in F^n \mid P(\overline{x}) = 0 \text{ for all } P \in S\}.$$

Recall that by our earlier convention a variety V over k is determined by a specific finite set of polynomials $P_1,\ldots,P_\ell \in k[X_1,\ldots,X_n]$.

(c) We write $I_F(V)$ for the ideal of $F[X_1,\ldots,X_n]$ generated by $P_1,\ldots,P_\ell$.

(d) The (affine) coordinate ring of a variety V over k is the ring $k[X_1,\ldots,X_n]/I(V(k))$, denoted by $k[V]$.

(e) A variety V is called irreducible if the ideal $I(V(k))$ is prime. □

REMARK. A variety V is irreducible iff we cannot write $V(k)$ as a proper union $V_1(k) \cup V_2(k)$, with V_1, V_2 varieties defined over k. cf. Hartshorne [3; Ch.I, §1] □

THEOREM 2.2. (Hilbert's nullstellensatz). Let V be a variety over an algebraically closed field k, and $P \in k[X_1,\ldots,X_n]$. Then the following are equivalent:

(i) $P(\overline{x}) = 0$ for every $\overline{x} \in V(k)$.

(ii) There is $n \geq 1$ such that $P^n \in I(V)$. □

In other words:

$$I(V(k)) = \text{Rad}\,(I(V)),$$

where $Rad(I) = \{P \in k[X_1,\ldots,X_n] \mid P^n \in I \text{ for some } n \geq 1\}$.
For a proof, see Lang [8; p. 256].

Of course, this theorem fails badly over $\mathbb{R}$ (or any real closed field):

EXAMPLE 2.3. Let V the variety over $\mathbb{R}$ defined by the polynomial $X^2 + 1$. Since $V(\mathbb{R}) = \emptyset$, then (i) of 2.2 is true for any P, while (ii) frequently fails. □

Turning to more interesting examples, we state a fundamental, and highly non-trivial, result about complex varieties:

THEOREM 2.4. Let V be an irreducible variety over $\mathbb{C}$. Then $V(\mathbb{C})$ is a connected subset of $\mathbb{C}^n$ (here $\mathbb{C}$ is considered with its standard, Euclidean topology). □

The proof is given in Shafarevich [14; pp. 320-32].

This result also fails over the real field:

2.5. EXAMPLES. (a) The hyperbola of equation $YX-1 = 0$, which clearly is irreducible, consists of two disjoint branches in $\mathbb{R}^2$. (b) Similarly, the cubic $Y^2 - X^3 + X$, irreducible by Einsenstein's criterion (cf. Lang [8; p. 128]), has two connected components:

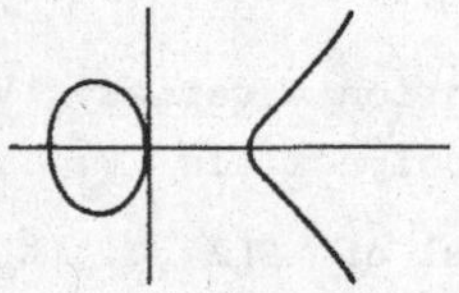

These examples remain irreducible over $\mathbb{C}$. □

The examples below require the notion of dimension of a variety. For our present purposes the intuitive notions suffice; but as we will develop the analogous "real" theory later, we recall here the customary algebraic definitions corresponding to those intuitive notions.

DEFINITION 2.6. Let A be a commutative ring with unit.

(a) The (Krull) dimension of A, dim A, is the supremum of the integers n such that there is a chain

$$P_o \subsetneq P_1 \subsetneq \cdots \subsetneq P_n$$

of length n + 1 of prime ideals of A (dim A = ∞ if there are chains of this type of unbounded length).

Let V be a variety over a field k.

(b) The (global) dimension of V, dim (V), is the dimension of the coordinate ring $k[V]$.

(c) The <u>(local dimension of V at a point $\bar{x}\in V(k)$</u>, $\dim(V,\bar{x})$, is the dimension of $k[V]_{M_{\bar{x}}}$, the localized of the ring $k[V]$ at the maximal ideal $M_{\bar{x}}$ of the point $\bar{x}$:

$$M_{\bar{x}} = \{{}^{P}/_{I(V(k))} \mid P\in k[X_1,\ldots,X_n] \text{ and } P(x) = 0\}. \qquad \square$$

There are several equivalent characterizations of these notions; we mention the following:

<u>THEOREM 2.7.</u> Let k be a field, A an integral k-algebra of finite type and K the field of fractions of A. Then:

$$\dim A = \text{transcendence degree of } K \text{ over } k. \qquad \square$$

In particular, if V is an irreducible variety over k and $k(V)$ denotes the field of fractions of $k[V]$ (called the <u>function field</u> of V), then we have:

$$\dim(V) = \text{transcendence degree of } k(V) \text{ over } k.$$

A proof of Theorem 2.7 can be found in Matsumura [9; Ch.5, §14].

The fundamental result on the dimension of varieties in classical algebraic geometry is the following:

<u>THEOREM 2.8.</u> Let V be an irreducible variety over an algebraically closed field k. Then

$$\dim(V) = \dim(V,\bar{x}) \quad \text{for all } \bar{x}\in V(k). \qquad \square$$

The proof is given in Atiyah-Macdonald [1; pp. 124-125].

This result also fails over the field of real numbers:

<u>2.9. EXAMPLES.</u> (a) The irreducible cubic $Y^2 - X^3 + X^2$ has an isolated point at the origin:

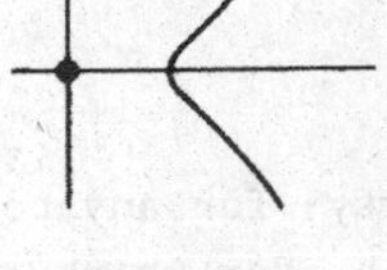

It is (intuitively) obvious that the local dimension at the origin is zero, while the same dimension at any point of the main branch is one.

(b) An example of the same phenomenon where the set $V(\mathbb{R})$ is, in addition, connected, is obtained by considering Whitney's "umbrella" (or Cartan's, if you are French), given by the irreducible polynomial $Y^3-(X^2+Y^2)Z$, whose graph looks as follows:

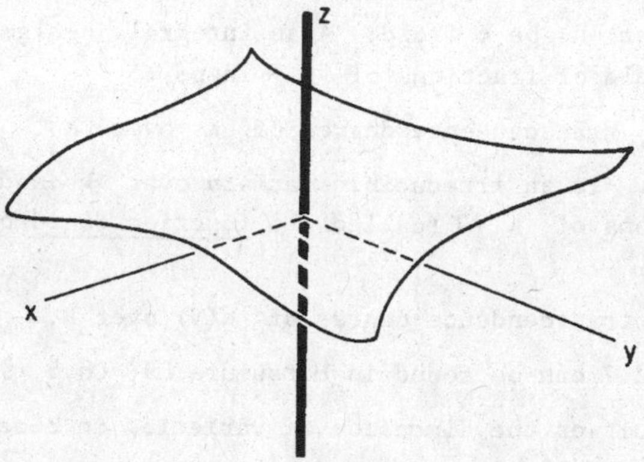

The theory of dimension for real varieties is discussed in §8 below.

§3. QUANTIFIER ELIMINATION.

The theory RCF of real closed fields is most simply defined as the first-order theory of the field $\mathbb{R}$ of real numbers in the language L of unitary ordered rings, built from the symbols $+, -, \cdot, 0, 1, <$ and logical symbols (but no symbol for the multiplicative inverse). In order to derive useful mathematical information it is necessary to have good axioms for the theory and some algebraic machinery, notably the existence and uniqueness of the real closure. The most efficient set of axioms is:

(1) The axioms for ordered fields.

(2) The intermediate value property: for any polynomial $P(X)$ in one variable, and for any points $a<b$ for which $P(a)$, $P(b)$ are non-zero and of opposite sign, $P(X)$ has a root in the interval (a,b).

Applied to polynomials of the form X^2-a, the intermediate value property implies that non-negative elements have square roots, and hence that real closed fields have a unique ordering which is definable in the language of rings. Nevertheless, for reasons which will soon be clear, we retain the symbol $<$.

DEFINITION 3.1. Let k, K be ordered fields, K an ordered extension of k. K is a (the) real closure of k if K is real closed and algebraic over k. □

The existence of a real closure is straightforward (Zorn). Its uniqueness up to (unique) k-isomorphism requires a more delicate touch, usually obtained by use of Sturm's algorithm. However, there are proofs which instead use the intermediate value property and Rolle's theorem (cf. Gross-Hafner [49]), or even other algebraic tools (see Knebusch [50]). Details concerning this result, and an elementary algebraic information on real close fields can be found in basic algebra texts. such as Jakobson [4, Ch.VI], Jakobson [5], Lang [8; Ch.IX] or Ribenboim [13; Ch. IX].

EXAMPLE 3.2. We recall an example of real closed field of great importance in real algebraic geometry: the field P(K) of (formal) Puiseux series with coefficients in a real closed field K. This is the field of Laurent series with fractional exponents, i.e. formal expressions of the type

$$\sum_{n=N}^{\infty} a_n X^{n/p}$$

where p is an integer ≥ 1, $N \in \mathbb{Z}$ and $a_N \neq 0$. The algebraic operations are defined as in the case of formal power series; the reader can figure out the explicit definitions as an exercise.

The real closedness of P(K) follows easily from that of K and the following:

THEOREM 3.3 (Newton). If k is an algebraically closed field of characteristic zero, then so is P(k). □

This is the fundamental theorem of the classical theory of plane algebraic curves. The standard proof, by the "Newton polygon" method, is given in Walker [15; Ch. IV, §3]; model-theoretic proofs also exist (an argument of this type gives a direct proof of the real closedness of P(K) as well). The key point here is that P(k) is the algebraic closure of k(X). For a geometric interpretation of the real closure of the field K(X) with its various orders, when K is real closed, see Brumfiel [65; pp. 183-184]. We will use this example in §9 below. □

The main result concerning the theory RCF is:

THEOREM 3.4. (Tarski; approx. 1930). Let $\phi(v_1,\ldots,v_n)$ be a first-order formula of the language L having $v_1,\ldots,v_n$ as free variables. Then there is a quantifier-free formula $\psi(v_1,\ldots,v_n)$ with the same free variables, such that

$$RCF \vdash \forall v_1,\ldots,v_n(\phi(v_1,\ldots,v_n) \leftrightarrow \psi(v_1,\ldots,v_n)).$$

In order to prove this result we will need the following model-theoretic criterion, which is a simple consequence of the compactness theorem of first-order logic:

<u>3.5.MODEL-THEORETIC CRITERION.</u> Let T be an arbitrary first-order theory formulated in a language L. Then the following are equivalent:

(1) T admits quantifier-elimination in the language L (i.e., the statement of Theorem 3.4 holds for T, instead of RCF).

(2) For every pair of models $\mathfrak{A}, \mathfrak{B}$ of T, every L-substructure $\mathbb{C} \subseteq \mathfrak{A}$, $\mathbb{C} \subseteq \mathfrak{B}$, and every conjunction $\psi(v,\bar{c})$ of atomic formulas of L and their negations, with one free variable and parameters $\bar{c}$ in $\mathbb{C}$, we have:

(*) $\mathfrak{A} \models \exists v\psi(v,\bar{c})$ implies $\mathfrak{B} \models \exists v\psi(v,\bar{c})$. □

For details see Chang-Keisler [33; 3.1.17] or Dickmann [72;Ch.II, §1].

<u>PROOF OF THEOREM 3.4.</u> By the model-theoretic criterion we need to prove (*) when $\mathfrak{A} = K_1$, $\mathfrak{B} = K_2$ are real closed fields and $\mathbb{C} = A$ is an ordered subring of both K_1 and K_2.

Trivial manipulations show that in the present case a formula $\psi(v,\bar{c})$ as above is of the form

$$\bigwedge_{i=1}^{m} P_i(v) = 0 \ \wedge \ \bigwedge_{j=1}^{t} Q_j(v) > 0,$$

with $P_i, Q_j \in A[X]$.

Let k be the field of fractions of A, which is an ordered subfield of both K_1 and K_2, and F_i its real closure in $K_i (i = 1,2)$. Thus, we have the situation

$$\begin{array}{ccc} & F_1 & \subseteq K_1 \\ k & \begin{array}{c}\rotatebox{45}{$\subseteq$}\\ \rotatebox{-45}{$\subseteq$}\end{array} & \\ & F_2 & \subseteq K_2. \end{array}$$

By uniqueness of the real closure there is a k-isomorphism $\sigma : F_1 \to F_2$.

In order to prove (*) it suffices to show:

(**) If $K_1 \models \exists v\ \psi(v,\bar{c})$, then there is $b \in F_1$ such that $F_1 \models \psi[b,\bar{c}]$. Indeed, we would have then, $F_2 \models \psi[\sigma(b),\bar{c}]$; since this formula is also valid in K_2, we conclude that $K_2 \models \exists v\ \psi(v,\bar{c})$.

Let us assume $K_1 \models \exists v\ \psi(v,\bar{c})$ and let $a \in K_1$ be such that $K_1 \models \bigwedge_i P_i(a) = 0 \wedge \bigwedge_j Q_j(a) > 0$.

CASE 1. $m > 0$ and one of the polynomials P_i is of degree >0. Then a is algebraic over k, which implies $a \in F_1$.

CASE 2. Either $m = 0$ or all the polynomials P_i are of degree 0. Let R be the (finite) set of all roots (in K_1) of the non-trivial polynomials amongst $Q_1,\dots,Q_t$. The elements of R are algebraic over k, and hence $R \subseteq F_1$. Let (α,β) be the interval of K_1 containing a determined by two successive elements of R (α or β may be $-\infty$ or $+\infty$). Since $Q_1,\dots,Q_t$ are positive at a, they are positive on the whole interval (α,β), and hence also positive on $(\alpha,\beta) \cap F_1$. Since $\alpha,\beta \in F_1$, this set is non-empty; choose $b \in (\alpha,\beta) \cap F_1$. □

3.6. COMMENTS. The foregoing proof of the quantifier elimination theorem is by far the shortest and most elegant, but it is not constructive. For most applications, however, the theorem is quite sufficient as stated; many applications require only the much weaker transfer principles below.

Constructive proofs of Theorem 3.4 do exist. Indeed, the first published proof, Tarski [40], is of this type; see also Kreisel-Krivine [37; Ch. IV] and Collins [35].

The primitive recursive algorithms for finding a quantifier-free equivalent of a given formula provided by the constructive proofs cannot be implemented in practice, except possibly for very particular formulas. Fischer-Rabin [36] showed that any such algorithm is necessarily of exponential complexity: they construct L-formulas of arbitrarily large length n for which no algorithm can compute a quantifier-free equivalent in less than 2^n steps. For further information in this direction, cf. Monk [39] and Collins [35].

Nevertheless, algorithmic proofs frequently yield additional information; in our case a useful corollary of such type of proof is the following:

PROPOSITION 3.7. For any L-formula ϕ there is a quantifier free equivalent modulo RCF, ψ, such that the coefficients of the polynomials occurring in ψ are polynomial functions over $\mathbb{Z}$ of the polynomials occurring in ϕ. □

Next we consider two important consequences of quantifier elimination.

THEOREM 3.8. (First transfer principle). Let ϕ be a sentence (i.e. a formula without free variables) of the language L. For any pair of real closed fields K_1, K_2, we have:

$$K_1 \models \phi \qquad \text{iff} \qquad K_2 \models \phi \qquad \square$$

The proof is trivial. The standard name of this result among logicians is "completeness of the theory RCF".

The transfer principles stated below are far more powerful than the preceeding one; they are also more useful, as they apply to formulas with parameters. The first is actually a reformulation of quantifier elimination (see 3.5 above), while the second is known as the "model-completeness of the theory RCF".

THEOREM 3.9. (a) Let $\langle k,\leq\rangle$ be an ordered field, and K_1, K_2 real closed ordered extension of $\langle k,\leq\rangle$. Then for every L-formula $\phi(v_1,\ldots,v_n)$ and every $a_1,\ldots,a_n \in k$, we have

$$K_1 \models \phi[a_1,\ldots,a_n] \qquad \text{iff} \qquad K_2 \models \phi[a_1,\ldots,a_n].$$

In particular:

(b) (Second transfer principle). The equivalence above holds whenever K_1, K_2 are real closed fields such that $K_1 \subseteq K_2$, and $a_1,\ldots,a_n \in K_1$.

PROOF. Since we may take ϕ quantifier-free, this is clear. $\square$

The following is an equivalent formulation of the second transfer principle:

PROPOSITION 3.10. Every unitary ring homomorphism between real closed fields, $f: K_1 \to K_2$, is an elementary map; in other words, for every L-formula $\phi(v_1,\ldots,v_n)$ and $a_1,\ldots,a_n \in K$, we have

$$K_1 \models \phi[a_1,\ldots,a_n] \qquad \text{iff} \qquad K_2 \models \phi[f(a_1),\ldots,f(a_n)].$$

In particular, an inclusion of real closed fields is an elementary inclusion.

3.11. COMMENTS. (i) It is well-known that quantifier elimination is very sensitive to the choice of language. For example, any theory admits quantifier elimination in a suitable language; but, as a rule, such language is of no mathematical interest. The interest of Theorem 3.4 lies precisely in that the language L is intimately connected with information of mathematical significance.

(ii) Quantifier elimination is an exceptional phenomenon. For example, Macintyre-McKenna-van den Dries [38] and van den Dries [41] have shown that RCF is the only theory of (not necessarily commutative) ordered rings with unit admitting quantifier elimination in the language L.

On the other hand, transfer principles are a much more common occurrence: Mc Kenna has shown the existence of $2^{\aleph_0}$ model-complete theories of ordered fields.

(iii) Quantifier elimination implies that any definable subset of $\mathbb{R}$ is a union of intervals. Van den Dries [87] has shown that this latter property <u>alone</u> already has some significant consequences normally associated with quantifier elimination techniques, and conjectures that this property is preserved if the exponential function is adjoined to the language. The study of properties of this type arises in the investigation of:

(iv) <u>Tarski's problem.</u> In his monograph [40; p. 45] Tarski asked whether quantifier elimination or some of its consequences could be extended to the theory of the structure $<\mathbb{R},+,\cdot,<,\exp,0>$. This problem has been taken up again in the last few years, and valuable new information has been obtained (we know, for example that quantifier elimination cannot be extended). Nevertheless, the problem is incomparably more difficult than its analog for polynomials. For a discussion and further references, see van den Dries [87]. □

§4. APPLICATIONS OF QUANTIFIER ELIMINATION.

A. Elementary properties of semi-algebraic sets and functions.

The following notion is generally considered to provide the analog in real algebraic geometry of the notion of "constructible set" in ordinary algebraic geometry.

<u>DEFINITION 4.1.</u> Let $K \models RCF$. A semi-algebraic (s.a.) subset of K^n, $n \geq 1$, is a finite boolean combination of sets of the form $\{\overline{x} \in K^n \mid Q(\overline{x}) > 0\}$, where $Q \in K[X_1, \ldots, X_n]$. □

Equivalently, s.a. sets are the subsets of K^n defined by formulas of the form

$$\bigvee_i \left(\bigwedge_j P_{ij}(v_1, \ldots, v_n) = 0 \wedge \bigwedge_k Q_{ik}(v_1, \ldots, v_n) > 0\right)$$

(finite conjunctions and disjunctions), with $P_{ij}, Q_{ik} \in K[X_1, \ldots, X_n]$, or in other words, by quantifier-free formulas in the language L, with parameters in K. Invoking the quantifier elimination theorem 3.4 we obtain at once the fundamental relation:

(*) Semi-algebraic = Definable

(i.e. parametrically definable) which establishes the connection between first-order logic and real algebraic geometry (or semi-algebraic geometry, as it is also called).

Once this relationship is understood, a number of fundamental results follow instantaneously:

THEOREM 4.2. Let $K \models RCF$. The class of s.a. sets is closed under the following operations:

(i) finite boolean operations;

(ii) closure and interior (in the euclidean topology for K^n);

(iii) projection (of K^m into K^n, say);

(iv) cartesian product.

PROOF. Immediate using (*). For example, the closure $\overline{S}$ of a definable set $S \subseteq K^n$ is defined by :

$$\overline{x} \in \overline{S} \quad \text{iff} \quad \forall \varepsilon > 0 \ \exists \ \overline{y} \in S (d(\overline{x},\overline{y}) < \varepsilon)$$

(d denotes the Euclidean distance). □

DEFINITION 4.3. Let K be a real closed field, $m,n \geq 1$, and S a subset of K^n. A function $F : S \to K^m$ is called semi-algebraic (s.a.) if its graph

$$Gr(F) = \{<\overline{x}, F(\overline{x})> \mid \overline{x} \in S\}$$

is a s.a. subset of K^{n+m}. □

Note that the domain $S = Dom(F)$ of a s.a. function is necessarily a s.a. set, since it is the projection over K^n of the graph of F.

THEOREM 4.4. Let $K \models RCF$.

(a) The set of K-valued s.a. functions defined on a s.a. subset of K^n is a ring under the standard, pointwise defined operations.

(b) The composition of s.a. functions is s.a.

(c) The image of a s.a. set under a s.a. function is s.a.

PROOF. As above; write down the standard definitions of the relevant operations (which are first-order), and use (*). □

The proof of these elementary facts without quantifier elimination is usually a headache (this is hardly surprising, since some of these facts are actually equivalent to quantifier elimination). One can avoid the trivial proofs using quantifier elimination by giving non-trivial proofs, if desired (in fact, this is done, sometimes).

4.5. EXAMPLE. As an example going beyond the standard examples of elementary analysis, consider, for a fixed s.a. set $S \subseteq K^n$, the function $d_S : K^n \to K$ defined by:

$$d_S(\overline{x}) \ = \ \text{the (Euclidean) distance from } \overline{x} \text{ to } S;$$

d_S is a continuous s.a. function. General properties of these functions, yielding non-trivial information, will be studied in §9. □

B. THE REAL NULLSTELLENSATZ.

A fundamental feature of algebraic geometry over an algebraically closed field, k, is the one-one correspondence between k-points of a variety V and (proper) maximal ideals of the coordinate ring k[V] established by Hilbert's nullstellensatz. This correspondence fails in the real case : the polynomial $X^2 + Y^2 + 1$ does not have real points, while its coordinate ring is not reduced to 0 and hence does have (proper) maximal ideals.

In order to develop algebraic geometry over $\mathbb{R}$, it is of the utmost importance to restore a correspondence of this kind, that is, to determine which type of ideals correspond to points. This amounts to proving a "real" version of the nullstellensatz, which we will presently do.

The key notion comes from the observation that the ideal $I(\bar{a})$ of real polynomials vanishing at a point $\bar{a} \in \mathbb{R}^n$ has the following property:

$$\sum_{i=1}^{n} P_i^2 \in I(\bar{a}) \quad \text{implies} \quad P_1,\dots,P_n \in I(\bar{a}).$$

Generalizing this observation we introduce the following notion:

DEFINITION 4.6. Let $\langle k, \leq \rangle$ be an ordered field, A a ring containing k and $I \subseteq A$ an ideal.

(a) We say that I is real over k iff

$$\sum_{i=1}^{n} p_i a_i^2 \in I, \text{ with } a_i \in A \text{ and } p_i \in k,\ p_i > 0, \text{ implies } a_1,\dots,a_n \in I.$$

We will also need the following weaker notion:

(b) I is called semi-real over k iff -1 is not of the form $\sum_i p_i a_i^2/_I$ with $p_i \in k$, $p_i > 0$, and $a_i \in A$.

(c) If $I = \{0\}$ we call A real (respectively, semi-real) over k. □

4.7. REMARKS. (a) If I is a prime ideal and $A/_I$ is of characteristic $\neq 2$, then I is real over k iff $A/_I$ has an ordering extending that of k iff -1 is not of the form $\sum_i p_i x_i^2$ with $p_i \in k$, $p_i > 0$, and x_i in the field of fractions of $A/_I$.

(b) I real over k implies I radical.

(c) Let $\langle F, \leq \rangle$ be an ordered extension of the ordered field $\langle k,\leq\rangle$, and $S \subseteq F^n$. Then the ideal $I_k(S)$ of polynomials over k vanishing at S (cf. § 2.1) is real over k.

(d) If A is a noetherian ring and I is a proper ideal of A, then I is real over k iff I is radical and is the intersection of finitely many prime ideals real over k.

(e) The example $I = (X^2 + Y^2)$, $A = k[X,Y]$, shows that semi-real is a notion strictly weaker than real. □

THEOREM 4.8. (Real nullstellensatz). Let $\langle k,\leq\rangle$ be an ordered field, $\bar{k}$ its real closure, and $I \subseteq k[X_1,\ldots,X_n]$ an ideal. Then

$$I \text{ is real over } k \quad \text{iff} \quad I = I_k(V_{\bar{k}}(I)).$$

PROOF. The implication from right to left is just the remark 4.7 (C). For the converse we need only prove the inclusion $I_k(V_{\bar{k}}(I)) \subseteq I$, since the other inclusion is trivial.

Let I be generated by $Q_1,\ldots,Q_r \in k[X_1,\ldots,X_n]$.
The condition $Q \in I_k(V_{\bar{k}}(I))$ is expressed by :

$$(*)\quad \bar{k} \models \forall\bar{v}[\bigwedge_{i=1}^{r} Q_i(v) = 0 \rightarrow Q(\bar{v}) = 0\,].$$

Using remark 4.7(d) we can find a representation $I = \bigcap_{j=1}^{\ell} P_j$ where the ideals P_j are prime and real over k. Then the condition $Q \in I$ is equivalent to the conjunction of the conditions

$$(**)\quad k[X_1,\ldots,X_n]/P_j \models Q({}^{X_1}/_{P_j},\ldots,{}^{X_n}/_{P_j}) = 0$$

for $j = 1,\ldots,\ell$. Since each of the rings $k[X_1,\ldots,X_n]/P_j$ has an ordering extending that of k (by 4.7(a)), we may consider the real closure L_j of its field of fractions with one of these orderings. In particular, we have $\bar{k} \subseteq L_j$; by Proposition 3.10 this inclusion is elementary. Hence the formula (*) holds in L_j as well. Applying (**) to $Q_1,\ldots,Q_r$ we get

$$L_j \models \bigwedge_{i=1}^{r} Q_i({}^{X_1}/_{P_j},\ldots,{}^{X_n}/_{P_j}) = 0.$$

Now, using (*) with $\bar{v} = \langle {}^{X_1}/_{P_j},\ldots,{}^{X_n}/_{P_j}\rangle$ we conclude that

$$L_j \models Q\,({}^{X_1}/_{P_j},\ldots,{}^{X_n}/_{P_j}) = 0,$$

which implies that $Q \in P_j$. Since this holds for $j = 1,\ldots,\ell$, we get $Q \in I$, as contended. □

Every ideal is contained in a smallest, possibly improper, real ideal; namely:

$$\bigcap \{J \mid J \text{ is real over } k \text{ and } I \subseteq J \subseteq A\}.$$

DEFINITION 4.9. We shall call real radical of I, $\sqrt[R]{I}$, the smallest ideal of A, real over k and containing I. □

The real radical admits a purely algebraic characterization:

PROPOSITION 4.10.

$a \in \sqrt[R]{I}$ iff there are $n, m \in \mathbb{N}$, $m > 0$, $p_1, \ldots, p_n \in k^+$ and $b_1, \ldots, b_n \in A$ such that

$$a^{2m} + \sum_{i=1} p_i b_i^2 \in I.$$ □

This result is proved in Krivine [51], Dubois-Efroymson [47] and Dickmann [72; Ch.III].

We have the following consequences of the real nullstellensatz:

COROLLARY 4.11. Let $\langle k, \leq \rangle$ be an ordered field, $\overline{k}$ its real closure and $I \subseteq k[X_1, \ldots, X_n]$ an ideal. Then:

$$I_k(V_{\overline{k}}(I)) = \sqrt[R]{I}.$$

PROOF. By the nullstellensatz $I_k(V_{\overline{k}}(\sqrt[R]{I})) = \sqrt[R]{I}$, and by the preceeding Proposition $V_{\overline{k}}(I) = V_{\overline{k}}(\sqrt[R]{I})$.

COROLLARY 4.12. With the notation of the preceeding Corollary we have:

$$1 \in \sqrt[R]{I} \quad \text{iff} \quad V_{\overline{k}}(I) = \emptyset.$$ □

We restate this corollary in more geometric language:

COROLLARY 4.12 bis. (Weak real nullstellensatz). Let V be a variety over an ordered field $\langle k, \leq \rangle$ and let $\overline{k}$ be the real closure of $\langle k, \leq \rangle$. Then the following are equivalent:

(i) V has a $\overline{k}$-point, i.e. $V(\overline{k}) \neq \emptyset$.

(ii) The ideal $I(V)$ is semi-real over k. □

(Note that an ideal I is semi-real iff $\sqrt[R]{I}$ is proper).

COROLLARY 4.13. Let $\langle k, \leq \rangle$ and $\overline{k}$ be as a Corollary 4.11, and let V be a variety over k. The map $\overline{a} \mapsto I(\overline{a})/_{I(V(k))}$, for $\overline{a} \in V(\overline{k})$, establishes a one-one correspondence between $\bar{k}$-points of V and ideals of $k[V]$ which are maximal among ideals real over k.

PROOF. We need to show:

(i) $I(\bar{a})$ contains $I(V(k))$ and is maximal among ideals of $k[X_1,\ldots,X_n]$ real over k; and

(ii) Every ideal of this type is of the form $I(\bar{a})$ for some $\bar{a}\in V(\bar{k})$.

PROOF of (i). $I(\bar{a}) \supseteq I(V(k))$ because $\bar{a}\in V(\bar{k})$. Remark 4.7(c) shows that $I(\bar{a})$ is real over k. Finally, if $I(\bar{a}) \subsetneq I$ with I real over k, then $V_{\bar{k}}(I) \subsetneq \{\bar{a}\}$, i.e. $V_{\bar{k}}(I) = \emptyset$; hence $1\in I$ by Corollary 4.12.

PROOF of (ii). Let $M \supseteq I(V(k))$ be maximal among ideals real over k. By 4.8, 4.11 and 4.12 we get $V_{\bar{k}}(M) \neq \emptyset$. Let $\bar{a}\in V_{\bar{k}}(M)$. Since $M \supseteq I(V(k))$ then $\bar{a}\in V(\bar{k})$. Also $M \subseteq I(\bar{a})$, and maximality implies the equality. □

Exercise. Prove that in a ring of the form $k[V]$ every ideal which is maximal among real ideals, is maximal.

Historical rehabilitation. The authorship of the real nullstellensatz is usually attributed to Dubois [46] and Risler [57]. However, Krivine proved it long before, as well as Proposition 4.10; see his paper [51].

C. THE SIMPLE POINT CRITERION.

The weak real nullstellensatz 4.12 bis expresses in geometric terms the algebraic condition "I(V) is a semi-real ideal". We are interested in finding a geometric expression for the closely realted notion "I(V) is a real ideal".

It turns out that this condition has a very interesting geometric content: it says that V has a non-singular $\bar{k}$-point ($\bar{k}$ = the real closure of $\langle k,\leq\rangle$). This is what we will prove below.

When the variety V is a hypersurface -i.e. it is given by a single polynomial $F(X_1,\ldots,X_n)$ - we know from elementary geometry that a point $\bar{a}\in V(k)$ is called non-singular (or simple) if at least one of the derivation $\frac{\partial F}{\partial X_i}$ $(i = 1,\ldots,n)$ does not vanish at $\bar{a}$. The correct definitives in the general case is as follows: assume that V is given by polynomials $P_1,\ldots,P_\ell\in k[X_1,\ldots,X_n]$, and consider the Jacobian matrix,

$$J(P_1, \ldots, P_\ell) = \begin{pmatrix} \frac{\partial P_1}{\partial X_1} & \cdots & \frac{\partial P_\ell}{\partial X_1} \\ \vdots & & \\ \frac{\partial P_1}{\partial X_n} & \cdots & \frac{\partial P_\ell}{\partial X_n} \end{pmatrix}$$

This is an $n \times \ell$ matrix with entries in $k[X_1, \ldots, X_n]$. For a fixed s, $1 \le s \le \min\{\ell, n\}$, let M_i^s $(i = 1, \ldots, t_i)$ denote the $s \times s$ minors of $J(P_1, \ldots, P_\ell)$; we have $M_i^s \in k[X_1, \ldots, X_n]$. Let r denote the largest s such that some $M_i^s \notin I(V)$; r is just the rank of $J(P_1, \ldots, P_\ell)$ calculated modulo $I(V)$.

<u>DEFINITION 4.14.</u> A point $\bar{a} \in V(\bar{k})$ of a variety V over k defined by $P_1, \ldots, P_\ell$ is called <u>non-singular</u> (or <u>simple</u>) iff the rank of $J(P_1, \ldots, P_\ell)(\bar{a})$, the Jacobian matrix at $\bar{a}$, equals r. □

It is easily seen that a point $\bar{a}$ is non-singular iff the tangent space to the variety at $\bar{a}$ has dimension $n-r$ (cf. Shafarevich [14; pp. 74,77]).

Our main result is:

<u>THEOREM 4.15.</u> Let $\langle k, \le \rangle$ be an ordered field, $\bar{k}$ its real closure, and I a prime ideal of $k[X_1, \ldots, X_n]$. Then the variety defined by I has a simple $\bar{k}$-point iff the ideal I is real over k.

<u>PROOF.</u> Let $A = k[X_1, \ldots, X_n]/_I$ and K be the field of fractions of A.

(a) We prove first the implication from right to left, which is a simple consequence of the second transfer principle.

By assumption K has an order $\le$ extending the order of k; let L denote the real closure of $\langle K, \le \rangle$. Assume that the polynomials $P_1, \ldots, P_\ell$ generate I. With r denoting the rank of $J(P_1, \ldots, P_\ell)$ in A, as above, we have:

$$K \models \bigwedge_{j=1}^{\ell} P_j({}^{X_1}/_I, \ldots, {}^{X_n}/_I) = 0 \wedge \bigvee_{i=1}^{t_r} M_i^r({}^{X_1}/_I, \ldots, {}^{X_n}/_I) \neq 0.$$

In particular, the statement

$$(*) \qquad \exists v_1, \ldots, \exists v_n \left(\bigwedge_{j=1}^{\ell} P_j(v_1, \ldots, v_n) = 0 \wedge \bigvee_{i=1}^{t_r} M_i^r(v_1, \ldots, v_n) \neq 0 \right),$$

holds in L. Since the parameters of $(*)$ lie in k and (up to k-isomorphism) we have $\bar{k} \subseteq L$, then $(*)$ also holds in $\bar{k}$. This means, precisely, that the variety defined by I has a simple $\overline{k}$-point.

(b) The proof of the other implication requires some high-powered commutative algebra, we only sketch it.

Let $\overline{a} \in V_{\overline{k}}(I)$ be a simple point and $M = I(\overline{a})/_I$ its maximal ideal in A. A (deep) result from commutative algebra tells us that, under the present assumptions, the following holds:

(†) The localized A_M of A at the ideal M is a regular (local)ring.

(For the definition of a regular local ring and the proof of (†), see Atiyah-Macdonald [1; Ch. 11].)

Next observe that the evaluation map at $\overline{a}$, $ev_{\overline{a}} : A \to \overline{k}$, defined by $ev_{\overline{a}}(^Q/_I) = Q(\overline{a})$, is a surjective ring homomorphism with kernel M; hence $A/_M \cong \overline{k}$.

On the other hahd, the map $^A/_M \to {}^{A_M}/_{MA_M}$ which sends $^x/_M$ into $^x/_{MA_m}$ is an isomorphism (easy verification). Hence

$$^{A_M}/_{MA_M} \cong {}^A/_M \cong \overline{k} .$$

It follows that the residue field of A_M is orderable. Now, another basic result from real commutative algebra tells us:

(††) If B is a regular local ring whose residue field is orderable, then B is real.

For a proof, see Lam [54; Prop. 2.7].

From (††) we conclude that A_M is real; this implies that A itself is real, as one can easily verify. □

The general case, when the variety is not necessarily irreducible, is easily derived from the preceeding theorem. We will not prove this, but in order to state the result we need to know that a proper ideal in a Noetherian ring is contained in only finitely many minimal prime ideals (that is, minimal among the prime ideals containing it); cf. Atiyah-Macdonald [1; Thm. 7.13 and Ch. 4].

THEOREM 4.16. Let V be a variety over an ordered field $\langle k, \leq \rangle$, $\overline{k}$ its real closure, and $J_1, \ldots, J_m$ the minimal prime ideals of $k[X_1, \ldots, X_n]$ containing the ideal $I(V)$. Then, the following are equivalent:

(i) The ideal $I(V)$ is real over k.

(ii) $I(V)$ is a radical ideal and each variety $V_{\overline{k}}(J_i)$, $i = 1, \ldots, m$, has a simple point. □

For details, see Lam [54; Prop. 2.9 and Thm. 6.10].

D. HILBERT'S 17th problem.

In its original formulation this is the problem of knowing whether a rational function with rational coefficients, $R \in \mathbb{Q}(X_1,\ldots,X_n)$, such that $R(\overline{x}) \geq 0$ at every $\overline{x} \in \mathbb{R}^n$ for which the denominator does not vanish, is a sum of squares in $\mathbb{Q}(X_1,\ldots,X_n)$.

A positive answer was given by Artin [19] in 1927, based on the path-breaking work of Artin-Schreier [42], [43]. In the 1950's Robinson [22], [23] gave an alternative (and generalized) proof using the second transfer principle 3.9. It should be emphasized that the ideas underlying Robinson's technique apply as well to other situations where the methods of Artin-Schreier are inapplicable; for example, to polynomials over real closed rings (see Dickmann [83]), or over fields with higher level orderings (see Becker - Jacob [81]), or even over p-adically closed fields (see Prestel-Roquette [86]). In a certain sense, one may say that Robinson's proof "trivializes" that part of the work of Artin-Schreier which deals properly with Hilbert's 17th problem.

Robinson's proof was the first truly mathematical application of quantifier elimination. The first chapter of Delzell's thesis [20] contains an extensive historial account of Hilbert's 17th problem.

Since the subject has been widely treated in the existing literature, we confine ourselves to the essential points.

THEOREM 4.17. Let $\langle k,\leq\rangle$ be an ordered field, $\overline{k}$ its real closure, and $R \in k(X_1,\ldots,X_n)$ a rational function with coefficients in k. Assume that $R(\overline{x}) \geq 0$ for every $\overline{x} \in \overline{k}^n$ for which its denominator does not vanish. Then

$$R = \sum_i p_i R_i^2$$

with $p_i \in k$, $p_i > 0$ and $R_i \in k(X_1,\ldots,X_n)$.

PROOF. Assume R does not have this form. The standard Artin-Schreier criterion of positivity (cf. Dickmann [72; Prop. I, 1.5] or Ribenboim [13; Thm. IX.2]) implies at once that $k(X_1,\ldots,X_n)$ has an order $\leq$ extending that of k, for which $R < 0$. Note also that if $R = {}^P/_Q$ with $P, Q \in k[X_1,\ldots,X_n]$, then $Q \neq 0$ and $R < 0$ iff $RQ^2 < 0$.

Let us denote by L the real closure of $\langle k(X_1,\ldots,X_n),\leq\rangle$. The statement

(*) $$\exists v_1,\ldots, \exists v_n (RQ^2(v_1,\ldots,v_n) < 0),$$

of the language L , with parameters in k, is valid in L (for $v_i = X_i$). We can assume, without loss of generality, that $L \supseteq \bar{k}$. Then (*) is true in $\bar{k}$, showing that $R(\bar{x}) < 0$ for some $x \in \bar{k}^n$, a contradiction. □

McKenna [21] has done a conclusive study of several questions related to this result.

A useful application of Theorem 4.17 in real commutative algebra is the following result, first proved by Dubois-Efroymson [47]:

THEOREM 4.18. (Change of sign criterion). Let $\langle k,\leq\rangle$ be an ordered field, $\bar{k}$ its real closure, and $Q, Q_1,\dots,Q_r \in k[X_1,\dots,X_n]$ monic polynomials such that $Q = Q_1\cdot\ldots\cdot Q_r$ and the Q_i's are irreducible. Then the following are equivalent:

(a) The principal ideal (Q) is real over k.

(b) Each Q_i changes sign in $\bar{k}^n$ and the Q_i's are pairwise distinct. □

Theorem 4.17 gives an easy proof of the implication from (a) to (b); the argument appears in Dickmann [72; Thm. II.6.5]. A model-theoretic proof, due to Prestel, appears in Elman-Lam-Wadsworth [48; §4. bis]. Other proofs can be found in Lam [54; Thm. 6.11] and Ribenboim [56].

E. FURTHER APPLICATIONS.

The results proved in the preceeding paragraphs do not exhaust the list of basic results obtained by direct application of model-theoretic methods. We metnion here, without proofs, three results originally proved by Lang [52] using other methods; they form the core of the theory of real algebraic function fields.

PROPOSITION 4.19. (The homomorphism theorem). Let $\langle k,\leq\rangle$ be an ordered field, $\bar{k}$ its real closure, and A an integral domain which is a k-algebra of finite type. If the field of fractions of A has an order extending that of k(i.e. is real over k), then there is a k-algebra homomorphism $f:A \to \bar{k}$. □

This result is, in fact, a particular case of the weak real nullstellensatz 4.12 bis, and the reader can prove it as an exercise; cf. Colliot-Thélène [44], Dickmann [72; Ch.II] or Lam [54; §5], for details.

THEOREM 4.20. (The embedding theorem). Let K be a real close field, L an orderable extension of K of (finite) transcendence degree n, and F a real closed extension of K of transcendence degree $\geq$ n, over K. Then L can be K-embedded into F. □

Proof of this theorem can be found in Lam [53; §6], [54, Thm. 6.15].

THEOREM 4.21. (The homomorphism extension theorem). Let $\langle k,\leq\rangle$ be an ordered field, A a subring of k, $g:A \to L$ a homomorphism of ordered rings with values in a real closed field L. Then g extends to an ordered ring homomorphism $g':B \to L'$, where B is a convex valuation subring of k containing A, and L' a real closed field extending L. □

This important result is a consequence of the amalgamation property of real closed fields, a property which follows from model-completeness. Lang [52] states the theorem in terms of real places. A corresponding existence theorem follows as corollary. A proof of Theorem 4.21 is given in Dickmann [72; Thm. II 7.10].

§5. THE STRUCTURE OF SEMI-ALGEBRAIC SETS.

The main topic of this section is the separation theorem, an important technical tool in real algebraic geometry, mainly used to show that the topological notion of connectedness for s.a. sets can be formulated in the first-order language L of real closed fields. Among its corollaries one obtains:

- The finiteness of the number of connected components of a s.a. subset of $\mathbb{R}^n$ (Theorem 5.8 below).
- A refinement of Tarski's theorem 3.4 known as "open quantifier elimination" (Theorem 5.9).

DEFINITION 5.1. Let k be an ordered field and $P_1,\ldots,P_\ell \in k[X_1,\ldots,X_n]$

(a) We call sign condition (resp. strict sign condition) on $P_1,\ldots,P_\ell$ any formula of the form:

$$\sigma(v_1,\ldots,v_n): \quad \bigwedge_{i=1}^{\ell} P_i(v_1,\ldots,v_n) \; ?_i \; 0$$

where each $?_i$ is one of the signs $\geq, >, \leq, <$ (resp. $>$, $<$) or $=$

(b) Given a sign condition $\sigma(v_1,\ldots,v_n)$ as above, we call enlarged sign condition associated to σ the formula

$$\hat{\sigma}(v_1,\ldots,v_n): \quad \bigwedge_{i=1}^{\ell} P_i(v_1,\ldots,v_n) \; \hat{?}_i \; 0.$$

where $\hat{?}_i$ is $\geq$ or $\leq$ if $?_i$ is $>$ or $<$, respectively, and is $?_i$ in all other cases. □

We begin our discussion by proving the easy case of the separation theorem, viz., for one-variable polynomials.

5.2. THOM'S LEMMA. Let $P_1,\ldots,P_\ell$ be a finite set of polynomials in $\mathbb{R}[X]$ closed under (non-constant) derivation. Then for every sign condition $\sigma(v)$ on $P_1,\ldots,P_\ell$, the set

$$A_\sigma \models \{x \in \mathbb{R} \mid \mathbb{R} \models \sigma[x]\}$$

is either empty or an interval (possibly reduced to one point or infinite).

PROOF. Induction on ℓ, the case $\ell = 1$ being trivial. Assume true for ℓ and suppose, renaming if necessary, that $P_{\ell+1}$ is of maximal degree. This has the effect of making $P_1, \dots, P_\ell$ closed under derivation, so that the induction hypothesis applies and the set

$$A' = \bigcap_{i=1}^{\ell} \{x \in \mathbb{R} \mid P_i(x) \ ?_i \ 0\}$$

is empty or an interval.

The derivative of $P_{\ell+1}$ - which is among $P_1, \dots, P_\ell$ - is of constant sign on A'; $P_{\ell+1}$ is, then, monotone or constant on A', and the lemma follows easily using the intermediate value property. □

The foregoing proof clearly applies to any real closed field. We mention the following interesting corollaries:

COROLLARY 5.3. Let $<k, \leq>$ be an ordered field, $\bar{k}$ its real closure, $P \in k[X]$ a one-variable polynomial. Then two distinct roots α_1, α_2, of P in $\bar{k}$ can be separated by a sign condition on one of the derivatives of P, i.e. there is i, $1 \leq i < \deg(P)$ such that $P^{(i)}(\alpha_1)$, $P^{(i)}(\alpha_2)$ have different signs (one of them can be zero). If P is irreducible over k, then i can be chosen so that $P^{(i)}(\alpha_1).P^{(i)}(\alpha_2)<0$. □

COROLLARY 5.4. Let $<k, \leq>$ be an ordered field, $\bar{k}$ its real closure, α an element algebraic over k with minimal polynomial F, and $\alpha_1 < \dots < \alpha_r$ all the roots of F in $\bar{k}$. For each $i = 1, \dots, r$ let us define an ordering $\leq_i$ on $k(\alpha)$ by the condition

$$0 \leq_i P(\alpha) \quad \text{iff} \quad P(\alpha_i) \geq 0 \ \ (\text{in } \bar{k}).$$

Then these orders extend the order of k, are pairwise distinct, and any order on $k(\alpha)$ extending the order of k is one of them. □

Returning to the separation theorem, the notion appropriate for a generalization of Thom's lemma to polynomials in several variables stems from the following (trivial) observations:

- intervals are exactly the connected subsets of $\mathbb{R}$;
- The closure of a set of form $A = \{x \in \mathbb{R} \mid \mathbb{R} \models \sigma[x]\}$, where σ is a sign condition, is the set

$$\bar{A} = \{x \in \mathbb{R} \mid \mathbb{R} \models \hat{\sigma}[x]\}$$

defined by the enlarged sign condition $\hat{\sigma}$ associated to σ.

Thus we introduce

DEFINITION 5.5. A finite family $P_1,\ldots,P_\ell \in \mathbb{R}[X_1,\ldots,X_n]$ is called a separating family iff for every sign condition σ on $P_1,\ldots,P_\ell$:

(i) The set

$$A_\sigma = \{x \in \mathbb{R}^n \mid \mathbb{R} \models \sigma[\bar{x}]\}$$

is either empty or connected.

(ii) If $A_\sigma \neq \emptyset$, then $\overline{A_\sigma}$ (=closure of A_σ) = $\{x \in \mathbb{R}^n \mid \mathbb{R} \models \hat{\sigma}[x]\}$. □

Thus, Thom's lemma says that any finite family of one-variable polynomials closed under derivation is separating.

The announced generalization is this:

THEOREM 5.6. (Separation theorem). Every finite family of polynomials, $P_1,\ldots,P_\ell \in \mathbb{R}[X_1,\ldots,X_n]$, can be completed into a separating family $P_1,\ldots,P_\ell, P_{\ell+1},\ldots,P_{\ell+s}$.

Furthermore

(a) The number s of additional polynomials is a primiteve recursive function of ℓ, n and d = maximum of the total degrees of the polynomials $P_1,\ldots,P_\ell$.

(b) There is a primitive recursive function $f(n,\ell,d)$ giving an upper bound on the total degrees of the new polynomials $P_{\ell+j}$ $(j=1,\ldots,s)$.

(c) The coefficients of the polynomials $P_{\ell+1},\ldots,P_{\ell+s}$ are polynomial functions over $\mathbb{Z}$ of the coefficients of $P_1,\ldots,P_\ell$. □

The proof is not easy. A word of caution: adjoining only the partial derivatives of $P_1,\ldots,P_\ell$ does not produce, in general, a separating family; counterexample: the polynomial

$(X \sin\alpha - Y \cos\alpha)(X \sin\beta - Y \cos\beta) + 1$, with $0 < \alpha < \beta < \frac{\pi}{2}$ (Houdebine [76]).

All known proofs proceed by induction on the number n of variables, using several of the following ingredients:

- Quantifier elimination;
- The continuity of the functions of $\bar{x}$ which give, in the increasing order of $\mathbb{R}$, the real part of the (complex) roots of a polynomial equation of the form

$$P_n(\bar{x}) \quad Y^n + \ldots + P_o(\bar{x}) = 0,$$

$P_o,\ldots,P_n \in \mathbb{R}[X_1,\ldots,X_n]$, at values $\bar{x} \in \mathbb{R}^n$ such that $P_n(\bar{x}) \neq 0$ (cf. Gillman-Jerison [85; Thm. 13.3(a)]);

- The following result, known as the "cylindrical decomposition" theorem:

THEOREM 5.7. Given a s.a. set $U \subseteq \mathbb{R}^n$ and a polynomial $P \in \mathbb{R}[X_1,\ldots,X_n]$ there is a partition of U into finitely many s.a. sets $A_1,\ldots,A_m$, such that for each $i = 1,\ldots,m$, either one of the following properties holds:

(1) $P(\overline{x},y)$ has constant sign (>0, <0 or $=0$) for all $\overline{x} \in A_i$ and $y \in \mathbb{R}$; or :

(2) There are continuous s.a. functions $\alpha_1^i,\ldots,\alpha_{\ell_i}^i : A_i \to \mathbb{R}$ ($\ell_i \geq 1$), such that:

(a) $\alpha_1^i(\overline{x}) < \ldots < \alpha_{\ell_i}^i(\overline{x})$ for all $\overline{x} \in A_i$;

(b) for each $\overline{x} \in A_i$, $\alpha_1^i(\overline{x}),\ldots,\alpha_{\ell_i}^i(\overline{x})$ are all the real roots of the polynomial $P(\overline{x},Y)$. □

Note that, in particular, for $\overline{x} \in A_i$ the polynomial $P(\overline{x},Y)$ does not change sign in any of the intervals $(-\infty, \alpha_1^i(\overline{x}))$, $(\alpha_j^i(\overline{x}), \alpha_{j+1}^i(\overline{x}))$ for $j = 1,\ldots,\ell_i - 1$, $(\alpha_{\ell_i}^i(\overline{x}), \infty)$.

Historical remarks (a) Thom's lemma was, indeed, first noticed by Thom; it first appeared in Łojasiewicz [28; p. 69]. The generalization to several variables is due to Efroymson [27], where the proof is (at best) incomplete. Later Coste/Coste-Roy [69] gave a correct (and complete) proof using a technique introduced by Cohen [34] to prove quantifier elimination for real closed fields. The most efficient proof is due to Houdebine [76].

(b) Cylindrical decomposition is sometimes used, disguised in different ways, as an alternative approach to quantifier elimination; for example, see Cohen [34], Collins [35], Coste [68], Brumfiel [65; Appendix].

We turn now to some important consequences of the separation theorem.

THEOREM 5.8. (Finiteness theorem). Any s.a. subset S of $\mathbb{R}^n$ has a finite number of connected components, each of which is a s.a. set.

PROOF. Since S is a finite union of sets defined by formulas of the form

$$\sigma(v_1,\ldots,v_n) : \bigwedge_{j=1}^{m} P_j(v_1,\ldots,v_n) = 0 \wedge \bigwedge_{k=m+1}^{\ell} P_k(v_1,\ldots,v_n) > 0$$

with $P_1,\ldots,P_\ell \in \mathbb{R}[X_1,\ldots,X_n]$, it suffices to prove the result for sets of this form.

Complete $P_1,\ldots,P_\ell$ to a separating sequence $P_1,\ldots,P_\ell,P_{\ell+1},\ldots,P_{\ell+s}$. By condition (i) of 5.5, the connected components of S are unions of non-empty sets of the form

$$\{\overline{x} \in \mathbb{R}^n \mid \mathbb{R} \models \sigma[\overline{x}] \wedge \delta[\overline{x}]\}$$

where $\delta(v_1,\ldots,v_n)$ is a sign condition on $P_{\ell+1},\ldots,P_{\ell+s}$. There are 3^s possible δ' s. □

The next result, known as "open quantifier elimination" is a particularly important tool for real algebraic geometry, specially for the investigation of the real spectrum (cf. §§6.C and 7 below). Its significance was pointed out by Brumfiel [65; Unproved Prop. 8.2] even before it was proved, independently, by Recio [78] in 1977, Coste/Coste-Roy [69] in 1979, and Delzell [20] in 1980.

The result gives an answer to the following question : assume we are given an <u>open</u> s.a. subset U of $\mathbb{R}^n$ (or, more generally, of K^n, with $K \models RCF$) ; what can we say about the quantifier-free representation of U?; does there exist, for example, a representation of the form

$$\bigvee_i \bigwedge_j P_{ij}(v_1,\ldots,v_n) > 0 \quad ?$$

Note that irredundant equalities may, <u>a priori</u>, appear in an specific representantion of an open s.a. set; for example :

$$U = \{\langle x,y\rangle \mid x^2 + y^2 < 1 \wedge y \neq 0\} \cup \{\langle x,y\rangle \mid y = 0 \wedge x^2 < {}^1/_4\}.$$

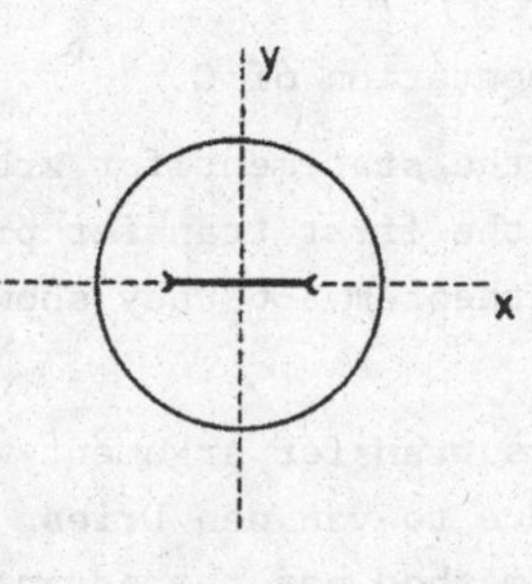

The same set, U, can also be defined by replacing the equality $y = 0$ by the inequality $y^2 < {}^1/_9$, for instance; but how can we guarantee that ad-hoc devices of this type will work in <u>all</u> cases?.

<u>THEOREM 5.9. (Open quantifier elimination)</u>. Let U be an open s.a. subset of K^n, where $K \models RCF$. Then there are polynomials $P_{ij} \in K[X_1,\ldots,X_n]$ such that

$$U = \bigcup_i \bigcap_j \{\overline{x} \in K^n \mid P_{ij}(\overline{x}) > 0\}.$$

Let us derive this result from the separation theorem in the case $K = \mathbb{R}$:

<u>PROOF OF THEOREM 5.9.</u> (case $K = \mathbb{R}$). We prove the dual statement: every closed s.a. set $C \subseteq \mathbb{R}^n$ has a definition of the form

$$\bigvee_i \bigwedge_j P_{ij}(v_1,\ldots,v_n) \geq 0 .$$

By quantifier elimination we know that C has a definition of the form

$$(*) \qquad \bigvee_k (\bigwedge_\ell P_{k\ell}(\overline{v}) = 0 \wedge \bigwedge_m Q_{km}(\overline{v}) > 0).$$

By the separation theorem this family of polynomials can be extended to a separating family; modulo a change of notation we may then assume that the polynomials $P_{k\ell}$, Q_{km} form a separating family. Let A_k be the subset of $\mathbb{R}^n$ defined by the corresponding disjunct of $(*)$, which we may suppose non-empty; by condition (ii) of Definition 5.5 we have:

$$\overline{A_k} = \{\overline{x} \in \mathbb{R}^n \mid \bigwedge_\ell P_{k\ell}(\overline{x}) = 0 \wedge \bigwedge_m Q_{km}(\overline{x}) \geq 0\}.$$

Since $C = \bigcup_k A_k$ and (finite) unions commute with closure, we obtain:

$$C = \overline{C} = \bigcup_k \overline{A_k} ,$$

which gives the desired representation of C. □

Coste/Coste-Roy [62] derived the statement for arbitrary $K \models RCF$ from the case $K = \mathbb{R}$ by use of the first transfer principle : using points (a) and (b) of the separation theorem 5.6 they show that the statement of Theorem 5.9 is first-order.

Instead of working through this transfer argument we outline below a direct model-theoretic proof due to van den Dries [79] which bypasses the separation theorem. This method has the advantage of solving a number of related problems which cannot be solved by the preceeding technique.

Van den Dries remarked that (the dual of) Theorem 5.9 just says that any formula defining a closed set $C \subseteq C^n$ is equivalent to a <u>positive</u> quantifier-free L-formula in any real closed field extending K (i.e., equivalent modulo the theory $T = RCF +$ the quantifier-free diagram of K). The model-theoretic criterion to be used in this case is easily derived from the compactness theorem.

5.10 <u>MODEL-THEORETIC CRITERION.</u> Let T be a theory with language L and $\phi(v_1,\dots,v_n)$ and L-formula. Then the following are equivalent:
(1) There is a positive, quantifier-free L-formula $\psi(v_1,\dots,v_n)$ such that $T \vdash \phi \leftrightarrow \psi$.

(2) Given $\mathfrak{A}, \mathfrak{B} \models T$, an L-substructure $\mathbb{C} \subseteq \mathfrak{A}$ and an L-homomorphism $f\colon \mathbb{C} \to \mathfrak{B}$, then

$$\mathfrak{A} \models \phi[\bar{c}] \quad \text{implies} \quad \mathfrak{B} \models \phi[f(\bar{c})]$$

for any $\bar{c} \in \mathbb{C}^n$. □

<u>PROOF OF THEOREM 5.9.</u> (general case). We check condition (2) of the model-theoretic criterion. Given the situation

$$(*)\qquad \begin{array}{ccc} F & & \\ \cup & & \\ A & \xrightarrow{\;f\;} & L \\ \cup & & \cup \\ K & & K \end{array}$$

with $F, L, K \models RCF$, A an ordered subring of F and f a homomorphism of unitary ordered rings such that $f \restriction K = id$, we have to show that

$$F \models \phi[\bar{a}] \quad \text{implies} \quad L \models \phi[f(\bar{a})],$$

where $\bar{a} \in A^n$.

Before proceeding with the proof, remark that we can make the following additional assumptions on the given situation (*):

(1) Replacing, if necessary, the field L by a larger real closed field, we can assume that A is a <u>convex</u> valuation ring of F. This is just Lang's homomorphism extension theorem 4.21.

(2) We can assume that A is a local ring with maximal ideal $Ker(f)$. Otherwise, replace A by its localization at $Ker(f)$.

(3) By changing K, if necessary, we may further assume that $f[K] = f[A]$. This is seen as follows. By Zorn, let R be a maximal subfield of A containing K. We have to show that R is real closed and $f[R] = f[A]$.

(a) R is algebraically closed in A (by maximality).

(b) R is real closed.

Since A is a convex subring of $F \models RCF$, the intermediate value property holds for polynomials $P \in A[X]$. Now, if $Q \in R[X]$, $Q \neq 0$, changes sign between a and b ($a, b \in R$, $a < b$), then, viewed as a polynomial in $A[X]$, Q has a root $c \in A$, $a < c < b$. Hence our claim follows from (a).

(c) $f[R]$ is a real closed field.

(d) $f[R] = f[A]$.

Otherwise, let $x \in A$ be such that $f(x) \notin f[R]$. By (c), $f(x)$ is trascendental over $f[R]$. It follows at once, using (2), that $Q(x)$ is invertible in A, whenever $Q \in R[X]$, $Q \neq 0$; hence $R(x) \subseteq A$, contradicting the maximality of R.

Now we can complete the proof of Theorem 5.9. Let $\bar{a} \in A^n$ be such that $F \models \phi[\bar{a}]$. Using (3) choose $\bar{b} \in K^n$ such that $f(b_i) = f(a_i)$ for $i=1,\ldots,n$. Since $f \restriction K$ is injective it suffices to show that $K \models \phi[\bar{b}]$.

By assumption the first-order statement " $\neg\phi$ is open" holds in K. If $\bar{b} \in (\neg\phi)^K$, then there is $\varepsilon \in K$, $\varepsilon > 0$, such that

(**) $$\forall v_1,\ldots,v_n \left(\sum_{i=1}^{n} (v_i - b_i)^2 < \varepsilon \longrightarrow \neg\phi(v_1,\ldots,v_n)\right)$$

holds in K; by transfer it also holds in F.

Since A is convex in F, it follows that the maximal ideal Ker(f) of A <u>is convex in A</u> (cf. Cherlin-Dickmann [82; Lemma 4]). As $\mathrm{Ker}(f) \cap K = \{0\}$ we conclude that $y < \varepsilon$ for all $y \in \mathrm{Ker}(f)$. Since $a_i - b_i \in \mathrm{Ker}(f)$, then $\sum_{i=1}^{n} (a_i - b_i)^2 < \varepsilon$. Condition (**) implies, then, that $F \models \neg\phi[\bar{a}]$, contradicting the assumption of the theorem. □

<u>COMMENTS.</u> Applying the same model-theoretic principle to algebraically closed fields, van den Dries [79] gets a simple proof of the completeness of projective varieties, a basic result in classical algebraic geometry (for the geometric meaning of this result, see Shafarevich [14; Ch. I, §5]). The author has used it in [73] to prove the following result, a refinement of Theorem 5.9, which answers a question of Bröcker [59; p. 261]:

<u>PROPOSITION 5.11.</u> Let $\langle k, \leq \rangle$ be an ordered field, K its real closure, and U an open s.a. subset of K^n. Then there is a finite set of polynomials $P_{ij} \in k[X_1,\ldots,X_n]$ such that

$$U = \bigcup_i \bigcap_j \{\bar{x} \in K^n \mid P_{ij}(\bar{x}) > 0\}.$$ □

REMARKS. (a) This section contains only a few of the most basic results on the topology of s.a. sets; much more is known. The reader interested in pursuing this line of enquiry may consult Coste [68; Prop. 3.5 and §IV, §V], Hardt [75], Mather [77].

(b) As a part of his research on Tarski's problem (cf. 3.11 (iv)) van den Dries has embarked on the project of showing that the major topological theorems of real algebraic geometry are consequences of the structure of the parametrically definable subsets of $\mathbb{R}$. He has conjectured that similar results hold for $<\mathbb{R}, +, \cdot, <, \exp, 0>$; cf. [87].

§6. THE REAL SPECTRUM.

A. INTRODUCTION.

One of the milestones of classical algebraic geometry was the introduction of the prime spectrum of a commutative ring by O. Zariski. When this construction is applied to the coordinate ring of a variety V over an algebraically closed field k, the space Spec (k[V]) contains the set V(k) of k-points of V with its Zariski topology as a dense subset- in the disguised form of the set of maximal ideals. Thus, we obtain a manageable topological space not too far removed from the original variety which gives valuable information about its geometry. This point of view leads to Grothendieck's theory of schemes, a far-reaching reformulation and generalization of classical algebraic geometry. Cf. Shafarevich [14], Hartshorne [3] or Mumford [10].

An obstacle to the development of algebraic geometry over the field of real numbers was the lack, until recently, of an appropriate analogue of the prime spectrum. The discovery of such an instrument by M. Coste and M.-F. Coste-Roy meant, therefore, a fundamental step forward in the study of real varieties.

To be sure, notions akin to the real spectrum of a ring have been known for quite some time; for example:

- the space of orderings of a field with the Harrison topology; see Prestel [55] and Lam [53];
- the subspace of real prime ideals of Spec(A); see Dubois [45].

However, none of these ancestors suitably reflected the wealth of geometric information contained in the rings arising in real algebraic geometry.

The crucial question in setting up a theory of the real spectrum is to decide which objects are to be its elements. Through a sequence of attempts initially motivated by topos-theoretic considerations, Coste/ Coste-Roy [69], [61], [62] realized that the appropriate choice of objects was a hybrid made of pairs of the form

(*) <prime ideals P of A; total orderings on the quotient ring $A/_P$>.

In order to give a motivation for this choice, recall that the members of the prime spectrum, i.e. the prime ideals of a given ring A, describe its homomorphisms onto integral domains (exactly the subrings of algebraically closed fields) modulo the natural equivalence relation corresponding to commutative diagrams

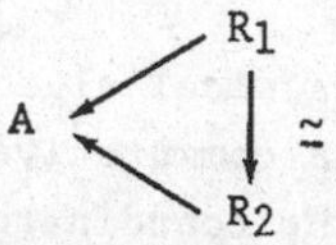

Substructures of real closed fields (in the natural language L for RCF) are ordered domains, and a moment's reflection shows that the pair (*) classify the corresponding class of epimorphisms.

Let us now show how the presentation of the objects (*) can be simplified in order to facilitate formal work with them. This is done by giving an axiomatic characterization of the set $\{a\in A \mid {}^a/_P \geq 0\}$ for each pair $\langle P,\leq\rangle$ of a (proper) prime ideal P of A and an order $\leq$ on $A/_P$.

DEFINITION 6.1. Let A be a commutative ring with unit. A prime precone is a subset α of A with the following properties:

(i) $x,y\in\alpha$ implies $x+y\in\alpha$;

(ii) $x^2\in\alpha$ for all $x\in A$;

(iii) $-1\notin\alpha$;

(iv) $xy\in\alpha$ iff $(x\in\alpha \wedge y\in\alpha) \vee (-x\in\alpha \wedge -y\in\alpha)$. □

Conversely, each prime precone α of A determines a pair $\langle P,\leq\rangle$ as described above, by setting:

$$P = \alpha\cap -\alpha$$
$$0 \leq a/_P \quad \text{iff} \quad a\in\alpha \qquad \text{for } a\in A.$$

Notation. Given a prime precone α of A we shall denote by $k(\alpha)$ the real closure of the fraction field of $A/_{\alpha\cap-\alpha}$ with the order determined by α, and by $\pi_\alpha : A \to k(\alpha)$ the corresponding canonical map. □

We shall now worry about the topology of the real spectrum, $\mathrm{Spec}_R(A)$. The topology of the prime spectrum, Spec (A), is generated by the base $\{D(a) \mid a \in A\}$, where

$$D(a) = \{P \in \mathrm{Spec}(A) \mid a \notin P\}$$

is the set of all P such that $a/_P \neq 0$.

It is natural, then, to give $\mathrm{Spec}_R(A)$ the topology generated by the sets $H(a)$ of prime precones α attributing a definite sign to $\pi_\alpha(a)$, say strictly positive sign. Since the family of such sets is not closed under intersection we are obliged to consider finite sequences of a's. We are now ready for the formal definition of the real spectrum:

<u>DEFINITION 6.2.</u> Let A be a commutative ring with unit. The <u>real spectrum</u> of A, $\mathrm{Spec}_R(A)$, is the set of all prime precones of A with the topology generated by the sets

$$H(a_1,\dots,a_n) = \{\alpha \in \mathrm{Spec}_R(A) \mid -a_1 \notin \alpha \wedge \dots \wedge -a_n \notin \alpha\}$$

for all finite sequences $a_1,\dots,a_n \in A$. □

Before studying the basic properties of the real spectrum let us compute a few examples.

<u>EXAMPLE 6.3.</u> If $A = k$ is a field, the only prime ideal is $\{0\}$ and $\mathrm{Spec}_R(k)$ is the space of orderings of k, mentioned above. □

<u>EXAMPLE 6.4.</u> The real spectrum of $\mathbb{R}[X]$, the coordinate ring of the real line $\mathbb{R}$.

Note that only real prime ideals (cf. Definition 4.6) are relevant in computing the real spectrum.

Since the prime ideals of $\mathbb{R}[X]$ are $\{0\}$ or of the form (Q) with irreducible Q, the changing sign criterion 4.18 tells us that (Q) is real iff changes sign. Since $\mathbb{R}$ is real closed, then Q has to be linear, say $Q = X-r$. Now, $\mathbb{R}[X]/_{(X-r)} \simeq \mathbb{R}$, which has only one order.

Thus, each $r \in \mathbb{R}$ gives rise to one point $\alpha_r \in \mathrm{Spec}_R(\mathbb{R}[X])$. The corresponding prime precone is

$$\alpha_r = \{F \in \mathbb{R}[X] \mid F(r) \geq 0\}.$$

Thr prime ideal $\{0\}$ gives rise to one point of the real spectrum for each ordering of $\mathbb{R}[X]$. We know (cf. Brumfiel [65; §7.5] or Dickmann [72; §I.5]) that the orderings of $\mathbb{R}[X]$ are determined by the cut that the element X defines in $\mathbb{R}$; these are as follows:

- X larger (smaller) than every element of $\mathbb{R}$, which gives rise to the following prime precones:

$\alpha_{+\infty} = \{F \in \mathbb{R}[X] \mid \text{There is } a \in \mathbb{R} \text{ such that } F \geq 0 \text{ on } (a,\infty)\}$,

$\alpha_{-\infty} = \{F \in \mathbb{R}[X] \mid \text{There is } b \in \mathbb{R} \text{ such that } F \geq 0 \text{ on } (-\infty,b)\}$.

- For each $r \in \mathbb{R}$, X is infinitesimally larger (smaller) than r, i.e. $r < X < (r,\infty)$ ($(-\infty,r) < X < r$, respectively). The corresponding prime precones are:

  $\alpha_{r^+} = \{F \in \mathbb{R}[X] \mid \text{There is } \varepsilon > 0 \text{ such that } F \geq 0 \text{ on } (r,r+\varepsilon)\}$,

  $\alpha_{r^-} = \{F \in \mathbb{R}[X] \mid \text{There is } \varepsilon > 0 \text{ such that } F \geq 0 \text{ on } (r-\varepsilon,r)\}$.

Observe that the only inclusion relations between these precones are $\alpha_{r^+} \subseteq \alpha_r$ and $\alpha_{r^-} \subseteq \alpha_r$ for $r \in \mathbb{R}$. Thus, we can draw the picture of $\mathrm{Spec}_R(\mathbb{R}[X])$:

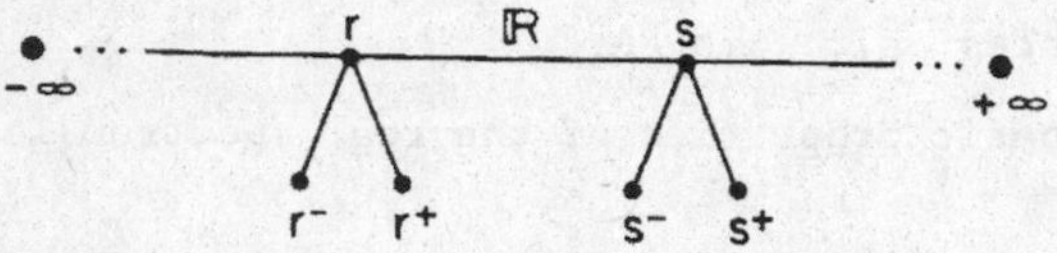

As a useful comparison, let us draw a picture of the prime spectrum of $\mathbb{R}[X]$. Here the points are the prime ideals of $\mathbb{R}[X]$: one corresponding to each polynomial $X-r$, for $r \in \mathbb{R}$; one corresponding to each polynomial $X^2 + aX + b$, with $a, b \in \mathbb{R}$, $a^2 - 4b < 0$; and the prime ideal $\{0\}$. The latter is contained in all the others, but no inclusion relation holds otherwise. The picture is this:

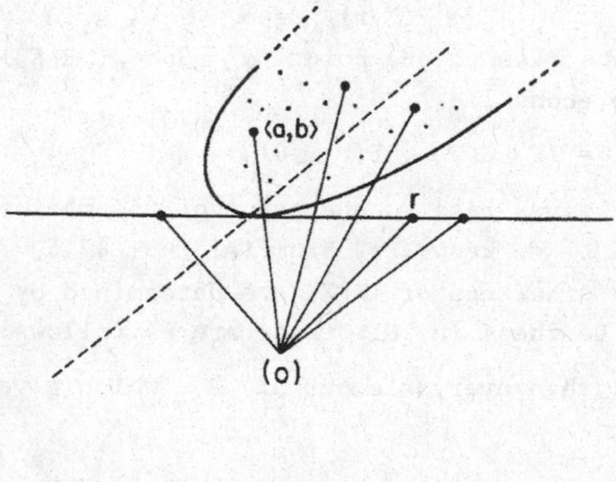

Note that in both the prime and the real spectrum, the set theoretic inclusion $\alpha \subseteq \beta$ means that β is in the closure of the singleton $\{\alpha\}$ (in geometric jargon : β is an specialization of α , and α a generization of β); this is easily checked from the definition of the respective topologies. In particular, $\overline{\{0\}}$ = Spec ($\mathbb{R}[X]$),i.e. the point $\{0\}$ is dense in Spec ($\mathbb{R}[X]$). Geometers say that the ideal (0) is a generic point of the variety $\mathbb{R}$ (apparently this notion goes back to the Italian school of 19th century geometers).

The real spectrum of $\mathbb{R}[X]$ does not have a generic point. But the points r^+, r^-, which together with r, form the closure of $\{r\}$ are interpreted as a sort of "generic points" of the half-lines $(r, +\infty)$ and $(-\infty, r)$ respectively.

Exercise. Compute $\mathrm{Spec}_R(\mathbb{R}[X]_{(x-r)})$, for $r \in \mathbb{R}$. □

EXAMPLE 6.5. In order to give an inkling on how the real spectrum reflects geometric properties of algebraic varieties, let us look at an example having more "singularities" than the preceeding one. We consider the plane curve C of equation $Y^2 - X^3 - X^2$.

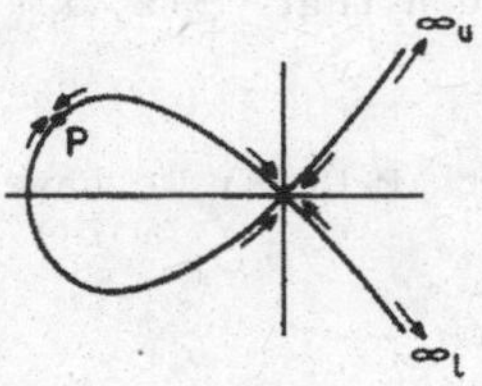

As before, each point $P \in C(\mathbb{R})$ gives rise to just one point $\alpha_P \in \mathrm{Spec}_R(\mathbb{R}[c])$, since $\mathbb{R}[C]/_{M_P} \simeq \mathbb{R}$ and $\mathbb{R}$ has only one order (M_P denotes the maximal ideal at P).

Since the polynomial defining C is irreducible, $\{0\}$ is a prime ideal of $\mathbb{R}[C]$; furthermore, it is easily checked to be a real ideal (exercise). The orderings of $\mathbb{R}[C]$ are computed as follows. The ring $\mathbb{R}[C]$ is the quadratic extension of $\mathbb{R}[X]$ in which the equation $Y^2 = X^3 + X^2$ holds; in particular, $X^3 + X^2 \geq 0$, i.e. $X \geq -1$. Hence, each ordering of $\mathbb{R}[X]$ satisfying this restriction gives rise to two orderings of $\mathbb{R}[C]$ by choosing the sign of Y. In geometric terms, the following prime preco es are thus generated:

(A) First fix a point $P = (a,b) \in C(\mathbb{R})$, $P \neq (-1,0)$, $P \neq (0,0)$. If $b > 0$, we have

$$\alpha_{P+} = \{F/_{I(C)} \mid F \in \mathbb{R}[X,Y] \text{ and there is } \varepsilon < 0 \text{ such that } F(x,\sqrt{x^3 + x^2}) \geq 0 \text{ on } (a,a+\varepsilon)\}$$

$$\alpha_{P-} = \{F/_{I(C)} \mid F \in \mathbb{R}[X,Y] \text{ and there is } \varepsilon > 0 \text{ such that } F(x,\sqrt{x^3 + x^2}) \geq 0 \text{ on } (a-\varepsilon,a)\}.$$

In order to check that these sets are in fact prime precones of $\mathbb{R}[C]$, only the implication from left to right in Definition 6.1 (iv) requires verification, as the other conditions are trivial. For this it suffices to see that if $F \notin I(C)$, then the function $F(x, \sqrt{x^3 + x^2})$ has finitely many zeros. Since the elements of $\mathbb{R}[C]$ are polynomials of the form $YP(X) + Q(X)$, our contention is clear as the equation $(X^3+X^2)\,P(X)^2-Q(X)^2 = 0$ has to be satisfied. (Alternatively, Bezout's theorem (cf. Walker [15 ;p. 59 ff.) could have been used to check this point).

The prime precones corresponding to the case $b < 0$ are obtained by replacing $-\sqrt{x^3 + x^2}$ for $\sqrt{x^3 + x^2}$ above.

(B) If $P = (-1,0)$, we have

$$\alpha_{P+} = \left\{F/_{I(C)} \mid \text{There is } \varepsilon > 0 \text{ such that } F(x,\sqrt{x^3 + x^2}) \geq 0 \text{ on } (-1,-1+\varepsilon)\right.,$$

and similarly for $-\sqrt{x^3 + x^2}$.

Thus, at each non-singular point $P \in C(\mathbb{R})$ we have a situation similar to that of Example 6.4:

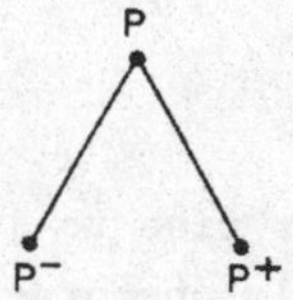

The points P^-, P^+ are interpreted as the "generic" points of the half-branches determined by P on the curve $C(\mathbb{R})$; see Figure 7.

(C) Next we have two "points at infinity" corresponding to the upper and lower infinite half-branches:

$$\alpha_{\infty,u} = \{F/_{I(C)} \mid \text{There is } a > 0 \text{ such that } F(x,\sqrt{x^3 + x^2}) \geq 0 \text{ on } (a,\infty)\},$$

and similarly $\alpha_{\infty,\ell}$ corresponding to $-\sqrt{x^3+x^2}$.

(D) Finally, at the singular point $P = (0,0)$ we have:

$$\alpha_{P_r^+} = \{F/_{I(C)} \mid \text{There is } \varepsilon > 0 \text{ such that } F(x, \sqrt{x^3+x^2}) \geq 0 \text{ on } (0,\varepsilon)\},$$

$$\alpha_{P_\ell^+} = \{F/_{I(C)} \mid \text{There is } \varepsilon > 0 \text{ such that } F(x, \sqrt{x^3+x^2}) \geq 0 \text{ on } (-\varepsilon,0)\},$$

and two more points, $\alpha_{P_r^-}$, $\alpha_{P_\ell^-}$ given by $-\sqrt{x^3+x^2}$ instead of $\sqrt{x^3+x^2}$.

As an exercise the reader may check that there is no inclusion relation between any two of these sets. Obviously these points specialize on α_P, and they correspond to the four half-branches of the curve $C(\mathbb{R})$ through the origin (see Figure 7).

In order to complete the analysis of this example we would have to show that

(i) There are no real prime ideals in $\mathbb{R}[C]$ other than $\{0\}$ and the maximal ideals M_P at each point $P \in C(\mathbb{R})$;

(ii) for each of these prime ideals there are no prime precones other than those explicitly constructed above.

A simple proof of point (i) goes as follows. Assume $J \neq \{0\}$ is a real prime ideal of $\mathbb{R}[C]$ and let $G_1,\ldots,G_\ell$ be its generators, where $G_i(X,Y) = YP_i(X) + Q_i(X)$. By Corollary 4.12, $V_{\mathbb{R}}(J) \neq \emptyset$. As in (A) above, the points $(x,y) \in V_{\mathbb{R}}(J)$ satisfy the equations $(x^3+x^2)\,P_i(x)^2 - Q_i(x)^2 = 0$ $(i = 1,\ldots,\ell)$, and hence $V_{\mathbb{R}}(J)$ is finite, say $V_{\mathbb{R}}(J) = \{P_1,\ldots,P_k\}$. By the real nullstellensatz 4.8 J consists of all polynomials vanishing at $P_1,\ldots,P_k$; but this ideal is never prime if $k \geq 2$. Hence $V_{\mathbb{R}}(J)$ consists of one point, P, and clearly $J = M_P$. Finally, point (ii) is clear by the argument preceeding (A) above. □

REMARKS. (a) The correspondence between points of the real spectrum and "oriented half-branches" illustrated by the preceeding examples is a general fact, true of any variety (of any dimension) over any real closed field. The proof of this requires a sophisticated algebraic machinery based on the analysis of valuations; cf. Coste/Coste-Roy [62; §7, §8].

(b) In both the preceeding examples the length of specialization chains of the real spectrum is at most 2. This is a manifestation of the fact that the (local) dimension at each point of the curves under consideration is 1, as we will se in §8 below. □

B. ELEMENTARY PROPERTIES.

PROPOSITION 6.6. Let A be a commutative ring with unit.

(1) The basic open sets $H(a_1,\dots,a_n)$, $a_1,\dots,a_n \in A$, of $\mathrm{Spec}_R(A)$ are quasi-compact, i.e. compact in the usual sense but not necessarily Hausdorff.

In particular:

(2) $\mathrm{Spec}_R(A)$ is quasi-compact.

(3) The irreducible closed subsets of $\mathrm{Spec}_R(A)$ are the closure of a unique point.

In particular:

(4) $\mathrm{Spec}_R(A)$ is a T_o-space.

(5) Let $\alpha, \beta, \gamma \in \mathrm{Spec}_R(A)$. If $\beta, \gamma \in \overline{\{\alpha\}}$, then $\beta \subseteq \gamma$ or $\gamma \subseteq \beta$.

REMARK. A closed set is called irreducible if it is not the union of two closed proper subsets. The closure of a point is clearly irreducible.

PROOF. (1) This is easily proved by applying the compactness theorem to an appropriate set of sentences with parameters in A of the firs-order theory (in the language of unitary rings plus an additional unary predicate P) whose axioms are:

- The axioms for commutative rings with unit.
- The axioms for P defining a prime precone (cf. Definition 6.1).

We leave the proof as an exercise for the reader.

(2) First check that the set

$$\alpha = \{a \in A \mid H(-a) \cap F = \emptyset\}$$

is a prime precone. Then, use the equivalence $\beta \in \overline{\{\alpha\}}$ iff $\alpha \subseteq \beta$ to show that $F = \overline{\{\alpha\}}$.

In order to show the uniqueness of such point α, it is easiest to check that $\mathrm{Spec}_R(A)$ is T_o:

$$\alpha \neq \beta \quad \text{implies} \quad \alpha \notin \overline{\{\beta\}} \quad \text{or} \quad \beta \notin \overline{\{\alpha\}}$$

(5) If $\beta \nsubseteq \gamma$ and $\gamma \nsubseteq \beta$, get $b \in \beta - \gamma$ and $c \in \gamma - \beta$. Since one of b-c or c-b is in α, and $\alpha \subseteq \beta, \gamma$, we obtain either $b=(b-c)+c \in \gamma$ or $c = (c-b)+b \in \beta$, a contradiction. □

6.7. REMARK. (Functorial properties of the real spectrum). The correspondence which assigns to each commutative unitary ring its real spectrum is a contravariant functor from the category of such rings with homomorphisms into the category of topological spaces (moreover, of spectral spaces, see Definition 6.9) with continuous functions as morphisms. This simply means that to each momomorphism of unitary rings $f:A \to B$ it is canonically associated a continuous map

$$\mathrm{Spec}_R f : \mathrm{Spec}_R(B) \to \mathrm{Spec}_R(A),$$

defined by

$$(\mathrm{Spec}_R f)(\beta) = f^{-1}[\beta] \qquad \text{for } \beta \in \mathrm{Spec}_R(B).$$

It is clear that for $a_1,\ldots,a_n \in A$ we have:

$$(\mathrm{Spec}_R f)^{-1}[H(a_1,\ldots,a_n)] = H(f(a_1),\ldots,f(a_n)),$$

which shows that $\mathrm{Spec}_R f$ is continuous and, moreover, that the inverse image of a compact open subset of $\mathrm{Spec}_R(A)$ is compact open. □

In this connection note the following:

<u>FACT 6.8.</u> With notation as above, let $\beta \in \mathrm{Spec}_R(B)$ and $\alpha = (\mathrm{Spec}_R f)(\beta)$. Then there is a (unique) ring monomorphism $\phi : k(\alpha) \to k(\beta)$ making the following diagram commute:

$$\begin{array}{ccc} A & \xrightarrow{\ f\ } & B \\ {\scriptstyle \pi_\alpha}\downarrow & & \downarrow{\scriptstyle \pi_\beta} \\ k(\alpha) & \xrightarrow[\ \phi\ ]{} & k(\beta) \end{array}$$

The map ϕ is elementary.

The easy proof is left as an exercise.

It is useful to recast the content of Proposition 6.6 in the following language:

<u>DEFINITION 6.9.</u> A topological space X is called an <u>spectral space</u> iff

(i) X is quasi-compact.

(ii) X has a base of open quasi-compact sets closed under intersection.

(iii) Every irreducible closed subset of X is the closure of a unique point. □

The Stone duality between Boolean algebras and Boolean (= compact, Hausdorff, totally disconnected) spaces can be extended to a duality between the category of distributive lattices with homomorphisms and the category of spectral spaces with continuous maps such that the inverse image of a compact open set is compact. To each spectral space it is associated the lattice of its compact open subsets. Conversely, to each distributive lattice it is associated the space of its prime filters with the spectral topology (defined exactly as for the spectrum of a ring).

The fundamental result about this class of space is:

THEOREM 6.10. (Hochster). A spectral space is homeomorphic to the (prime) spectrum of a ring. □

For a proof see Hochster [63] or Laffon [7].

In particular, the real spectrum of a ring A is homeomorphic to the prime spectrum of another ring. In the case where A is the coordinate ring of a variety over a real closed field, the ring B can be computed explicitly, as we will see later (Corollary 9.10).

C. CONSTRUCTIBLE SETS.

DEFINITION 6.11. A subset of $\mathrm{Spec}_R(A)$ is called constructible if it is a Boolean combination of basic open sets. □

Quantifier elimination shows at once that the constructible sets coincide with the definable sets in the following sense:

PROPOSITION 6.12. A set $C \subseteq \mathrm{Spec}_R(A)$ is constructible iff there is an L-sentence with parameters in A, $\phi_C = \phi_C(a_1,\ldots,a_n)$, such that

$$C = \{\alpha\in \mathrm{Spec}_R(A) \mid k(\alpha) \models \phi_C[\pi_\alpha(a_1),\ldots,\pi_\alpha(a_n)]\}.$$ □

The (easy) proof is left as an exercise for the reader.

It is clear that the constructible sets form a basis for a topology on $\mathrm{Spec}_R(A)$, called the constructible topology. This topology is obviously finer than the spectral topology and is compact Hansdorff.

The main property of constructible sets is:

6.13. THE REAL CHEVALLEY THEOREM. Let A be a ring, $B=A[X_1,\ldots,X_n]/_I$ a finitely presented A-algebra (i.e. the ideal I is finitely generated), and $f:A\to B$ the canonical morphism. Then $\mathrm{Spec}_R f$ transforms constructible sets into constructible sets.

PROOF. Let C be a constructible subset of $\mathrm{Spec}_R(B)$ given by

$$C = \{\beta\in \mathrm{Spec}_R(B) \mid k(\beta) \models \phi_C[\pi_\beta(Q_1/_I),\ldots, \pi_\beta(Q_m/_I)]\},$$

where $Q_1,\ldots,Q_m \in A[X_1,\ldots,X_n]$. Let I be generated by $P_1,\ldots,P_\ell \in$

$A[X_1,\dots,X_n]$. Then we have the equality:

$$(\mathrm{Spec}_R f)\,[C] = \{\alpha\in\mathrm{Spec}_R(A) \mid k(\alpha) \models \exists y_1,\dots,\exists y_n \bigwedge_{i=1}^{\ell} (\pi_\alpha P_i)(y_1,\dots,y_n)=0 \wedge$$

$$\wedge\ \phi_C(\pi_\alpha Q_1(y_1,\dots,y_n),\dots,\pi_\alpha Q_m(y_1,\dots,y_n))]\},$$

where for $F\in A[X_1,\dots,X_n]$, $\Pi_\alpha F$ denotes the polynomial whose coefficients are the images of the coefficients of F.

The inclusion $\subseteq$ follows easily from Fact 6.8. For the other inclusion, if $y_1,\dots,y_n \in k(\alpha)$ satisfy the given formula, then the correspondence

$$a \longmapsto \pi_\alpha(a) \qquad \text{for } a\in A,$$

$$X_{i/I} \longmapsto y_i \qquad i = 1,\dots,n,$$

extends to a ring homomorphism $g:B \to k(\alpha)$. This morphism gives a prime precone of B,

$$\beta = g^{-1}\,[k(\alpha)^+] = \{F/_I \mid k(\alpha) \models (\pi_\alpha F)(y_1,\dots,y_n) \geq 0\},$$

such that $\alpha = (\mathrm{Spec}_R f)(\beta)$ and $\beta\in C$. □

The study of topological properties of the map $\mathrm{Spec}_R f$ is of central interest in real algebraic geometry. This study is frequently based on an elegant combination of logical and geometrico-topological techniques. For example, if one needs to show that a certain constructible set $C \subseteq \mathrm{Spec}_R(A)$, given by an L-formula $\phi_C(a_1,\dots,a_n)$, is open, logic helps by reducing the problem to showing that the set

$$(*) \qquad \{\bar{x}\in\mathbb{R}^n \mid \mathbb{R} \models \phi_C\,[\bar{x}]\}$$

is open in $\mathbb{R}^n$ (of course, $\mathbb{R}$ can be replaced by any other real closed field). The analytic techniques available in the reals often are of help in proofs of this kind.

Indeed, if the set (*) is open, the open quantifier elimination theorem 5.9 implies that the formula $\phi_C(v_1,\dots,v_n)$ (without parameters) is equivalent in the theory RCF to one of the form

$$\bigvee_i \bigwedge_{j=1}^{n_i} P_{ij}(v_1,\dots,v_n) > 0$$

with $P_{ij}\in\mathbb{Z}\,[X_1,\dots,X_n]$. It follows that

$$C = \bigcup_i H(P_{i1}(a_1,\ldots,a_n),\ldots,P_{in_i}(a_1,\ldots,a_n))$$

and, hence, that C is open.

As an illustration of the use of this technique we prove the following result due to Elman-Lam-Wadsworth [48].

THEOREM 6.14. (The open mapping theorem). Let K,F be orderable fields, where F is a finitely generated extension of K, and let $i:K \to F$ denote the inclusion map. Then $\mathrm{Spec}_R i$ is an open map.

PROOF. Obviously it suffices to show that $(\mathrm{Spec}_R i)[0]$ is open in $\mathrm{Spec}_R(K)$, whenever 0 is a basic open set of $\mathrm{Spec}_R(F)$. Note that $(\mathrm{Spec}_R i)(\alpha) = \alpha \cap K$ for $\alpha \in \mathrm{Spec}_R(F)$.

We may assume $F = K(x_1,\ldots,x_n,a)$, where $x_1,\ldots,x_n$ are a transcendence base of F over K and a is algebraic over $K(x_1,\ldots,x_n)$. Obviously it suffices to prove the theorem in the cases $F = K(a)$ and $F = K(X)$.

CASE 1. $F = K(a)$, a algebraic over K.

Let $f \in K[X]$ be the minimal polynomial of a. A non-empty basic subset of $\mathrm{Spec}_R(F)$ is of the form:

$$H(P_1(a),\ldots,P_\ell(a)) = \{\alpha \in \mathrm{Spec}_R(F) \mid \overline{\langle F,\alpha\rangle} \models \bigwedge_{i=1}^{\ell} P_i(a) > 0\}$$

where $P_i \in K[X]$ and $P_i(a) \neq 0$. Then $f \nmid P_i$ and we may assume that $\deg(P_i) < \deg(f)$.

Let $X = (\mathrm{Spec}_R i)[H(P_1(a),\ldots,P_\ell(a))]$. For $\beta \in \mathrm{Spec}_R(K)$ the following are equivalent:

(i) $\beta \in X$

(ii) β extends to an order α of F such that $\langle F,\alpha\rangle \models \bigwedge_{i=1}^{\ell} P_i(a) > 0$.

Since $K(a) \simeq K[X]/(f)$, the sign-changing criterion 4.18 tells us that the following conditions are equivalent for any order β of K:

- β extends to an order of F;
- f changes sign in $\overline{\langle K,\beta\rangle}$.

By definition $\overline{\langle K,\beta\rangle} = k(\beta)$, and if α is an order of F extending β we have $\langle F,\alpha\rangle \subseteq k(\beta)$. Hence (ii) is equivalent to :

(iii) f changes sign in $k(\beta)$ and $k(\beta) \models \bigwedge_{i=1}^{\ell} P_i(a) > 0$.

This condition is first-order. Let:

$$\Phi(x,y,a_o,\ldots,a_{n-1}): \qquad x<y \wedge f(x)\, f(y)<0,$$

$$\Psi(a,b_o,\ldots,b_{m-1}): \qquad \bigwedge_{i=1}^{\ell} P_i(a)>0,$$

where $a_o,\ldots,a_{n-1}$ are the coefficients of f, and $b_o,\ldots,b_{m-1}$ those of $P_1,\ldots,P_\ell$. By continuity, the parameter-free formulas $\Phi(x,y,v_o,\ldots,v_{n-1})$ $\Psi(z,w_o,\ldots,w_{m-1})$ define open subsets of $\mathbb{R}^{n+2}$ and $\mathbb{R}^{n+1}$, respectively. Hence, the formula

$$\phi(v_o,\ldots,v_{n-1}): \qquad \exists xy\ \Phi(x,y,v_o,\ldots,v_{n-1})$$

defines an open subset of $\mathbb{R}^n$. By the remark preceeding the theorem, the constructible set

$$\{\beta\in \mathrm{Spec}_R(K) \mid k(\beta) \models \phi(a_o,\ldots,a_{n-1}) \wedge \Psi(a,b_o,\ldots,b_{m-1})\}$$

is open. By the equivalence shown above this set equals X.

CASE 2. $F = K(X)$.

By trivial manipulations (cf. 4.17) a basic open subset of $\mathrm{Spec}_R(F)$ can be written in the form:

$$H(P_1,\ldots,P_\ell) = \{\alpha\in \mathrm{Spec}_R(F) \mid \overline{\langle F,\alpha\rangle} \models \bigwedge_{i=1}^{\ell} P_i>0\}$$

for non-zero polynomials $P_i \in K[X]$. Putting $X = (\mathrm{Spec}_R i)[H(P_1,\ldots,P_\ell)]$ we have, as before, the equivalence between :

(i') $\beta\in X$,

(ii') β extends to an order α of F such that $\overline{\langle F,\alpha\rangle} \models \bigwedge_{i=1}^{\ell} P_i>0$,

where $\beta\in \mathrm{Spec}_R(K)$. Below we prove that these conditions are equivalent to:

(iii') $\overline{\langle K,\beta\rangle} \models \exists x \bigwedge_{i=1}^{\ell} P_i(x)>0$.

With this equivalence established, the proof is completed as above for (iii') defines an open condition, hence an open constructible subset of $\mathrm{Spec}_R(K)$.

Let $L = \overline{\langle K,\beta\rangle}$. Clearly (ii') implies (iii'), as $L \prec \overline{\langle F,\alpha\rangle}$.

Conversely, (iii') implies that all the P_i's are positive on an interval $(a,a+\varepsilon)$ of L. Hence the set α_{a^+} defined in Example 6.4 (with L replacing $\mathbb{R}$) defines an order of $L(X)$ extending β and making the P_i's positive. This order induces an order on F with the properties required in (ii'). □

REMARK. The map $\text{Spec}_R i$ is also closed, since the real spectrum of an orderable field is Hausdorff (exercise). As a matter of fact, something much more general is proved by Coste/Coste-Roy [62;Thm. 6.2] with the techniques used above:

THEOREM 6.15. (The closed mapping theorem). Let A, B be rings and $f: A \to B$ a homomorphism such that B is integral over $f[A]$. Then $\text{Spec}_R f: \text{Spec}_R(B) \to \text{Spec}_R(A)$ is a closed map. $\qquad\qquad\Box$

For still another application of the same technique, see Roy [32; §2].

§7. AFFINE VARIETIES OVER REAL CLOSED FIELDS.

Now we shall study the interplay between the geometry of affine varieties over real closed fields -in particular, over $\mathbb{R}$- and the topology of the real spectra of their coordinate rings. Throughout this section varieties are equipped with the euclidean topology derived from the order topology in the base field, and spectra are equipped with their spectral topology (cf. Definition 6.2).

Observe that for any ordered base field $\langle K, \leq \rangle$ and any variety V over K, there is an obvious embedding

$$\alpha : V(K) \longrightarrow \text{Spec}_R(K[V])$$

given by

$$\bar{x} \longmapsto \alpha_{\bar{x}} = \{ {}^Q/_I \mid Q \in K[X_1, \ldots, X_n] \text{ and } Q(\bar{x}) \geq 0 \}$$

(we write I instead of $I(V(K))$).

7.1 FACT. The map α is injective and continuous.

PROOF. Injectivity follows easily by considering linear polynomials. Continuity follows from the equality

$$\alpha^{-1}[H({}^{Q_1}/_I, \ldots, {}^{Q_M}/_I)] = V(K) \cap \bigcap_{i=1}^{m} Q_i^{-1}[(0, \infty)],$$

which is checked without problem. $\qquad\qquad\Box$

Since the family of sets of the form $V(K) \cap \bigcap_{i=1}^{m} Q_i^{-1}[(0, \infty)]$ for $m \in \mathbb{N}$ and $Q_1, \ldots, Q_m \in K[X_1, \ldots, X_n]$ clearly is a basis for the topology of $V(K)$, 7.1 says, furthermore, that the image of $V(K)$ is a subspace of $\text{Spec}_R(K[V])$; therefore, we may (and will) identify $V(K)$ with its image by α. Henceforth we also assume that K is real closed.

THEOREM 7.2. With the convention above, restriction to $V(K)$, $S \longmapsto S \cap V(K)$, defines a bijective map between:

(i) Constructible subsets of $\text{Spec}_R(K[V])$ and s.a. subsets of $V(K)$.

(ii) Open constructible subsets of $\mathrm{Spec}_R(K[V])$ and open s.a. subsets of $V(K)$.

PROOF. (i) Assume C is constructible, defined by the formula $\phi_C(^{Q_1}/_I,\ldots,^{Q_m}/_I)$, with $Q_1,\ldots,Q_m \in K[X_1,\ldots,X_n]$ (see Proposition 6.12). Then the equality

(*) $$C \cap V(K) = \{\overline{x} \in V(K) \mid K \models \phi_C[Q_1(\overline{x}),\ldots,Q_m(\overline{x})]\},$$

shows that $C \cap V(K)$ is s.a. By induction on the (Boolean) structure of C one gets reduced to showing (*) when C is basic open, $C = H(^{Q_1}/_I,\ldots,^{Q_m}/_I)$. In this case one may take $\phi_C(v_1,\ldots,v_m): \bigwedge_{i=1}^{m} v_i > 0$, and then (*) is just the equality appearing in the proof of 7.1.

It is clear that the map $C \mapsto C \cap V(K)$ takes on all s.a. subsets of $V(K)$ as values.

In order to see that it is injective, assume that C and C' are constructible subsets of $\mathrm{Spec}_R(K[V])$ defined by formulas $\phi_C(^{Q_1}/_I,\ldots,^{Q_m}/_I)$ and $\phi_{C'}(^{F_1}/,\ldots,^{F_r}/_I)$ respectively, and that $C \cap V(K) = C' \cap V(K)$. By (*) this equality translates as:

(**) $$K \models \forall \overline{x}[\bigwedge_{j=1}^{\ell} P_j(\overline{x}) = 0 \rightarrow (\phi_C(Q_1(\overline{x}),\ldots,Q_m(\overline{x})) \leftrightarrow \phi_{C'}(F_1(\overline{x}),\ldots,F_r(\overline{x})))],$$

where $P_1,\ldots,P_\ell$ are polynomials generating the ideal I of V. This is a formula with parameters in K (the coefficients of the polynomials). On the other hand, for every $\alpha \in \mathrm{Spec}_R(K[V])$, $k(\alpha)$ is a real closed field containing K, and hence (**) holds in $k(\alpha)$. Specializing (**) to $X_i = \pi_\alpha(^{X_i}/_I)$, $i = 1,\ldots,n$, we have $P_j(\pi_\alpha(^{X_1}/_I),\ldots,\pi_\alpha(^{X_n}/_I) =$ $= \pi_\alpha(^{P_j}/_I) = 0$, and hence:

$$k(\alpha) \models \phi_C(\pi_\alpha(^{Q_1}/_I),\ldots,\pi_\alpha(^{Q_m}/_I)) \longleftrightarrow \phi_{C'}(\pi_\alpha(^{F_1}/_I),\ldots,\pi_\alpha(^{F_r}/_I)).$$

In view of Proposition 6.12, this means:

$$\alpha \in C \text{ iff } \alpha \in C' \quad \text{for } \alpha \in \mathrm{Spec}_R(K[V]),$$

i.e., $C = C'$.

(ii) Since the identification map α is continuous, it is clear that $C \cap V(K)$ is open, whenever C is. The fact that every open s.a. subset of $V(K)$ is of the form $C \cap V(K)$ for some open constructible set C is an immediate consequence of open quantifier elimination (Theorem 5.9). □

COROLLARY 7.3. $V(K)$ is dense in $\mathrm{Spec}_R(K[V])$.

<u>Notation.</u> Given a s.a. subset S of $V(K)$, we denote by $\tilde{S}$ the unique constructible subset of $\mathrm{Spec}_R(K[V])$ such that $S = \tilde{S} \cap V(K)$.

<u>COROLLARY 7.4.</u> The map $S \longmapsto \tilde{S}$ induces a one-one correspondence between the connected components of $V(\mathbb{R})$ and those of $\mathrm{Spec}_R(\mathbb{R}[V])$. In particular, $\mathrm{Spec}_R(\mathbb{R}[V])$ has a finite number of connected components and is locally connected.

<u>PROOF.</u> Let $U_1,\dots,U_r$ be the connected components of $V(\mathbb{R})$ (Theorem 5.8). We want to show that $\tilde{U}_1,\dots,\tilde{U}_r$ are the connected components of $\mathrm{Spec}_R(\mathbb{R}[V])$. By Theorem 7.2 these sets are clopen and form a partition of $\mathrm{Spec}_R(\mathbb{R}[V])$. We only need to show that they are connected. Assume $\tilde{U}_i = C \cup C'$, with C,C' open (in $\tilde{U}_i$), non-empty and disjoint. Since the $\tilde{U}_j$'s are open, then C,C' are clopen (in $\mathrm{Spec}_R(\mathbb{R}[V])$); hence compact. Therefore C,C' are finite unions of basic open sets, and hence constructible. By Theorem 7.2 again,

$$U_i = (C \cap V(\mathbb{R})) \cup (C' \cap V(\mathbb{R}))$$

is a partition of U_i in non-empty open subsets, a contradiction.

The proof of local connectedness is left as an exercise. □

<u>Exercise.</u> (a) Prove the statement of Corollary 7.4 with a s.a. set $S \subseteq V(\mathbb{R})$, and $\tilde{S}$ replacing $V(\mathbb{R})$, and $\mathrm{Spec}_R(\mathbb{R}[V])$, respectively.

(b) The operation $S \longmapsto \tilde{S}$ commutes

(i) with the finite Boolean operations;

(ii) with closure and interior;

(iii) with images and inverse images by morphisms of algebraic varieties over K. (Cf. Hartshorne [3; Ch.I] for the notion of morphism of algebraic variety).

(c) If $U \subseteq V(\mathbb{R})$ is open s.a., then $\tilde{U}$ is the largest open subset W of $\mathrm{Spec}_R(\mathbb{R}[V])$ such that $W \cap V(\mathbb{R}) = U$.

Corollary 7.4 depends essentially on the fact that $V(\mathbb{R})$ has finitely many connected components, a property that only the real numbers enjoy amongst real closed fields. In fact:

<u>Exercise.</u> Prove that if K is a real closed field $\neq \mathbb{R}$, then the connected component of one point of K^n is the singleton of that point. □

However, using transfer on an appropriate $\mathcal{L}$-formula - namely one that for a fixed constructible set C expresses the property "C is connected" -, Coste/Coste-Roy [62; Thm. 5.5] show:

PROPOSITION 7.5. Let K be a real closed field and V a variety over K. Then $\mathrm{Spec}_R(K[V])$ has a finite number of connected components which are constructible sets. The same is true of any constructible subset of $\mathrm{Spec}_R(K[V])$. □

What kind of partition do the connected components of $\mathrm{Spec}_R(K[V])$ induce on $V(K)$?. It turns out that the members of this partition are precisely the components for the following notion, defined for s.a. sets $S \subseteq V(K)$:

> S is called s.a.-connected if it cannot be split into two disjoint, non-empty, s.a. open sets.

This notion, which clearly coincides with the standard notion of connectedness in the case $K = \mathbb{R}$, has a deep geometrical meaning. In order to see this, let us consider the restriction of the two-component cubic of Example 2.5(b) to the field $\overline{\mathbb{Q}}$ of real algebraic numbers. Manifestly, this is still a two-component variety, although it has many "holes". Clearly, the topology of $\overline{\mathbb{Q}}$ cannot reflect this property (see Exercise above), while the notion of s.a.-connectedness does.

It is well-known that the notion of path-connectedness coincides in $\mathbb{R}^n$ with that of connectedness. The first of these notions can be generalized to any real closed base field (this was done by Delfs [70]). Remarkably, this generalized notion turns out to be equivalent to that of s.a.-connectedness.

Delfs and Knebusch [71] have introduced a theory of "restricted topological spaces" intended to provide a frame in which the "semi-algebraic" versions of some topological notions (e.g. that of s.a.-connectedness) may be cast in much the same terms in which the corresponding standard topological notions are formulated in the frame of general, point-set topology. See also Bröcker [59; § 1].

§8. DIMENSION.

We give in this section a brief summary, without proofs, of the theory of dimension for affine varieties over a real closed base field K, and for s.a. subsets of K^n. This theory was developed by Coste/Coste-Roy [62; § 8]. The algebraic notions of dimension used in classical geometry were briefly reviewed in § 2.

For most of the present section we will assume that the polynomials $P_1,\dots,P_\ell \in K[X_1,\dots,X_n]$ determining our variety V generate a real ideal $I(V)$ over K; we will say that the variety V is real. From a geometrical point of view this is no restriction at all: it suffices to replace $P_1,\dots,P_\ell$ by a (finite) set of generators of the ideal $\mathcal{I}(V(K))$ which, by Corollary 4.11, is real; the latter obviously generate the same set of K-points. Moreover, the rings $K[V]$ and $K[X_1,\ldots,X_n]/I(V)$ have the same real spectrum.

A. GLOBAL DIMENSION.

The notion of prime precone gives, a priori, a new way of measuring dimensions:

DEFINITION 8.1. Let A be a commutative ring with unit, and V an affine variety over a real closed field K.

(a) The real dimension of A, $\dim_R A$, is the supremum of the integers n such that there is a strict chain

$$\alpha_o \subsetneq \alpha_1 \subsetneq \cdots \subsetneq \alpha_n$$

of prime precones of A ($\dim_R A = \infty$ if there are such chains of unbounded length).

(b) The real dimension of V, $\dim_R(V)$, is defined to be the real dimension of the coordinate ring $K[V]$. □

Since $\alpha \cap -\alpha = \beta \cap -\beta$ and $\alpha \subseteq \beta$ imply $\alpha = \beta$ $(\alpha,\beta \in \mathrm{Spec}_R(A))$, every strict chain of prime precones of A induces a strict chain of real prime ideals of A. Hence, $\dim_R A \le \dim A$. However, for real varieties these two quantities are equal:

PROPOSITION 8.2. Let V be a real irreducible affine variety over a real closed field K. Them $\dim_R K[V]$ equals the transcendence degree of $K(V)$ over K. □

(Cf. Theorem 2.7).

In addition it follows that:

COROLLARY 8.3. For K and V as in Proposition 8.2, the following quantities are equal to $\dim_R(V)$:

(a) The combinatorial dimension of $V(K)$.

(b) The supremum of the length of strict chains of real prime ideals in $K[V]$. □

The combinatorial dimension of $V(K)$ is the supremum of the lengths of strict chains of closed irreducible subsets of $V(K)$ with the Zariski topology (cf. Hartshorne [3;Ch. 1]); or, in other words, of strict chains of irreducible subvarieties of $V(K)$. Proposition 8.2 is used in proving the equality of this quantity with $\dim_R(V)$. The equality between quantities (a) and (b) follows from the Real Nullstellensatz 4.8: the map $I \mapsto V_K(I)$ is a bijective correspondence between real prime ideals of $K[V]$ and irreducible subvarieties of $V(K)$.

Note that the preceeding results are false for non-real varieties: if V is given by the polynomial $X^2 + Y^2$, so that $V(\mathbb{R}) = \{<0,0>\}$, then $\dim_R(V) = 0$ since the only prime precone of $A = \mathbb{R}[X,Y]/_I$, $I = (X^2 + Y^2)$ is $\alpha = \{ F/_I \mid F(0,0) \geq 0\}$. However A contains the chain $(o) \subsetneq (X/_I, Y/_I)$ of real prime ideals.

The results above show that, as far as measuring <u>global dimensions</u> is concerned, the use of prime precones yields the same results as the tools of classical commutative algebra. However, prime precones provide the means of constructing a theory of <u>local dimension</u> capable of explaining the phenomena of "fall of dimension" observed in the examples of §2; this cannot be done with the classical tools.

B. <u>LOCAL DIMENSION.</u>

<u>Definition 8.4.</u> Let V be an affine variety over a real closed field K, and $\overline{x} \in V(K)$. The <u>(local) real dimension of V at $\overline{x}$,</u> $\dim_R(V,\overline{x})$, is the supremum of the integers n such that there is a strict chain

$$\alpha_o \subsetneq \alpha_1 \subsetneq \cdots \subsetneq \alpha_n = \alpha_{\overline{x}}$$

of prime precones of K[V] ending in $\alpha_{\overline{x}}$(= the prime precone corresponding to $\overline{x}$; cf. §7). □

Comparing this definition with Definition 2.6(c) one may wonder whether $\dim_R(V,x)$ coincides with the real dimension of the ring $K[V]_{M_{\overline{x}}}$. This is not true in general, but we have:

<u>PROPOSITION 8.5.</u> $\dim_R(V,\overline{x}) = \dim_R(K[V]_{M_{\overline{x}}})^h$,

where A^h denotes the Henselization of a local ring A. □

The proof of this result requires some non-trivial arguments developed by Coste/Coste-Roy [62]. For the construction of the Henselization of a local ring, see Lafon [6] or Nagata [11].

Next we state the central geometric theorem on local dimension:

<u>THEOREM 8.6.</u> Let V be an affine real irreducible variety over a real closed field K, and $\overline{x} \in V(K)$. Them $\dim_R(V) = \dim_R(V,\overline{x})$ iff $\overline{x}$ belongs to the closure (in the euclidean topology) of the set of non-singular points of V(K). □

The proof is done in Coste/Coste-Roy [62;Thm. 8.9]. This result is the analogue of Theorem 2.8 for real varieties. Looking back at the Examples 2.9 we can see now that our notion assigns the correct dimension to the origin in both cases : 0 in the first example, 2 in the second.

C. THE DIMENSION OF SEMI-ALGEBRAIC SETS.

One of the remarkable features of the local theory of real dimension is that it assigns a dimension (both locally and globally) not only to varieties but, more generally, to s.a. sets. In particular, it provides a notion of dimension for the connected components of a variety.

DEFINITION 8.7. Let K be a real closed field, $S \subseteq K^n$ a s.a. set, and $\overline{x} \in S$. The real dimension of S at $\overline{x}$, $\dim_R(S,\overline{x})$, is the supremum of the length, of strict chains of prime precones in $\tilde{S}$ ending in $\alpha_{\overline{x}}$.□

PROPOSITION 8.8. $\sup\{\dim_R(X,\overline{x}) \mid \overline{x} \in S\}$ is equal to the (real) dimension of the closure of S in K^n with the Zariski topology. □

It follows that the natural notion of (global) real dimension for a s.a. set is that of the real dimension of its Zariski closure; this is well-defined, for such a closure is a variety by definition.

PROPOSITION 8.9. Let $S \subseteq K^n$ be a s.a. set. Then the function

$$\overline{x} \longmapsto \dim_R(S,\overline{x}),$$

defined on S, is upper semi-continuous. For a given integer $k \geq 1$, the set

$$\{\overline{x} \in S \mid \dim_R(S,\overline{x}) < k\}$$

is s.a., open in S. □

Looking at Example 2.9(b), we see that Proposition 8.9 gives the "right" result; the set $\{\overline{x} \in V(\mathbb{R}) \mid \dim_R(V,\overline{x}) = 1\}$ is open in $V(\mathbb{R})$, as it should be, since it coincides with $\{\langle x,y,Z\rangle \mid x=y=0 \wedge z \neq 0\}$.

COROLLARY 8.10. If $S \subseteq K^n$ is sla. and $\overline{x} \in S$, then there is a s.a. neighborhood U of $\overline{x}$ in S such that $\dim_R(S,\overline{x})$ equals the (real) dimension of the Zariski closure of U in K^n. □

§9. CONTINUOUS SEMI-ALGEBRAIC FUNCTIONS.

The study of continuous s.a. functions has only begun recently. The subject is still largely unexplored, as algebraic geometers have in the past concentrated on the study of analytic s.a.(=Nash) functions, which lie closer to the geometrical phenomena and have better algebraic properties. Nevertheless, the investigations carried out so far underline the increasingly important role of continuous s.a. functions. Furthermore, the study of these functions leads naturally to that of Nash functions.

We begin by introducing a class of functions which, in most interesting cases, turns out to coincide with that of (continuous) s.a. functions.

DEFINITION 9.1. Let $S \subseteq \mathbb{R}^n$ be a s.a. set and $f: S \to \mathbb{R}$ a function. We say that f is globally algebraic (over polynomials) iff there is $\ell \geq 1$ and polynomials $P_o, \ldots, P_\ell \in \mathbb{R}[X_1, \ldots, X_n]$, with some P_i non-zero, such that the equation

$$P_\ell(\overline{x})\ f(\overline{x})^\ell + P_{\ell-1}(\overline{x})\ f(\overline{x})^{\ell-1} + , \ldots, + P_o(\overline{x}) = 0$$

is verified for all $\overline{x} \in S$. □

PROPOSITION 9.2. Let $f: S \to \mathbb{R}$ be a function defined on a s.a. set $S \subseteq \mathbb{R}^n$. Then:

(1) If f is s.a., then f is globally algebraic.

(2) If f is continuous and globally algebraic, and S is open, then f is s.a.

PROOF. (1) By trivial manipulations the graph of f, Gr(f), has a definition of the form:

(*) $$\bigvee_i (P_i(\overline{x},y) = 0 \ \wedge \ \bigwedge_k Q_{ik}(\overline{x},y) > 0)$$

with $P_i, Q_{ik} \in \mathbb{R}[X_1, \ldots, X_n, Y]$. Furthermore, we can assume that each disjunct defines a non-empty set.

It suffices to show that each disjunct of (*) contains a non-trivial polynomial equation, i.e. that a polynomial P_i of degree ≥ 1 in Y does actually occur. For then, setting $P(\overline{X},Y) = \prod_i P_i(\overline{X},Y)$, we have $P(\overline{x}, f(\overline{x})) = 0$ for all $\overline{x} \in S$.

Assume that one disjunct contains no non-trivial polynomial equation. If this disjunct holds at $\langle \overline{x}_o, y_o \rangle$, then (*) shows that f is not single-valued at $\overline{x}_o$, a contradiction.

(2) We shall use now the cylindrical decomposition theorem 5.7 with $P(\overline{X},Y) = P_\ell(\overline{X})Y^\ell + \ldots + P_o(\overline{X})$ a non-zero polynomial annihilated by f on S. Under the present hypothesis, the first alternative of Theorem 5.7 may only occur when the sign of P is zero. Therefore, we can arrange a partition $A_o, A_1, \ldots, A_m$ of S into s.a. sets, so that:

(j) $P_o(\overline{x}) = \ldots = P_\ell(\overline{x}) = 0$ for all $\overline{x} \in A_o$,

and for each $i = 1, \ldots, m$, we have:

(jj) There are continuous s.a. functions $\alpha_1^i, \ldots, \alpha_{\ell_i}^i : A_i \to \mathbb{R}$, $\ell_i \geq 1$, giving exactly the real roots of the polynomial $P(\overline{x},Y)$, for all $\overline{x} \in A_i$. In particular, $P(\overline{x},Y)$ has constant sign $\neq 0$ in each of the intervals

$(-\infty, \alpha_1^i(\overline{x})), \ldots, (\alpha_j^i(\overline{x}), \alpha_{j+1}^i(\overline{x})), \ldots, (\alpha_{\ell_i}^i(\overline{x}), +\infty)$. By Theorem 5.8 there is no loss of generality in assuming that $A_1, \ldots, A_m$ are connected. Since $f(\overline{x})$ is a real root of $P(\overline{x}, Y)$, by (jj) it coincides, for $\overline{x} \in A_i$, with one of the $\alpha_j^i(\overline{x})$, say $j = s_i$. Now, the connectedness of A_i and the continuity of f imply that s_i is the same for all $\overline{x} \in A_i$ (exercise).

The foregoing argument shows that the graph of $f \upharpoonright A_1 \cup \ldots \cup A_m$ is defined by the formula:

$$\bigvee_{i=1} \phi_i(\overline{x}) \wedge \text{"y is the } s_i^{\text{th}} \text{ root of } P(\overline{x}, Y)\text{"},$$

where ϕ_i is a formula defining A_i.

Next we need to worry about the definability of $f \upharpoonright A_o$. This is just an argument of continuity. In fact, (j) shows that $A_o \subseteq S \cap V(\mathbb{R})$, where V is the variety defined by $P_o, \ldots, P_\ell$. In particular, A_o has empty interior (in $\mathbb{R}^n$), as $V(\mathbb{R})$ has this property. Since S is open, A_o has empty interior in S, and each neighborhood (in S) of a point $\overline{x} \in A_o$ intersects $A_1 \cup \ldots \cup A_m$. By continuity we have

$$f(\overline{x}) = \lim_{\substack{\overline{y} \to \overline{x} \\ \overline{y} \in A_1 \cup \ldots \cup A_m}} f(\overline{y})$$

Since $f \upharpoonright A_1 \cup \ldots \cup A_m$ is s.a. and the definition of limit is first-order, $f \upharpoonright A_o$ is also s.a. □

There is a notion of <u>minimal polynomial</u> for globally algebraic functions. This is a consequence of the following algebraic result:

<u>PROPOSITION 9.3.</u> Let R be a ring, A a unique factorization subring of R, and $b \in R$ an element algebraic over A. Then the ideal $I_b = \{P \in A[Y] \mid P(b) = 0\}$ is principal. If R is an integral domain, then I_b is also prime. □

The proof, which is just a variant of standard arguments, appears in Palais [30; §3] and in Dickmann [72; Prop. V.3.1]. This result applies to the case under consideration by setting:

R = the ring of real-valued , continuous, s.a. functions on a s.a. set $S \subseteq \mathbb{R}^n$ (henceforth denoted $C(U)$);

A = the ring $\mathbb{R}[X_1, \ldots, X_n]$.

<u>DEFINITION 9.4.</u> The <u>minimal polynomial</u> of a continuous s.a. (or,

more generally, a globally algebraic) function f on S is defined to be a generator of the ideal

$$I_f = \{P \in \mathbb{R}[\overline{X},Y] \mid P(\overline{x},f(\overline{x})) = 0 \text{ for all } \overline{x}\in S\}$$

such that the g.c.d. of its coefficients is 1. □

The minimal polynomial is, of course, a polynomial of lowest degree in I_f, and is unique, but <u>not necessarily irreducible</u>; for example, the minimal polynomial of the absolute value function on $\mathbb{R}$ is $(Y-X)(Y+X)$. However, it has the following properties:

<u>PROPOSITION 9.5.</u> Let $P \in \mathbb{R}[\overline{X},Y]$ be the minimal polynomial of $f \in C(S)$, and let $P = P_1,\ldots,P_\ell$ be a decomposition of P into irreducible factors. Then:

(a) The P_i are distinct.

(b) If $F_i = {}^{P}/_{P_i}$ and $U_i = \{\overline{x}\in S \mid F_i(\overline{x},f(\overline{x})) \neq 0\}$, then the U_i are pairwise disjoint, open s.a. subsets of S, and P_i vanishes identically on U_i.

(c) $S - \bigcup_{i=1}^{\ell} U_i$ has empty interior (in $\mathbb{R}^n$); in particular, if S is open, then $S \subseteq \bigcup_{i=1}^{\ell} \overline{U_i}$. □

For a proof, see Brumfiel [65; Prop. 8.13.15].

The proofs of some of the basic properties of continuous s.a. functions which we will consider below, use a technique depending on the fact that the real roots of a polynomial equation $P(X,Y) = 0$ <u>in one variable X</u> have a convergent Puiseux series expansion (cf. Example 3.2). Precisely:

<u>PROPOSITION 9.6.</u> Let $P \in \mathbb{R}[X,Y]$ be a polynomial in two variables and ρ a real valued function defined on an interval $[x_o,x_o+\varepsilon)$, for some $\varepsilon>0$, and such that

$$P(x,\rho(x)) = 0 \qquad \text{for all } x \in [x_o,x_o+\varepsilon).$$

Then ρ has an absolutely convergent Puiseux series expansion

$$\rho(x) = \sum_{k=N}^{\infty} a_k(x-x_o)^{k/p}$$

for all x in some interval $(x_o,x_o+\delta)$, $0<\delta\le\varepsilon$. Here $p\ge 1$, $N\in\mathbb{Z}$, $a_k\in\mathbb{R}$ for all $k\ge N$, and $a_N\neq 0$. In particular, any globally algebraic function ρ defined on $[x_o,x_o+\varepsilon)$ has such a Puiseux series expansion. □

The existence of a formal solution is a consequence of Theorem 3.3; see Walker [15; Ch. IV, § 3]. It is a remarkable fact that a formal Puiseux series which is a solution of an equation $P(X,Y)=0$ is automatically convergent. A neat discussion of the question of convergence appears in Chenciner [2; Ch. VIII, §8.6]. See also Hormander [18; Appendix].

As an application of this technique we prove:

THEOREM 9.7 (The Łojasiewicz inequality). Let $C \subseteq \mathbb{R}^n$ be a closed s.a. set an f, g continuous s.a. functions defined on C such that:

(i) For all $\varepsilon > 0$, $\{\bar{x} \in C \mid |g(\bar{x})| \geq \varepsilon\}$ is compact.

(ii) $Z(f) \subseteq Z(g)$ (where $Z(f) = f^{-1}[0]$).

Then there are constants $c, r > 0$ such that $|f| \geq c|g|^r$ on C.

PROOF. Let

$$H = \{\langle u,v\rangle \in \mathbb{R}^2 \mid \exists\, \bar{x} \in C\ (u = |g(\bar{x})| \wedge v = |f(\bar{x})|)\}.$$

Clearly H is a s.a. set contained in the positive quadrant of $\mathbb{R}^2$.

Then H is given by a disjunction of conjunctions of sign conditions on certain polynomials, say $P_1(u,v),\ldots,P_t(u,v)$. By Proposition 9.6 there is $\varepsilon > 0$ such that all the real roots of these polynomials have absolutely convergent Puiseux series expansions in $(0,\varepsilon)$.

Let us consider the set $H \cap ((0,\varepsilon) \times \mathbb{R})$. If it is empty, either $g = 0$ or $|g| \geq \varepsilon$ on C. In the last case, C is compact (by assumption (i)) and the result follows easily by considering the minimum of $|f|$ and the maximum of $|g|$ in C.

Assume, then, that $H \cap ((0,\varepsilon) \times \mathbb{R}) \neq \emptyset$. This set is bounded below on $(0,\varepsilon)$ by one of the roots of one of the polynomials $P_1,\ldots,P_t$, say $v(u)$. This means that for $\bar{x} \in C$ we have:

$$0 < |g(\bar{x})| < \varepsilon \quad \text{implies} \quad |f(\bar{x})| \geq v(|g(\bar{x})|), \quad \text{and}$$

$$v(|g(\bar{x})|) = \inf\,(H \cap (\{|g(\bar{x})|\} \times \mathbb{R}).$$

Observe also that $v > 0$ on $(0,\varepsilon)$; for if $u \in (o,\varepsilon)$, then $\{\bar{x} \in C \mid |g(\bar{x})| = u\}$ is a compact set (by(i)) on which $f \neq 0$ (by(ii)); it follows that $|f|$ has a minimum $\neq 0$ on this set, which equals, $v(u)$; hence $v(u) > 0$.

If v is bounded below by $\alpha > 0$ on $(0,\varepsilon)$, then an easy argument shows that $|f| \geq c|g|$ for some $c > 0$.

Let us assume that v is not bounded away from 0, i.e. $\lim_{u \to 0^+} v(u) = 0$.

Using the Puiseux series expansion of v we get:

$$(*)\qquad v(u) = \sum_{k=N}^{\infty} a_k u^{k/p} = a_N u^{N/p} (1 + \sum_{k=N+1}^{\infty} a_k/a_N\, u^{k-N/p})$$

for $u\in(0,\varepsilon)$. Let $\gamma(u)$ denote the series expnasion in the last term; since its exponents are positive, then $\lim_{u\to 0^+} \gamma(u) = 0$. Choose δ, $0<\delta\le\varepsilon$, so that $1+\gamma\ge 1/2$ on $(0,\delta)$. Since $v(u)>0$, it follows from (*) that $a_N>0$. The fact that $v(u)\xrightarrow[u\to 0^+]{} 0$ implies that $N>0$. Putting $c' = a_N/2$ and $r = N/p$ we have $|f|\ge c'|g|^r$ on $\{\bar{x}\in C \mid 0\le|g(x)|<\delta\}$. The argument used at the beginning of the proof shows that $|f|\ge c''\,|g|^r$ for some $c''>0$ on the compact set $\{\bar{x}\in C \mid |g(x)|\ge\delta\}$. Put $c=\min\{c',c''\}$. □

REMARKS. (a) The idea of the proof above goes back to Hörmander [18; Appendix; Lemma 2.1]

(b) When the set C is compact, the statement of Theorem 9.7 is in fact first-order, as one may take r rational, cf. Dickman [72; Prop.V 3.4]. By transfer, the inequality is valid in each real closed field, for C closed and bounded. Delfs [70; Lemma 3.2] gives an elementary proof valid for arbitrary real closed fields. □

In the remainder of this section we sum up other results obtained by application of the same technique, and study their effect on the structure of the rings $C(S)$.

We shall denote by $C^K(S)$ the ring of K-valued s.a. functions on a s.a. set $S\subseteq K^n$, K a real closed field, which are continuous in the euclidean topology.

The following result is proved by Carral-Coste [60], first for $\mathbb{R}$ and then, by transfer, for any real closed field K.

PROPOSITION 9.8. Let $S\subseteq K^n$ be a locally closed s.a. set, $f\in C^K(S)$ and $g\in C^K(S-Z(f))$. Then there is $m\ge 1$ such that the s.a. function $f^m g$, prolonged by 0 on $Z(f)$, is continuous. □

COROLLARY 9.9. Let S be as in 9.8 and $f,g\in C^K(S)$ be such that $Z(g)\subseteq Z(f)$. Then there is $m\ge 1$ such that g divides f^m in $C^K(S)$. □

COROLLARY 9.10. Let S be as in 9.8. Then $\mathrm{Spec}(C^K(S))$ is homeomorphic to $\tilde{S}$. In particular, if V is an affine variety over K, then $\mathrm{Spec}\,(C^K(V(K)))$ is homeomorphic to $\mathrm{Spec}_R(K[V])$.

PROOF. This is an application of the duality between spectral spaces and distributive lattices mentioned in §6.B. It suffices to prove that the lattices of compact open subsets of Spec $(C^K(S))$ and of $\tilde{S}$ are isomorphic.

By Theorem 7.2 the latter is isomorphic to the lattice of open s.a. subsets of S.

The former is simply $\{D(f) \mid f \in C^K(S)\}$, since Corollary 9.9 implies that $D(f) \cup D(g) = D(f^2 + g^2)$. The map $D(f) \longmapsto \{\bar{x} \in S \mid f(\bar{x}) \neq 0\}$ establishes the required isomorphism : it is injective by Corollary 9.9, and it is surjective since for a given open s.a. set $U \subseteq S$, the function d_{S-U} (= distance to S-U) is continuous. □

COROLLARY 9.11. Let $S \subseteq K^n$ be a locally closed s.a. set. Then

$$\dim (C^K(S)) = \dim_R(S). \qquad \Box$$

This corollary shows that the rings $C^K(S)$, $C(S)$, are radically different from the rings of arbitrary continuous functions : the Krull dimension of the latter is one or infinite, whatever the underlying space; inclusion chains of prime ideals in this case are of length 1 or at least $2^{\aleph_1}$ (cf. Gillman - Jerison [85; Thm. 14.19]). Thus, we see that rings of continuous s.a. functions are well-behaved objects which reflect geometric properties of the underlying spaces.

The results above have an effect on the structure of the ideals of $C^K(S)$.

COROLLARY 9.12. Let S be as in Proposition 9.8, and let I be an ideal of $C^K(S)$. The following are equivalent:

(i) I is real.

(ii) I is radical.

(iii) I is a z-ideal (i.e. $Z(f) = Z(g)$ and $g \in I$ imply $f \in I$). □

The residue rings $C^K(S)/_P$, where P is a prime ideal, have the following properties, similar to those holding in rings of arbitrary continuous:

PROPOSITION 9.13. Let $S \subseteq K^n$ be a s.a. set, and P a prime ideal of $C^K(S)$.

(i) The relation

$$f/_P \geq 0 \quad \text{iff} \quad \text{there is } g \in C^K(S) \text{ such that } g \geq 0 \text{ on S and } f/_P = g/_P,$$

defines a total ordering on $C^K(S)/_P$.

(ii) The ring $C^K(S)/_P$ has the following properties:

(a) It is a local ring (i.e. P is contained in exactly one maximal ideal).

(b) Every non-negative element has a square root.

(c) Every monic polynomial of odd degree has a zero.

In particular:

(iii) If M is a maximal ideal, then $C^K(S)/_M$ is a real closed field. □

The proof, given in Dickmann [84], is a "definable" version of an argument known in the case of rings of (arbitrary) continuous functions; see Gillman-Jerison [85; Thm. 13.4].

A result of geometric nature concerning the residue rings $C(C(\mathbb{R}))/_P$, where C is an algebraic curve over $\mathbb{R}$ and P is a prime ideal, is proved in Dickmann [84]; it establishes a link with the notion of real closed ring, introduced in Cherlin-Dickmann [82].

COMMENT. One may consider classes of continuous s.a. functions obta-ned by imposing further "regularity" conditions; many conditions of this type are familiar in analysis. As far as we know, nothing has been done in this direction beyond the study of Nash functions. Semi-algebricity may not always be a natural condition in connection with other requirements; for example, r-fold continuously differentiable s.a. functions for r finite may not be a natural class from the point of view of differential geometry.

However, the following is important:

PROPOSITION 9.14. Let f be a C^∞ (i.e. infinitely differentiable) realvalued function defined on an open, connected s.a. subset of $\mathbb{R}^n$. Then f is analytic. □

The proof is implicit in Brumfiel [65; Prop. 8.13.16].

§10. NASH FUNCTIONS.

DEFINITION 10.1. Let U be an open s.a. subset of $\mathbb{R}^n$. A function $f: U \to \mathbb{R}$ is called a Nash function if it is s.a. and analytic on U. We denote by $N(U)$ the ring of Nash functions defined on U. □

These functions, first considered by Nash [29], constitute a tool of prime importance in real algebraic geometry. The point is, as Bochnak-Efroymson observe, [25; p. 214], that Nash functions have the good algebraic properties of polynomials, but better geometric properties.

There is a vast literature dealing with Nash functions. Bochnak-Efroymson [26] is an introduction to the subject, while Bochnak-Efroymson [25] is a comprehensive survey of the algebraic theory of $N(U)$ and its subrings. Many of the basic results were first collected in Łojasiewicz [28], which contains a wealth of material. Valuable information can also be found in Roy [32] and Palais [30]; the points of view of these two papers are very different from the one adopted here.

In this survey we shall only consider Nash functions defined on connected open domains in $\mathbb{R}^n$. This will be quite sufficient for our purposes, although much of the theory below applies to more general domains of definition (see Bochnak-Efroymson [25]).

A. BASIC ALGEBRAIC PROPERTIES.

Many of the good algebraic properties of Nash functions follow from: 10.2. Fundamental fact. If U is open and connected, then $N(U)$ is an integral domain.

PROOF. Let $f, g \in N(U)$ be such that $fg = 0$, i.e. $Z(f) \cup Z(g) = U$. One of these sets, say $Z(f)$, has non-empty interior. Since U is connected $f = 0$ on U by the principle of analytic continuation (Dieudonné [17; 9.4.2]) □

An immediate consequence is :

COROLLARY 10.3. Let f be an analytic function on U. Then f is Nash iff it is locally algebraic; i.e. for every $\overline{x_o} \in U$ there is a neighborhood V of $\overline{x_o}$ and a polynomial $P \in \mathbb{R}[X_1,\ldots,X_n,Y]$, $P \neq 0$, such that $P(\overline{x},f(\overline{x})) = 0$ for all $\overline{x} \in V$. □

By Proposition 9.3 the minimal polynomial of a Nash function is irreducible. This has a number of simple but important algebraic consequences; we mention the following:

COROLLARY 10.4. Let $f: U \times \mathbb{R} \to \mathbb{R}$ be a non-zero Nash function and $g: U \to \mathbb{R}$ an analytic function satisfying

$$f(\overline{x},g(\overline{x})) = 0 \qquad \text{for all } \overline{x} \in U.$$

Then g is Nash.

PROOF. Let $P \in \mathbb{R}[X_1,\ldots,X_n,X_{n+1},Y]$ be the minimal polynomial of f. Since $f \neq 0$, then $P \neq Y$ and hence P is not divisible by Y; it follows that $Q(X_1,\ldots,X_{n+1}) = P(X_1,\ldots,X_{n+1}, 0)$ is not the zero polynomial. We also have:

$$Q(\overline{x},g(\overline{x})) = P(\overline{x},g(\overline{x}),0) = P(\overline{x},g(\overline{x}), f(\overline{x},g(\overline{x}))) = 0$$

for all $\overline{x} \in U$. □

COROLLARY 10.5. Let $F \in N(U)[Y]$ be a non-zero polynomial and g an analytic function on U satisfying $F(g) = 0$. Then g is Nash. □

COROLLARY 10.6. The ring $N(U)$ is differentially stable: if $F \in N(U)$ and $i = 1,\ldots,n$, then $\frac{\partial f}{\partial x_i} \in N(U)$.

PROOF. If f satisfies the polynomial equation $P(\overline{x},f(\overline{x})) = 0$, then $\frac{\partial f}{\partial x_i}$ satisfies the equation

$$\frac{\partial P}{\partial X_i}(\overline{x},f(\overline{x})) + \frac{\partial P}{\partial Y}(\overline{x},f(\overline{x})) \cdot \frac{\partial f}{\partial X_i}(\overline{x}) = 0$$

which has Nash coefficients. Hence it is Nash by 10.5. □

Another consequence of 10.4 is:

PROPOSITION 10.7. (Implicit function theorem). Let $U \subseteq \mathbb{R}^{n+1}$ be open s.a. and $\overline{a} \in \mathbb{R}^n$, $b \in \mathbb{R}$ be such that $\langle \overline{a},b\rangle \in U$. Let $f:U \to \mathbb{R}$ be a Nash function such that $f(\overline{a},b) = 0$ and $\frac{\partial f}{\partial X_{n+1}}(\overline{a},b) \neq 0$. Then there is a s.a. neighborhood V of $\overline{a}$ in $\mathbb{R}^n$ and a Nash function $g:V \to \mathbb{R}$ such that $g(\overline{a}) = b$ and $f(\overline{x},g(\overline{x})) = 0$ for all $\overline{x} \in V$.

PROOF. By the implicit function theorem for analytic functions (see Dieudonné [17; 10.2.4]) there is V, which we may take s.a., and an analytic solution g as above; g is Nash by 10.4. □

Palais [30; §1] shows that this statement is equivalent to more general versions of the implicit function theorem, for a variety of situations including, of course, Nash functions.

The following algebraic property is a consequence of Corollary 10.5:

PROPOSITION 10.8. The ring $N(U)$ is integrally closed; that is, if a quotient f/g of Nash functions satisfies a monic polynomial equation with coefficients in $N(U)$, then g divides f in $N(U)$.

PROOF. The ring of analytic functions on U is known to be integrally closed (cf. Dickmann [72; Ch. V]); hence f/g is analytic on U. By 10.5, f/g is Nash. □

Now we mention, without proof, an algebraic property of crucial importance.

THEOREM 10.9. The ring $N(U)$ is noetherian. □

The original proof, due to Risler [31], is basically of algebraic nature. A proof using complexification techniques is sketched in Bochnak-Efroymson [25; Thm. 3.1]. We shall use later the following consequence:

COROLLARY 10.10. Let A be a subring of $N(U)$ containing the polynomials, and I an ideal of A. Then there are $f_1,\ldots,f_k \in I$ such that $Z(I) = \bigcap_{i=1}^{k} Z(f_i)$.

PROOF. The ideal $I \cdot N(U)$ generated by I in $N(U)$ is finitely generated. Each of these generators is a linear combination of members of I, say $f_1,\ldots,f_k$. The conclusion follows at once. □

The algebraic properties considered above are valid for all (open) domains U. On the contrary, unique factorization in $N(U)$ is an algebraic property which depends essentially on the geometry of the domain U; namely, on the triviality of its first cohomology group (cf. Bochnak-Efroymson [25; §4], where further references are given, and Risler [31] for simple examples).

B. NASH FUNCTIONS AND REAL ALGEBRAIC GEOMETRY.

In order to understand the relevance of Nash functions for real algebraic geometry, we underline the basic fact that in the classical theory of algebraic curves, the notion of a branch is defined in terms of analytic parametrizations; see Walker [15; Ch. IV, §2].
For example, the branches of the curve $Y^2 - (X^3 + X^2)$ considered in Example 6.5, are given by the functions
$b_i : (-1, +\infty) \to \mathbb{R}$, $i = 1,2$, defined as follows:

$$b_i(x) = \begin{cases} (-1)^i \sqrt{X^3 + X^2} & \text{for } -1 < X \leq 0 \\ (-1)^{i+1} \sqrt{X^3 + X^2} & \text{for } 0 \leq X \end{cases}$$

(where $\sqrt{\ }$ denotes the positive value of the square root). These functions are analytic branches (exercise), and they are obviously s.a.

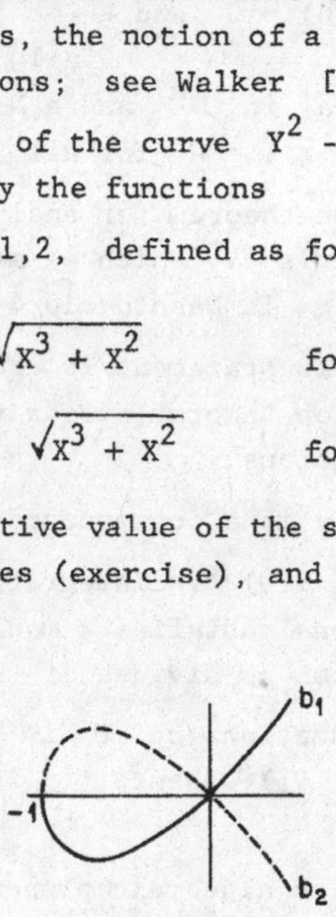

The s.a. function $f:[-1, +\infty) \to \mathbb{R}$ defined by

$$f(x) = \min \{y \in \mathbb{R} \mid y^2 = x^3 + x^2\}$$

is continuous but not analytic, thus not a branch.

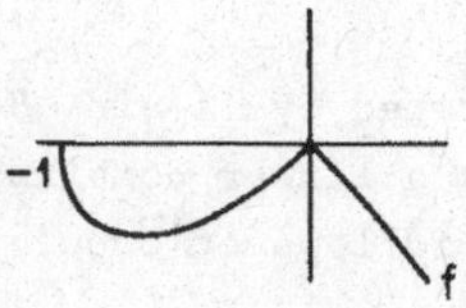

As a second example consider the curve $Y^2 - X^3 + X^2$ of Example 2.5 (a). Its only branch is given by the map

$$b(y) = \text{ the unique } x \text{ such that } y^2 = x^3 - x^2, \text{ and } x > 1/2$$

defined on $\mathbb{R}$ (we have to give it in this form in order to have a function).

It is a remarkable fact that the description of Nash functions illustrated by the preceeding examples gives their general form. Every Nash function, in any number of variables is, in a suitable sense, a "branch" of an algebraic variety. (Compare with the description of continuous s.a. given by Proposition 9.2). The precise result is as follows:

THEOREM 10.11. (Artin-Mazur). Let U be an open, connected, s.a. subset of $\mathbb{R}^n$ and $f:U \to \mathbb{R}$. The following are equivalent:

(i) f is Nash.

(ii) There is $q \geq n+1$, an irreducible affine variety V given by polynomials in q variables, a continuous s.a. map $s = \langle s_1,\dots,s_q \rangle : U \to V(\mathbb{R})$ and a polynomial $P \in \mathbb{R}[X_1,\dots,X_q]$ such that:

(a) V is étale over $\mathbb{R}$.

(b) $(pr \restriction V(\mathbb{R})) \circ s = id_U$, where $pr : \mathbb{R}^q \to \mathbb{R}^n$ denotes the projection onto the first n coordinates.

(c) $f = P \circ s$.

The proof is far outside the scope of this survey; see Artin-Mazur [24] and Roy [32] for details and further uses of this characterization.

Condition (a) has a deep geometric significance, but the mere presentation of it requires the use of heavy machinery from commutative algebra; see Raynaud [12]. Since this condition will not be used explicitly, we omit further explanations, but point out for later use the following consequences of conditions (a) - (c) above:

10.12 FACT. With the notations of Theorem 10.11, we have:

(j) The functions $s_1,\dots,s_q$ are Nash.

(jj) The map $pr \restriction V(\mathbb{R})$ is locally injective. □

The following exercise gives a clue as to why Nash functions do not appear is classical algebraic geometry.

Exercise. Let $f:\mathbb{C}^n \to \mathbb{C}$ be a holomorphic (=complex analytic) function. If f is globally algebraic over polynomials (in the sense of Definition 9.1), then f is a polynomial. [Hint: use Liouville's theorem, Dieudonné [17; 9.11.1].]

C. THE SEPARATION THEOREMS.

Now we will consider the problem of finding a simple class of s.a. functions having the following separation property: given two disjoint closed s.a. sets C_1, C_2, there is a function f in the class such that $f \upharpoonright C_1 > 0$ and $f \upharpoonright C_2 < 0$.

It is known that polynomials are not sufficient to separate closed s.a. sets; the sets of Figure 11 provide a counter-example. It turns out that Nash functions - even Nash functions of a particularly simple form

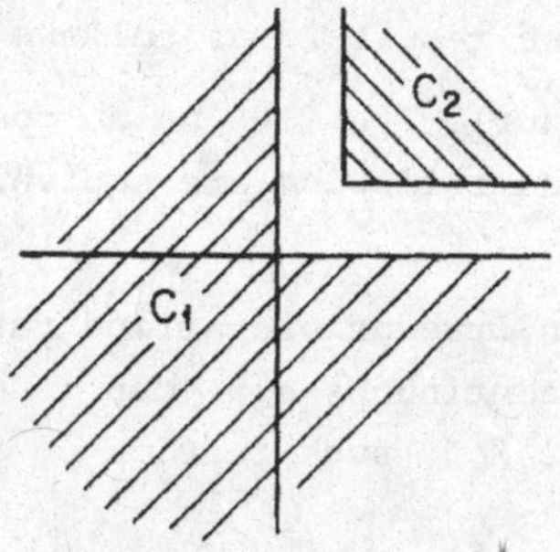

- have the required separation property.

DEFINITION 10.13. Let $U \subseteq \mathbb{R}^n$ be open s.a. We define $R(U)$ to be the smallest subring of $C(\mathbb{R}^n)$ containing the polynomials and such that:

(i) If $f \in \sum R(U)^2$ is such that $f > 0$ on $\mathbb{R}^n$, then $1/f \in R(U)$.

(ii) If $f \in \sum R(U)^2$ is such that $f > 0$ on U, then $\sqrt{f} \in R(U)$. □

The ring $R(U)$ can be constructed inductively as follows: for a ring $B \subseteq C(\mathbb{R}^n)$, let $B^{(1)}$ denote the ring generated over B by all functions $\sqrt{f}$, with $f \in \sum B^2$ and $f > 0$ on U. Let

$$B^{(o)} = B$$
$$B^{(n+1)} = (B^{(n)})^{(1)}$$
$$B^{(\infty)} = \bigcup_{n \in \omega} B^{(n)}$$

Let S_B denote the multiplicative subset of B of all functions $f \in \sum B^2$ such that $f > 0$ on $\mathbb{R}^n$. Then $R(U)$ is the localization of $\mathbb{R}[X_1,\ldots,X_n]^{(\infty)}$ at the set $S_{\mathbb{R}[X_1,\ldots,X_n]^{(\infty)}}$.

Observe that the functions in $R(U)$ are continuous on $\mathbb{R}^n$ and Nash on U; condition 10.13 (ii) is introduced in order to make this true, since $\sqrt{f}$ is not analytic at any zero of f.

THEOREM 10.14. Let C_1, C_2 be disjoint closed s.a. subsets of $\mathbb{R}^n$. Then there is $f \in R(\mathbb{R}^n)$ - even $f \in (\mathbb{R}[\overline{X}]_{S_{\mathbb{R}[\overline{X}]}})^{(1)}$ - such that $f \upharpoonright C_1 > 0$ and $f \upharpoonright C_2 < 0$. □

PROPOSITION 10.15. Let $U \subseteq \mathbb{R}^n$ be open s.a. Then there is $f \in R(U)$ such that $f \upharpoonright U > 0$ and $f \upharpoonright (\mathbb{R}^n - U) = 0$. □

A relative separation theorem is obtained from 10.14 and 10.15:

THEOREM 10.16. Let $U \subseteq \mathbb{R}^n$ be open s.a. and C_1, C_2 be closed s.a. subsets of U. Then there is $f \in R(U)$ such that $f \upharpoonright C_1 > 0$ and $f \upharpoonright C_2 < 0$. □

Various proofs of these results, due to Mostowski, can be found in Efroymson [27; §1] and Bochnak - Efroymson [25; §5]. As a corollary we have the following result of Risler [31]:

COROLLARY 10.17. Let P be a prime ideal of $N(U)$, where $U \subseteq \mathbb{R}^n$ is an open, connected s.a. set. Then $Z(P)$ is connected.

PROOF. By Corollary 10.10 there are $g_1, \ldots, g_r \in P$ such that $Z(P) = \bigcap_{i=1}^{r} Z(g_i)$. Hence $Z(P)$ is s.a. and has finitely many connected components (Theorem 5.8). If $Z(P)$ is not connected, there are disjoint s.a. subsets C_1, C_2 of $Z(P)$, closed in $\mathbb{R}^n$, such that $C_1 \cup C_2 = Z(P)$. Let $f \in R(U)$ be such that $f \upharpoonright C_1 > 0$ and $f \upharpoonright C_2 < 0$. Let us now define

$$h_i = \sqrt{\sum_{i=1}^{r} g_i^2 + f^2} + (-1)^i f \qquad \text{for } i = 1,2.$$

Then $h_1 h_2 = \sum_{i=1}^{r} g_i^2 \in P$, but $h_1 \notin P$ because $h_1 \upharpoonright C_1 = 0$ and $h_1 \upharpoonright C_2 > 0$. Similarly $h_2 \notin P$, contradicting that P is prime. □

REMARK. The proof shows that the result holds for any subring A of $N(U)$ which contains $\sqrt{f}$, whenever $f \in A$ and $f > 0$ on U.

§11. THE SUBSTITUTION THEOREM; "STELLENSÄTZE".

In this paragraph we prove another central result in the theory of Nash functions: the substitution theorem. We use it later to derive a number of "stellensätze" for the rings $N(U)$ and many of its subrings.

A. THE SUBSTITUTION THEOREM.

If K is a real closed field containing $\mathbb{R}$, $S \subseteq \mathbb{R}^n$ is a s.a.

set and $f:S \to \mathbb{R}$ a s.a. function, it is clear that the formulas (with parameters in $\mathbb{R}$) defining S and $Gr(f)$ are interpretable in K, and define, respectively, a s.a. subset of K^n, denoted by S^K, and a K-valued function on S^K, denoted by f^K. The substitution theorem says that any ring homomorphism $\phi: N(U) \to K$ is the evaluation homomorphism at some point of U^K. Precisely:

THEOREM 11.1 (The substitution theorem). Let $U \subseteq \mathbb{R}^n$ be an open, connected s.a. set, K a real closed field containing $\mathbb{R}$, and $\phi: N(U) \to K$ an $\mathbb{R}$-algebra homomorphism. Then :

(i) $\langle\phi(\pi_1),\ldots,\phi(\pi_n)\rangle \in U^K$, where $\pi_1,\ldots,\pi_n$ denote the projection maps, $\pi_i(x_1,\ldots,x_n) = x_i$.

(ii) For every $f \in N(U)$, $\phi(f) = f^K(\phi(\pi_1),\ldots,\phi(\pi_n))$.

PROOF. Observe that (i) gives a sense to (ii). The proof proceed in the following steps:

(I) We prove (ii) for $f \in R(U)$. Since any such f is defined on $\mathbb{R}^n$, this makes sense even in the absence of (i).

(II) We prove (i).

(III) We prove (ii) for arbitrary $f \in N(U)$.

STEP I. Condition (ii) is clear for polynomials. By the inductive construction of $R(U)$ it suffices to prove that for $h \in C(\mathbb{R}^n)$, $h \geq 0$:

(a) if (ii) holds for h and $h \upharpoonright U > 0$, then (ii) holds for $f = \sqrt{h}$;
(b) if (ii) holds for h and $h > 0$, then (ii) holds for $f = {}^1/_h$.

PROOF of (a). Since $f^2 = h$, we have

$$\phi(f)^2 = \phi(h) = h^K(\phi(\overline{\pi})) = f^K(\phi(\overline{\pi}))^2,$$

where $\phi(\overline{\pi}) = \langle\phi(\pi_1),\ldots,\phi(\pi_n)\rangle$. In order to conclude that $\phi(f)=f^K(\phi(\overline{\pi}))$ it suffices to show that both sides are non-negative. The right-hand side is ≥ 0 for we have assumed that $f \geq 0$. But we also have:

(*) $\phi(f) > 0$ for any $f \in N(U)$ such that $f > 0$ on U.

Indeed, as we have ${}^1/_{\sqrt{f}} \in N(U)$, then $\phi(f)\cdot\phi({}^1/_{\sqrt{f}})^2 = 1$; this implies that $\phi(f) > 0$.

The proof of (b) is similar.

STEP II. By Proposition 10.15 let $f \in R(U)$ be such that $f \upharpoonright U > 0$ and $f \upharpoonright (\mathbb{R}^n - U) = 0$. Then the first-order statement

(**) $$\forall\overline{x}\ (\overline{x} \in U \longleftrightarrow f(\overline{x}) > 0)$$

holds in $\mathbb{R}$, and hence in K. By (*) we have $\phi(f) > 0$, and by (I),

$f^K(\phi(\overline{\pi})) > 0$. It follows from (**) that $\phi(\overline{\pi}) \in U^K$.

STEP III. Now we use the Artin-Mazur characterization of Nash functions (Theorem 10.11). Therefore, let $q \geq n+1$, $P_1, \ldots, P_\ell \in \mathbb{R}[X_1, \ldots, X_q]$ be polynomials defining a variety V, $s = \langle s_1, \ldots, s_q \rangle : U \to V(\mathbb{R})$ a continuous s.a. function and P a polynomial, so that conditions 10.11 (a)-(c) – and hence also 10.12 (j), (jj) – hold.

Clearly, it suffices to prove (ii) for the functions $s_1, \ldots, s_q$. Put $W = pr^{-1}[U]$ and $X = V(\mathbb{R}) \cap W$.

The set $s[U]$ obviously is connected, and we prove next that it is clopen in X (hence a connected component of X). Indeed, if $\overline{y}$ belongs to the closure of $s[U]$ in X, then $\overline{y} = \lim s(\overline{u_n})$ for some sequence $\overline{u_n} \in U$. By continuity and $pr \circ s = id_U$ (10.11(b)) we conclude that $s(pr(\overline{y})) = \lim s(u_n) = \overline{y}$, i.e. $\overline{y} \in s[U]$. Hence $s[U]$ is closed in X. In order to see that $s[U]$ is open in X, let $\overline{u} \in U$, and using 10.12 (jj) choose an open set $O \subseteq \mathbb{R}^q$ so that $s(\overline{u}) \in O \cap X$ and pr is injective on $O \cap X$. Using 10.11(b) it is checked at once that $O \cap X \subseteq s[U]$, which proves our contention.

Now we invoke the separation theorem 10.16 to get $h \in R(W)$ such that $h \restriction s[U] > 0$ and $h \restriction (X - s[U]) < 0$. It follows from this situation that the first-order statement

(***) $\forall \overline{x} \in U \ \forall \overline{z} \in \mathbb{R}^q \ [\overline{z} \in V \cap W \wedge h(\overline{z}) > 0 \wedge pr(\overline{z}) = \overline{x} \to \overline{z} = s(\overline{x})]$

holds in $\mathbb{R}$. By transfer it holds in K. Specializing to $\overline{z} = \langle \phi(s_1), \ldots, \phi(s_q) \rangle$ and $\overline{x} = \langle \phi(\pi_1), \ldots, \phi(\pi_n) \rangle = \phi(\overline{\pi})$ (which is in U^K by (i)), the consequent of (***) is the equality $\phi(s_i) = s_i^K(\phi(\overline{\pi}))$, $i = 1, \ldots, q$, which we want to prove. It suffices, then, to check that the premises of (***) hold in K.

- $pr^K(\overline{z}) = \overline{x}$

Immediate from $pr \circ s = id_U$ (10.11(b)).

- $h^K(z) > 0$.

By (I), $\phi(h \circ s) = (h \circ s)^K (\phi(\overline{\pi})) = h^K(z)$. Since $h > 0$ on $s[U]$, then (*) shows that $\phi(h \circ s) > 0$.

- $\overline{z} \in (V \cap W)^K$.

Since $\overline{x} = pr^K(\overline{z}) \in U^K$, then $\overline{z} \in W^K$. As the polynomials $P_1, \ldots P_\ell$ define the variety V and $s[U] \subseteq V(\mathbb{R})$, we have

$$\mathbb{R} \models \forall \overline{y} \in U \bigwedge_{j=1}^{\ell} P_j(s(\overline{y})) = 0,$$

that is, the Nash functions $P_j \circ s$ vanish on U; therefore

$$0 = \phi(P_j \circ s) = P_j^K(\phi(\bar{s})) = P_j^K(\bar{z}) \quad \text{for } j = 1,\ldots,\ell. \quad \square$$

The substitution theorem is due to Efroymson [27]. The present proof is due to Coste [67]; for a different proof see Bochnak-Efroymson [25; §7]. As shown in this paper, the result applies as well to a wide class of subrings of $N(U)$.

B. APPLICATIONS : "STELLENSÄTZE" ; COMMUTATIVE ALGEBRA OF NASH FUNCTIONS.

Now we show how the substitution theorem can be used to derive, by a uniform method, all the "stellensätze" holding for rings of Nash functions. These derivations are purely algebraic – in fact, they are a sophisticated elaboration of Robinson's method to prove Hilbert's nullstellensatz – and work for any ring of s.a. functions for which a substitution theorem is available.

A "stellensätz" is a result establishing an equivalence between a condition of the form

$$\mathbb{R} \models \forall \bar{x} \in U [\bigwedge_i g_i(\bar{x}) \geq 0 \wedge \bigwedge_j h_j(\bar{x}) > 0 \wedge \bigwedge_k f_k(\bar{x}) = 0 \rightarrow f(\bar{x})\,?\,0]$$

where $g_1,\ldots,g_r$, $h_1,\ldots,h_p$, $f_1,\ldots,f_q$, $f \in N(U)$ and $?$ is a sign condition ($>$, $\geq$ or $=$) and, on the other hand, an equation

$$P(g_1,\ldots,g_r,\ h_1,\ldots,h_p,\ f_1,\ldots,f_q,\ f) = 0$$

where P is a polynomial with coefficients in $N(U)$. Here, one or more of r, p, q may be zero.

Thus we have:

THEOREM 11.2. (General stellensatz for Nash functions). Let $g_1,\ldots,g_r$, $h_1,\ldots,h_p$, $f_1,\ldots,f_q$, $f \in N(U)$.

(A) The following are equivalent:

(1) $R \models \forall \bar{x} \in U [\bigwedge_i g_i(\bar{x}) \geq 0 \wedge \bigwedge_j h_j(\bar{x}) > 0 \wedge \bigwedge_k f_k(\bar{x}) = 0 \rightarrow f(\bar{x}) > 0]$.

(2) There are $t, s, u_k \in N(U)$ and $\varepsilon_j \in \{0,1\}$ $(1 \leq j \leq p,\ 1 \leq k \leq q)$ where s,t are of the form $F(g_1,\ldots,g_r,\ h_1,\ldots,h_p)$, with F a polinomial whose coefficients are squares in $N(U)$, such that

$$f \cdot t = s + \prod_j h_j^{\varepsilon_j} + \sum_k f_k u_k .$$

(B) Similarly, a condition of type (A.1) with $f(\bar{x}) > 0$ replaced by $f(\bar{x}) \geq 0$ is equivalent to a polynomial equation

$$f\cdot t = (\prod_j h_j^{\varepsilon_j})\ f^{2\ell} + s + \sum_k f_k u_k,$$

for some $\ell \geq 0$ and s, t, u_k, ε_j as is (A).

(C) A condition of type (A.1) with $f(\bar{x}) = 0$ instead of $f(x) > 0$ is equivalent to

$$(\prod_j h_j^{\varepsilon_j})\ f^{2\ell} + s + \sum_k f_k u_k = 0$$

with $\ell \geq 1$ and s, u_k, ε_j as in (A).

Each of the equivalences (A), (B), (C) is derived from a corresponding "formal stellensätz". For example, the result needed for (A) is:

<u>PROPOSITION 11.3.</u> (Formal positivstellensatz). Let A be a commutative ring with unit, and:

(a) S a sub-semi-ring of A(= a subset of A closed under sum and product, but not necessarily under difference) containing the squares.

(b) M a multiplicative subset of A containing 1.

(c) I an ideal of A.

(d) $f \in A$.

Then, the following are equivalent:

(1') For every ring homomorphism $\phi : A \to K$, with K a real closed field, we have:

$$\phi[S] \geq 0, \quad \phi[M] > 0 \text{ and } \phi[I] = 0 \text{ imply } \phi(f) > 0.$$

(2') There are $s, t \in S$, $m \in M$ and $u \in I$ such that

$$f\cdot t = s + m + u$$

<u>PROOF of : Proposition 11.3 implies Theorem 11.2(A).</u>

The implication (A.2) implies (A.1) is checked without difficulty.

(A.1) implies (A.2). As it is obvious, we apply 11.3 with $A = N(U)$ and:

S = the sub-semi-ring generated by the g_i's, the h_j's and the squares.
M = the multiplicative subset generated by the h_j's.

I = the ideal generated by the f_k's.

It suffices to derive condition (1') of 11.3 from condition (A.1) of Theorem 11.2. This is done using the substitution theorem, as follows: Let ϕ be a homomorphism verifying the assumptions of (1'); thus we have:

$$\phi(g_i) \geq 0, \qquad \phi(h_j) > 0, \qquad \phi(f_k) = 0 \quad \text{for all } i,j,k.$$

By the substitution theorem we get:

$$g_i^K(\phi(\overline{\pi})) \geq 0, \qquad h_j^K(\phi(\overline{\pi})) > 0, \qquad f_k^K(\phi(\overline{\pi})) = 0$$

The assumption (A.1) then yields $f^K(\phi(\overline{\pi})) > 0$, and by substitution again, we conclude $\phi(f) > 0$. □

The "formal stellensätze" are, in turn, a consequence of:

<u>11.4 The pivotal lemma.</u> Let A be a commutative ring with unit and I, T, N be subsets of A such that :

(a) T is closed under multiplication.

(b) $I \subseteq T$ is an ideal of A.

(c) i) $-1 \in N$.

ii) $x, y \in N$ implies $-(xy) \in N$.

iii) $(-N) . \Sigma TA^2 \subseteq \Sigma TA^2$.

Then, the following are equivalent:

(1") $N \cap \Sigma TA^2 \neq \emptyset$.

(2") For every prime ideal P of A containing I, if L denotes the fraction field of $A/_P$, we have:

$$N/_P \cap \Sigma \; T/_P \cdot L^2 \neq \emptyset. \qquad \square$$

We omit the purely algebraic proof of this result, which can be found in Dickmann [72; Ch. V]. This is a generalization of Colliot-Thélene [44; Lemma 1.bis], who proves it in the case $I = (0)$ and $N = \{-1\}$. However, we give the

<u>PROOF OF : Lemma 11.4 implies Proposition 11.3.</u> The implication (2') implies (1') in 11.3 is evident.

(1') implies (2'). Assume not (2'); this just says:

(*) $$-M \cap ((S-f\,S) + I) = \emptyset.$$

We apply Lemma 11.4 with $N = -M$ and $T = (S=f\cdot S) + I$, leaving as an easy but tedious exercise for the reader the task of checking that the assumptions (a)-(c) of 11.4 are fulfilled.

The equality (*) amounts to the negation of condition (1") of 11.4. Then, by (2"), there is a prime ideal P containing I such that

(**) $$N/_P \cap \Sigma \; T/_P \cdot L^2 = \emptyset.$$

Since $1 \in M$, then $-1 \notin \Sigma \; T/_P \; L^2$. This means that L admits an order, $\leq$, which makes $T/_P \geq 0$. Let K be the real closure of $\langle L, \leq \rangle$ and $\phi : A \to K$ the canonical homomorphism, $\phi(a) = a/_P$. Then we have $\phi[T] \geq 0$, which implies:

(***) $\phi[S] \geq 0, \quad \phi(f) \leq 0 \quad \text{and} \quad \phi[M] \geq 0,$

as $S \subseteq T$, $-f \in T$ by definition and $M \cdot \Sigma TA^2 \subseteq \Sigma TA^2$ by (c.iii).

By (**) we get $\phi(x) \neq 0$ for $x \in M$, i.e.

(****) $\phi[M] > 0.$

Since $I \subseteq P$ we have

(*****) $\phi[I] = 0$

Conditions (***), (****) and (*****) contradict (1'), which proves Proposition 11.3. □

We mention a few amongst the numerous corollaries of Theorem 11.2.

PROPOSITION 11.5 (Real nullstellensätz for Nash functions). Given an ideal I of $N(U)$ and $f \in N(U)$, the following are equivalent:

(i) f vanishes on $Z(I)$.

(ii) $f \in \sqrt[R]{I}$.

In particular, I is real iff for every $f \in N(U)$, $f \upharpoonright Z(I) = 0$ implies $f \in I$.

PROOF. The implication (ii) implies (i) is trivial. Conversely, by Corollary 10.10 there are $f_1, \ldots, f_t \in I$ such that $Z(I) = \bigcap_k Z(f_k)$; hence condition (i) is equivalent to

$$R \models \forall \bar{x} \in U\, [\bigwedge_k f_k(\bar{x}) = 0 \rightarrow f(x) = 0].$$

By Theorem 11.2(c) (with $r, p = 0$) this is equivalent to

$$f^{2\ell} \in I \qquad \text{for some} \quad \ell \geq 1,$$

which implies $f \in \sqrt[R]{I}$. □

REMARK. The proof shows that $\sqrt[R]{I}$ coincides with Rad(I), the radical of I. In particular we have:

COROLLARY 11.6. Let $U \subseteq \mathbb{R}^n$ be an open connected s.a. set. Every maximal ideal of $N(U)$ is real, and the map:

$$\bar{a} \longmapsto M_{\bar{a}} = \{f \in N(U) \mid f(\bar{a}) = 0\}$$

establishes a one-one correspondence between points of U and maximal ideals of $N(U)$.

PROOF. A maximal ideal is always radical; by the remark above, it is also real. Since M is proper, 11.5 also shows that $Z(M) \neq \emptyset$. Let $\bar{a} \in Z(M)$; then $M \subseteq M_{\bar{a}}$ and, by maximality, $M = M_{\bar{a}}$.

PROPOSITION 11.7. (Solution of Hilbert's 17th problem for Nash functions).

Let $U \subseteq \mathbb{R}^n$ be an open, connected s.a. domain and $f \in N(U)$. If $f \geq 0$ on U, then f is a sum of squares in the fraction field of $N(U)$.

PROOF. By Theorem 11.2(B) (with $r = p = q = 0$), we get $f \cdot t = f^{2\ell} + s$, where ℓ is an integer and s, t sums of squares in $N(U)$. □

Further results in the same vein are reported in Bochnak-Efroymson [25].

As remarked earlier, Theorem 11.2 holds for any ring of real-valued s.a. functions for which the substitution theorem holds. Trivially, this is the case for the polynomial ring $\mathbb{R}[X_1,\ldots,X_n]$. Thus we get back, with a uniform proof, generalized versions of earlier semi-algebraic stellensatze: the real nullstellensatz 4.8 (as in Proposition 11.5) when there are no g_i's or h_j's but we allow the f_k's; the equivalence (c) of Theorem 11.2, when there are no f_k's but we allow the g_i's and the h_j's, is the semi-algebraic real nullstellensätz of Stengle [58]; when there are no f_k's or h_j's but we allow the g_i's, the equivalence 11.2(B) is Stengle's nicht-negativstellensätz.

REFERENCES

I. Algebra; commutative algebra; algebraic geometry: basic texts.

[1] Atiyah, M. & Macdonald, I.G. Introduction to Commutative Algebra, Addison-Wesley Publ.Co. (1969).

[2] Chenciner, A. Courbes Algebriques Planes, Publ.Math.Univ. Paris VII, (1980)

[3] Hartshorne, R. Algebraic Geometry, Springer-Verlag (1977).

[4] Jakobson, N. Lectures in Abstract Algebra, vol.III, Van Nostrand Inc., (1964).

[5] ________ Basic Algebra, 2 vols. W.I. Freeman (1974).

[6] Lafon, J.P. Anneaux Henséliens, Univ.Sao Paulo, (1968).

[7] ________ Algebre Commutative, Hermann, Paris (1977).

[8] Lang, S. Algebra, Addison-Wesley Publ.Co., (1971).

[9] Matsumura, H. Commutative Algebra, W.Benjamin (1970).

[10] Mumford, D. Introduction to Algebraic Geometry, mimeographed.

[11] Nagata, M. Local Rings, Interscience Publishers, (1962).

[12] Raynaud, M. Anneaux Locaux Henséliens, Lect.Notes Math. 169 Springer-Verlag (1970).

[13] Ribenboim, P. L'arithmétique des corps, Hermann, Paris (1972).

[14] Shafarevich, I.R. Basic Algebraic Geometry, Springer-Verlag (1977).

[15] Walker, R. Algebraic Curves, Springer-Verlag (1978).

[16] Zariski, O. & Samuel, P. Commutative Algebra, 2 vols., Van Nostrand Inc., (1958), (1960).

II. Analysis : basic texts.

[17] Dieudonné, J. Foundations of Modern Analysis, Academic Press (1960).

[18] Hörmander, L. Linear Partial Differential Operators, Springer-Verlag (1963).

See also : [85].

III. Hilbert's 17th problem.

[19] Artin, E. Uber die Zerlegung definiter Funktionen in Quadrate, Abh.Math.Sem.Univ.Hamburg 5(1927), pp.100-115.

[20] Delzell, C. A constructive, continuous solution to Hilbert's 17th problem and other results in semi-algebraic geometry, Doctoral dissertation, Stanford Univ. (1980).

[21] McKenna, K. New facts about Hilbert's seventeenth problem, in Model Theory and Algebra, A Memorial Tribute to A. Robinson, Lect.Notes Math 498(1975), pp.220-230.

[22] Robinson, A. On ordered fields and definite functions. Math. Ann. 130(1955), pp. 257-271.

[23] ________ Further remarks on ordered fields and definite functions, Math.Ann. 130(1956), pp. 405-409.

See also: [25], [72], [74], [81], [83], [86].

IV. Nash functions.

[24] Artin, M. & Mazur, B. On periodic points, Ann. Math. 81(1965), pp.82-99.

[25] Bochnak, J. & Efroymson, G. Real algebraic geometry and the 17th Hilbert problem, Math.Ann. 251(1980), pp.213-241.

[26] ________ An Introduction to Nash Functions, in [66], pp.41-54.

[27] Efroymson, G. Substitution in Nash Functions. Pacific J.Math. 63(1973), pp.137-145.

[28] Łojasiewicz,S. Ensembles semi-analytiques, mimeographed, I.H.E.S. (1965).

[29] Nash,J. Real algebraic manifolds, Ann.Math. 56(1952)pp. 405-421.

[30] Palais,R. Equivariant, real algebraic, differential topology. I. Smothness categories and Nash manifolds, notes, Brandels Univ. (1972).

[31] Risler, J.J. Sur l'anneau des fonctions de Nash globales, Ann. Sci.Ecole Norm.Sup. 8(1975), pp.365-378.

[32] Roy,M.F. Faisceau structural sur le spectre réel et fonctions de Nash, in [66], pp. 406-432.

See also : [66], [67], [72].

V. Quantifier elimination; model theory.

[33] Chang,C.C.& Keisler,H.J. Model Theory, North-Holland (1973).

[34] Cohen,P.J. Decision procedures for real and p-adic fields, Comm.Pure Appl.Math. 22(1969), 131-151.

[35] Collins, G.E Quantifier elimination for real closed fields by cylindrical algebraic decomposition, in Automata Theory and Formal Languages, 2nd G.I. Conf., Kaiserslautern (1975), Springer-Verlag pp. 134-184.

[36] Fischer, M.J.& Rabin,M.O. Super-Exponential Complexity of Presburger's Arithmetic, M.I.T. MAC Tech.Memo. 43, (1974).

[37] Kreisel, G. & Krivine,J.L. Elements of Mathematical Logic (Model Theory) North-Holland Publ.Co.,(1967).

[38] Macintyre, A., McKenna,K. & van den Dries,L. Elimination of quantifiers in algebraic structures, Adv.Math. 47(1983), pp. 74-87.

[39] Monk,L. Effective-recursive decision procedures. Doctoral Dissertation, Univ. California, Berkeley (1975).

[40] Tarski,A. A decision method for elementary algebra and geometry 2nd. revised ed., Berkeley, Los Angeles (1951).

[41] van den Dries,L A linearly ordered ring whose theory admits quantifier elimination is a real closed field, Proc. Am.Math.Soc. 79(1980),pp. 97-100.

See also: [22], [23], [55], [72], [82], [86].

VI. Real closed fields; real commutative algebra.

[42] Artin,E.& Schreier,O. Algebraische Konstruktion reeler Körper,, Abh. Math.Sem.Univ.Hamburg 5(1926),pp. 85-99.

[43] ______ Eine Kennzeichnung der reell abgeschlossenen Korper, Abh.Math.Sem.Univ.Hamburg 5(1927),pp.225-231.

[44] Colliot-Thélène, J.L. Variantes du Nullstellensätz réel et anneaux formellement réels, in [66], 98-108.

[45] Dubois,D.W. A note on D.Harrison's theory of preprimes, Pacific J.Math. 21(1967),pp. 15-19.

[46] ______ A nullstellensätz for ordered fields, Arkiv für Math. 8 (1969), pp.111-114.

[47] Dubois,D.W.& Efroymson,G. Algebraic theory of real varieties, in Studies and essays presented to Yu-Why Chen on his 60th birthday, Taiwan Univ. (1970).

[48] Elman,R., Lam,T.Y. & Wadsworth,A. Orderings under field extensions, J.Reine Angew. Math. 306(1979),pp. 7-27.

[49] Gross,H. & Hafner,P. Über die Eindeutigkeit des reellen Abschlusses eine angeordneten Körpers, Comm.Math.Helvetici 44(1969), pp. 491-494.

[50] Knebusch,M. On the uniqueness of real closures and the existence of real places, Comm.Math.Helvetici 47(1972), pp.260-269.

[51] Krivine,J.L. Anneaux préordonnés, J.Anal.Math. 21(1964),pp. 307-326.

[52] Lang,S. The theory of real places, Ann.Math. 57(1953), pp. 378-351.

[53] Lam,T.Y. The theory of ordered fields, in Ring Theory and Algebra III, B.McDonald(ed), M.Dekker,(1980), pp.1-152.

[54] Lam,T.Y. An introduction to real algebra, mimeographed, Univ.Calif.Berkeley (1983).

[55] Prestel,A. Lectures on Formally Real Fields, IMPA, Rio de Janeiro (1975).

[56] Ribenboim,P. Le thérème des zéros pour les corps ordonnés, in Sem.Alg.et Th.Nombres, Dubriel-Pisot, 24e année, 1970-71, exp. 17.

[57] Risler, J.J. Une caráctérisation des idéaux des variétés algébriques réelles, C.R.A.S.(1970), pp. 1171-1173.

[58] Stengle,G. A Nullstellensatz and a Positivestellensatz in semi-algebraic geometry, Math.Ann.(1974),pp. 87-97.

See also: [13], [20], [21], [22], [23], [65], [66], [72], [74].

VII. Spectra.

[59] Brocker,L. Real spectra and ditribution of signatures, in [66], pp. 249-272.

[60] Carral,M. & Coste,M. Normal Spectral Spaces and their Dimensions, J.Pure Appl. Algebra 30(1983),pp. 227-235.

[61] Coste,M. & Coste-Roy,M.F. Spectre de Zariski et spectre réel: du cas classique an cas réel, mimeographed, Univ. Catholique Louvain, Rapport 82 (1979).

[62] ______ La topologie du spectre réel in [74], pp. 27-59.

[63] Hochster,M. Prime ideal structure in commutative rings, Trans.Amer.Math.Soc. 142(1969),pp. 43-60.

[64] Knebusch,M. An invitation to real spectra, in Proc.Conf.Quadratic Forms. Hamilton, Ontario (1983) (To appear).

See also: [32],[69],[72].

VIII. Topology of semi-algebraic sets.

[65] Brumfiel, G.W. Partially Ordered Rings and Semi-Algebraic Geometry. London Math.Soc.Lect.Notes 37(1979), Cambridge Univ. Press.

[66] Colliot-Thélene,J.L. Coste,M.,Mahé,L & Roy,M.F. (eds). Géométric Algébrique Réelle et Formes Quadratiques, Lect.Notes Math. 959 ,Springer-Verlag (1982).

[67] Coste,M. Ensembles semi-algébriques et fonctions de Nash, Prépublications mathématiques 18(1981),Univ. Paris Nord.

[68] ______ Ensembles semi-algébriques, in [66], 109-138.

[69] Coste,M. & Coste-Roy,M.F. Topologies for real algebraic geometry, in Topos-Theoretic Methods in Geometry, A. Kock (ed). Aarhus Univ. (1979), pp. 37-100.

[70] Delfs,H. Kohomologie affiner semi-algebraischer Räume, Doctoral dissertation, Univ. Regensburg (1980).

[71] Delfs,H. & Knebusch,M. Semialgebraic topology over a real closed field. II. : Basic theory of semialgebraic spaces, Math. Zeit. 178(1981), pp. 175-213.

[72] Dickmann,M.A. Logique, algèbre réelle et géométrie réelle, manuscript notes, Univ. Paris VIII (1983) (To appear).

[73] ________ Sur les ouverts semi-algébriques d'une clôture réelle, Portugalia Math. 40(1984).

[74] Dubois,D.W.& Recio,T. (eds), Ordered fields and real algebraic geometry, Comtemporary Math. 8, Amer.Math.Soc. Providence, R.I. (1982).

[75] Hardt,R. Semi-algebraic local triviality in semi-algebraic mappings, Amer.J.Math. 102 (1980),pp. 291-302.

[76] Houdebine, J. Lèmme de séparation, mimeographed, Rennes (1980).

[77] Mather,J. Stratifications and mappings, in Dinamical Systems, M.Peixoto (ed), Acad.Press (1973),pp. 195-232.

[78] Recio,T. Una descomposición de un conjunto semialgebraico, Actas V Reunión de Matemáticos de Expresión Latina, Mallorca (1977).

[79] Van den Dries, L. An application of a model-theoretic fact to (semi-) algebraic geometry, Indag.Math. 44(1982),pp.397-401.

[80] Whitney,H. Elementary structure of real algebraic varieties, Ann. Math. 66(1957),pp. 545-556.

See also: [18], [20], [27], [28], [47], [57], [58], [62].

VIII. Various.

[81] Becker,E. & Jacob,B. Rational points on algebraic varieties over a generalized real closed field: a model-theoretic approach, mimeographed (1983),33 pp.

[82] Cherlin,G. & Dickmann,M.A. Real Closed Rings. II: Model theory, Ann.Pure Appl.Logic 25(1983),pp.213-231.

[83] Dickmann,M.A. On polynomials over real closed rings, in Model Theory of Algebra and Arithmetic, L. Pacholski,J. Wierzejewski, A.J. Wilkie(eds), Lect.Notes Math.

834 , Springer Verlag (1980), pp. 117-135.

[84] Dickmann, M.A. A property of the continuous semi-algebraic functions defined on a real curve, (To appear).

[85] Gillman, L. & Jerison, M. Rings of Continuous Functions, van Nostrand Reinhold Co. (1960).

[86] Prestel, A. & Roquette, P. Formally p-adic Fields, Lect. Notes Math. 1050 Springer-Verlag (1984).

[87] Van den Dries, L. Remarks on Tarski's problem concerning < $\mathbb{R}$, +, · , exp >, manuscript (1981) pp. 25.

See also : [24].

ON THE SPACE $(\lambda)^\kappa$

Carlos A. Di Prisco
I.V.I.C.
Departamento de Matemáticas
Apartado 1827
Caracas 1010A, Venezuela

Wiktor Marek
Dept. Computer Science
University of Kentucky
Lexington, KY 40506-0027
U.S.A.

§1. The space $(\lambda)^\kappa = \{P \subseteq \lambda \mid \text{order type of } P = \kappa\}$ appears naturally in the study of large cardinals and is specifically related to the so called huge cardinals. As it is shown in [S.R.K.], a cardinal κ is huge with target λ (c.f. also [B.D.T.]) if there is a κ-complete normal non-trivial ultrafilter on $(\lambda)^\kappa$ containing the sets of the form $\hat{p} = \{P \in (\lambda)^\kappa \mid p \subseteq P\}$ for $p \in P_\kappa(\lambda)$.

Since any ultrafilter on the set $[\lambda]^\kappa = \{P \subseteq \lambda \mid |P| = \kappa\}$ with the properties mentioned above (κ-complete, normal and fine) concentrates on $(\lambda)^\kappa$, the existence of such ultrafilter on $[\lambda]^\kappa$ is also equivalent to hugness of κ with target λ.

In [DP.M2] we constructed $F_{\kappa,\lambda}$, the least κ-complete, fine, normal non-trivial filter on $[\lambda]^\kappa$, which permitted us to complete the following table

Large cardinal property	Normal Measure on κ (κ measurable)	Normal Measure on $P_\kappa(\lambda)$ (κ λ-supercompact)	Normal measure on $[\lambda]^\kappa$ ($\kappa \to (\lambda)$)
Related combinatorial structure.	Closed unbounded subsets of κ	Closed unbounded subsets of $P_\kappa(\lambda)$	$F_{\kappa,\lambda}$

This raises the following question, could the table be also completed putting a combinatorial structure on $(\lambda)^\kappa$ instead of $F_{\kappa,\lambda}$?. The purpose of this note is to show that there is no such possibility: there is no combinatorial "absolute" notion on $(\lambda)^\kappa$ approximating normal measures on $(\lambda)^\kappa$. In this way we present results which stress the differences between $(\lambda)^\kappa$ and $[\lambda]^\kappa$ studied by Mignone in [Mig.]. One should note that the difference pointed by Mignone pertains to the situation in which the axiom of determinateness is accepted, whereas in this paper we work under the axiom of choice. We would like to thank M. Magidor for providing us with the exact statement of Theorem 4.

Let us recall some results and definitions from [DP.M2]: Let $\kappa<\lambda$ be regular uncountable cardinals. If $X \subseteq P_\kappa(\lambda)$ is a closed unbounded subset of $P_\kappa(\lambda)$, let $A_X = \{P\in[\lambda]^\kappa | P$ is the union of an increasing κ-chain of elements of $X\}$. The sets of the form $\{A_X | X$ closed unbounded subset of $P_\kappa(\lambda)\}$ generate a κ-complete, normal, fine, nontrivial filter on $[\lambda]^\kappa$ called $F_{\kappa,\lambda}$.

It is interesting to note that if in the definition of A_X we use directed sets of elements of X instead of κ-chains we obtain the same filter ([DP.M2]). The following propositions are proved in [DP.M2].

PROPOSITION 1. If $V = L$ then $(\lambda)^\kappa$ is not $F_{\kappa,\lambda}$-stationary. □

PROPOSITION 2. $F_{\kappa,\lambda}$ is the least κ-complete, fine, normal, nontrivial filter on $[\lambda]^\kappa$. □

From these propositions we obtain the following

METATHEOREM: If ZFC is consistent then ZFC $\nvdash$ "There is a κ-complete, normal, fine, nontrivial filter on $(\lambda)^\kappa$".

PROOF. Assume the contrary, so ZFC + V=L proves as well that there is a nontrivial, κ-complete, fine, normal filter F on $(\lambda)^\kappa$. Define a filter G on $[\lambda]^\kappa$ as follows

$$X\in G \iff X\cap(\lambda)^\kappa \in F.$$

Then G is a κ-complete, fine, normal, nontrivial filter on $[\lambda]^\kappa$ and therefore by proposition 2, $F_{\kappa,\lambda} \subseteq G$. But $(\lambda)^\kappa \in G$ and so $(\lambda)^\kappa$ is $F_{\kappa,\lambda}$-stationary contradicting proposition 1. □

The next statement clarifies the situation.

PROPOSITION 3. The following statements are equivalent:

a) There exists a κ-complete, normal, fine, nontrivial filter on $(\lambda)^\kappa$.

b) $F_{\kappa,\lambda} \restriction (\lambda)^\kappa$ is such a filter.

c) $(\lambda)^\kappa$ is $F_{\kappa,\lambda}$-stationary.

PROOF.

a) ⇒ b): The argument of the metatheorem shows that $(\lambda)^\kappa$ is $F_{\kappa,\lambda}$-stationary and therefore $F_{\kappa,\lambda}$ induces a filter on $(\lambda)^\kappa$. The other implications follow easily. □

The results in [DP.M2] suggest that perhaps we have that either $(\lambda)^\kappa$ is $F_{\kappa,\lambda}$-stationary or $F_{\kappa,\lambda}$ is the CLUB filter on $P_{\kappa^+}(\lambda)$ restricted to $[\lambda]^\kappa$, or equivalently that either $(\lambda)^\kappa$ is $F_{\kappa,\lambda}$-stationary or $\check{\kappa}$ belongs to $F_{\kappa,\lambda}$ (where $\check{\kappa} = \{P\in[\lambda]^\kappa | \kappa\subseteq P\}$). (In fact, the first part of the disjunction holds if κ is huge with target λ and the second occurs if $V = L$ or, as it will be shown in the next section if $\lambda = \kappa^+$).

This alternative does not hold in general but we have something close to it:

THEOREM 4. Given regular cardinals $\kappa<\lambda$ then either $\check{\kappa}\in F_{\kappa,\lambda}$ or there is $\kappa<\lambda'\leq\lambda$ such that $(\lambda')^\kappa$ is $F_{\kappa,\lambda'}$ stationary.

PROOF. Suppose $\check{\kappa}\notin F_{\kappa,\lambda}$ then, its complement in $[\lambda]^\kappa$, i.e. the set $S = \{P \subseteq [\lambda]^\kappa | \kappa\not\subseteq P\}$, is stationary. Considering the function $f:S \to \lambda$ defined by $f(P)$ = first element of P after the first gap, and using the normality of $F_{\kappa,\lambda}$ we see that there is an $F_{\kappa,\lambda}$- stationary $S_1 \subseteq S$ such that for all $P\in S_1, f(P)$ is a fixed value α. Since $F_{\kappa,\lambda}$ is fine (i.e. contains all cones $\check{p} = \{P\in[\lambda]^\kappa | p\subseteq P\}$ for $p\in P_\kappa(\lambda)$) this α cannot belong to κ. Thus $\alpha\geq\kappa$. So, if an element P of S' has order type $>\kappa$, its κth element must be above κ. Therefore, if $(\lambda)^\kappa$ is not $F_{\kappa,\lambda}$ stationary then a stationary subset S_2 of S_1 contains only elements of order type greater than κ. Define $g:S_2 \to \lambda$ by $g(P) = \kappa^{th}$ element of P. There is a stationary subset $S_3 \subseteq S_2$ and a $\lambda'<\lambda$ such that for every $p\in S_3$, $g(P) = \lambda'$.

From this we conclude that $(\lambda')^\kappa$ is $F_{\kappa,\lambda'}$ stationary. This is so because given $X \subseteq P_\kappa(\lambda')$ closed and unbounded the set $X' \subseteq P_\kappa(\lambda)$ defined by $X' = \{p\in P_\kappa(\lambda) | p\cap\lambda'\in X\}$ is closed and unbounded in $P_\kappa(\lambda)$ so $A_{X'}\in F_{\kappa,\lambda}$. The set $S_3\cap A_{X'}$ is thus $F_{\kappa,\lambda}$-stationary.
For $p\in S_3\cap A_{X'}$, $P\cap\lambda'$ has order type κ and so $P\cap\lambda'\in A_{X'}\cap(\lambda)^\kappa$. □

The possibility that $(\lambda)^\kappa$ is not $F_{\kappa,\lambda}$ stationary and $\check{\kappa}$ is not in $F_{\kappa,\lambda}$ (but $(\gamma)^\kappa$ is $F_{\kappa,\gamma}$-stationary for some $\gamma<\lambda$) actually occurs. We give an example due to J. Henle.

Consider $\kappa<\gamma$ and suppose that $\kappa \to (\gamma)$, and let $\lambda=\gamma+\gamma$. The set $A = \{P\in[\lambda]^\kappa | \gamma+[P\cap\gamma]=P-\gamma\}$ (i.e. A is the set of $P\in[\lambda]^\kappa$ such that $P\cap\gamma$ is "just like" $P-\gamma$). This set is in $F_{\kappa,\lambda}$ because the set $X = \{p\in P_\kappa(\lambda) | \gamma+(p\cap\gamma)=p-\gamma\}$ is club and $A_X = A$. Thus $(\lambda)^\kappa$ is not $F_{\kappa,\lambda}$-stationary. We also have that $\check{\kappa}\notin F_{\kappa,\lambda}$ since if $\check{\kappa} \in F_{\kappa,\lambda}$ then there is a closed and unbounded set $X \subseteq P_\kappa(\lambda)$ such that $\check{\kappa} \supseteq A_X$. But since $\kappa<\lambda\approx\gamma$ there is a bijection f from λ to γ which is the identity on κ. The set $Y = \{f''p | p\in X\}$ is a closed and unbounded subset of $P_\kappa(\gamma)$ such that A_Y is in $F_{\kappa,\gamma}$ and is contained in $\check{\kappa}$. But this contradicts that $\kappa\to(\gamma)$ since in this case $(\gamma)^\kappa$ is $F_{\kappa,\gamma}$-stationary.

§2. We consider in this section the case in which $\lambda = \kappa^+$.

PROPOSITION 1. Let X be a closed unbounded subset of $P_\kappa(\kappa^+)$. Then $A_X \cap \mathrm{Ord}$ is an unbounded subset of κ^+.

<u>PROOF.</u> Given $\gamma<\kappa^+$, let $p\in X$ be such that $\gamma\in p$. We will construct in stages a directed subset of X whose union is an ordinal in κ^+ greater than γ.

<u>STAGE 1</u>: For each ordinal α in $\cup p-p$ let p_α be an element of X containing it. Put $D_1=\{p_\alpha | \alpha\in\cup p-p\}\cup\{p\}$.

<u>STAGE 2:</u> Close D_1 to obtain $D_2 \supseteq D_1$ such that D_2 is a directed subset of X. (Note we can obtain such a directed set D_2 such that $|D_2|=|D_1|+\aleph_0$).

<u>STAGE 2n+1:</u> For every $\alpha\in\sup(\cup D_{2n})-\cup D_{2n}$ let p_α be a set in X containing it. Put $D_{2n+1}=D_{2n}\cup\{P_\alpha | \alpha\in\sup(\cup D_{2n})-\cup D_{2n}\}$.

<u>STAGE 2n+2:</u> Close D_{2n+1} to obtain a directed subset of X, D_{2n+2}, containing D_{2n+1}.

Finally we put $D = \bigcup_{n\in\omega} D_n$.

Clearly D is a directed subset of X such that $\cup D$ is an ordinal in κ^+ above γ. □

It follows from the lemma 2.6 of [DP.M2] that for each $X\subseteq P_\kappa(\kappa^+)$, X closed and unbounded, $A_X\cap\kappa^+$ is a closed subset of κ^+, thus we have shown that $F_{\kappa,\kappa^+}\restriction\kappa^+ \subseteq \text{CLUB}_{\kappa^+}$. On the other hand, $\text{CLUB}_{\kappa^+} \subseteq F_{\kappa,\kappa^+}\restriction\kappa^+$ since this is a κ^+-complete normal fine filter on κ^+ (it is fine because given any $\alpha<\kappa^+$, if $p\in P_\kappa(\kappa^+)$ contains α, $\hat{p}$ is a closed unbounded subset of $P_\kappa(\kappa^+)$ and $A_{\hat{p}}\restriction\text{ord} \subseteq \kappa^+-\alpha$).

Therefore $F_{\kappa,\kappa^+}\restriction\kappa^+=\text{CLUB}_{\kappa^+}$ and so $(\kappa^+)^\kappa$ is not F_{κ,κ^+} stationary. On the other hand, $\hat{\kappa}\in F_{\kappa,\kappa^+}\restriction\kappa^+$ because if X is closed unbounded then the set $\hat{\kappa}=\kappa^+-\kappa$ contains $(A_{X\cap\{\hat{\kappa}\}})\cap\kappa^+$.

§3. The metatheorem of section 1 shows that if we want to produce (in ZFC) a filter on $(\lambda)^\kappa$ which provides an "absolute" counterpart to a measure on $(\lambda)^\kappa$, we must drop one of the three usual properties: κ-completeness, normality, or fineness. The easiest to give up seems to be fineness, specially if we can maintain unboundedness of the elements of our filter. (That is to say: even if we do not require cones to be in the filter, they must be stationary with respect to it). Another condition we wish to maintain is that, if $\kappa\to(\lambda)$ the elements of the filter must belong to any normal measure witnessing this fact.

We will define such a filter modifying a construction from [DP.M2].

We say that $X \subseteq P_\kappa(\lambda)$ is end-unbounded if for all $p\in P_\kappa(\lambda)$ there is $q\in X$ such that $p \subseteq_e q$ (where $p \subseteq_e q$ means that q is an end extension of p).

Clearly every end-unbounded set is unbounded. The cones $\hat{p}=\{q|p\subseteq q\}$ are not end-unbounded (with the exception of $\hat{\phi}=P_\kappa(\lambda)$). Call SUPERCLUB the filter generated on $P_\kappa(\lambda)$ by the end-unbounded and closed subsets of $P_\kappa(\lambda)$.

PROPOSITION. The filter SUPERCLUB is κ-complete, normal and non-trivial. It is not fine but all its elements are unbounded subsets of $P_\kappa(\lambda)$.

PROOF. Let us first recall a definition from [DP.M1]. E-CLUB is the filter on $P_\kappa(\lambda)$ generated by sets which are end-unbounded and closed with respect to unions of end extension chains of length $<\kappa$. This filter is κ-complete and normal but not fine ([DP.M1]). It is easy to verify that SUPERCLUB=CLUB $\cap$ E-CLUB and thus we have only to prove that this intersection is non empty. For this, note that the sets $\hat{\hat{\alpha}} = \{p\epsilon P_\kappa(\lambda)|\text{order type of } p>\alpha\}$ are in SUPERCLUB. □

Since $\cap\{\hat{\hat{\alpha}}:\alpha\epsilon\kappa\}=\phi$, SUPERCLUB is not κ^+ complete. SUPERCLUB is not the least κ-complete, normal filter in $P_\kappa(\lambda)$ containing $\{\hat{\hat{\alpha}}|\alpha<\kappa\}$. In fact, the filter generated by the sets which are closed and unbounded by final segments is strictly contained in SUPERCLUB (we say that $X\subseteq P_\kappa(\lambda)$ is unbounded by final segments if for any $p\epsilon P_\kappa(\lambda)$ there is $\alpha<\lambda$ such that $p\cup(\alpha-\cup p)\epsilon X$).

Using the ideas of [DP.M2] we define now a κ-complete normal non-trivial filter on $(\lambda)^\kappa$. Given $X\subseteq P_\kappa(\lambda)$, let

$$A''_X=\{P\epsilon(\lambda)^\kappa| \text{ There is an end extension chain } \{p_\xi\}_{\xi<\kappa}\subseteq X \text{ such that } \cup p_\xi=P\}$$

PROPOSITION 2. If X is end-unbounded then $A''_X\neq\phi$. □

Let now $G_{\kappa,\lambda}$ the the filter on $(\lambda)^\kappa$ generated by $\{A''_X|X \text{ superclub}\}$.

THEOREM 3. $G_{\kappa,\lambda}$ is a κ-complete, normal, non-trivial filter in $(\lambda)^\kappa$.

PROOF. The κ-completeness and normality are easily verified. To see that $G_{\kappa,\lambda}$ is non-trivial, in the sense that it does not contain small subsets of $(\lambda)^\kappa$, we prove the following lemma.

LEMMA. If $B\subseteq(\lambda)^\kappa$ is such that $|\lambda-\cup B|=\lambda$ then B is not $G_{\kappa,\lambda}$-stationary.

PROOF. It is enough to show that there is a superclub set X_B such that $A''_{X_B}\cap B=\phi$. We put $X_B=\{p\epsilon P_\kappa(\lambda)| \text{ for all } P\epsilon B(p-P\neq\phi)\}$.

Clearly X_B is closed, and since $\lambda - \cup B$ is cofinal in λ, it is super-unbounded. □

Therefore, if $\kappa<\lambda$ then $G_{\kappa,\lambda}$ contains the complements of all subsets of $(\lambda)^\kappa$ of cardinality $<\lambda$. □

We will now show that the elements of $G_{\kappa,\lambda}$ are of measure one with respect to normal measures on $(\lambda)^\kappa$.

<u>LEMMA.</u> If X is closed and unbounded then $A''_X = A_X \cap (\lambda)^\kappa$.

<u>PROOF.</u> The inclusion $\subseteq$ is obvious. For the other one, assume that P belongs to $A_X \cap (\lambda)^\kappa$. Then $P = \bigcup_{\xi<\kappa} p_\xi$ where $\langle p_\xi \rangle_{\xi<\kappa}$ is a chain of elements of X. Since we can assume that $(P_\xi)_{\xi<\kappa}$ is a continuous chain and $\overline{P} = \kappa$ it is not difficult to construct a subchain of $\langle p_\xi \rangle_{\xi<\kappa}$ which is an end extension chain. □

<u>COROLLARY 4.</u> If $A \in G_{\kappa,\lambda}$, $\kappa \to (\lambda)$ and μ is a witnessing normal measure, then $A \in \mu$.

<u>PROOF.</u> There is an $X \in$ SUPERCLUB such that $A''_X \subseteq A$. But then $A_X \in \mu$ and $(\lambda)^\kappa \in \mu$ so $A''_X \in \mu$. □

Finally let us mention that if X_ξ is closed and unbounded for each $\xi<\lambda$,

$$\mathop{\Delta}_{\xi<\lambda} A''_{X_\xi} = A''_{\mathop{\Delta}_{\xi<\lambda} X_\xi}$$

Note that both sides could be empty unless the sets X_ξ are superclub.

REFERENCES

[B.D.T.] Barbanel, J., Di Prisco, C.A. & Tan, I.B. Many-times huge and superhuge cardinals. J. Symbolic Logic 49(1984), pp. 112-122.

[DP.M1] DiPrisco, C.A. & Marek, W. Some properties of stationary sets. Dissertationes Mathematicae CCXVIII(1982), pp.1-37.

[DP.M2] Di Prisco, C.A. & Marek, W. A filter on $[\lambda]^\kappa$ Proc.Amer.Math.Soc. 90(1984), pp. 591-598.

[Mig] Mignone, R.J. On the ultrafilter characterization of huge cardinals. Proc.Amer.Math.Soc. 90(1984), pp. 585-590.

[S.R.K.] Solovay, R., Reinhardt, W. Kanamori, A. Strong axioms of infinity and elementary embeddings. Annals of Mathematical Logic 13(1978), pp. 73-116.

This work was partially supported by CONICIT grants S1-1129 & Capt 15-84

THE MODEL EXTENSION THEOREMS FOR $\mathbb{J}_3$-THEORIES

Itala M.L. D'Ottaviano
Universidade Estadual de Campinas
Instituto de Matematica, Estatística e Ciencia da Computação
Caixa Postal 6155
13100 Campinas, SP, Brasil

Two generalized versions of Keisler's classical Model Extension Theorem are obtained for $\mathbb{J}_3$-theories. These theories are three-valued with more than one distinguished truth-value, reflect certain aspects of modal logics and can be paraconsistent. $\mathbb{J}_3$-theories were introduced in the author's doctoral dissertation.

§1. INTRODUCTION. A theory T is said to be inconsistent if it has as theorems a formula and its negation; and it is said to be trivial if every formula of its language is a theorem.

A logic is paraconsistent if it can be used as the underlying logic for inconsistent but nontrivial theories (see [1] and [2]).

In a previous paper (see [4]) we introduced a three-valued propositional system, $\mathbb{J}_3$, with two distinguished truth-values, which is paraconsistent and reflects some aspects of certain types of modal logics.

In another paper (see [5]) we axiomatized $\mathbb{J}_3$ and established relations between this calculus and several known logical systems, as, for example, intuitionism. We especially emphasized the close analogy between $\mathbb{J}_3$ and Lukasiewicz' three-valued propositional calculus Ł$_3$.

We also introduced the corresponding three-valued first-order $\mathbb{J}_3$-Theories in whose languages may appear other equalities in addition to ordinary identity, and which extend the first-order predicate calculus $\mathbb{J}_3^*=$. We proved the Completeness Theorem and the Compactness Theorem for $\mathbb{J}_3$-theories.

The model theory for $\mathbb{J}_3$-theories reflects much of the classical model theory. We were able to obtain in our doctoral dissertation generalized versions of the following classical results: Keisler Model Extension Theorem, Łoś-Tarski Theorem, Chang-Łoś-Suszko Theorem, Tarski Cardinality Theorem, Löwenheim - Skolem Theorem, Joint-Consistency Theorem,

Craig Interpolation Lemma, Beth-Padoa Definability Theorem, the Quantifier Elimination Theorem and many of the usual theorems on complete theories and categoricity.

In some cases, as for example the Model Extension Theorem and the Definability Theorem, we proved more than one generalization of the classical results, all of them compatible both with the many-valued aspects and the modal aspects of $\mathbb{J}_3$-theories.

Our aim here is to present two generalized versions of Keisler's classical Model Extension Theorem, asserting both the conditions to extend a structure $\mathfrak{A}$ for a $\mathbb{L}_3$-language L of T to a model of the $\mathbb{J}_3$-theory T.

In §2, after giving some basic notions and the axiomatization for $\mathbb{J}_3$-theories, we emphasize several results and theorems which are necessary for the development of the last part of the paper.

In §3, after adapting the concepts of isomorphism and elementary equivalence, we define Γ-substructures and discuss the case of certain special sets of formulas Γ.

After this, we present the two Model Extension Theorems for $\mathbb{J}_3$-theories.

Both theorems generalize the classical Model Extension Theorem of Keisler and their proofs involve specific characteristics of many-valued and modal logics, reflecting the existence of more than one distinguished truth-value in the matrices defining $\mathbb{J}_3$-theories.

Finally, we give a three-valued version of Łoś-Tarski Theorem, which characterizes universal-open $\mathbb{J}_3$-theories; Tarski Lemma, on the union of elementary chains; and Chang-Łoś-Suszko Theorem, characterizing universal-existential $\mathbb{J}_3$-theories.

In this paper, definitions, theorems and proofs, when similar to the corresponding classical case, are omitted.

For definitions and theorems we use the nomenclature of [13].

We have also extended some of the above results about $\mathbb{J}_3$-theories to $\mathbb{J}_n$-theories, $3 \leq n \leq \aleph_0$.

The results about Joint non-Trivialization, Definability Theorems, Complete Theories, Quantifier Elimination Theorems, Categoricity and the mentioned results about $\mathbb{J}_3$-theories will appear elsewhere.

The content of this work is part of the results of our doctoral dissertation.

§2. FIRST ORDER $\mathbb{L}_3$-THEORIES. The symbols of a first-order $\mathbb{L}_3$-language are the individual variables, the function symbols, the predicate symbols, the primitive connectives $\neg$, $\vee$ and ∇, the quantifiers $\exists$ and $\forall$, and the parentheses.

The identity = must be among the predicate symbols of every $\mathbb{L}_3$-language. In particular cases, other predicate symbols can be considered as equalities.

We use x,y,z and w as syntactical variables for individual variables; f and g, for function symbols; p and q, for predicate symbols, and c for constants.

The definitions of term and atomic formula are the usual ones for first-order languages; $a,b,\ldots$,etc, are syntactical variables for terms.

The definition of formula is also the usual plus the condition: if A is a formula, then ∇A is a formula; A,B,C, etc, are syntactical variables for formulas.

By a $\mathbb{L}_3$-language we understand a first-order language (in the sense of [13]) whose logical symbols are the ones mentioned above.

The truth-functions; $H_\vee$, H_∇ and $H_\neg$ are defined by the following tables:

A \ B ($A \vee B$)	0	½	1
0	0	½	1
½	½	½	1
1	1	1	1

A	∇A
0	0
½	1
1	1

A	$\neg$A
0	1
½	½
1	0

The set of truth-values $\{0,½,1\}$ and the set of distinguished truth-values $\{½,1\}$ are denoted by "V" and "V_d" respectively.
The following abbreviations will be used:

$A \mathbin{\&} B =_{def} \neg(\neg A \vee \neg B)$

$\Delta A =_{def} \neg\nabla\neg B$

$\neg^* A =_{def} \neg\nabla A$

$A \rightarrowtail B =_{def} \nabla\neg A \vee B$

$A \supset B =_{def} \neg\nabla A \vee B$

$A \twoheadrightarrow B =_{def} (A \rightarrowtail B) \mathbin{\&} (\neg B \rightarrowtail \neg A)$

$A \equiv B =_{def} (A \supset B) \mathbin{\&} (B \supset A)$

$\neg$ is called weak negation or simple negation, $\neg *$ strong negation, $\supset$ basic implication, $\equiv$ basic equivalence.

We present the table of some of the non primitive connectives:

$\neg^* A$

A	$\neg^* A$
0	1
½	0
1	0

ΔA

A	ΔA
0	0
½	0
1	1

$A \supset\!\!\rightarrow B$

A \ B	0	½	1
0	1	1	1
½	½	1	1
1	0	½	1

$A \supset B$

A \ B	0	½	1
0	1	1	1
½	0	½	1
1	0	½	1

$A \equiv B$

A \ B	0	½	1
0	1	0	0
½	0	½	½
1	0	½	1

Free occurrence of a variable, open formula, closed formula, variable free term and closure of a formula are defined as in [13].

We let $b_{x_1,\ldots,x_n}[a_1,\ldots,a_n]$ be the term obtained from b by replacing all occurrences of $x_1,\ldots,x_n$ by the terms $a_1,\ldots,a_n$ respectively; and we let $A_{x_1,\ldots,x_n}[a_1,\ldots,a_n]$ be the formula obtained from A by replacing all free occurrences of $x_1,\ldots,x_n$ by $a_1,\ldots,a_n$ respectively.

Whenever either of these is used, it will be implicitly assumed that $x_1,\ldots,x_n$ are distinct variables and that, in the case of $A_{x_1,\ldots,x_n}[a_1,\ldots,a_n]$, a_i is substitutible for x_i, $i = 1,\ldots,n$.

DEFINITION 2.1. A structure $\mathfrak{A}$ for the first-order $\mathbb{L}_3$-language consists of :

i) a nonempty set $|\mathfrak{A}|$ called the universe of $\mathfrak{A}$;

ii) for each n-ary function symbol f of L, an n-ary function $f_{\mathfrak{A}}$ from $|\mathfrak{A}|^n$ to $|\mathfrak{A}|$;

iii) for each n-ary predicate symbol p of L, other than $=$, an n-ary function $p_{\mathfrak{A}}$ from $|\mathfrak{A}|^n$ to V.

As in [13], we construct the language $L(\mathfrak{A})$; and define a value $\mathfrak{A}(a) \in V$ for each variable free term a of $L(\mathfrak{A})$, and $\mathfrak{A}$-instance of a formula A.

We use i and j as syntactical variables for the names of individuals of $\mathfrak{A}$.

DEFINITION 2.2. The truth-value $\mathfrak{A}(A)$ for each closed formula A in $L(\mathfrak{A})$ is given by:

i) if A is $a = b$, then $\mathfrak{A}(A) = 1$ if $\mathfrak{A}(a) = \mathfrak{A}(b)$; on the contrary, $\mathfrak{A}(A) = 0$;

ii) if A is $p(a_1,\ldots,a_n)$, where p is not =, then $\mathfrak{A}(A) = p_{\mathfrak{A}}(\mathfrak{A}(a_1),\ldots,\mathfrak{A}(a_n))$;

iii) if A is $\neg B$, then $\mathfrak{A}(A)$ is $H_{\neg}(\mathfrak{A}(B))$;

iv) if A is ∇B, then $\mathfrak{A}(A)$ is $H_{\nabla}(\mathfrak{A}(B))$;

v) if A is $B \vee C$, then $\mathfrak{A}(A)$ is $H_{\vee}(\mathfrak{A}(B), \mathfrak{A}(L))$;

vi) if A is $\exists xB$, then $\mathfrak{A}(A) = \max\{\mathfrak{A}(B_x[i]/i \in L(\mathfrak{A})\}$;

vii) if A is $\forall xB$, then $\mathfrak{A}(A) = \min\{\mathfrak{A}(B_x[i]\ i \in L(\mathfrak{A})\}$.

DEFINITION 2.3. A formula A of L is valid in $\mathfrak{A}$ if, and only if for every $\mathfrak{A}$-instance A' of A, $\mathfrak{A}(A')$ belongs to V_d.

DEFINITION 2.4. A first-order $\mathbb{J}_3$-theory is a formal system T such that:

i) the language of T, L(T), is a $\mathbb{L}_3$-language;

ii) the axioms of T are the logical axioms of L(T) and certain further axioms, the nonlogical axioms;

iii) the logical axioms and rules of T are the following, with the usual restrictions:

AXIOM 1 : $\Delta(A \supset\!\!\rightarrow (B \supset\!\!\rightarrow A))$

AXIOM 2 : $\Delta((A \supset\!\!\rightarrow B) \supset\!\!\rightarrow ((B \supset\!\!\rightarrow C) \supset\!\!\rightarrow (A \supset\!\!\rightarrow C)))$

AXIOM 3 : $\Delta((\neg A \supset\!\!\rightarrow \neg B) \supset\!\!\rightarrow (B \supset\!\!\rightarrow A))$

AXIOM 4 : $\Delta(((A \supset\!\!\rightarrow \neg A) \supset\!\!\rightarrow A) \supset\!\!\rightarrow A)$

AXIOM 5 : $\Delta(\Delta(A \supset\!\!\rightarrow B) \supset\!\!\rightarrow \Delta(\Delta A \supset\!\!\rightarrow \Delta B))$

AXIOM 6 : $\forall x(x = x)$

AXIOM 7 : $x = y \supset (A[x] \equiv A[y])$

AXIOM 8 : $A_x[a] \supset \exists xA$

AXIOM 9 : $\forall xA \supset A_x[a]$

AXIOM 10 : $\exists xA \equiv \neg \forall x \neg A$

AXIOM 11 : $\forall xA \equiv \neg\exists x \neg A$

AXIOM 12 : $\neg\exists xA \equiv \forall x \neg A$

AXIOM 13 : $\neg\forall xA \equiv \exists x \neg A$

AXIOM 14 : $\nabla\exists xA \equiv \exists x \nabla A$

AXIOM 15 : $\nabla\forall xA \equiv \forall x \nabla A$

RULE R_1 : $\dfrac{A, \Delta(A \supset\!\!\rightarrow B)}{B}$

RULE R_2 : $\dfrac{\nabla A}{A}$

RULE R_3 ($\exists$-Introduction): $\dfrac{A \supset C}{\exists xA \supset C}$

RULE R_4 ($\forall$-Introduction): $\dfrac{C \supset A}{C \supset \forall xA}$

A is a Theorem of T, in symbols : $\vdash_T A$, is defined in the standard way.

Observe that it is possible to define $\mathbb{L}_3$-languages, using the Łukasiewicz' connectives $\neg$ and $\supset\!\!\rightarrow$, instead of $\vee$, ∇ and $\neg$.

So, there is a close analogy between Łukasiewicz's three-valued logic $\pounds_3$ (see [3], [7], [8] and [9]) and the underlying logic for $\mathbb{J}_3$-theories.

<u>THEOREM 2.1.</u> $\mathbb{J}_3$ is a non-conservative extension of $\pounds_3$ with connectives $\neg$ and $\supset\!\!\rightarrow$.

<u>THEOREM 2.2.</u> $\mathbb{J}_3$ is a conservative extension of the classical propositional calculus with connectives $*$, , &, $\supset$ and $\equiv$.

The initial notion of equivalence $\equiv$ has to be gradually fortified (see [5] and [6]), in order to obtain a relation compatible with the fact that the matrices defining $\mathbb{J}_3$ have more than one distinguished truth-value.

In the case of classical logic, the equivalence $\equiv$ behaves as a congruence relation with respect to the other logical symbols. Unfortunately this is not the case for $\mathbb{J}_3$-theories, since it is possible to have $\vdash_T A \equiv B$ and $\nvdash_T \neg A \equiv \neg B$.

Hence, we introduce a stronger equivalence, $\equiv^*$, called strong equivalence in $\mathbb{J}_3^=$- theories and which is a $\mathbb{J}_3^=$-congruence relation.

<u>DEFINITION 2.5.</u> $A \equiv^* B =_{def} (A \equiv B)$ & $(\neg A \equiv \neg B)$.

For this strong equivalence we obtain an Equivalence Theorem for

$\mathbb{I}_3$-theories (see [5] and [6]).

The following closure theorem and the theorem on constants for $\mathbb{I}_3$-theories are identical to the classical ones.

THEOREM 2.3. If A' is the closure of A, then $\vdash_T A$ if, and only if, $\vdash_T A'$.

If T' is a $\mathbb{I}_3$-theory obtained from T by adding new constants, then for every formula A of T and every sequence $e_1, \ldots e_n$ of new constants, $\vdash_T A$ if, and only if,

$$\vdash_{T'} A_{x_1,\ldots,x_n}[e_1,\ldots,e_n].$$

We observe that if Γ is a set of formulas in the theory T, then T[Γ] is the theory obtained from Γ by adding all of the formulas in Γ as new nonlogical axioms.

THEOREM 2.4. (Reduction Theorem for non-Trivialization): Let Γ be a non empty set of formulas in a $\mathbb{I}_3$-theory. Then the extension T[Γ] is trivial if, and only if, there is a theorem of T which is a disjunction of negations of closures of formulas of type ∇A, with A in Γ.

To finish this section, we discuss how to convert a formula A into a formula in prenex form.

DEFINITION 2.6. If A is a formula in a $\mathbb{L}_3$-language L, the prenex operations on A are:

a) the classical prenex operations, with the usual restrictions;

b) replacement of a part ∇QxB of A by Qx∇B, where Qx is either ∃x or ∀x.

By the above definition we can obtain the prenex form of every formula A, based on the following result.

THEOREM 2.5. If A' is obtained from A with A in L(T), by a prenex operation, then $\vdash_T A \equiv^* A'$.

$\mathbb{I}_3$-theories are three-valued theories with more than one-distinguished truth-value, they can be paraconsistent and reflect certain aspects of modal type logics.

The details of a further study, the proofs of the theorems mentioned in this paragraph and other results about $\mathbb{I}_3$-theories can be found in [5] and [6].

§3. THE THEORY OF MODELS.

The theory of models for $\mathbb{I}_3$-theories reflects a great deal of classical model theory.

In some cases, there is more than one generalization of the classical theorems, all of them compactible with the characteristics of $\mathbb{I}_3$-theories.

In the following, we give the two versions of the classical Keisler Model Extension Theorem for $\mathbb{I}_3$-theories.

The proofs of both theorems reflect the specific characteristics of many-valued and modal logics, and they specially involve the existence of more than one distinguished truth-value.

We shall use the notations and the conventions of [13], with the appropriate adaptations.

DEFINITION 3.1. Let $\mathfrak{A}$ and $\mathbb{B}$ be structures for the $\mathbb{L}_3$-language L. An isomorphism of $\mathfrak{A}$ and $\mathbb{B}$ is a bijective mapping ϕ from $|\mathfrak{A}|$ to $|\mathbb{B}|$ such that for $a_1,\ldots,a_n$ in $|\mathfrak{A}|$:

i) $\phi(f_{\mathfrak{A}}(a_1,\ldots,a_n)) = f_{\mathbb{B}}(\phi(a_1),\ldots,\phi(a_n))$, for every function symbol f of L;

ii) $p_{\mathfrak{A}}(a_1,\ldots,a_n) = p_{\mathbb{B}}(\phi(a_1),\ldots,\phi(a_n))$, for every predicate symbol p of L.

If ϕ is a mapping from $|\mathfrak{A}|$ to $|\mathbb{B}|$ and i is the name of an individual a of $|\mathfrak{A}|$, we use i^{ϕ} to designate the name of the individual $\phi(a)$ of $\mathbb{B}$. If u is an expression of $L(\mathfrak{A})$, u^{ϕ} is the expression obtained from u by replacing each name i by i^{ϕ}.

THEOREM 3.1. Let ϕ be an isomorphism of $\mathfrak{A}$ and $\mathbb{B}$. Then $\phi(\mathfrak{A}(a)) = \mathbb{B}(a^{\phi})$ for every variable-free term a of $L(\mathfrak{A})$, and $\mathfrak{A}(A) = \mathbb{B}(A^{\phi})$ for every closed formula A of $L(\mathfrak{A})$.

The following definitions are similar to the classical ones.

DEFINITION 3.2. Two estructures $\mathfrak{A}$ and $\mathbb{B}$ for L are elementarily equivalent, in symbols $\mathfrak{A} \simeq \mathbb{B}$, if the same closed formulas of L are valid in $\mathfrak{A}$ and $\mathbb{B}$.

As in the classical case, if $\mathfrak{A}$ and $\mathbb{B}$ are elementarily equivalent, then they are models of the same $\mathbb{I}_3$-theories. But if $\mathfrak{A} \simeq \mathbb{B}$ and A is a closed formula of L, it is not possible to conclude that $\mathfrak{A}(A) = \mathbb{B}(A)$.

DEFINITION 3.3. An embedding of $\mathfrak{A}$ in $\mathbb{B}$ is an injective mapping ϕ from $|\mathfrak{A}|$ to $|\mathbb{B}|$ such that conditions (i) and (ii) of Definition 3.1 hold for all nonlogical symbols f and p of L and all $a_1,\ldots,a_n$ in $|\mathfrak{A}|$.

DEFINITION 3.4. We say that $\mathfrak{A}$ is a substructure of $\mathbb{B}$, or $\mathbb{B}$ is an extension of $\mathfrak{A}$, when $|\mathfrak{A}|$ is a subset of $|\mathbb{B}|$ and the identity mapping from $|\mathfrak{A}|$ to $|\mathbb{B}|$ is an embedding from $\mathfrak{A}$ in $\mathbb{B}$, that is:

i) $f_{\mathfrak{A}}(a_1,\ldots,a_n) = f_{\mathbb{B}}(a_1,\ldots,a_n)$, for every f;

ii) $p_{\mathfrak{A}}(a_1,\ldots,a_n) = p_{\mathbb{B}}(a_1,\ldots,a_n)$, for every p.

DEFINITION 3.5. In the conditions of the definition 3.4, if $\mathfrak{A}$ and $\mathbb{B}$ are models of some $\mathbb{L}_3$-theory T, we say that $\mathfrak{A}$ is a submodel of $\mathbb{B}$.

THEOREM 3.2. Let $\mathfrak{A}$ and $\mathbb{B}$ be structures for L, and let ϕ be a mapping from $|\mathfrak{A}|$ to $|\mathbb{B}|$. Then ϕ is an embedding of $\mathfrak{A}$ in $\mathbb{B}$ if, and only if, $\mathfrak{A}(A) = \mathbb{B}(A^{\phi})$, for every variable free formula A of $L(\mathfrak{A})$.

Let Γ be a set of formulas in the $\mathbb{L}_3$-language L and let $\mathfrak{A}$ be a structure for L. As in the classical model Theory, $\Gamma(\mathfrak{A})$ designates the set of $\mathfrak{A}$-instances of formulas in Γ.

DEFINITION 3.6. Let $\mathfrak{A}$ and $\mathbb{B}$ be structures for L, such that $|\mathfrak{A}|$ is a subset of $|\mathbb{B}|$, and let Γ be a set of formulas in L. We say that $\mathfrak{A}$ is a Γ-substructure of $\mathbb{B}$ and that $\mathbb{B}$ is a Γ-extension of $\mathfrak{A}$ if $\mathfrak{A}(A) \in V_d$ implies $\mathbb{B}(A) \in V_d$, for every formula A in $\Gamma(\mathfrak{A})$.

THEOREM 3.3. Let Γ be a set of formulas in L such that for every formula A in Γ, $\neg\nabla A$ is in Γ. If $\mathfrak{A}$ is a Γ-substructure of $\mathbb{B}$, then $\mathfrak{A}(A) \in V_d$ if, and only if, $\mathbb{B}(A) \in V_d$ for every formula A in $\Gamma(\mathfrak{A})$.

PROOF. If $\mathfrak{A}(A) = 0$, then $\mathbb{B}(\neg\nabla A) = 1$. □

If Γ is a set of formulas such that for every formula A in Γ, $\neg A$ belongs to Γ, it does not seem possible to obtain the result above. In fact, under such conditions, if $\mathfrak{A}(A) = 0$ we can conclude only that $\mathbb{B}(A) = \frac{1}{2}$ or $\mathbb{B}(A) = 0$.

On the other hand, we can prove that $\mathfrak{A}(A) = \frac{1}{2}$ implies $\mathbb{B}(A) = \frac{1}{2}$, and $\mathbb{B}(A) = 1$ implies $\mathfrak{A}(A) = 1$.

It seems impossible also to prove the above result when Γ is a set of formulas such that for every formula A in Γ, ∇A is in Γ.

But, if for every formula A in Γ, $\neg A$ and ∇A belong to Γ, then we can prove a stronger result.

THEOREM 3.4. Let Γ be a set of formulas in L such that for every formula A in Γ, $\neg A$ and ∇A are in Γ. If $\mathfrak{A}$ is a Γ-substructure of $\mathbb{B}$, then $\mathfrak{A}(A) = \mathbb{B}(A)$ for every formula A in $\Gamma(\mathfrak{A})$.

PROOF. By Theorem 3.3, as $\mathfrak{A}(A) = \frac{1}{2}$ implies $\mathfrak{B}(A) = \frac{1}{2}$, it is sufficient to prove that $\mathfrak{A}(A) = 1$ implies $\mathfrak{B}(A) = 1$ for every formula A in $\Gamma(\mathfrak{A})$. □

THEOREM 3.5. (1) If Γ is the set of all open formulas in L, $\mathfrak{A}$ and $\mathfrak{B}$ are structures for L and $\mathfrak{A}$ is a Γ-substructure of $\mathfrak{B}$, then $\mathfrak{A}$ is a substructure of $\mathfrak{B}$ and $\mathfrak{B}$ is an extension of $\mathfrak{A}$.

(2) If Γ is the set of all formulas in L and $\mathfrak{A}$ is a Γ-substructure of $\mathfrak{B}$, then $\mathfrak{A}$ and $\mathfrak{B}$ are elementarily equivalent.

DEFINITION 3.7. If Γ is the set of all formulas in an $\mathbb{L}_3$-language L and $\mathfrak{A}$ is a Γ-substructure of $\mathfrak{B}$, we say that $\mathfrak{A}$ is an elementary substructure of $\mathfrak{B}$, or that $\mathfrak{B}$ is an elementary extension of $\mathfrak{A}$.

THEOREM 3.6. If $\mathfrak{A}$ is a elementary substructure of $\mathfrak{B}$, then $\mathfrak{A}$ is elementarily equivalent to $\mathfrak{B}$ and $\mathfrak{A}(A) = \mathfrak{B}(A)$ for every closed formula A in L.

Furthermore, in the case of $\mathbb{L}_3$-languages, it is convenient to observe that if $\mathfrak{A}$ is a substructure of $\mathfrak{B}$ and elementarily equivalent to $\mathfrak{B}$, there can be closed formulas A in $L(\mathfrak{A})$ such that $\mathfrak{A}(A)$ and $\mathfrak{B}(A)$ are distinguished values, but different ones.

THEOREM 3.7 If $x = e$ belongs to the set Γ of formulas in L and $\mathfrak{A}$ is a Γ-substructure of $\mathfrak{B}$, then $\mathfrak{A}(e) = \mathfrak{B}(e)$.

If $\mathfrak{A}$ is a structure for L, we define the Γ-diagram theory of $\mathfrak{A}$ as in the classical case.

DEFINITION 3.8. Let $\mathfrak{A}$ and $\mathfrak{B}$ be structures for L with $\mathfrak{A} \subset \mathfrak{B}$. The structure $\mathfrak{B}_{\mathfrak{A}}$ for the $\mathbb{L}_3$-language $L(\mathfrak{A})$ is an expansion of $\mathfrak{B}$ such that, if i is the name for an individual a of $|\mathfrak{A}|$ then $\mathfrak{B}_{(\mathfrak{A})}(i) = a$.

DEFINITION 3.9. The Γ-diagram theory of $\mathfrak{A}$, $D_\Gamma(\mathfrak{A})$, is the $\mathbb{J}_3$-theory whose language is $L(\mathfrak{A})$ and whose non-logical axioms are the formulas A in $\Gamma(\mathfrak{A})$ such that $\mathfrak{A}(A)$ belongs to V_d.

LEMMA. (Diagram Lemma) : Let Γ be a set of formulas in L, and let $\mathfrak{A}$ and $\mathfrak{B}$ be structures for L such that $|\mathfrak{A}|$ is a subset of $|\mathfrak{B}|$. Then $\mathfrak{A}$ is a Γ-substructure of $\mathfrak{B}$ if, and only if, $\mathfrak{B}_{\mathfrak{A}}$ is a model of $D_\Gamma(\mathfrak{A})$.

DEFINITION 3.10. A set Γ of formulas of L is almost-regular if every formula of the form $x = y$ or $x \neq y$ is in Γ, and if for every formula A in Γ, every formula of the form $A[x_1,\ldots,x_n]$ is in Γ.

THEOREM 3.8. (Model Extension Theorem I): Let $\mathfrak{A}$ be a structure for L, T a $\mathcal{J}_3$-theory with language L and Γ an almost-regular set of formulas in L. Then $\mathfrak{A}$ has a Γ-extension which is a model of T if, and only if, every theorem of T which is a disjunction of strong negations of formulas in Γ is valid in $\mathfrak{A}$.

PROOF. Suppose that such a Γ-extension $\mathfrak{B}$ exists. If $\vdash_T \neg{}^*A_1 \vee \ldots \vee \neg{}^*A_n$, with $A_1,\ldots,A_n$ formulas in Γ, and $\neg{}^*A_1 \vee \ldots \vee \neg{}^*A_n$ is not valid in $\mathfrak{A}$, then there are $\mathfrak{A}$-instances $A'_1,\ldots,A'_n$ of $A_1,\ldots,A_n$, such that $\mathfrak{A}(A'_i)$ belongs to V_d, for i, $1 \le i \le n$. Therefore $\mathfrak{B}(\neg{}^*A'_i) = 0$ for every $i, 1 \le i \le n$. Hence $\mathfrak{B}(\neg{}^*A'_1 \vee \ldots \vee \neg{}^*A'_n) = 0$, what is impossible, since $\mathfrak{B}$ is a model of T.

On the other hand, suppose that the condition holds. Let T' be the $\mathcal{J}_3$-theory obtained from T by adding all the names of $L(\mathfrak{A})$ as new constants; and let be T" obtained from T' by adding all the non-logical axioms of $D_\Gamma(\mathfrak{A})$ as new axioms. We shall show that T" is non-trivial. If not, then by the Reduction Theorem for non-Trivialization there is a formula of type $\neg{}^*A'_1 \vee \ldots \vee \neg{}^*A'_n$, with $A'_1,\ldots,A'_n$ in $\Gamma(\mathfrak{A})$, such that $\vdash_{T'} \neg{}^*A_1 \vee \ldots \vee \neg{}^*A'_n$ and $\mathfrak{A}(A'_i)$ belongs to V_d for $i = 1,\ldots,n$.

Hence by the Theorem on Constants, $\vdash_T \neg{}^*A_1 \vee \ldots \vee \neg{}^*A_n$, where A_i results from A'_i by replacing the names by new variables. Then, by the almost-regularity of Γ, we have that $A_1,\ldots,A_n$ belong to Γ and so $\neg{}^*A_1 \vee \ldots \vee \neg{}^*A_n$ is valid in $\mathfrak{A}$. Therefore $\mathfrak{A}(A'_i) = 0$ for some i, which is a contradiction.

By the Completeness Theorem T" has a model $\mathfrak{B}'$ and from here on the proof becomes identical to the classical one. That is, we prove that the restriction $\mathfrak{B}$ of $\mathfrak{B}'$ to L is a Γ-extension of $\mathfrak{A}$, which is a model of T. □

COROLLARY. Let Γ be an almost regular set of formulas in L and let Λ be a set of formulas containing every formula $\forall x_1 \ldots \forall x_n A$, where A is a disjunction of strong negations of formulas in Γ. If $\mathfrak{A}$ is a structure for L and $\mathfrak{B}$ is a Λ-extension of $\mathfrak{A}$, then there is a Γ-extension C of $\mathfrak{B}$ which is an elementary extension of $\mathfrak{A}$.

PROOF. Let Γ' be the set of formulas $A[i_1,\ldots,i_n]$ in $L(\mathfrak{A})$.

If $A_1,\ldots,A_n$ are in Γ' and $\vdash_{D_{\Gamma'}(\mathfrak{A})} \neg{}^*A_1 \vee \ldots \vee \neg{}^*A_n$, then by the Diagram Lemma $\mathfrak{A}_{\mathfrak{A}}$ is a model of $D_\Gamma(\mathfrak{A})$. So $\mathfrak{A}_{\mathfrak{A}}(B) \in V_d$ and $\mathfrak{A}_{\mathfrak{A}}(B) = \mathfrak{A}(B)$, where B is the closure of $\neg{}^*A_1 \vee \ldots \vee \neg{}^*A_n$. As B is an axiom of $D_\Lambda(\mathfrak{A})$, then by the Diagram Lemma $\mathfrak{B}_{\mathfrak{A}}(B)$ is distinguished. Hence, by Closure Theorem $\neg{}^*A_1 \vee \ldots \vee \neg{}^*A_n$ is valid in $\mathfrak{B}_{\mathfrak{A}}$.

Therefore there is a Γ'-extension $\mathfrak{C}'$ of $\mathfrak{B}_{\mathfrak{A}}$, which is a model of $D_{\Gamma'}(\mathfrak{A})$ and the proof follows as in the classical case. □

If in the statement of Theorem 3.8, we had "disjunction of negations of formulas in Γ", then we could not prove it by the same method.

Given an almost regular set Γ of formulas in $L(T)$ and a structure $\mathfrak{A}$ for $L(T)$, if we suppose that there is a model $\mathfrak{B}$ of T which is a Γ-extension of $\mathfrak{A}$, then we can not prove that $\vdash_T \neg A_1 \vee \ldots \vee \neg A_n$ implies $\mathfrak{A} \models \neg A_1 \vee \ldots \vee \neg A_n$. In fact, if there are $\mathfrak{A}$-instances $A'_1, \ldots, A'_n$ of formulas $A_1, \ldots, A_n$ in Γ, such that $\mathfrak{B}(A'_i)$ belongs to V_d for every i, $1 \le i \le n$, we cannot conclude that $\mathfrak{B}(\neg A'_1 \vee \ldots \vee \neg A'_n) = 0$, on account of the characteristics of the basic negation $\neg$ of $\mathbb{J}_3$-theories.

On the other hand, if we suppose that $\vdash_T \neg A_1 \vee \ldots \vee \neg A_n$ implies $\mathfrak{A} \models \neg A_1 \vee \ldots \vee \neg A_n$ and if we construct the theories T' and T'', as in Theorem 3.8, then we can not prove that T'' is non-trivial. In fact, if $\vdash_T \neg A'_1 \vee \ldots \vee \neg A'_n$ and $\mathfrak{A} \models \neg A_1 \vee \ldots \vee \neg A_n$, with $A_1, \ldots, A_n$ in Γ and $A'_1, \ldots, A_n$ in $\Gamma(\mathfrak{A})$, we cannot conclude that $\mathfrak{A}(A'_i) = 0$ for some $i, 1 \le i \le n$.

By the observations above, in order to obtain a model extension theorem which generalizes classical Keisler Model Extension Theorem and which uses the primitive negation $\neg$ of $\mathbb{J}_3$-theories, we have to consider almost regular sets with special characteristics.

<u>DEFINITION 3.11</u>. A set Γ of formulas of L is ∇-regular if Γ is almost regular and if for every formula A in Γ, ∇A belongs to Γ.

<u>DEFINITION 3.12</u>. A set Γ of formulas of L is Δ-regular if Γ is almost regular and if for every formula A in Γ, ΔA belongs to Γ.

<u>DEFINITION 3.13</u>. A set Γ of formulas of L is regular if Γ is ∇-regular and Δ-regular.

<u>THEOREM 3.9.</u> Let $\mathfrak{A}$ be a structure for L, T a $\mathbb{J}_3$-theory with language L and let Γ be a ∇-regular set of formulas of L. If every theorem of T which is a disjunction of negations of formulas in Γ is valid in $\mathfrak{A}$, then $\mathfrak{A}$ has a Γ-extension which is a model of T.

<u>PROOF.</u> We construct the $\mathbb{J}_3$-theories T' and T'' like in the proof of Theorem 3.8.

If T'' is trivial, by the Reduction Theorem for non-Trivialization, there are formulas $A'_1, \ldots, A'_n$ in $\Gamma(\mathfrak{A})$ such that $\vdash_T \neg{}^*A'_1 \vee \ldots \vee \neg{}^*A'_n$

and $\mathfrak{A}(A_i') \in V_d$, $i = 1,\ldots,n$. By the Theorem on Constants, as in Theorem 3.8, we can obtain formulas $A_1,\ldots,A_n$ such that $\vdash_T \neg *A_1 \vee \ldots \vee \neg *A_n$. By hypothesis, $\mathfrak{A} \models \neg\nabla A_1 \vee \ldots \vee \neg\nabla A_n$. Therefore, we have $\mathfrak{A}(A_i') = 0$ for some $i, i = 1,\ldots,n$, and A_i' in $\Gamma(\mathfrak{A})$.

That is a contradiction, so T'' has a model $\mathfrak{B}'$.

The proof then follows as in Theorem 3.8. □

THEOREM 3.10. Let $\mathfrak{A}$ be a structure for L, T a $\mathbb{JI}_3$-theory with language L and let Γ be a Δ-regular set of formulas of L. If $\mathfrak{A}$ has a Γ-extension which is a model of T, then every theorem of T which is a disjunction of negations of formulas in Γ is valid in $\mathfrak{A}$.

PROOF. If $\vdash_T \neg A_1 \vee \ldots \vee \neg A_n$, with the formulas $A_1,\ldots,A_n$ in Γ, and $\neg A_1 \vee \ldots \vee \neg A_n$ is not valid in $\mathfrak{A}$, then there are $\mathfrak{A}$-instances $A_1',\ldots,A_n'$ of $A_1,\ldots,A_n$ such that $\mathfrak{A}(\neg A_i') = 0$, for every i, $1 \le i \le n$. So, $\mathfrak{A}(\Delta A_i') = 1$, for every i, $1 \le i \le n$.

If $\mathfrak{B}$ is a Γ-extension of $\mathfrak{A}$ which is a model of T, we have that $\mathfrak{B}(A_i')$ and $\mathfrak{B}(\neg A_i')$ belong to V_d, for every i, $1 \le i \le n$. Hence $\mathfrak{B}(\neg A_1' \vee \ldots \vee \neg A_n') = 0$, which is a contradiction. Therefore, $\vdash_T \neg A_1 \vee \ldots \vee \neg A_n$ implies $\mathfrak{A} \models \neg A_1 \vee \ldots \vee \neg A_n$. □

THEOREM 3.11. (Model Extension Theorem II). Let $\mathfrak{A}$ be a structure for L, T a $\mathbb{JI}_3$-theory with language L and let Γ be a regular set of formulas in L. Then $\mathfrak{A}$ has a Γ-extension which is a model of T if, and only if, every theorem of T which is a disjunction of negations of formulas in Γ is valid in $\mathfrak{A}$.

COROLLARY 1. Let Γ be a regular set of formulas in L and let Λ be a set of formulas containing every formula $\forall x_1 \ldots \forall x_n A$, where A is a disjunction of negations of formulas in Γ. If $\mathfrak{A}$ is a structure for L and $\mathfrak{B}$ is a Γ-extension of $\mathfrak{A}$, then there is a Γ-extension $\mathfrak{C}$ of $\mathfrak{B}$ which is an elementary extension of $\mathfrak{A}$.

COROLLARY 2. Let $\mathfrak{A}$ be a structure for L. T a $\mathbb{JI}_3$-theory with language L and let Γ be an almost regular set of formulas in L, such that $\neg A$ and ∇A belong to Γ for every formula A in Γ. Then, $\mathfrak{A}$ has a Γ-extension which is a model of T if, and only if, every theorem of T which is a disjunction of negations of formulas in Γ is valid in $\mathfrak{A}$.

The Model Extension Theorem I (Theorem 3.8) seems to be natural for $\mathbb{JI}_3$-theories, on account of the role played by $\neg *$ and the Reduction Theorem for non Trivialization in $\mathbb{JI}_3$-theories.

The Extension Theorem II (Theorem 3.11) seems to be more compatible with the main characteristics of $\mathbb{I}_3$-theories, besides more meaningful for $\mathbb{I}_3$-theories and general many-valued theories and model logics.

But it seems not to be possible to obtain each one of the two versions from the other. That is, it seems that Theorem 3.8, does not imply Theorem 3.11, and vice-versa.

It is convenient to observe that if Γ is an almost regular set of formulas in L such that $\neg A$ and ΔA belong to Γ for every formula A in Γ, then we can prove a result similar to Corollary 2 to Theorem 3.11, though Γ can be not regular.

We yet observe that, in the proofs of some of the following theorems, as the case of Łoś-Tarski Theorem, Chang-Łoś-Suszko Theorem and also in the Joint non-Trivialization Theorem for $\mathbb{I}_3$-theories, when we apply one of the model extension theorems, we need only a sufficient condition to Γ-extend a structure $\mathfrak{A}$ for L(T) to a model $\mathbb{B}$ of the $\mathbb{I}_3$-theory T.

In these cases, considering the special characteristics of the sets Γ, we can indistinctly apply Model Extension Theorem I, or Model Extension Theorem II.

Now we study three-valued versions of Łoś-Tarski Theorem and Chang-Łoś-Suszko Theorem.

<u>THEOREM 3.12.</u> (Łoś-Tarski Theorem): A $\mathbb{I}_3$-theory T is equivalent to an open $\mathbb{I}_3$-theory if, and only if, every substructure of a model of T is a model of T.

<u>PROOF.</u> If T is equivalent to an open $\mathbb{I}_3$-theory T', it suffices to show that every substructure $\mathfrak{A}$ of a model $\mathbb{B}$ of T' is a model of T'. By Theorem 3.2, every open formula valid in $\mathbb{B}$ is valid in $\mathfrak{A}$, so $\mathfrak{A}$ is a model of T'.

On the other hand, we first prove that if Γ is the set of open formulas in T, if Γ' is the set of formulas in Γ which are theorems of T, and if every structure for L(T) in which all the formulas of Γ' are valid is a model of T, then T is equivalent to a $\mathbb{I}_3$-theory whose nonlogical axioms are in Γ. Then, it suffices to show that if every open theorem of T is valid in $\mathfrak{A}$, then $\mathfrak{A}$ is a model of T. So, as Γ is a regular set, by the Model Extension Theorem II, $\mathfrak{A}$ has a Γ-extension which is a model of T. □

<u>DEFINITION 3.13.</u> A sequence $\mathfrak{A}_1, \mathfrak{A}_2, \ldots$ of structures for L is a chain if for each n, $\mathfrak{A}_{n+1}$ is an extension of $\mathfrak{A}_n$.

DEFINITION 3.14. Given a chain of structures for L, the union of the chain is the structure $\mathfrak{A}$ whose universe is the union of the universes of the $\mathfrak{A}_n$; if $a_1,\dots,a_k$ are in this union, then there is an n such that all of $a_1,\dots,a_k$ are individuals of $\mathfrak{A}_n$ and we set

$$f_{\mathfrak{A}}(a_1,\dots,a_k) = f_{\mathfrak{A}_n}(a_1,\dots,a_k)$$

$$p_{\mathfrak{A}}(a_1,\dots,a_n) = p_{\mathfrak{A}_n}(a_1,\dots,a_k).$$

DEFINITION 3.15. If $\mathfrak{A}$ is the union of the chain $\mathfrak{A}_1, \mathfrak{A}_2, \dots$ and A is a formula of L, then A is valid in $\mathfrak{A}$ if $\mathfrak{A}_n(A')$ belongs to V_d, for every $\mathfrak{A}_n$-instance A' of A, n = 1,2... .

DEFINITION 3.16. An elementary chain is a chain $\mathfrak{A}_1, \mathfrak{A}_2, \dots$, such that for each n, $\mathfrak{A}_{n+1}$ is an elementary extension of $\mathfrak{A}_n$.

THEOREM 3.13. (Tarski's Lemma): If $\mathfrak{A}_1, \mathfrak{A}_2, \dots$ is an elementary chain, then the union $\mathfrak{A}$ of the chain is an elementary extension of each $\mathfrak{A}_n$.

PROOF. As in the classical case, we must show that if A is a closed formula in $L(\mathfrak{A}_n)$, then $\mathfrak{A}_n(A) = \mathfrak{A}(A)$. We use induction on the length of A.

If A is atomic, the result is obvious.

If A is of type $\neg B$ or $B \vee C$, the result follows from the induction hypothesis and the definition tables of $H_\neg$ and $H_\vee$.

If A is of type ∇B, we observe that $\mathfrak{A}_n(\nabla B) = 0$ if, and only if, $\mathfrak{A}(B) = 0$; $\mathfrak{A}_n(\nabla B) = 1$ if, and only if, $\mathfrak{A}(B) = 1$ or $\mathfrak{A}(B) = \frac{1}{2}$.

Let A be of type $\exists xB$. $\mathfrak{A}(\exists xB) = 0$ if, and only if, $\mathfrak{A}_n(B_x[i]) = 0$ for each i in $L(\mathfrak{A}_n)$; if $\mathfrak{A}(\exists xB) = \frac{1}{2}$, then $\mathfrak{A}(B_x[i]) = \frac{1}{2}$ for some i in $L(\mathfrak{A})$, we choose k such that $k > n$ and i is a name in $L(\mathfrak{A}_k)$, and we have that $\mathfrak{A}_k(B_x[i]) = \frac{1}{2}$; if $\mathfrak{A}(\exists xB) = 1$ we proceed similarly. □

THEOREM 3.14. (Chang-Łoś-Suszko Theorem): A Π_3-theory T is equivalent to a Π_3-theory whose non-logical axioms are existential if, and only if, the union of every chain of models of T is a model of T.

PROOF. As in the classical case, if T is equivalent to a Π_3-Theory T' whose non logical axioms are existential, then if suffices to prove that the union $\mathfrak{A}$ of a chain $\mathfrak{A}_1, \mathfrak{A}_2, \dots$ of models of T' is a model of T'. Let $\exists x_1 \dots \exists x_n A$ be an $\mathfrak{A}$-instance of a non logical axiom of T'. For large enough k, $\mathfrak{A}_k(A[i_1,\dots,i_n]) \in V_d$ for some $i_1,\dots,i_n$ in $L(\mathfrak{A}_k)$. Hence $\mathfrak{A}(\exists x_1 \dots \exists x_n A) \in V_d$ and $\mathfrak{A}$ is a model of T'.

Now suppose that every union of a chain of models of T is a model of T. It also suffices to show that if $\mathfrak{A}$ is a structure in which every existential theorem of T is valid, then $\mathfrak{A}$ is a model of T. We proceed as in the classical case by constructing a chain $\mathfrak{A}_1, \mathfrak{A}_2, \ldots$ such that $\mathfrak{A}_1 = \mathfrak{A}$, $\mathfrak{A}_{2n}$ is a model of T and $\mathfrak{A}_{2n+3}$ is an elementary extension of $\mathfrak{A}_{2n+1}$.

If $\mathfrak{B}$ is the union of the chain, as $\mathfrak{A}$ and $\mathfrak{B}$ are elementarily equivalent, then $\mathfrak{A}$ is a model of T.

The second part of the proof is possible because the set of all universal formulas of L(T) is regular, in the sense of Definition 3.13. □

REFERENCES

[1] ARRUDA, A.I. A survey of paraconsistent logic, Mathematical Logic in Latin America, North-Holland, Amsterdam, (1980), pp. 1-41.

[2] ARRUDA, A.I. Aspects of the historical development of paraconsistent logic (To appear)

[3] BORKOWSKI,L. (ed.) Selected Works of J. Łukasiewicz, North-Holland, Amsterdam (1970).

[4] D'OTTAVIANO, I.M.L. & da Costa,N.C.A. Sur un problème de Jaśkowski, C.R. Acad.Sc.Paris, 270A (1970), pp. 1349-1343.

[5] D'OTTAVIANO, I.M.L. The completeness and compactness of a three-valued first-order logic. To appear in Mathematical Logic, V Latinamerican Symposium on Mathematical Logic, Marcel-Dekker.

[6] ____________ Sobre uma teoría de modelos trivalente (Thesis) Universidade Estadual de Campinas, Campinas (1982).

[7] ŁUKASIEWICZ,J. On the principle of contradiction in Aristotle, Review of Metaphysics XXIV (1971), pp. 485-509.

[8] ____________ Philosophische Bermerkungen zu mehrwertigen systemen des Aussagenkalküls, C.R.Soc.Sci.Lett. Varsovie 23, (1930), pp. 51-57 (Translation to English in [3] , pp. 153-178).

[9] ŁUKASIEWICZ,J. & Tarski, A. Untersuchungen über den Aussagenkalküll, C.R.Soc. Sci.Lett.Varsovie 23 (1930) pp. 39-50 (Translation to English in [3] , pp 131-152).

[10] RASIOWA,H. An algebraic approach to non-classical logics, North Holland, Amsterdam (1974).

[11] RESCHER, N. Many-valued logics, McGraw-Hill, N.York (1969).

[12] ROSSER, J.B. & Turquette,A. Many-valued logics, North-Holland, Amsterdam (1952).

[13] SHOENFIELD,J.R. Mathematical Logic. Addison Wesley, Reading (1967).

COMPLETENESS THEOREMS FOR THE GENERAL THEORY OF STOCHASTIC PROCESSES

Sergio Fajardo**
University of Wisconsin
Department of Mathematics
Madison, Wisconsin 53706
U.S.A.

§0. INTRODUCTION.

Adapted Probability Logic, denoted L_{ad}, is a logic adequate for the study of stochastic processes. In this paper we show how to axiomatize in L_{ad} the basic notions of the general theory of stochastic processes (i.e. stopping time, martingale, adapted, progressively measurable, optional and predictable stochastic processes) and prove a "completeness" theorem for each such notion.

The study of stochastic processes has been for the last forty years one of the main topics of research in probability theory. The development of the theory of martingales and Markov processes has brought many new ideas into this field. Concepts such as filtration of σ-algebras and stopping times are now of fundamental importance after the work of Levy, Doob, Ito, Chung, Hunt and others. Along with these ideas the need for more advanced tools has appeared. The "General theory of Processes" is the branch of probability that has as its subject of study all these new objects. Among the many concepts studied by "general theorists" (besides the basic notions mentioned above) are: stopping times of different types, local martingales, semimartingales, generalized stochastic integrals, stochastic differential equations, etc. Good sources of information about these topics are [D], [DM_1], [DM_2], [MP] and the Lecture Notes in Mathematics published regularly by the members of the Strasbourg school of probability

---

** Current Address : Department of Mathematics
University of Colorado
Campus Box 426
Boulder, Colorado 80309
(U.S.A.)

After the introduction of the Loeb measure in [Lo] and Anderson's nonstandard construction of Brownian motion [An], nonstandard analysts have developed a new approach for the study of stochastic integration (see [K_1], [Pa], [Li], [HP], [Pe]) and stochastic integral equations ([K_1], [HP], [C]) that is more intuitive and seems to be very well suited for direct applications.

What does logic have to do here? Hoover and Keisler realized that many of the concepts in [K_1] could be naturally expressed in a language with integral quantifiers and conditional expectation operators, and that some of the properties of the hyperfinite adapted spaces introduced in that paper were general model theoretic theorems in the logic constructed with this language. Thus, adapted probability logic was born. This logic first appeared in [K_2]. Later Rodenhausen in his thesis [R] proved, among other things, a completeness theorem for L_{ad}. Hoover and Keisler in [HK] introduced different notions of elementary equivalence for L_{ad} and using a model theoretic approach gave direct applications to probability theory.

Keisler in [K_3] tied L_{ad} with the two previously known probability logics L_{AP} and $L_{A\int}$ (see [K_3], [K_4], [H_1], [H_2], [H_3], [H_4]) by introducing a new logic, the so called Probability Logic with Conditional Expectation L_{AE}. L_{ad} was then reintroduced as a two sorted form of L_{AE} and in this way Keisler in [K_3], [K_5] gave a new (simpler) proof of Rodenhausen's completeness theorem for L_{ad}. Keisler [K_3] is an up-to-date account of the development of the different probability logics including L_{ad}. For more on L_{AE} see Fajardo [F].

A description of the contents of this paper is as follows : Section 1 contains an introduction to adapted probability logic that "almost" does not require a previous knowledge of the other probability logics, but Keisler's survey paper [K_3] is strongly recommended. In section 2 we show how to axiomatize the concept of stopping time and prove our first completeness theorem. In section 3 we give a set of axioms and prove a corresponding completeness theorem that simultaneously covers adapted, progressively measurable and optional stochastic processes. We also give a different set of axioms for predictable processes and prove its completeness. Finally we present a completeness theorem for martingales.

The reader is assumed to be familiar with the basic concepts of probability theory such as random variable and conditional expectation which can be found in any elementary textbook in probability (for example

[As]). We do not expect readers to know the general theory of processes, so all the basic definitions are presented. Elliot [E] is a good introductory book covering the concepts studied here. We do not use nonstandard analysis and only require some familiarity with basic model theory.

§1. DEFINITIONS AND BACKGROUND:

We introduce adapted probability Logic L_{ad} following Keisler's original presentation in $[K_2]$. An alternative way of defining L_{ad} as a two-sorted form of logic with conditional expectation $L_{\mathbb{A}E}$ was given recently in Keisler $[K_3]$. We follow the former approach since it allows us to go directly into the basics of L_{ad} without assuming a previous knowledge about the other probability logics. Basic notions from adapted probability logic such as adapted model, interpretation and adapted elementary equivalence (weak and strong) together with a complete set of axioms for L_{ad} are given. We also present all the basic definitons from the general theory of processes that are relevant to our exposition. Those readers familiar with Keisler $[K_3]$ will realize that we do not worry about admissible fragments $L_{\mathbb{A}ad}$ of L_{ad}. Instead, we just work with $\mathbb{A}$ = HC (the hereditarily countable sets), but all results generalize in the obvious way to fragments with $\mathbb{A} \subseteq$ HC and $\omega \in \mathbb{A}$. Throughout this paper we assume that so-called "usual conditions" of a filtration of σ-algebras (see definition 1.1 below). These are, as the name suggests, the kind of assumptions that probabilists working with the type of structures that we are going to be dealing with usually make. A theory of processes that does not assume these conditions has been partially developed but it is at a very early stage and it does not yet have important applications. For more on this subject see $[DM_1]$.

DEFINITION 1.1

(a) An Adapted Probability Space or Stochastic Base is a structure $\underline{\Omega} = (\Omega, (F_t)_{t\in[0,1]}, P)$ where:

(i) P is a complete probability measure on Ω.

(ii) $(F_t)_{t\in[0,1]}$ is an increasing filtration of σ-algebras of P-measurable sets satisfying the Usual Conditions, i.e.:

(1) F_0 is Complete: all the P-null sets are contained in F_0 and

(2) $(F_t)_{t\in[0,1]}$ is Right Continuous : For all $t \in [0,1]$ $F_t = \bigcap_{s>t} F_s$.

(b) Let M be a Polish space (i.e. complete separable metric space). A function $X:\Omega \times [0,1] \to M$ is an M-valued Stochastic Process on Ω if X is measurable with respect to the product measure $P \times \beta$ where β is the Borel measure on $[0,1]$. Sometimes we write $X(t)$ instead of $X(w,t)$ and $X = (X(t))_{t \in [0,1]}$.

(c) Let X and Y be stochastic processes on Ω taking values on the same space M. Then we say:

(i) X is Indistinguishable from Y if for almost all $w \in \Omega$ for all $t \in [0,1]$: $X(w,t) = Y(w,t)$

(ii) X is a Modification of Y if for all $t \in [0,1]$: $X(t) = Y(t)$ a.s.

Now we give the formal definition of adapted probability logic.

DEFINITION 1.2.

(a) The Logical Symbols of L_{ad} are:

(i) Variables:

(1) Countably many time variables: $t_1, t_2, \ldots, s_1, s_2, \ldots$

(2) One Probabilistic variable: w

(ii) Connectives:

(1) First Order connectives: $\neg, \wedge$.

(2) Function connectives : the function symbol $\underline{F}$ for each $n \in \mathbb{N}$ and $F \in C(\mathbb{R}^n)$ = the set of continuous functions from $\mathbb{R}^n$ into $\mathbb{R}$.

(iii) Integral quantifier symbol: $\int$

(iv) Conditional Expectation operator symbol: $E[\,|\,]$.

(b) The Non-Logical Symbols of L_{ad} are stochastic process symbols $(X_i(\ldots): i \in \mathbb{N})$ with two arguments, the first argument for the probabilistic variable and the second for time variables. Most proofs are carried out using just one stochastic process symbol X, but everything is easily seen to work for countably many stochastic processes.

DEFINITION 1.3.

(a) The set of L_{ad}-terms is defined inductively as follows:

(i) For each $r \in \mathbb{Q}^+$, $[X(w,t)\ulcorner r]$ is an atomic term (see remark 1.9).

(ii) Each time variable t is a term.

(iii) Each real $r \in \mathbb{R}$ is a term.

(iv) If τ is a term, so are $\int \tau\, dw$, $\int \tau\, dt$ and $E[\tau | s]\,(w,t)$.

(v) If $\tau_1, \ldots, \tau_n$ are terms and $F \in C(\mathbb{R}^n)$ then $\underline{F}(\tau_1, \ldots, \tau_n)$

is a term.

(b) Free and Bound variables in a term are defined as usual, with the integral $\int \tau dx$ binding x, in $E[\tau|s](w,t)$ all occurrences of s are bound and w and t are free, and in $E[\tau|s](w,s)$ the first occurrence of s is bound and the second is free. Sometimes we shorten this expression to $E[\tau|s]$ or $E[\tau|s](s)$. A closed term is a term with no free variables.

(c) The set of L_{ad}-Formulas is the least set such that

(i) For each L_{ad}-term τ, $[\tau \geq 0]$ is an Atomic formula

(ii) If ϕ is a formula so is $\neg \phi$.

(iii) If ϕ is a countable set of formulas with finitely many free variables then $\wedge \phi$ is a formula. A Sentence is a formula with no free variables.

(d) There is one natural way of extending the definition of L_{ad}-formula given in (c). We can add probability quantifiers $(P\bar{t} \geq r)$ with $r \in \mathbb{Q} \cap [0,1]$ and allow formulas of the form $(P\bar{t} \geq r)\ \phi$ in our language. As for the Logic with Integral Quantifier $L_{\omega_1 \int}$, this addition does not increase the power of L_{ad} (see Section 3.6 of $[K_3]$ for a rigorous proof of this fact) but sometimes it makes things smoother. We will not hesitate to use these quantifiers in situations where their presence helps the reader to have a better understanding.

Remark 3.9 suggests other possible extension of the definition of L_{ad}-formula.

DEFINITION 1.4.

An Adapted Probability Structure (Model) for L_{ad} is a structure $A = (\Omega, (F_t)_{t \in [0,1]}, X, P) = (\underline{\Omega}, X)$ where X is a real valued stochastic process on the adapted probability space $\underline{\Omega}$.

Before we can give the definition of the notion of interpretation of L_{ad} in a model, we need to introduce the main concepts of the general theory of processes.

DEFINITION 1.5.

Let $\underline{\Omega} = (\Omega, (F_t)_{t \in [0,1]}, P)$ be a fixed adapted probability space and X a real valued stochastic process on $\underline{\Omega}$.

(a) X is Right Continuous with Left Limits (r.c.l.l.) if for almost all $w \in \Omega$, the path $X(w,\cdot) : [0,1] \to \mathbb{R}$ is right continuous with left limits.

(b) X is Adapted with respect to $(F_t)_{t \in [0,1]}$ if for all $t \in [0,1]$ the random variable $X(t)$ is F_t-measurable.

(c) X is a Martingale if the following two conditions hold:

(1) For each t, X_t is integrable

(2) For all $s \leq t$, $E(X_t | F_s) = X_s$ a.s.

(d) X is Progressively Measurable with respect to $(F_t)_{t\varepsilon[0,1]}$ if for all $t\varepsilon[0,1]$ the restriction of X to $\Omega \times [0,t]$ is $F_t \times B([0,t])$-measurable, where $B([0,t])$ is the σ-algebra of Borel sets in $[0,t]$. $A \subseteq \Omega \times [0,1]$ is said to be a Progressive set if the indicator function $I_A(w,t)$ is a progressively measurable stochastic process. The Progressive σ-algebra M associated to $(F_t)_{t\varepsilon[0,1]}$ is the σ-algebra on $\Omega \times [0,1]$ generated by the progressive sets.

(e) A random variable $S:\Omega \to [0,1]$ is a Stopping Time with respect to $(F_t)_{t\varepsilon[0,1]}$ if for all $t \varepsilon [0,1]$ $\{w\varepsilon\Omega : S(w) < t\} \varepsilon F_t$. (Stopping times are sometimes called Optional Times).

(f) The σ-algebra O on $\Omega \times [0,1]$ generated by the $P \times \beta$-measurable sets B such that for each $t \varepsilon [0,1]$ $B_t = \{w:(w,t)\varepsilon B\}$ belongs to F_t and $B_t = \bigcap_{s>t} B_s$, is called the Optional σ-algebra with respect to $(F_t)_{t\varepsilon[0,1]}$

(g) X is Optional if it is measurable with respect to O.

(h) The σ-algebra P on $\Omega \times [0,1]$ generated by the sets of the form $A \times \{0\}$ with $A\varepsilon F_0$ and $A \times [s,t]$ with $A\varepsilon \bigcup_{r<s} F_r$ is called the Predictable σ-algebra with respecto to $(F_t)_{t\varepsilon[0,1]}$.

(i) X is Predictable if it is measurable with respect to P.

NOTE: In the probability literature there are other ways of defining the Optional and Predictable σ-algebras. See for example the corresponding sections in [D], [DM_1] and [MP]. The definition of Optional σ-algebra that we have given is taken from [K_3] and we think it is easier to understand for the reader not familiar with stochastic processes.

The following theorem gives an equivalent way of introducing the Optional and Predictable σ-algebras and establishes the relationship between them. A proof can be found in [DM_1].

THEOREM 1.6.

Let $\underline{\Omega} = (\Omega, (F_t)_{t\varepsilon[0,1]}, P)$ be an adapted probability space, then:

(a) The Predictable (Optional) σ-algebra on $\Omega \times [0,1]$ is generated by the real valued continuous (right continuous) (F_t)-adapted stochastic processes on $\underline{\Omega}$.

(b) $P \subseteq \mathcal{O} \subseteq M$. (These inclusions are usually strict).

DEFINITION 1.7.

(Keisler [K_3]): An Interpretation of L_{ad} in an L_{ad} model A is a function that assigns to each L_{ad}-term $\tau(w,t)$ a $P \times \beta^n$ -measurable function (β^n is the Borel measure on $[0,1]^n$) τ^A such that:

(i) If τ is $[X(w,t)\ulcorner r]$ then $\tau^A(w,t)=\begin{cases} r & \text{if } X(w,t) \geq r \\ -r & \text{if } X(w,t) \leq -r \\ X(w,t) & \text{otherwise} \end{cases}$

(ii) If τ is a time variable then $\tau^A(a) = a (a\varepsilon[0,1])$.

(iii) If τ is $\int\theta dw$ then $\tau^A = \int\tau^A dP$.

(iv) If τ is $\int\theta dt$ then $\tau^A = \int\theta^A d\beta$.

(v) For each term $(\vec{t},w,s)$ the following conditions hold:

(a) $(E[\tau(\vec{b},w,s)|s](w,a))^A$ is $\mathcal{O} \times B^n$-measurable, where B^n is the Borel σ-algebra on $[0,1]^n$, n is the length of $\vec{t}$ and $\mathcal{O}$ is the Optional σ-algebra.

(b) For each n-tuple $\bar{b}$ we have:

1- $(E[\tau(\vec{b},w,s)|s](w,a))^A = E[\tau^A(\vec{b},\cdot,\cdot)|\mathcal{O}](w,a)$ $P \times \beta$ a.s.

2- For all $a\varepsilon[0,1]$,
$(E[\tau(\vec{b},w,s)|s](w,a))^A = E[\tau^A(b,\cdot,a)|F_a](w)$ (P a.s.)

LEMMA 1.8. (Keisler [K_3]): Every model A has an interpretation of L_{ad}. For each term $\tau(w,\vec{s})$ and all $\vec{a}$ in $[0,1]$, any two interpretations agree at $\tau(w,\vec{a})$ for P-almost all w. In particular, if w is not free in $\tau(\vec{a})$ then any two interpretations in A agree at $\tau(\vec{a})$ for all $\vec{a}$ in $[0,1]$.

REMARK 1.9. We can see from the definition of interpretation of L_{ad} in a model that the interpretation of each term is a bounded stochastic process. The process X may be unbounded but the atomic terms $[X(w,t)\ulcorner r]$ truncate X at r and consequently all the terms, built from the atomic terms, are bounded.

DEFINITION 1.10. (Keisler [K_3]):

(i) The Logic L_{ad} has the following axiom schemes:

(0) Axiom schemes for two-sorted $L_{A\int}$ (see [K_3])

(1) Axioms for the conditional expectation operator:

a - $E[\tau(u,\vec{t})|u](w,s) = E[\tau(v,\vec{t})|v](w,s)$ where u and v do not occur in $\vec{t}$.

b - $\int E[\sigma|t](w,t) \cdot \tau dw = \int E[\sigma|t](w,t) \cdot E[\tau|t](w,t)dw$

(2) Axioms for the filtration and stochastic processes:

a - $\int \underline{F}(t)dt = r$. For any $F:[0,1] \to \mathbb{R}$ continuous with $\int_0^1 F(x)dx = r$

b - $s \le t \to E[\tau|s] = E[E[\tau|s]|t]$.

c - $\bigwedge_m \bigvee_n \int\int\int |\tau(w,s) - \tau(w,t)| \cdot \max(0,1-|s-t|\cdot n)\,ds\,dt\,dw$

$$\le \frac{1}{m\cdot n}$$

(ii) L_{ad} has the following Rules of Inference:

a - Modus ponens: $\phi, \phi \to \psi \vdash \psi$

b - Conjunction: $\{\phi \to \psi | \psi \in \Gamma\} \vdash \phi \to \bigwedge \Gamma$

c - Generalization:

1-$\phi \to [\tau(w,s,\vec{t}) \ge 0] \quad \vdash \phi \to \int \tau(w,s,\vec{t})dw \ge 0$

where w is not free in ϕ.

2-$\phi \to [\tau(w,s,\vec{t}) \ge 0] \quad \vdash \phi \to \int \tau(w,s,\vec{t})ds \ge 0$.

where s is not free in ϕ.

THEOREM 1.11.

(Rodenhausen's completeness and soundness for L_{ad}). A countable set ϕ of L_{ad}-sentences has a model if and only if ϕ is consistent with L_{ad}.

PROOF. See [K_3] and [K_5] □

To conclude this section we present two notions of elementary equivalence for adapted models. They were introduced by Hoover and Keisler in [HK] in a probabilistic context and their importance for L_{ad} is explained in [K_3]. Here we just present the definitions and some properties without proofs.

DEFINITION 1.12. (Hoover and Keisler [HK]) Let $\mathcal{A}$ and $\mathcal{B}$ be models.

(a) $\mathcal{A}$ is said to be weakly L_{ad}-Equivalent to $\mathcal{B}$, denoted by $\mathcal{A} \overset{w}{\equiv} \mathcal{B}$, if for every L_{ad}-sentence θ: $\mathcal{A} \models \theta$ if and only if $\mathcal{B} \models \theta$.

(b) $\mathcal{A}$ is strongly L_{ad}-Equivalent to $\mathcal{B}$, denoted by $\mathcal{A} \overset{s}{\equiv} \mathcal{B}$, if for each $\vec{a}$ in $[0,1]$ and formula $\phi(\vec{t})$ of L_{ad} in which w is not free: $\mathcal{A} \models \phi[\vec{a}]$ if and only if $\mathcal{B} \models \phi[\vec{a}]$.

THEOREM 1.13 (Hoover and Keisler [HK])

Let $\mathcal{A}$ and $\mathcal{B}$ be models. The following are equivalent:

(i) $\mathcal{A} \overset{w}{\equiv} \mathcal{B}$

(ii) There is a set $T \subseteq [0,1]$ of measure one such that for each $\vec{a}$ in T and formula $\phi(\vec{s})$: $A \models \phi[\vec{a}]$ if and only if $B \models \phi[\vec{a}]$.

(iii) For each L_{ad}-term $\tau(\vec{s})$ with no integrals over time variables $\tau^A(\vec{a}) = \tau^B(\vec{a})$ for almost all $\vec{a}$ in $[0,1]$

<u>THEOREM 1.14.</u> <u>(Hoover and Keisler</u> [HK]):

(a) $A \overset{s}{\equiv} B \Rightarrow A \overset{w}{\equiv} B$

(b) If X and Y are stochastic processes on $\underline{\Omega}$ and Y is a modification of X then $(\underline{\Omega},X) \overset{s}{\equiv} (\underline{\Omega},Y)$.

(c) $A \overset{s}{\equiv} B$ if and only if for each L_{ad}-term $\tau(\vec{s})$ with no integrals over time variables and all $\vec{a}$ in $[0,1]$, $\tau^A(\vec{a}) = \tau^B(\vec{a})$.

(d) If $T \subseteq [0,1]$ has measure one and $A \overset{s}{\equiv} B$ with time parameters restricted to T then $A \overset{w}{\equiv} B$.

§2. <u>A COMPLETENESS THEOREM FOR STOPPING TIMES</u>. "Just as the seemingly trivial definition of the derivative contains in germ all of the calculus, and it its discovery may have involved as much genius as the whole development that followed it, the seemingly trivial notion of stopping time (due to Doob) is the cornerstone of the "General Theory of Processes" ". With this statement, Dellacherie and Meyer in [DM_1] (page 115) initiate the study of stopping times (sometimes called optional times). Thus it seems appropriate that in this paper where we intend to study the general theory of processes using adapted probability logic, we dedicate some few lines to this basic concept. This is precisely the content of this section. The definition of stopping time (def 1.5.d) is very simple and it immediately suggests a natural set of axioms. We first have to observe that stopping times are random variables defined on Ω and not stochastic processes on $\Omega \times [0,1]$. We do not work with $L_{\mathbb{A}\int}$, the probability logic designed for the study of random variables, because stopping times take into account the time structure of the adapted probability space. Instead of artificially using a stochastic process symbol $X(w,t)$ and giving an axiom that characterizes those stochastic processes that are time independent (i.e. random variables), we take as primitive symbol the random variable symbol $S(w)$. The definition of L_{ad}-terms (definition 1.3) is modified in the obvious way with $[S(w)\upharpoonright r]$ in place of $[X(w,t)\upharpoonright r]$. All the results stated in section 1 remain valid with this modification. Let us first introduce some notation so that we can follow the usual probabilistic practice.

NOTATION.

(a) From now on we are going to shorten the expression $[X(w,t)\lceil r]$ and $[S(w)\lceil r]$ to $X_r(t)$ and S_r respectively.

(b) Given L_{ad}-terms ϕ and ψ we write $\phi = \psi$ a.s. instead of the L_{ad}-formula $\int|\phi - \psi|dw \leq 0$ and $\psi \leq \phi$ a.s. instead of $\int(\psi - \min(\phi,\psi))dw \leq 0$.

DEFINITION 2.1. A model $A = (\Omega,S) = (\Omega,S,(F_t)_{t\in[0,1]},P)$ is said to be a Stopping Time model if S is a stopping time with respect to $(F_t)_{t\in[0,1]}$.

DEFINITION 2.2. Let ST be the following set of axioms in L_{ad}:

(ST_1) Axioms for L_{ad} (see definiton 1.9)

(ST_2) $\bigwedge_{r\in\mathbb{Q}^+} (0 \leq S_r \leq 1 \text{ a.s.})$

(ST_3) $\bigwedge_{r\in\mathbb{Q}^+} (Pt \geq 1)(E[\min(S_r,t)|t](t) = \min(S_r,t)\text{a.s.})$

(Intuitively, for almost all t the random variable $S_r\wedge t$ is F_t-measurable).

THEOREM 2.3 (Soundness) Let $A = (\Omega,S)$ be a stopping time model then $A \models$ ST.

PROOF. It follows trivially from the definition of stopping time. □

THEOREM 2.4. (Completeness) Let ϕ be a countable set of sentences in L_{ad}. If $\phi \cup$ST is L_{ad}-consistent, then ϕ has a stopping time model.

PROOF. If we go back to Keisler's proof of the completeness theorem for L_{ad}(see [K_3] and [K_5]), it is easy to see that by adjoining axioms (ST_2) and (ST_3) to (ST_1) we can get a model $A = (\Omega,S)$ of ϕ having the properties written in parentheses. This model is "almost" the one that we want but we have to make sure that these properties hold everywhere and not only almost surely. We proceed as follows: From (ST_2) we can find $U \subseteq \Omega$ of measure one such that

(1) For all $w \in U$ and all $r\in\mathbb{Q}^+$, $S_r(w) \in [0,1]$.

If we recall that for all $w\in\Omega$ $S(w)= \lim_{r\to\infty} S_r(w)$ then using (1) the function $S'\colon \Omega \to 0,1$ defined by

$$S'(w) = \begin{cases} S(w) & \text{if } w \in U \\ 1 & \text{if } w \notin U \end{cases}$$

has the following properties:

(2) $\begin{cases} \text{(i)} & S' \text{ is measurable (the probability space is complete)} \\ \text{(ii)} & S' = S \text{ a.s.} \end{cases}$

<u>CLAIM.</u> S' is a stopping time and $A' = (\Omega, S')$ is a model of ϕ: In order to prove this we just have to check that for all $t \in [0,1]$ the random variable $S' \wedge t$ is F_t-measurable. By (ST_3) we know that there exists $T \subseteq [0,1]$ of measure one such that:

(3) For all $t \in T, S \wedge t$ is F_t-measurable.

We can strengthen (3) to

(4) For all $t \in [0,1]$, $S \wedge t$ is F_t-measurable.

This is easy to see: First observe that we can assume $0 \in T$, since the function $S \wedge 0$ is trivially F_0-measurable. Now let's consider $t \in [0,1] \setminus T$. Since $[0,1] \setminus T$ has measure zero, for each $n \in \mathbb{N}$ we can find $m_n \in T$ such that

(5) $\begin{cases} \text{(i)} & \text{for each } n \in \mathbb{N}\ m_n < m_{n+1} < t \text{ and} \\ \text{(ii)} & \lim_{n \to \infty} m_n = t \end{cases}$

For each $w \in \Omega$ we have $S(w) \wedge t = \lim_{n \to \infty} S(w) \wedge m_n$. By (3) we know that for each n the function $S \wedge m_n$ is F_{m_n}-measurable and by (5.i) we have that for each n, $S \wedge m_n$ is F_t-measurable; therefore $S \wedge t$ is also F_t-measurable as we wanted to show.

Finally, given the fact that the filtration $(F_t)_{t \in [0,1]}$ satisfies the usual conditions, we can conclude that (4) holds with S' in place of S. This proves that S' is a stopping time and by (2.ii.) the model A' is weakly-elementarily equivalent to and so $A' \models \phi$. □

§3. <u>COMPLETENESS THEOREM FOR STOCHASTIC PROCESSES.</u> Given a theory T in L_{ad}, when does T have an adapted model $A = (\underline{\Omega}, X)$ with X a predicctable process on $\underline{\Omega}$? In this section (theorem 3.12) we answer this question; in other words, we prove a completeness theorem for predictable processes. Similar questions can be asked by replacing predictable by optional or progressively measurable, and corresponding completeness theorems can be proved. The main difficulty in finding correct axiomatization of these notions arises from the fact that the notions of predictable, optional and progressive processes are given in terms of measurability conditions with respect to σ-algebras on the product space $\Omega \times [0,1]$ but the conditional expectation operator for L_{ad} only "talks" about σ-algebras defined on Ω. Therefore there are no "obvious" ways of axiomatizing these notions in L_{ad}. Using tools from Probability theory and the model theory of adapted probability logic we can find the correct axioms and prove the completeness theorems. To conclude this section we present a completeness theorem for martingales. In handling this notion it is, in principle, easy to write down the "obvious" axioms but we immediately realize that a universal quantifier over time variables is needed. (See remark 3.9 for some comments about this problem); nevertheless, we are able to avoid this obstacle by making use of a simple probabilistic construction. Here are the details.

DEFINITION 3.1. (Skorokhod [Sk]): Let $\underline{\Omega}$ be an adapted probability structure and X a stochastic process taking values in a Polish space $(M,B(M))$ where $B(M)$ is the Borel σ-algebra on M. S is Countably Generated in $(M,B(M))$ if $\sigma(\{X(t): t \in [0,1]\})$ is contained in the completion under P of a countably generated σ-algebra.

Now we state the main results from probability theory that we use in this section.

THEOREM 3.2. (Skorokhod [Sk]) Let $X = (X(t))_{t\ [0,1]}$ be a countably generated M-valued stochastic process. The following are equivalent:

(a) X has a predictable modification (i.e. there exists Y on $\underline{\Omega}$ predictable such that for all t, $Y(t) = X(t)$ a.s.)

(b) For all $t \in [0,1]$, $X(t)$ is F_{t^-} measurable, where $F_{t-} = \sigma(\bigcup_{s<t} F_s)$.

THEOREM 3.3. (Dellacherie and Meyer [DM_2]): If X is a measurable adapted stochastic process on $\underline{\Omega}$ then there exists an optional process Y on $\underline{\Omega}$ such that for all $t \in [0,1]$, $X(t) = Y(t)$ a.s. (i.e. X has an optional modification).

REMARK 3.4. The obvious question is: are there enough countably generated stochastic processes? In an earlier version of this paper we gave an argument showing that the optional processes are countably generated. The referee has pointed out that it is easy to prove that any measurable process is countably generated.

The importance of Skorkhod's theorem 3.2 for us is that it gives a a characterization of "predictable modification" in terms of a measurability condition with respect to the σ-algebras $(F_{t^-})_{t\in[0,1]}$. The following theorem is going to show us a natural way of expressing this condition in the language of adapted probability logic.

THEOREM 3.5. Let $\underline{\Omega} = (\Omega, (F_t)_{t\in[0,1]} P)$ be an adapted probability structure and $X = (X(t))_{t\in[0,1]}$ an integrable stochastic process on $\underline{\Omega}$ (i.e. for all $t\in[0,1]$, $\int X(w,t)dP < \infty$). The following are equivalent:

(a) $X(t)$ is F_{t-}-measurable

(b) $\bigwedge_m \bigvee_n \bigwedge_{p\geq n} |E(X(t)|F_{t-1/p}) - X(t)| < 1/m$ a.s.

PROOF. Condition (b) is just a "suggestive" way of writing:

(1) $$E[X(t)|F_{t-1/n}] \xrightarrow[n\to\infty]{} X(t) \quad \text{a.s.}$$

For each $n \in \mathbb{N}$ let:

$$F_{(n)} = \begin{cases} F_{t-1/n} & \text{if } t \geq 1/n \\ F_0 & \text{otherwise} \end{cases} \quad \text{and } X_{(n)} = E[X(t)|F_{(n)}]$$

Clearly, the sequence $(X_{(n)})_{\overrightarrow{n\in\mathbb{N}}}$ is a discrete martingale with respect to the increasing family of σ-algebras $(F_{(n)})_{n\in\mathbb{N}}$. Since $X(t)$ is integrable, we can apply a well known martingale convergence theorem due to Levy (see for example [DM_2] Thm. 31, page 28). With our notation the theorem says:

(3) $$X_{(n)} = E[X(t)|F_{(n)}] \underset{n\to\infty}{} E[X(t)|\sigma(\bigcup_{n\in\mathbb{N}} F_{(n)})] \quad \text{a.s.}$$

Using this fact our theorem can be easily proved as we now indicate.

Suppose $X(t)$ is F_{t-}-measurable. This by definition, means:

$E[X(t)|F_{t-}] = X(t)$ a.s. with $F_{t-} = \sigma(\bigcup_{s<t} F_s) = \sigma(\bigcup_{n\in\mathbb{N}} F_{(n)})$,

therefore by (3) and (a) we have

$X_{(n)} = E[X(t)|F_{(n)}] \xrightarrow[n\to\infty]{} X(t)$ a.s. and this is (b) (i.e. (1)).

Suppose (b) holds, then $X_{(n)} \to X_{(t)}$ a.s. and by (3) we can conclude $E[X(t)|F_t] = X(t)$ a.s. □

NOTE. Observe the importance of the integrability condition on the stochastic process $X = (X(t))_{t\in[0,1]}$ in the previous theorem. In particular, if X is bounded then it is integrable.

<u>DEFINITION 3.6.</u> A model $A = (\underline{\Omega}, X)$ is said to be a <u>Predictable Model</u> if the stochastic process X is predictable with respect to $(F_t)_{t\varepsilon[0,1]}$. Similarly we can define adapted, progressively measurable and optional models. By Theorem 1.6, every predictable model is optional, every optional model is progressively measurable and every progressively measurable model is adapted.

LEMMA 3.7. Let $\underline{\Omega} = (\Omega, (F_t)_{t\varepsilon[0,1]}, P)$ be an adapted probability structure.

(a) If $A = (\underline{\Omega}, X)$ is an adapted model then there exists an optional stochastic process Y on $\underline{\Omega}$ such that the optional model $B = (\underline{\Omega}, Y)$ is strongly L_{ad}-equivalent to A.

(b) If $X(t)$ is F_{t-}-measurable then $X(t)$ is F_t-measurable.

<u>PROOF.</u>

(a) Recall that in our definition of stochastic process (def 1.b) we only deal with measurable stochastic processes. Then if X is adapted, Theorem 3.3 provides an optional modification Y of X and Thm 1.14.b shows that $B = (\underline{\Omega}, Y) \overset{s}{\equiv} A = (\underline{\Omega}, X)$.

(b) Trivial. By definition, $F_{t-} \subseteq F_t$.

We can now present the first completeness theorem. It basically says: "If X is adapted this is enough in order to get an optional model". At first sight it may seem strange that from an adapted model we can go to an optional model (compare the definitions), but the force behind this is Theorem 3.3 which is a non-trivial probabilistic result.

<u>THEOREM 3.8.</u> (<u>Soundness and Completeness for Adapted, Progressively Measurable and Optional Processes</u>): Let T be a countable set of sentences in L_{ad}.

(a) If T has an adapted model then

$$T \cup \{ \bigwedge_{r\varepsilon\mathbb{Q}^+} (Pt \geq 1)(E[X_r(t)|t](t) = X_r(t) \text{ a.s.})$$

is consistent in L_{ad}.

(b) If

$$T \cup \{ \bigwedge_{r\varepsilon\mathbb{Q}^+} (Pt \geq 1)(E[X_r(t)|t](t) = X_r(t) \text{ a.s.})$$

is consistent in L_{ad} then T has an Optional model.

<u>PROOF.</u>

(a) Immediate from the definition of adapted model.

(b) If we add the axiom

$$T \cup \{ \bigwedge_{r \in \mathbb{Q}^+} (Pt \geq 1)(E[X_r(t)|t](t) = X_r(t) \quad a.s.)$$

to the list of axioms for L_{ad} (definition 1.9), Keisler's proof of the completeness theorem for L_{ad} (see [K_3] and [K_5]) can be naturally extended in order to get a model $A = (\underline{\Omega}, X)$ of T, such that:

(1) For each $r \in \mathbb{Q}^+$, for almost all $t \in [0,1]$, $X_r(t)$ is F_t-measurable. From (1) we can find $U \subseteq [0,1]$ of measure one such that:

(2) For each r $\mathbb{Q}^+$ for each t U, $X_r(t)$ is F_t-measurable.

Let $X' = (X'(t))_{t \in [0,1]}$ be defined as follows: If $t \in [0,1]$ and $w \in \Omega$ then let $X'(w,t) = \lim_{r \to \infty} E[X_r(t)|F_t](w)$ if the limit exists and 0 otherwise.

<u>CLAIM.</u> The model $A' = (\underline{\Omega}, X')$ is an adapted model of T: It is easy to see that X' is adapted since for each $r \in \mathbb{Q}^+$ $E[X_r(t)|F_t]$ is F_t-measurable and this property is preserved by passing to the limit.

By the definition of the truncated process $(X_r(t))_{t \in [0,1]}$ we know that for each $w \in \Omega$ and each $t \in [0,1]$:

(3) $$\lim_{r \to \infty} X_r(w,t) = X(w,t).$$

Then by (2) and (3) for all $t \in U$ and almost all $w \in \Omega$:

$$X'(w,t) = \lim_{r \to \infty} E[X_r(t)|F_t](w) = \lim_{r \to \infty} X_r(w,t) = X(w,t)$$

Therefore by proposition 1.13.d. $A' \overset{w}{\equiv} A$ and so $A' \models T$. Finally by lemma 3.7.a we can find an optional modification X'' of X' and the model $A'' = (\underline{\Omega}, X'')$ is such that $A'' \overset{s}{\equiv} A'$ and consequently A'' is an optional model of T as we want to show. □

<u>REMARK 3.9.</u> In the above proof we did not immediately get an adapted model of T due to the fact that with the integral quantifier we can only make sure that a given property holds for almost all $t \in [0,1]$ but not necessarily for all $t \in [0,1]$. A research problem proposed in Keisler [K_3] is to develop adapted probability logic with the addition of the universal quantifier ranging over time variables. If this were done, the argument of the previous theorem could be simplified and many properties of stochastic processes that hold for "all times" could be naturally expressed in this extension of adapted probability logic (see sec. 4.4 in [K_3]).

DEFINITION 3.10. Let P be the following set of axioms in L_{ad}.

(P_1) Axioms for L_{ad} (see definition 1.9)

(P_2)

$$(Pt\geq 1)\left(\bigwedge_{r\in\mathbb{Q}^+}\ \bigwedge_{m\in\mathbb{N}}\ \bigvee_{n\in\mathbb{N}}\ \bigwedge_{p\geq n}(Pu\geq 1)(t-1/p<u<t\rightarrow|E[X_r(t)|u](u)-X_r(t)<1/m \text{ a.s.})\right.$$

(intuitively, for almost all t, $X_r(t)$ is F_{t^-}-measurable).

THEOREM 3.11. (Soundness of P) The axioms of P are true in every predictable model.

PROOF. It follows from the soundness theorem for L_{ad} and theorem 3.2 and 3.4. □

We are now ready for our main completeness theorem.

THEOREM 3.12. (Completeness Theorem for Predictable Processes). Let T be a countable set of sentences in L_{ad}. If $T\cup P$ is consistent in L_{ad} then T has a predictable model.

PROOF. Let's first make an observation about $L_{\omega_1\int}$ (Logic with integral quantifier). Consider structures for $L_{\omega_1\int}$ of the form $A = (\Omega,(Y_n)_{n\in\mathbb{N}},Y,P)$ where the Y_n's and Y are unary random variables and add to the axioms of $L_{\omega_1\int}$ the axiom $\bigwedge_m \bigvee_n \bigwedge_{p\geq n} |Y_p - Y| < 1/m$ a.s.

Then Keisler's completeness theorem for $L_{\omega_1\int}$(see [K_3]) can be extended to obtain models A as above where $Y_n \xrightarrow[n\to\infty]{} Y$ a.s. in Ω.

The idea here is to introduce countably many new stochastic process symbols $(X^n)_{n\in\mathbb{N}}$ that are going to represent the processes $(E[X(t)|F_{t-1/n}])_{t\in[0,1],n\in\mathbb{N}}$. So, for each n let $X_r^n = (X_r^n(t))_{t\in[0,1]}$ be defined by

$$(1)\quad X_r^n(w,t) = \begin{cases} E[X_r(t)|F_{t-1/n}](w) & \text{if } t\geq 1/n \\ E[X_r(t)|F_0](w) & \text{if } t\leq 1/n \end{cases}$$

As we noticed in Theorem 3.5, the expression

$\bigwedge_{m\in\mathbb{N}}\ \bigwedge_{n\in\mathbb{N}}\ \bigwedge_{p\geq n} |E(X_r(t)|F_{t-1/p}) - X_r(t)| < 1/m$ a.s. just means

$X^n(t)\xrightarrow[n\to\infty]{} X_r(t)$ a.s.. Axiom (P_2) is the formal way of expressing this (for almost all t) in the language of L_{ad}. Therefore any model in which axiom (P_2) holds is such that:

(2) For all $r\in\mathbb{Q}^+$ for almost all $t\in[0,1]$ the random variable $X_r(t)$ is F_{t-}-measurable.

By the above remarks it is clear that we can get a model $A = (\underline{\Omega},X)$ of T such that (2) is true of X. Using a similar argument to the one given in the proof of theorem 8.b. we can find a process $X' = X'(t))_{t\in[0,1]}$ on $\underline{\Omega}$ such that:

(3) For all $t\in[0,1]$ the random variable $X'(t)$ is F_{t-}-measurable and

(4) $A' = (\underline{\Omega},X') \overset{w}{\equiv} A.$

By lemma 3.7 and theorem 3.3 we can find an optional modification $X'' = (X''(t))_{t\in[0,1]}$ of X' such that

(5) $A'' = (\underline{\Omega},X'') \overset{s}{\equiv} A'$

By the characterization (Theorem 1.13.c) of strong equivalence and (P_2) property (3) also holds in A''. Then by theorems 3.2 and 3.4 we can obtain a predictable modification $X''' = (X'''(t))_{t\in[0,1]}$ of X" such that the model $A''' = (\underline{\Omega},X''') \overset{s}{\equiv} A''$. Thus A''' is the desired predictable model of $T\cup P$. □

If we take a close look at the above proof we will see that the theory T plays a "cosmetic" role because its importance is only felt in Keisler's proof of the completeness theorem, for L_{ad} and we are just quoting his theorem. We can isolate the main model theoretic feature with the following

<u>THEOREM 3.13.</u> If $A \models$ "for almost all t, X_t is F_{t-}-measurable," then there is a predictable model B such that $A \overset{w}{\equiv} B$. □

Let's remark that all the other completeness theorems from this section can also be put in this form. We leave the details to the interested reader. The two previous theorems are stated in a form that is familiar to logicians but we can use the same proof in order to present some results that are along the lines of the probabilistic work of Hoover and Keisler in [HK]. We refer the reader to that paper where all the undefined notions that appear in the following theorem can be found.

<u>THEOREM 3.14.</u> Let X and Y be stochastic processes (on perhaps different adapted spaces) and suppose X is predictable, then

(a) If X and Y have the same adapted distribution ($X \overset{s}{\equiv} Y$ in our language) then Y has a predictable modification (but Y itself is not necessarily predictable).

(b) We can find a predictable stochastic process Z on the adapted Loeb space such that $X \overset{s}{\equiv} Z$.

PROOF.

(a) Just read carefully the proof of Theorem 3.12.

(b) Use (a) and the saturation property of the adapted Loeb space. □

Finally we study the notion of martingale which is of central importance in probability theory and crucial for the development of the general theory of processes. A thorough exposition of the theory of discrete and continuous time martingales can be found in [DM_2] . We are going to prove a completeness theorem that gives us martingales with time parameter [0,1) instead of [0,1]. The problem arises because 1 is an end-point on the right. After the proof of the theorem we indicate how to take care of this problem.

DEFINITION 3.15. Let M be the following set of axioms in L_{ad}:

(M_1) Axioms for L_{ad} (see Definition 1.9).

(M_2) $(Pt \geq 1) \bigwedge_n \bigvee_m \bigwedge_{q \geq p \geq m} (\int |X_q(t)-X_p(t)| dw < 1/n)$

(M_3) $\bigwedge_{r \in \mathbb{Q}^+} (Pst \geq 1)(s \leq t \rightarrow (E[X_r(t)|s](s) = X_r(s)\text{a.s.}))$

Intuitively, axiom (M_2) says that for almost all t the random variable X(t) is integrable.

THEOREM 3.16. Let T be a countable set of sentences in L_{ad}. If $T \cup M$ is consistent in L_{ad} then T has a martingale model.

PROOF. As in previous theorems the first step is easy: use M to extend Keisler's proof of the completeness theorem for L_{ad} in order to get a model $A = \langle \Omega, (X_t)_{t\in[0,1]}, (F_t)_{t\in[0,1]}, P\rangle$ of T such that:

(1) For almost all t, X_t is integrable

(2) For almost all $(t,s) \in [0,1) \times [0,1)$, if $s \leq t$ then $E(X_t|F_s) = X_s$ a.s.

We now show how to find a martingale $(Y_t)_{t\in[0,1)}$ living on $(\Omega, (F_t)_{t\in[0,1]}, P)$ such that $A \overset{w}{\equiv} B = \langle \Omega, (Y_t)_{t\in[0,1]}, (F_t)_{t\in[0,1]}, P\rangle$.

From (1) and (2) we can get $G \subseteq [0,1)$ of measure one such that:

(3) For all $t \in G$, X_t is integrable

(4) For all $t \in G$ the following holds in A:

$$(P_s \geq 1)(s \leq t \rightarrow E(X_t | F_s) = X_s \text{ a.s.}$$

It is easy to verify that $(X_t)_{t\in G}$ is a martingale. (3) gives us the integrability condition. So we just have to check:

For all $s,t\in G(s \leq t \rightarrow E(X_t|F_s) = X_s$ a.s.)

In order to see this let $s, t\in G$ with $s \leq t$. Using (4) we can find an $r\in G$ such that $s \leq t < r$ and

(5) $E(X_r|F_s) = X_s$ a.s.

(6) $E(X_r|F_t) = X_t$ a.s.

Thus we have:

$$E(X_t|F_s) \underset{a.s.}{=} E(E(X_r|F_t)|F_s) \text{ by (6),}$$

$$\underset{a.s.}{=} E(X_r|F_s) \text{ since } F_s \leq F_t,$$

$$\underset{a.s.}{=} X_s \text{ by (5).}$$

Now the definition of $(Y_t)_{t\in[0,1)}$. If $t \notin G$ find $s_t \in G$ such that $s_t > t$. Let $t\in[0,1)$

$$Y_t = \begin{cases} X_t & \text{if } t \in G \\ E(X_{s_t}|F_t) & \text{if } t \notin G \end{cases}$$

It is a simple exercise to show that $(Y_t)_{t\in[0,1)}$ is a martingale and we leave it to the reader. Finally observe that for all $t \in G$, $X_t = Y_t$ and since G has measure one theorem 1.14.d. tells us that $A \overset{w}{\equiv} B$ as we wanted to show. □

Let's indicate what to do if we want to have a martingale with $t\in[0,1]$. The idea is that in order to ensure the transition from $[0,1)$ to $[0,1]$ we have to have a uniformly integrable martingale.

Fortunately in L_{ad} we can express the fact that on a set $G \subseteq [0,1)$ of measure one $(X_t)_{t\in G}$ is uniformly integrable. We do it by adding the following axiom to the list of axioms of the definition 3.15:

$$(M_4) \quad \bigwedge_n \bigvee_m (Pt \geq 1)(\bigwedge_{q\geq m} (\int |X_q(t) - X_m(t)| dw < 1/n))$$

Once this axioms holds in our model (see above proof) we can choose from G an increasing sequence $(d_n)_{n\in\mathbb{N}}$ converging to 1 and define

$Y_1 = \lim_{n\to\infty} X_{d_n}$. It is now easy to show that $(Y_t)_{t\varepsilon[0,1]}$ is a martingale and we leave the details to the reader.

REFERENCES

[An] Anderson, R.M., A non-standard Representation of Brownian Motion and Itô Integration, Israel J.Math 25(1976), pp. 15-46.

[As] Ash, R., Real Analysis and Probability, Academic Press (1972)

[C] Cutland, N., On the Existence of Strong Solutions to Stochastic Differential Equations on Loeb Spaces, Z.W.V.G., 60(1982), pp. 335-357.

[D] Dellacherie, C., Capacites et Processus Stochastiques, Springer-Verlag, (1972).

[DM_1] Dellacherie, C. and Meyer, P.A., Probabilities and Potential, Vol. I, North-Holland, 1978.

[DM_2] __________, Probabilités et Potentiel, Vol. II, Hermann, 1980.

[E] Elliot, R., Stochastic Calculus and its Applications, Springer-Verlag, (1982).

[F] Fajardo, S., Probability Logic with Conditional Expectation in: Annals of pure and applied Logics (To appear).

[H_1] Hoover, D., Probability Logic, Ph.D. Thesis,Univ. of Wisconsin (1978)

[H_2] __________, Probability Logic, Ann.Math.Logic 14 (1978),pp. 287-313.

[H_3] __________, A Normal Form Theorem for $L_{w,p}$ with Applications, J.S.L. 47(1982), pp. 605-624.

[H_4] __________, A Probabilistic Interpolation Theorem. (To appear)

[HK] Hoover, D. and Keisler, H.J., Adapted Probability Distributions. T.A.M.S. (To appear).

[HP] Hoover, D. and Perkins, E., Nonstandard Construction of the Stochastic Integral and Applications to SDE's. I and II, T.A.M.S. 275(1983) ,pp. 1-36 and 37-58.

[K_1] Keisler, H.J., An Infinitesimal Approach to Stochastic Analysis. Memoirs Amer.Math.Soc. 297, Vol. 48(1984).

[K_2] Keisler, H.J., Hyperfinite Probability Theory and Probability Logic. Lecture Notes, Univ. of Wisconsin (Unpublished)(1982).

[K_3] __________, Probability Quantifiers, To appear in Abstract Model Theory and Logics of Mathematical Concepts, ed. by J. Barwise and F. Feferman, Springer-Verlag.

[K_4] __________, Hyperfinite Model Theory, pp. 5-110 in Logic Colloquium 76, North-Holland (1977).

[K_5] __________, A Completeness Proof for Adapted Probability Logic. (To appear)

[Li] Lindström, T., Hyperfinite Stochastic Integration, I,II, and III. Math.Scand. 46(1980), pp 265-333.

[Lo] Loeb, P., Conversion from Nonstandard to Standard Measure Space and Applications to Probability Theory, T.A.M.S. 211(1975), pp. 113-122.

[MP] Metiver, M. and Pellaumail, J., Stochastic Integration, Academic Press, 1980.

[Pa] Panetta, L., Hyperreal Probability Spaces: Some Applications of the Loeb Construction, Ph.D. Thesis, Univ. of Wisconsin (1978)

[Pe] Perkins, E., Stochastic Processes and Nonstandard Analysis. LNM 983.

[R] Rodenhausen, H., The Completeness Theorem for Adapted Probability Logic, Ph.D. Thesis, Heidelberg University (1982).

[Sk] Skorokhod, A.V., On Measurability of Stochastic Processes, Theory of Probability and Its Applications, Vol. XXV,(1980), pp.139-141.

I wish to thank my advisor, Professor H.J. Keisler for suggesting the subject of this paper and for his invaluable comments. This research was partially supported by a University of Wisconsin Research Assistantship.

A BARREN EXTENSION

J.M.Henle
Smith College
Northampton, Mass.
U.S.A.

A.R.D. Mathias
Peterhouse
Cambridge, England

W. Hugh Woodin
California Institute of Technology
Pasadena - California
U.S.A.

ABSTRACT. It is shown that provided $\omega \to (\omega)^\omega$, a well-known Boolean extension adds no new sets of ordinals. Under an additional assumption, the same extension preserves all strong partition cardinals. This fact elucidates the role of the hypothesis $V = L[R]$ in the Kechris-Woodin characterization of the axiom of determinacy.

§0. INTRODUCTION.

Let $B = \text{Power}(\omega)/\text{Fin}$ be the quotient of the Boolean algebra of all subsets of ω by the ideal Fin of finite sets, and $P = [\omega]^\omega/\text{Fin}$ the set of non-zero elements of B, with the induced partial ordering. We shall study the Boolean extension that results from using P as a notion of forcing: with a famous theorem [5] about the existence under AC of $(\omega_1,\omega_1{}^*)$ gaps in P in mind, we shall call this the Hausdorff extension.

Our underlying set theory is Zermelo-Fraenkel. We shall be working in contexts where the full axiom of choice is false; at times, we shall use DC, the axiom of dependent choices, or DCR, its weaker form, which states that if Q is a relation on Power(ω) such that $\forall p \exists q\, Q(p,q)$, then there is a map $f:\omega \to \text{Power}(\omega)$ such that $\forall n\, Q(f(n),f(n+1))$. Our notation of Boolean extensions and forcing follows that described in Mathias [15], 3.7 and 3.8. We write $[\kappa]^\lambda$ for the set of subset of κ of order type λ, and follow Supercontinuity [8] in our notation of partition relations.

It has long been known that under DCR the Hausdorff extension adds no new sets of integers: what it does add is a Ramsey ultrafilter on ω. (Cf [15], Theorem 4.2). For a recent discussion under AC of P, see Dordal [3].

Since without AC there are difficulties in the simultaneous choice of representatives of equivalence classes , it will be convenient to take as our forcing conditions infinite subsets p,q of ω, with the understanding that if p and q have finite symmetric difference, written $p \simeq q$, they force the same statements, and that p is a stronger condition than q if $p \setminus q$ is finite.

In section 1 we prove in the theory $ZF + \omega \to (\omega)^{\omega}$ that the Hausdorff extension is barren in the sense that every map in the extension from an ordinal into the ground model lies in the ground model: in particular, no new sets of ordinals are added. We characterize this latter property in terms of the Galvin-Prikry notion [6] of completely Ramsey families.

In section 2, we consider three set-theoretic principles, called LU, LSU and EP. LU and LSU are the weak and strong uniformisation principles discussed in Mathias [16], where it is shown that LSU is equivalent in ZF + DCR to $\omega \to (\omega)^{\omega}$ + LU, and that LSU is true if AD holds and V = L[R] or if ADR holds, or if V is Solovay's model for "all sets of reals are Lebesgue measurable". EP, which we derive in this section from LSU in the theory ZF + DCR, is, to use the topological terminology of Ellentuck [4], the assertion that the intersection of any well-ordered collection of co-meagre sets is co-meagre.

In section 3, we prove in the theory ZF + LU + EP that the Hausdorff extension preserves every partition property of the form $\kappa \to (\kappa)^{\lambda}\mu$, where $\kappa > \omega$, $2 \le \mu < \kappa$, and $0 < \lambda = \omega\lambda \le \kappa$.
Finally, in section 4, we comment on the implications of the results of section 3 for the recent characterization [12] of AD in L[R] by Kechris and Woodin, and discuss some problems related to our work: reference to this discussion is made in the text by the string [PROB].
Assumptions are given in full in the statements of theorems, but may be omitted in Lemmata and Propositions when the flow of the narrative demans it. The end of a proof is signalled by $\dashv$.

§1. THE BARRENNESS OF THE HAUSDORFF EXTENSION.

THEOREM 1.0. Let M be a transitive model of $ZF + \omega \to (\omega)^{\omega}$ and N its Hausdorff extension. Then M and N have the same sets of ordinals; moreover every sequence in N of elements of M lies in M.

PROOF. It will be sufficient to prove in the theory $ZF + \omega \to (\omega)^{\omega}$ that if $p_0 \Vdash f:\hat{\zeta} \to \hat{V}$, then for some $q_0 \subseteq p_o$, $q_o \Vdash f \in \hat{V}$.
Fix then such p_0, f, ζ and suppose that no such q_0 exists. Then for

each $p \in [p_0]^\omega$ there will be at least one ordinal $\xi < \zeta$ for which no $x \in V$ exists with $p \Vdash f(\hat{\xi}) \equiv \hat{x}$: define $\phi(p)$ to be the least such ξ.

For $p \subseteq \omega$, define $p_\ell \subseteq \omega$, $p_r \subseteq \omega$ by writing $\tilde{q}$ for the monotonic enumeration of $q \subseteq \omega$ and setting $\tilde{p}_\ell(n) = \tilde{p}(2n)$, $\tilde{p}_r(n) = \tilde{p}(2n+1)$.

Now define a partition $\pi:[p_0]^\omega \to 3$ by setting $\pi(p) = 0$, 1 or 2 according as $\phi(p_\ell)$ is less than, equal to, or greater than $\phi(p_r)$, and let $P \in [p_0]^\omega$ be homogeneous for π. Notice that $\pi(P) = 1$: for putting $k = \min P$ and $Q = P\setminus\{k\}$, we have $Q_\ell = P_r$, $Q_r = P_\ell\setminus\{k\}$; P_ℓ and $P_\ell\setminus\{k\}$ force the same statements, and so have the same value under ϕ; thus if $\pi(P) = 0$, $\pi(Q) = 2$; if 2, 0; but $\pi(P) = \pi(Q)$ by homogeneity, so $\pi(P)$ can only equal 1.

Put now $\nu = \phi(P)$. By definition of ϕ there are $q,r \in [P]^\omega$ and $x,y \in V$ with $x \neq y$, $q \Vdash f(\hat{\nu}) \equiv \hat{x}$ and $r \Vdash f(\hat{\nu}) \equiv \hat{y}$. Define $s \in [P]^\omega$ by choosing the first three elements of s from q, the fourth from r, then three from q, then one from r, and so on, so that $s_\ell \subseteq q$, $(s_r)_\ell \subseteq q$, $(s_r)_r \subseteq r$.

Since s_r is compatible with both q and r, and extends P, $\phi(s_r) = \nu$. Since $s_\ell \subseteq q$, $\phi(s_\ell) \geq \phi(q)$; but $\phi(q) > \nu$, since $q \subseteq P$ and q forces the value x for $f(\nu)$. So $\phi(s_\ell) > \phi(s_r)$ and thus, $\pi(s) = 2 \neq 1 = \pi(P) = \pi(s)$, an evident absurdity. ⊣

As in Happy Families [15], for $s \in [\omega]^{<\omega}$, we write $|s|$ for $\sup\{n+1 \mid n \in s\}$, and for $p \in [\omega]^\omega$ with $|s| \leq \cap p$, we write $[s,p]$ for $\{x \in [\omega]^\omega \mid s \subseteq x \subseteq s \cup p\}$. Thus $[0,p] = [p]^\omega$.

DEFINITION 1.1. A subset A of $[\omega]^\omega$ will be called invariant if ($p \in A$ and $p \approx p'$) always implies $p' \in A$; similarly a function F defined on (an invariant subset of) $[\omega]^\omega$ will be called invariant if ($p \in$ domain(F) and $p \approx p'$) always implies ($p' \in$ domain(F) and $F(p) = F(p')$).

DEFINITION 1.2. Following Galvin and Prikry [6] we call a subset P of $[\omega]^\omega$ completely Ramsey, or CR, if for all $\langle s,p\rangle$ there is a $q \in [p]^\omega$ with either $[s,q] \subseteq P$ or $[s,q] \cap P = 0$.

The CR sets are chosen in $\underset{=}{C}_H$ in the notation of [15], 1.3. The statement $\omega \to (\omega)^\omega$ is equivalent to the assertion that all subsets of $[\omega]^\omega$ are completely Ramsey.

DEFINITION 1.3. We shall call a subset P of $[\omega]^\omega$ CR^+ if for all $\langle s,p\rangle$ there is a $q \in [p]^\omega$ with $[s,q] \subseteq P$, and CR^- if for all $\langle s,p\rangle$ there is a $q \in [p]^\omega$ with $[s,q] \cap P = 0$.

The CR^- sets are those in $\underset{=}{I}_H$ in [15], 1.4, and are the meagre sets in the topology defined by Ellentuck in [4].

1.4 In a manner familiar to readers of [15], we shall want relativized versions of these concepts: we shall for example say that P is CR^+ on $[s,p]$ if for all $<t,q>$ with $s \subseteq t$ and $(t\backslash s) \cup q \subseteq p$, there is an r in $[q]^\omega$ with $[t,r] \subseteq P$.

We are now in a position to reformulate a weaker form of the conclusion of 1.0 as

PROPOSITION 1.5 (ZF) The following are equivalent:

(1.6) The Hausdorff extension adds no new sets of ordinals.

(1.7) Let ζ be an ordinal and $<C_\nu \mid \nu < \zeta>$ a sequence of invariant CR subsets of $[\omega]^\omega$. Then there is a $p \in [\omega]^\omega$ such that for each $\nu < \zeta$, C_ν is, relative to $[p]^\omega$, either CR^+ or CR^-.

PROOF. (1.6) → (1.7) : Fix $<C_\nu \mid \nu < \zeta>$ and let G be the generic filter in Power(ω) added by the Hausdorff extension. In $V[G]$ put

$$B = \{\nu < \zeta \mid p \in G(C_\nu, \text{ relative to } [p]^\omega, \text{ is } CR^+)\}$$

By (1.6), there is an $A \subseteq \zeta$, and a q such that $q \Vdash B \equiv \hat{A}$. We shall show that for all ν in A, C_ν is CR^+ relative to $[q]^\omega$, and that for all ν not in A, C_ν is CR^- relative to $[q]^\omega$, by proving the contrapositives of these statements.

Let $\nu < \zeta$. If C_ν is not CR^+ relative to $[q]^\omega$, then since C_ν is CR there is $[s,r] \subseteq [0,q]$ with $[s,r] \cap C_\nu$ empty, so by the invariance of C_ν, $[r]^\omega \cap C_\nu$ is empty, and so $r \Vdash \hat{\nu} \notin B$; hence $q \nVdash \hat{\nu} \in B$, so $q \nVdash \hat{\nu} \in \hat{A}$, and thus $\nu \notin A$.

If C_ν is not CR^- relative to $[q]^\omega$, there is $[s,r] \subseteq [0,q]$ with $[s,r] \subseteq C_\nu$, so by invariance $[r]^\omega \subseteq C_\nu$, and so $r \Vdash \hat{\nu} \in B$; hence $q \nVdash \hat{\nu} \notin B$, so $q \nVdash \hat{\nu} \notin \hat{A}$, and so $\nu \in A$.

(1.7) → (1.6): Suppose $\Vdash B \subseteq \hat{\zeta}$. Set $C_\nu = \{q \mid q \Vdash \hat{\nu} \in B\}$. Then each C_ν is invariant. Notice that C_ν is CR^+ relative to $[p]^\omega$ iff $p \Vdash \hat{\nu} \in B$, and C_ν is CR^- relative to $[p]^\omega$ iff $p \Vdash \hat{\nu} \notin B$: so a p satisfying the conclusion of (1.7) will force $B \in \hat{V}$. ⊣

REMARK 1.8. (1.7) is false without the hypothesis of invariance: for $n \in \omega$, let $C_n = \{p \mid n \leq \inf p\}$.

1.9 The conclusion of Theorem 1.0 may be derived from certain square-bracket partition relations: write $\omega \to [\omega]^\omega \lambda$, where $2 \leq \lambda \leq \omega$, to mean that for any $\psi:[\omega]^{\omega} \to \lambda$ there is an $x \in [\omega]^{\omega}$ such that for some $\nu < \lambda$ and all $y \in [x]^{\omega}$, $\psi(y) \neq \nu$. Kleinberg has observed that from a sequence $<\pi_n \mid 2 \le n < \omega>$ of partitions of $[\omega]^{\omega}$, where for each n, π_n is a counterexample to $\omega \to [\omega]^{\omega}_n$, a counterexample to $\omega \to [\omega]^{\omega}\omega$ may be constructed: thus assuming DC, or at least AC for countable families, $\omega \to [\omega]^{\omega}_{\omega}$ implies that for some $n < \omega$, $\omega \to [\omega]^{\omega} n+2$. It is not known whether $\omega \to (\omega)^{\omega}$ can be derived from any $\omega \to [\omega]^{\omega} n$ for $n \ge 3$ [PROB] but the conclusion of theorem 1.0 may be derived as follows.

Suppose $\omega \to [\omega]^{\omega} n$. Let β be a prime number very much larger than n, and for $p \in [\omega]^{\omega}$, $0 \le i < \beta$, let $(p)_i$ be $\{\tilde{p}(m\beta+i) \mid m \in \omega\}$. Thus if $q = p\setminus\{\tilde{p}(0)\}$, $(q)_i = (p)_{i+1}$, $(q)_{\beta-1} = (p)_0\setminus\{\tilde{p}(0)\}$.

Let p_0, f, ζ, ϕ be as in the proof of Theorem 1.0. Set

$$\psi(p) = \{i \mid i < \beta \text{ and } \forall j<\beta\, \phi((p)_j) \le \phi((p)_i)\}.$$

By (repeated) application of $\omega \to [\omega]^{\omega} n$, a $\bar{p} \in [p_0]^{\omega}$ may be found for which $\psi[[\bar{p}]^{\omega}]$ is of cardinality less than n. But then $\phi((p)_0) = \phi((p)_1) = \ldots = \phi((p)_{\beta-1})$ for any $p \in [\bar{p}]^{\omega}$, since if $\phi((p)_j) < \phi((p)_i)$ for some i,j less than β, by the primality of β and the invariance of ϕ, the β values

$\psi(p)$, $\psi(\tilde{p}\setminus\{p(0)\})$, $\psi(\tilde{p}\setminus\{\tilde{p}(0),\tilde{p}(1)\})$, .. ,

$\psi(p\setminus\{\tilde{p}(0),\tilde{p}(1),\ldots,\tilde{p}(\beta-1)\})$ will be distinct.

Now put $\nu = \phi(\bar{p})$. As before, there are $q, r \in [\bar{p}]^{\omega}$, x, $y \in V$, with $q \Vdash f(\hat{\nu}) \equiv \hat{x}$, $r \Vdash f(\hat{\nu}) \equiv \hat{y}$, and $x \neq y$. Find $s \subseteq q \cup r$ so that $(s)_0 \cap q$ is infinite, $(s)_0 \cap r$ is infinite, $(s)_1 \subseteq q$. Then $\phi((s)_0) = \nu$ while $\phi((s)_1) \ge \phi(q) > \nu$, contradicting the property of $\bar{p}$ established in the previous paragraph. ⊣

<u>REMARK 1.10.</u> It is a theorem of Vopěnka and Balcar [21] that if two transitive models, of which at least one is known to satisfy the axiom of choice, have the same sets of ordinals, they coincide. That they might differ if neither satisfies AC is due to Jech [9]: for an extension of that result, see Monro [17].

2. THE LARGENESS OF THE INTERSECTION OF CR^+ FAMILIES.

DEFINITION 2.0. A function $F:[\omega]^\omega \to \text{Power}(\omega)$ is called strongly continuous on $[s,p]$ if there is a tree $\langle t_u | u \in [p]^{<\omega}\rangle$ of finite subsets of ω such that for all $q \in [p]^\omega$ and all $k \in q$,

$$F(q) \cap (k+1) = t_{q\cap(k+1)}.$$

DEFINITION 2.1. A relation $R \subseteq [\omega]^\omega \times \text{Power}(\omega)$ is (strongly) uniformised on $[s,p]$ if there is some (strongly continuous) function F such that for all $q \in [s,p]$, $R(q,F(q))$.

DEFINITION 2.2. LU (LSU) is the assertion that for any relation R as above, such that $\forall p \exists y\, R(p,y)$, $\{x | R$ is (strongly) uniformised on $[0,x]\}$ is CR^+.

THEOREM 2.3. (ZF + DCR + LSU) Let θ be any ordinal and $\langle C_\nu | \nu < \theta\rangle$ a sequence of CR^+ families. Then $\bigcap_{\nu<\theta} C_\nu$ is CR^+.

PROOF. For θ countable, this is a theorem of ZF + DCR, and is due to Galvin and Prikry [6]. A proof may be given by the methods of Happy Families ([15]; see in particular Proposition 1.10).

Let us therefore suppose the theorem true for all sequences of length less than θ, and that it fails at θ for the sequence $\langle C_\nu | \nu < \theta\rangle$. By the minimality of θ and the first sentence of this proof, θ must be of uncountable cofinality.

Now put $D_\nu =$ df $\{p \in [\omega]^\omega | [p]^\omega \subseteq \bigcap_{\mu<\nu} C_\mu\}$. Then by Proposition 2.8 of Happy Families (for the case $A = [\omega]^\omega$, $B = [\omega]^\omega \setminus (\bigcap_{\mu<\nu} C_\mu)$), D_ν is CR^+ for each $\nu < \theta$, $D_\mu \supseteq D_\nu$ for $\mu < \nu < \theta$, $q \subseteq p \in D_\nu$ implies $q \in D_\nu$, and, by relativizing to some $[s,S]$ if necessary, we may further assume that $\bigcap_{\nu<\theta} D_\nu$ is empty.

Thus we may define $\chi(p)$, for $p \in [\omega]^\omega$, to be the least $\nu < \theta$ with $p \notin D_\nu$. The function $\chi:[\omega]^\omega \to \theta$ will have these properties:

$$q \subseteq p \to \chi(p) \le \chi(q),$$

$$\forall \nu<\theta \ \forall \langle s,q\rangle \ \exists p \in [s,q]\ (\chi(p) > \nu).$$

Define now $\psi(p) = \cup\{\chi(q) | q \approx p\}$ and $\phi(p) = \cap\{\chi(q) | q \approx p\}$. We assert that

$$\forall q \in [\omega]^\omega \exists p \in [q]^\omega (\psi(q) < \phi(p)).$$

To see that, put $\nu = \psi(q)$, construct sequences $n_0 < n_1 < n_2 < \ldots$ and $q = p_0 \supseteq p_1 \supseteq p_2 \supseteq \ldots$ such that for each $i, n_i = \min p_i$, and for each

$s \subseteq n_i+1$, $[s,p_{i+1}] \subseteq D_\nu$, and set $p = \{n_i | i < \omega\}$.

If $p' \approx p$, there is an n_k such that $p' \setminus n_k = p \setminus n_k$, so $p' \in [p' \cap (n_k+1), p_{k+1}]$, so $p' \in D_\nu$, and $\chi(p') > \nu$; consequently $\phi(p) > \nu$.

By LSU, there is a $\overline{p} \in [\omega]^\omega$ and a strongly continuous function $F:[\overline{p}]^\omega \to [\overline{p}]^\omega$ such that

$$\forall p \in [\overline{p}]^\omega (F(p) \subseteq p \quad \text{and} \quad \phi(F(p)) > \psi(p)).$$

Define the relation R on $[\overline{p}]^\omega$ by setting $R(p,q)$ iff $\exists q' \approx q\ \exists p' \approx p (F(p') = q')$: R will be well-founded since if $R(p,q)$,

$$\chi(q) \geq \phi(q) = \phi(q') > \phi(F(p')) > \psi(p') = \psi(p) \geq \chi(p).$$

Since F is continuous, R is Borel, and so by the Kunen-Martin theorem (Moschovakis [18], page 99, Theorem 2G.2; see also page 114, footnote 12), if we define $\rho_R(q) = \cup\{\rho_R(p)+1 | p \in [\overline{p}]^\omega$ and $F(p) = q\}$, then for some η less than ω_1 and all $q \in [\overline{p}]^\omega$, $\rho_R(q) < \eta$.

Notice that for any $\langle s,p \rangle$ with $s \cup p \subseteq \overline{p}$, there is a $q \in [s,p]$ with $\rho_R(q) > \rho_R(s \cup p)$. To see that, take $q = s \cup (F(p) \setminus |s|)$: then $s \cup p \approx p$ and $q \approx F(p)$, so $R(s \cup p, q)$, and so $\rho_R(q) > \rho_R(s \cup p)$.

Thus if for $\zeta < \eta$ we set $E_\zeta = \{p \in [\overline{p}]^\omega | \rho_R(p) = \zeta\}$, each E_ζ will be CR^- relative to $[0,\overline{p}]$. But then since η is countable $\bigcup_{\zeta<\eta} E_\zeta$ is CR^-, relative to $[0,\overline{p}]$; but that is absurd, as $[0,\overline{p}] = \bigcup_{\zeta<\eta} E_\zeta$. ⊣

DEFINITION 2.4. EP is the statement of Theorem 2.3, that the intersection of any well-ordered collection of CR^+ sets if CR^+.

PROPOSITION 2.5 (ZF) EP implies (1.7).

PROOF. If $\langle C_\nu | \nu < \zeta \rangle$ is as in (1.7), let $D_\nu = \{x | \forall s \subseteq \cap x \ ([s,x] \subseteq C_\nu$ or $[s,x] \cap C_\nu = 0\}$. Then each D_ν is CR^+: by EP, $\bigcap_{\nu<\zeta} D_\nu$ is not empty. Let P be a member: then P satisfies the conclusion of (1.7). ⊣

§3. THE PERSISTENCE OF PARTITION PROPERTIES.

PROPOSITION 3.0 (ZF + EP) Let κ be an ordinal and $\Phi:[\omega]^\omega \to [\kappa]^\kappa$ an invariant function. Then Φ is constant on some $[p]^\omega$.

PROOF. For $\nu < \kappa$, put $C_\nu = \{p | \nu \in \Phi(p)\}$. Then each C_ν is invariant, so by 1.7, which by 2.5 follows from EP, there is a $p \in [\omega]^\omega$ such that, relative to $[p]^\omega$, each C_ν is either CR^+ or CR^- : put

$D_\nu = \{q \in [p]^\omega | [q]^\omega \subseteq C_\nu\}$ in the first case and $= \{q \in [p]^\omega | [q]^\omega \cap C_\nu = 0\}$ in the second. Then each D_ν is CR^+ on $[p]^\omega$.

Let p be in the intersection, non-empty by EP, of $\{D_\nu | \nu < \kappa\}$. Then for each $q \in [p]^\omega$ and $\nu < \kappa$, $\nu \in \Phi(q)$ iff $\nu \in \Phi(p)$, since

$$\nu \in \Phi(p) \to p \in C_\nu \to D_\nu \subseteq C_\nu \to q \in C_\nu \to \nu \in \phi(q), \text{ and}$$
$$\nu \notin \Phi(p) \to p \notin C_\nu \to D_\nu \cap C_\nu = 0 \to q \notin C_\nu \to \nu \notin \Phi(q). \dashv$$

<u>REMARK 3.1.</u> The above Proposition may fail if Φ is not required to be invariant : for example, if Φ is an injection.

<u>PROPOSITION 3.2</u> (ZF + LU) Suppose that $0 < \lambda = \omega\lambda \leq \kappa$, $2 \leq \mu < \kappa$, $\kappa \to (\kappa)^\lambda \mu$, that there is a surjection $\psi: [\omega]^\omega \to [\kappa]^\kappa$ and that $\langle \pi_p | p \in [\omega]^\omega \rangle$ is a collection of partitions $\pi_p : [\kappa]^\lambda \to \mu$. Then there is a $p^* \in [\omega]^\omega$ and an invariant function $\Phi: [p^*]^\omega \to [\kappa]^\kappa$ such that for each $p \in [p^*]^\omega$, $\Phi(p)$ is homogeneous for π_p.

<u>PROOF.</u> For each p, define $\rho_p : [\kappa]^\lambda \to \mu$ by

$$\rho_p(x) = \pi_p({}_\omega x),$$

where, as in Supercontinuity [8], ${}_\omega x = \{\bigcup_{n\in\omega} \tilde{x}(\omega\zeta + n) | \zeta < \lambda\}$: note that as $\omega\lambda = \lambda$, ${}_\omega x$ is in $[\kappa]^\lambda$ whenever x is. Set

$$H_p = \{q \in [\omega]^\omega | \psi(q) \text{ is homogeneous for } \rho_p\}.$$

By LU there is a p^* and an $F: [p^*]^\omega \to [\omega]^\omega$ such that for all $q \in [p^*]^\omega$, $\psi(F(q))$ is homogeneous for ρ_p. Write $B(q) = \psi(F(q))$.

For each $q \in [p^*]^\omega$, define $C(q) \in [\kappa]^\kappa$ thus: let $C(q)(0)$ be the least ordinal greater than all $B(q')(0)$ for $q' \approx q$; let $C(q)(\nu)$ be the least ordinal ξ such that setting $\eta = \cup\{C(q)(\nu') | \nu' < \nu\}$, the interval $[\eta, \xi)$ contains at least one element of each $B(q')$ for $q' \approx q$.

The regularity of κ ensures the soundness of this definition. Note that this definition does not rely on an enumeration of any $\{q' | q' \approx q\}$, and therefore that $C(q) = C(q')$ whenever $q \approx q'$: so that C is invariant.

Now put $\Phi(q) = {}_\omega C(q)$. Then Φ is invariant; any $x \in [\Phi(q)]^\lambda$ is of the form ${}_\omega y$ for some $y \in [B(q)]^\lambda$, and hence $\pi_q(x) = \pi_q({}_\omega y) = \rho_q(y)$, which, by the homogeneity of $B(q)$, is independent of y and therefore of x. Hence $\Phi(q)$ is homogeneous for π_q as required. $\dashv$

THEOREM 3.3 (ZF + EP + LU) Suppose $0 < \lambda = \omega\lambda \leq \kappa$, $2 \leq \mu < \kappa, \kappa \to (\kappa)^{\lambda}_{\mu}$, and that there is a surjection $\psi : [\omega]^{\omega} \to\!\!\to [\kappa]^{\kappa}$. Then in the Hausdorff extension, $\kappa \to (\kappa)^{\lambda}\mu$.

PROOF. Note first that by 2.5 no new subsets of κ are added; consequently $[\kappa]^{\lambda}$ is the same whether interpreted in the ground model or in the extension, and thus may be written without ambiguity.

Note secondly that $\kappa \to (\kappa)^{\lambda}\mu{+}1$.

Suppose $p_0 \Vdash f : [\hat{\kappa}]^{\hat{\lambda}} \to \hat{\mu}$. We shall find a $p_2 \in [p_0]^{\omega}$ and an $A \in [\kappa]^{\kappa}$ such that $p_2 \Vdash f$ is constant on $[\hat{A}]^{\hat{\lambda}}$.

For each $p \in [p_0]^{\omega}$, define a partition $\pi_p : [\kappa]^{\lambda} \to \mu{+}1$ by

$$\pi_p(A) = \zeta \quad \text{if} \quad p \Vdash f(\hat{A}) \equiv \hat{\zeta}$$

$$\pi_p(A) = \mu \quad \text{otherwise.}$$

By 3.2, there is a $p_1 \in [p_0]^{\omega}$ and an invariant function $\Phi : [p_1]^{\omega} \to [\kappa]^{\kappa}$ such that for each $p \in [p_1]^{\omega}$, $\Phi(p)$ is homogeneous for π_p. By 3.1 there is some $p_2 \in [p_1]^{\omega}$ with Φ constant on $[p_2]^{\omega}$. Set $A = \Phi(p_2)$. We assert that $p_2 \Vdash \hat{A}$ is homogeneous for f.

If not, there will be $D, E \in [A]^{\lambda}$, $q \in [p_2]^{\omega}$, $\xi < \zeta < \mu$ with $q \Vdash (f(\hat{D}) \equiv \hat{\xi}$ and $f(\hat{E}) \equiv \hat{\zeta})$: so $\pi_q(D) = \xi$, $\pi_q(E) = \zeta$, and thus A is inhomogeneous for π_q. But $A = \Phi(p_2) = \Phi(q)$ which is homogeneous for π_q. ⊣

§4. PARTITION CARDINALS WITHOUT DETERMINACY.

An important ordinal in the study of AD is θ, the least ordinal > 0 onto which Power(ω) cannot be mapped.

LEMMA 4.0. If the Hausdorff extension is barren, then θ is the same whether calculated in the ground model or in the extension.

PROOF. Suppose $p \Vdash f : \widehat{[\omega]^{\omega}} \to\!\!\to \hat{\theta}$. For each pair (q,r) with $q \setminus p$ finite, put $\psi(q,r) = \xi$ if $q \Vdash f(\hat{r}) \equiv \hat{\xi}$, and $\psi(q,r) = 0$ otherwise. Then ψ is a surjection of $[\omega]^{\omega} \times [\omega]^{\omega}$ onto θ, a contradiction. ⊣

PROPOSITION 4.1 (AD + V = L[R]) If $0 < \omega\lambda = \lambda \leq \kappa$, $2 \leq \mu < \kappa$ and $\kappa \to (\kappa)^{\lambda}\mu$, then $\kappa \to (\kappa)^{\lambda}\mu$ in the Hausdorff extension.

PROOF. By a theorem of Kleinberg [14], κ is measurable, so by arguments to be found in [13], $\kappa < \theta$; hence, by a theorem of Moschovakis

([18], 7D.19, page 442), there is a surjection $\psi:[\omega]^{\omega} \to\to [\kappa]^{\kappa}$.

DC is provable in ZF + AD + V = L[R], by Kechris [10]; so is LSU, by Theorem 2.2 of [16]; by Theorem 2.3 above, EP then holds.

Thus all the hypothesis of 3.3 hold, so the conclusion follows. ⊣

THEOREM 4.2 If AD is consistent, so is DC + $\forall\lambda<\theta\ \exists\ \kappa(\lambda \le \kappa < \theta$ and κ is a strong partition cardinal) + "there is a Ramsey ultrafilter on ω".

PROOF. If AD holds, it stays true in L[R], as strategies are reals. Again, by [10], DC holds, so DC holds in the Hausdorff extension. As $\omega \to (\omega)^{\omega}$, (cf Theorem 2.2 of [16]), the extension is barren. By the results of Kechris, Kleinberg, Moschovakis and Woodin [11], there are arbitrarily large strong partition cardinals below θ in the ground model, so by 4.0 and 4.1, the same holds in the extension. A remark in the introduction completes the proof. ⊣

REMARK 4.3. The significance of 4.2 is that as the existence of an ultrafilter on ω implies the failure of AD, the hypothesis V = L[R] is an essential ingredient in the Kechris-Woodin [12] derivation of AD from the existence of arbitrary large strong partition cardinals below θ.

REMARK 4.4. The arguments of section 1 generalize with little change to obtain a new proof of Henle's result [7] that Spector forcing [20] at a strong partition cardinal κ is barren.

REMARK 4.5. If AD holds and V = L[R], then ω_1 and ω_2 are measurable, and $\omega \to (\omega)^{\omega}$, so in the Hausdorff extension, there is a Ramsey ultrafilter on ω, and ω_1 and ω_2 are still measurable: we may say that there are in this model two and a half contiguous measurables.

A model for that statement may also be found assuming something presumably much weaker than Con(AD): by an unpublished result of Woodin, a model of ZFC in which κ is λ-supercompact and $\lambda > \kappa$ is measurable admits a Boolean extension in which $\kappa = \omega_1$, $\lambda = \omega_2$, κ and λ are still measurable, and $\omega \to (\omega)^{\omega}$; in the Hausdorff extension of that model, there will be a Ramsey ultrafilter on ω, while the measures on κ and λ will remain measures as no new subsets of either will be added.

Several open problems are related to [16] and the present work:

PROBLEM 4.6. Can $\omega \to (\omega)^{\omega}$ be deduced from $\omega \to [\omega]^{\omega}n$ for any $n \ge 3$?

PROBLEM 4.7. Does AD imply that ω_1 is huge? Is there a huge measure on $[\omega_2]^{\omega_1}$?

PROBLEM 4.8. Is it a theorem of $ZF + DC + \aleph_1 \not\leq 2^{\aleph_0}$ or of $ZF + DC + \omega \to (\omega)^\omega$ that there are no MAD families coded on ω?

PROBLEM 4.9. Is there a $k \in \omega$ such that it is a theorem of ZF + DC that there cannot exist k contiguous measurables?

By a result of Kechris, if AD is consistent, k must be greater than three. k might be quite small, though: for limitations on contiguous large cardinals, see Apter [1] and Bull [2].

PROBLEM 4.10. Is EP a theorem of $ZF + DC + \omega \to (\omega)^\omega$?

The least θ for which 2.3 fails must be measurable.

PROBLEM 4.11. If κ and λ are strong partition cardinals with $\kappa < \lambda$, will κ remain strong in the Spector extension for λ, or vice versa?

The similarities between strong partition cardinals and ω, when $\omega \to (\omega)^\omega$, which are studied in [8], and the result mentioned in 4.4 above, suggest that Spector forcing should preserve something more than plain measurability. But there are limits: the analogue of 2.3, when ω is replaced by a strong partition cardinal κ, fails for $\theta = \kappa^+$.

PROBLEM 4.12. Does ADR imply that every set of reals is Souslin?

A theorem of Woodin states that ADR is provable in ZF + AD + "every set of reals in Souslin".

PROBLEM 4.13. How strong is the theory $ZF + DC + V = L[R]$ + "θ is a regular limit cardinal"? Does it prove the existence of $\alpha^\#$ for every real α?

The challenge in this question is to get θ a limit cardinal: it is a result of the Cabal that $ZF + DC + V = L[R]$ proves that θ is regular. In [13], Kechris and Woodin show that if $AD + V = L[R]$ holds, then θ cannot be weakly compact.

PROBLEM 4.14. It follows from the results of [10], that Con(ω_1 is a strong partition cardinal) follows from, e.g., Con(there are arbitrarily large strong partition cardinals), but the proof goes via determinacy. Is there a direct proof?

REFERENCES

[1] Apter, A.W. Some results on consecutive large cardinals, Annals of Pure and Applied Logic, 25(1983), 1-17.

[2] Bull, E.L. Successive large cardinals, Annals of Mathematical Logic, 15(1978), 161-191.

[3] Dordal, P.L. Independence Results concerning Combinatorial Properties of the Continuum, Thesis, Harvard, (1982).

[4] Ellentuck, E. A new proof that analytic sets are Ramsey, J. Symbolic Logic 39(1974), 163-165.

[5] Hausdorff, F. Summen von $\aleph_1$ Mengen, Fundamenta Mathematicae, 26(1936), 241-255.

[6] Galvin, F. & Prikry, K. Borel sets and Ramsey's theorem, J.Symbolic Logic 38(1973), 193-198.

[7] Henle, J.M. Spector forcing, J.Symbolic Logic 49(1984), 542-554.

[8] Henle, J.M. & Mathias, A.R.D. Supercontinuity, Mathematical Proc. Cambridge Philosophical Society 92(1982), 1-15.

[9] Jech, T. On models for set theory without AC, Mathematical Reviews, 43 # 6078.

[10] Kechris, A.S. The axiom of determinacy implies dependent choices in L[R], J. Symbolic Logic 49(1984), 161-173.

[11] Kechris, A.S., Kleinberg, E.M., Moschovakis, Y. N. & Woodin, W. H. The axiom of determinacy, strong partition properties, and non-singular measures, Cabal Seminar 77-79, Springer Lecture Notes in Mathematics, Volume 839, ed. by A.S. Kechris, D.A.Martin and Y.N. Moschovakis, 75-99.

[12] Kechris, A.S. & Woodin, W.H. Equivalence of partition properties and determinacy, Proc. Natl.Acad. Sci. United States of America, 80(1983), 1783-1786.

[13] Kechris, A.S.& Woodin, W.H. On the size of θ in L[R], (To appear).

[14] Kleinberg, E.M. Infinitary Combinatorics and the Axiom of Determinateness, Springer-Verlag Lectures Notes in Mathematics, Vol. 612(1977).

[15] Mathias, A.R.D. Happy Families, Annals of Mathematical Logic, 12(1977), 59-111.

[16] Mathias, A.R.D. A notion of forcing: some history and some applications, (To appear).

[17] Monro, G.P. Models of ZF with the same sets of sets of ordinals, Fundamenta Mathematicae, 80(1973), 105-110.

[18] Moschovakis, Y.N. Descriptive Set Theory, North Holland, Studies in Logic, vol. 100(1980).

[19] Oxtoby, J.C. Measure and Category. Springer-Verlag, New York, Heidelberg, Graduate Texts in Mathematics, vol. 2 Second Edition (1971).

[20] Spector, M. A measurable cardinal with a non-well-founded ultrapower, J. Symbolic Logic 45(1980), 623-628.

[21] Vopěnka, P. & Balcar, K. On complete models, Bull. l'Académie Polonaise des Sciences, Sér. Sci. Math. Astron. et Phys. 15(1967), 839-841.

PROOF FUNCTIONAL CONNECTIVES

E.G.K. Lopez-Escobar
University of Maryland
Department of Mathematics
College Park, MD. 20742
U.S.A.

A characteristic of the classical propositional connectives is that they are truth-functional, that is, the truth of a sentence depends only on the truth of its prime components. On the other hand the intuitionistic connectives are supposed to be much less dependent on semantical notions. Consequently, one avoids saying that a "sentence is true", rather one tends to say:

"the construction c proves (or justifies) the sentence A".

Nevertheless, some aspects of truth-functionality can still be found in that:

(1) a construction c either proves a sentence A or it does not ,

(2) the conditions for a construction to prove a compound sentence is given in terms of the conditions for the proof of the components; for example, c proves the conjunction $(A_0 \wedge A_1)$ iff c is a pair of constructions $\langle c_0, c_1 \rangle$ such that c_0 proves A_0 and c_1 proves A_1.

Thus for a conjunction to be provable it is necessary (and sufficient) that the conjuncts be provable. In other words, the fact that quite different constructions may prove a given formula is not exploited in traditional intuitionism.

G. Pottinger [1980] introduced a conjunction, which he called "strong conjunction", which requires more than the existence of constructions proving the conjuncts. According to Pottinger:

"The intuitive meaning of $\hat{\&}$ can be explained by saying that to assert $A \hat{\&} B$ is to assert that one has a reason for asserting A which is also a reason for asserting B".

Hence $\hat{\&}$ is one of the first, if not the first, connective which is truly proof-functional. This paper is an introduction to the "logic" of a proof-functional connective; in fact for the most part we shall consider the extension of the positive intuitionistic calculus of implication obtained by the addition of strong-conjunction $\hat{\&}$.

As with any "new" logic, there are five questions that immediately come to mind:

(1) Is there a reasonable, intuitive concept of validity for the sentences of the language?
(2) Is there a formal concept of validity for the sentences of the language?
(3) Is there a formal concept of derivation for the sentences of the language?
(4) What are the relations between (1), (2) and (3)?
(5) How does the new connective affect familiar mathematical theories (for example : elementary number theory)?

The paper is broken down into 5 sections corresponding to the above 5 questions.

§1. INTUITIVE VALIDITY.

1.1 THE LANGUAGE $\mathcal{L}$. The sentential language $\mathcal{L}$ is to have the following symbols:

p,q,r,...	for the sentential variables,
$\supset$	for the conditional connective,
&	for Pottinger's strong conjunction,
(,)	for auxiliary symbols.

1.2 AN INTUITIVE INTERPRETATION FOR THE INTUITIONISTIC CONNECTIVES. Let $\pi_A(c)$ express the (decidable) predicate "the construction c proves the sentence A" and $\pi(c,\theta(x))$ express the (decidable) predicate "the construction c proves the free-variable formula $\theta(x)$". Finally if c,d are constructions then d'c is result of applying d to c.

The Brouwer-Kreisel interpretation of the intuitionistic conditional is that:

(*) $\quad \pi_{A\supset B}(<c,d>)$ iff $\pi(c,(\pi_A(x) \supset \pi_B(d'x)))$.

And although (*) is not universally accepted, it certainly is, and was, a good point of departure for an intuitive interpretation for the intuitionistic conditional.

1.3 AN INTUITIVE INTERPRETATION FOR STRONG CONJUNCTION. The Brouwer-Kreisel interpretation for ordinary conjunction $\wedge$ is

$$\pi_{A\wedge B}(<c,d>) \text{ iff } \pi_A(c) \wedge \pi_B(d).$$

The latter suggests the following interpretation for strong conjunction $\&$:

$$\pi_{A\&B}(<c,c>) \text{ iff } \pi_A(c) \wedge \pi_B(c).$$

The following are equivalent interpretations (in an appropriate theory of constructions)

$$\pi_{A\&B}(c) \text{ iff } \pi_A(c) \wedge \pi_B(c),$$

$$\pi_{A\&B}(<c,d>) \text{ iff } \pi_A(c) \wedge \pi_B(d) \wedge c \cong d,$$

where $\cong$ is an equality relation on constructions.

1.3 AN INTUITIVE CONCEPT OF VALIDITY FOR SENTENCES OF $\not{L}$. A sentence A of $\not{L}$ is intuitively valid iff there is a construction c such that $\pi_A(c)$.

The following sentences are easily seen to be intuitively valid:

(1) $A \& (A \supset B) \supset B$

(2) $(A \supset B) \& (A \supset C) \supset (A \supset B \& C)$

(3) $(A \supset B \& C) \supset (A \supset B) \& (A \supset C)$

(4) $(A \supset C) \supset (A \& B \supset C)$

(5) $A \& B \supset (A \supset B)$

(6) $A \supset (B \supset C) \supset (A \& B \supset C)$

(7) $A \& B \supset A$

(8) $A \supset A \& A$

(9) $A \& B \supset B \& A$

(10) $((A \& B) \& C) \supset A \& (B \& C)$

(11) $\{(A \supset B \& C) \supset (A \supset B) \& (A \supset C)\} \& \{(A \supset B) \& (A \supset C) \supset (A \supset B \& C)\}$

We let $IVAL_{\not{L}}$ be the set of intuitively valid sentences of $\not{L}$.

§2. FORMAL VALIDITY.

2.1 APPLICATIVE ALGEBRAS AND CURRY ALGEBRAS. The formal semantics for $\not{L}$ is to be of the algebraic type. However instead of using the algebras of open sets, we plan to use algebras related to Curry's combinatory logic.

2.1.1 DEFINITION. An applicative algebra is an algebra $M = <M, \circ>$, where $\circ$ is a binary operation on M.

Given an applicative algebra $<M, \circ>$, then instead of writing "$\circ(a,b)$", we will write "(a,b)". A polynomial in $<M, \circ>$ is a term built up from indeterminates $x, y, z, \ldots$, elements of M and the application operator.

2.1.2 DEFINITION. A combinatory algebra is an applicative algebra $\langle M,\circ\rangle$ which is non-trivial (i e. contains at least two elements), and which is combinatory complete; that is, to each polynomial $h(x,y,\ldots,z)$ in $\langle M,\circ\rangle$ there corresponds an element $a \in M$ such that:

$$\forall x \forall y \ldots \forall z[(\ldots(a,x),y),\ldots,z) = h(x,y,\ldots,z)]$$

2.1.3 DEFINITION. A curry algebra is an algebra $C = \langle M,\circ,k,s\rangle$ such that $\langle M,\circ\rangle$ is an applicative algebra and

(1) $k \neq s$,

(2) $\forall x \forall y[((k,x),y) = x]$,

(3) $\forall x \forall y \forall z[(((s,x),y),z) = ((x,z),(y,z))]$.

It was one of the first theorems on combinatory logic that the Curry algebras are combinatory complete.

2.2 SATISFACTION IN CURRY ALGEBRAS.

Assume that $C = \langle M,\circ,k,s\rangle$ is a Curry algebra. A C-proof assignment is a function pa which maps the sentential variables to subsets of M. We extend pa to act on all sentences of $\not{L}$ by requiring that:

$$pa(A\supset B) = \{m \in M \mid \forall n[n \in pa(A) \rightarrow (m,n) \in pa(B)]\}$$
$$pa(A \& B) = pa(A) \cap pa(B).$$

The elements in pa(A) will be called the "pa-proofs of A (in the Curry algebra C)".

Loosely speaking, a sentence A of $\not{L}$ is to be formally valid iff $pa(A) \neq \emptyset$ for every Curry algebra C and C-proof assignment pa. However, because of the constructive (intuitionistic) character of the logic, a certain degree of uniformity is required. We achieve the required uniformity by using terms of a first order language $L_{\underset{\sim}{C}}$ suitable for both the Curry algebras and the proof-assignments. Or more specifically, $L_{\underset{\sim}{C}}$ is the first-order language (with equality: $\doteq$) which also contains

the individual constants: K, S,

the binary (infix) function symbol: $\cdot$,

the unary relation symbols: P,Q,R,...

Next, given a Curry algebra $C = \langle M,\circ,k,s\rangle$ and a C-proof assignment pa we form the first-order structure:

$C[pa] = \langle M,\circ,k,s,pa(p),pa(q),\ldots\rangle$. Clearly C[pa] is a structure associated to the language $L_{\underset{\sim}{C}}$.

In order to relate $pa(A)$ to satisfaction in $C[pa]$, we associate to each formula A of $\mathcal{L}$ a formula $\underset{\sim}{r}_A[x]$ of $\underset{\sim}{L}_C$ as follows:

if A is the sentential variable p, then $\underset{\sim}{r}_A[x] = P(x),\ldots,$

if $A = (B\&C)$, then $\underset{\sim}{r}_A[x] = \underset{\sim}{r}_B[x] \wedge \underset{\sim}{r}_C[x]$,

if $A = (B\supset C)$, then $\underset{\sim}{r}_A[x] = \forall y[\underset{\sim}{r}_B[y] \supset \underset{\sim}{r}_C[x\cdot y]]$.

A routine induction then gives us that for every sentence A of $\mathcal{L}$:

(*) m satisfies $\underset{\sim}{r}_A[x]$ in the structure $C[pa]$ iff $m \in pa(A)$.

Finally, given a closed term t of the first-order language $\underset{\sim}{L}_C$ and a Curry algebra C, t^C is the denotation of t in the algebra C.

We are now ready for the definition of formal validity:

2.2.1 <u>DEFINITION.</u> A sentence A of $\mathcal{L}$ is formally valid iff there is a closed term t of the first-order language $\underset{\sim}{L}_C$ such that for all Curry algebras C and all proof-assignments $pa: t^C \in pa(A)$. In view of the remark (*), the condition "$t^C \in pa(A)$" can be replaced by:

(**) The first-order sentence $\underset{\sim}{r}_A[t]$ is true in the structure $C[pa]$.

We let $FVAL_{\mathcal{L}}$ be the set of sentences of $\mathcal{L}$ which are formally valid.

2.3 AXIOMATIZABILITY OF THE FORMALLY VALID SENTENCES OF $\mathcal{L}$.

Let CA be the following set of sentences of $\underset{\sim}{L}_C$:

$$\neg K \doteq S,$$

$$\forall x\forall y[((K\cdot x)\cdot y) \doteq x],$$

$$\forall x\forall y\forall z[(((S\cdot x)\cdot y)\cdot z) \doteq ((x\cdot z)\cdot(y\cdot z))]$$

and THM_{CA} be the set of logical consequences of CA.

Gödel'a completeness theorem and (**) gives us that:

2.3.1 <u>THEOREM.</u> A formula A of $\mathcal{L}$ is formally valid iff there is a closed term t of $\underset{\sim}{L}_C$ such that $\underset{\sim}{r}_A[t] \in THM_{CA}$.

An immediate consequence of the above theorem is that $FVAL_{\mathcal{L}}$ is a recursively enumerable set. A variant of Craig's lemma gives us the following:

2.3.3 <u>THEOREM.</u> There is a recursive axiomatization for $FVAL_{\mathcal{L}}$.

<u>PROOF.</u> Let $S_0, S_1, \ldots, S_n, \ldots$ be an enumeration of $FVAL_{\mathcal{L}}$ by a (primitive) recursive function. Then for axioms take the following

set of sentences of $\mathcal{L}$:

$$\{(S_0 \& S_0), ((S_1 \& S_1) \& S_1), (((S_2 \& S_2) \& S_2) \& S_2), \ldots\}$$

Clearly the set is recursive. As rule of inference take the rule that allows one to conclude A from $((\cdots(A \& A) \& \ldots) \& A)$.

An application of the concept of formal validity is to show that strong conjunction is indeed different from conjunction. For example it can be shown that there are Curry algebras C and proof-assignments pa such that pa(A) is empty for the following sentences A:

(1) $p \supset (q \supset p \& q)$

(2) $(p \supset q) \supset ((p \supset r) \supset (p \supset q \& r))$

(3) $(p \& q \supset r) \supset (p \supset (q \supset r))$

(4) $(p \supset p) \& (p \supset (q \supset p))$.

§3. FORMAL DERIVABILITY.

3.1 MINIMAL REQUIREMENTS. Let $\Gamma \vdash A$ be a relation of formal derivability between a sentence A and a finite sequence Γ of sentences of $\mathcal{L}$. Then the minimal requirements we place on $\vdash$ are:

(R) $A \vdash A$, for all sentences of $\mathcal{L}$,

(T) if $\Gamma \vdash A$ and $A, \Gamma \vdash B$ then $\Gamma \vdash B$,

(M) if $\Gamma \vdash A$ then $\Delta \vdash A$ whenever every sentence that occurs in Γ also occurs in Δ.

Furthermore, since the conditional is the intuitionistic conditional we also require that

(D) $\Gamma, A \vdash B$ iff $\Gamma \vdash (A \supset B)$.

3.2 SEMANTICAL CONSEQUENCE. In view of the definition of formal validity (see 2.2.1), we propose the following definition for semantical consequence:

3.2.1 DEFINITION. The sentence A is a semantical consequence of the sequence $B_1, \ldots, B_r$ of sentences, in symbols: $B_1, \ldots, B_r \models A$, iff there is a term $t_{x_1, \ldots, x_r}$ of $\underset{\sim}{L}_C$ such that for all Curry algebras C and all C-proof assignments pa:

if $b_1 \in pa(B_1), \ldots, b_r \in pa(B_r)$, then $t^C_{b_1, \ldots, b_r} \in pa(A)$.

We shall say that the term $t_{x_1, \ldots, x_r}$ validates the pair $\langle\langle B_1, \ldots B_r\rangle, A\rangle$.

Call a pair $<\Gamma,A>$ a sequent. Define VALS = $\{<\Gamma,A> \mid \Gamma \models A\}$. Using the formulae $\underset{\sim}{r}_A[x]$ of section 2.2 one can then show that:

3.2.2 LEMMA. VALS is a recursively enumerable set.

3.3 A RECURSIVE AXIOMATIZATION FOR DERIVABILITY.

Let $<<B_{00},\ldots,B_{0r_0}>,A_0>$, $<<B_{10},\ldots B_{1r}>,A_1>,\ldots$ be an enumeration of VALS by a (primitive) recursive function. Then let AXMS be the following set of sequents:

$$\{<<(B_{00}\&B_{00}),\ldots,(B_{0r_0}\&B_{0r_0})>, (A_0\&A_0)>,<<((B_{10}\&B_{10})\&B_{10}),\ldots ((B_{1r}\&B_{1r})\&B_{1r})>,((A_1\&A_1)\&A_1)>,\ldots\}$$

Again, it is clear that AXMS is a recursive set of sequents.

Then let $\vdash$ be the smallest relation containing AXMS $\cup$ $\{<<A>,A> \mid A$ a sentence of $\not{L}\}$ and closed under (T), (M),(D),$(\& \Rightarrow)^*$ and $(\Rightarrow \&)^*$, where the last two rules are:

$$(\& \Rightarrow)^* \qquad \frac{\Gamma,(\ldots(((B\&B)\&B)\ldots\&B) \vdash A}{\Gamma, B \vdash A,}$$

$$(\Rightarrow \&)^* \qquad \frac{\Gamma \vdash (\ldots(((A\&A)\&A)\ldots\&A)}{\Gamma \vdash A,}$$

respectively.

After verifying that the above mentioned rules preserve semantical consequences one easily obtains the following completeness theorem:

3.3.1 THEOREM. For any sequent $<\Gamma,A>$:

$$\Gamma \vdash A \text{ iff } \Gamma \models A.$$

3.4 A SOUND AND NATURAL AXIOMATIZATION FOR DERIVABILITY.

The set of axioms of the axiomatization given in §3.3, although recursive, is of no practical use. We now present another axiomatization which is closer to Gentzen's Sequent calculus for the intuitionistic sentential calculus. In fact, at first sight, it may appear as simply Gentzen's axiomatization for the intuitionistic sentential calculus of the conditional and conjunction. The modification (for the introduction of & in the succedent) is in the applicability of the rule, not in the schema for the rule. Some persons may find the restriction of interest since it is a global restriction involving all the nodes above the node of the inference, and not just those immediately above.

Unfortunately a price has to be paid for such naturalness and although we can show that the axiomatization is sound, we have not yet succeeded in showing it is complete.

In order to further emphasize the relation to Gentzen's systems we will write the sequents in the form "$\Gamma \Rightarrow A$".

3.4.1 AXIOMA SCHEMA:

$$A \Rightarrow A.$$

3.4.2 CUT RULE OF INFERENCE:

$$\frac{\Gamma, A \Rightarrow B \quad \Gamma \Rightarrow A}{\Gamma \Rightarrow B}$$

3.4.3 STRUCTURAL RULES OF INFERENCE

(MON) $$\frac{\Gamma \Rightarrow A}{\Delta \Rightarrow A},$$

provided every sentence occuring in Γ also occurs in Δ.

(REP) $$\frac{\Gamma \Rightarrow A \quad \Gamma \Rightarrow A}{\Gamma \Rightarrow A}$$

3.4.4 RULES OF INFERENCE FOR THE CONDITIONAL.

$(\Rightarrow \supset)$ $$\frac{\Gamma, A \Rightarrow B}{\Gamma \Rightarrow A \supset B}$$ $(\supset \Rightarrow)$ $$\frac{\Gamma \Rightarrow A \quad \Gamma, B \Rightarrow C}{\Gamma, A \supset B \Rightarrow C}$$

3.4.5 RULES FOR STRONG CONJUNCTION.

$(\Rightarrow \&)$ $$\frac{\Gamma \Rightarrow A \quad \Gamma \Rightarrow B}{\Gamma \Rightarrow A \& B}$$ $(\& \Rightarrow)$ $$\frac{\Gamma, B \Rightarrow C}{\Gamma, A \& B \Rightarrow C} \qquad \frac{\Gamma, B \Rightarrow C}{\Gamma, B \& A \Rightarrow C}$$

As already remarked the rule $(\Rightarrow \&)$ is not universally applicable. Loosely speaking $(\Rightarrow \&)$ may be applied when the (sub) derivations of $\Gamma \Rightarrow A$ and $\Gamma \Rightarrow B$ look the same.

We now proceed to make precise the latter statement.

3.4.7 PRE-DERIVATIONS. A pre-derivation consists of a finite tree T (of "nodes") and two functions $ and ₰ defined on T. $\$_N$ is the sequent occuring at the node N, while $₰_N$ is the name of the rule schema followed in obtaining $\$_N$; in other words a pre-derivation in the language Ⱡ is to all intent and purposes a derivation (with analysis) of the intuitionistic sentential calculus of the conditional and usual conjunction.

Given a pre-derivation P = <T,$,₰>, then <T,₰> is called the logical structure of P. The reduced logical structure of P is the pair

$\langle T, \hat{R}\rangle$ where $\hat{R}$ is the function defined on T such that $\hat{R}_N = \mathcal{R}_N$ unless $\mathcal{R}_N$ is either (MON), (REP), ($\& \Rightarrow$) or ($\Rightarrow \&$) in which case $\hat{R}_N = 0$.

Then two pre-derivations $P_1 = \langle T_1, \$_1, \mathcal{R}_1\rangle$ and $P_2 = \langle T_2, \$_2, \mathcal{R}_2\rangle$ are equivalent, in symbols: $P_1 \equiv P_2$, iff their reduced logical structures are isomorphic.

3.4.8 DERIVATIONS. A pre-derivation $\langle T, \$, \mathcal{R}\rangle$ is a derivation when the following condition is met:

At each node N of T at which $\mathcal{R}_N = (\Rightarrow \&)$, the two sub-pre-derivations immediately above N are equivalent.

If $D = \langle T, \$, \mathcal{R}\rangle$ is a derivation and $\emptyset$ is the root of the tree T, then D is a derivation of the sequent $\$_\emptyset$. We write "$\Gamma \vdash A$" just in case that there is a derivation of the sequent $\Gamma \Rightarrow A$.

If Γ is empty and $\Gamma \vdash A$, then A is a (formally) derivable sentence of $\mathcal{L}$.

3.4.9 AN EXAMPLE OF A DERIVATION. First consider the following two derivations:

$$D_1 \qquad \dfrac{\dfrac{A \supset B \Rightarrow A \supset B}{(A\supset B)\ \&\ (A\supset C) \Rightarrow A \supset B} \qquad \dfrac{A \Rightarrow A \quad B \Rightarrow B}{A, A \supset B \Rightarrow B}}{(A\supset B)\ \&\ (A\supset C), A \Rightarrow B}$$

$$D_2 \qquad \dfrac{\dfrac{A \supset C \Rightarrow A \supset C}{(A\supset B)\ \&\ (A\supset C) \Rightarrow A \supset C} \qquad \dfrac{A \Rightarrow A \quad C \Rightarrow C}{A, A \supset C \Rightarrow C}}{(A\supset B)\ \&\ (A\supset C), A \Rightarrow C.}$$

A moments reflexion shows that they are equivalent. Thus the following is a derivation of $(A\supset B)\ \&\ (A\supset C) \supset (A\supset B\&C)$:

$$\dfrac{\dfrac{\dfrac{\overset{D_1}{(A\supset B)\ \&\ (A\supset C), A \Rightarrow B} \qquad \overset{D_2}{(A\supset B)\ \&\ (A\supset C), A \Rightarrow C}}{(A\supset B)\ \&\ (A\supset C), A \Rightarrow B\ \&\ C}}{(A\supset B)\ \&\ (A\supset C) \Rightarrow A \supset B\ \&\ C}}{\Rightarrow (A\supset B)\ \&\ (A\supset C) \supset (A\supset B\&C)}$$

3.4.10 SOUNDNESS THEOREM. An induction on the length of the derivation gives us that to each derivation D and each natural number n we can associate a term $t^D_{x_0,\ldots,x_{n-1}}$ such that :

(1) if D is a derivation of the sequent $S = B_0,\ldots,B_{n-1} \Rightarrow A$,

then $t^{D}_{x_0,\ldots,x_{n-1}}$ validate S

(2) if D_1, D_2 are equivalent derivations, then

$$t^{D_1}_{x_0,\ldots,x_{n-1}} = t^{D_2}_{x_0,\ldots,x_{n-1}}.$$

The term t^D can then be used to show that:

THEOREM. If $\Gamma \vdash A$, then $\Gamma \models A$.

3.4.11 A NORMAL FORM THEOREM FOR DERIVABILITY. Another advantage of the natural axiomatization given in §3.4 is that we can prove a normal form (cut-elimination) theorem for it. An interesting aspect of the proof of normalization is the essential use of the rules of repetition (the structural rule (MON) also includes the usual rule of repetition).

Let us call a derivation D in which there are no cut-formulae of the form $(A\&B)$ and $\&$-cut-free derivation.

The first thing we show is how given a derivation $D = \langle T, \$, R\rangle$ and a cut-node N of T with cut-formula $(A\&B)$, one can transform D to an equivalent derivation with cut-formula either A or B. Thus assume that around the node N, the derivation D is as follows:

$$\dfrac{\begin{matrix} D_1 \\ \Gamma, (A\&B) \Rightarrow C\end{matrix} \qquad \begin{matrix} D_2 \\ \Gamma \Rightarrow (A\&B)\end{matrix}}{\Gamma \Rightarrow C}$$
$$\vdots$$

The critical case is when the last rule of inference applied in D_1 $[D_2]$ is $(\& \Rightarrow)$ $[(\Rightarrow \&)$ respec.$]$. In such a situation we would have (for example):

$$\dfrac{\dfrac{\begin{matrix} D_1' \\ \Gamma, A \Rightarrow C\end{matrix}}{\Gamma, (A\&B) \Rightarrow C} \qquad \dfrac{\begin{matrix} D_2^1 \\ \Gamma \Rightarrow A\end{matrix} \quad \begin{matrix} D_2^2 \\ \Gamma \Rightarrow B\end{matrix}}{\Gamma \Rightarrow (A\&B)}}{\Gamma \Rightarrow C}$$
$$\vdots$$

Then we transform the above derivation into the derivation:

$$\cfrac{\cfrac{\begin{matrix} D_1' \\ \Gamma, A \Rightarrow C \end{matrix}}{\Gamma, A \Rightarrow C} \qquad \cfrac{\begin{matrix} D_2^1 & D_2^1 \\ \Gamma \Rightarrow A & \Gamma \Rightarrow A \end{matrix}}{\Gamma \Rightarrow A}}{\Gamma \Rightarrow C}$$
$$\vdots$$

Iterating the above procedure one can then transform a given derivation into a derivation which is $\mathcal{S}$-cut-free.

Then traditional methods of reducing a cut-formula of the form $(A \supset B)$ can be modified so as to apply to our axiomatization. Intertwining the two reductions one can then obtain the normalization theorem.

§4. RELATION BETWEEN THE CONCEPTS INTRODUCED.

4.1 THE CLASSICAL CASE. Let $IVAL_C$, $SVAL_C$, THM_C be the sets of intuitively valid sentences, set-theoretically valid sentences and provable sentences (in some natural axiomatization) of classical logic respectively. Then one usually argues as follows:

(.1) $IVAL_C \subseteq SVAL_C$,

because if a sentence is valid in all possible structures then it certainly is valid in all set-theoretical structures.

(.2) $THM_C \subseteq IVAL_C$,

because the axioms and rules were chose so as to be correct.

Combining (.1) and (.2) one then obtains

(.3) $THM_C \subseteq IVAL_C \subseteq SVAL_C$.

Then Gödel's completeness theorem, using a rather weak set-theory, gives the mathematical result:

$$SVAL_C \subseteq THM_C.$$

Combining the latter with (.3) then gives us:

(.4) $THM_C = IVAL_C = SVAL_C$.

We remark once again that in order to derive (.4) certain existential assumptions in set-theory were required.

4.2 THE CASE FOR THE (CONSTRUCTIVE) LANGUAGE $\not{L}$. First of all, instead of 4.1.1 we have

(.1) $FVAL_{\not{L}} \subseteq IVAL_{\not{L}}$,

because if we have a term (of combinatory logic) which validates a sentence A, then we do have an intuitive construction that proves A. The converse is by no means obvious.

The soundness theorem, combined with (.1) then gives us

(.2) $THM_{\mathcal{L}} \subseteq FVAL_{\mathcal{L}} \subseteq IVAL_{\mathcal{L}}$.

Now a completeness theorem for the axiomatization of §3.4 (we already know that $FVAL_{\mathcal{L}}$ is recursively axiomatizable) would give us only the mathematical result that

$$THM_{\mathcal{L}} = FVAL_{\mathcal{L}}.$$

Unfortunately the above result does not produce the analog of the classical 4.1.4. In order to obtain such a result (even in the presence of a mathematical completeness theorem) we need first to justify that

(.3) $IVAL_{\mathcal{L}} \subseteq FVAL_{\mathcal{L}}$

The latter could be immediately obtained if one could show that any intuitive construction could be represented by a term of combinatory logic; in other words, 4.2.3 would be a consequence of (some form of) Church's thesis.

If Church's thesis is to be involved, then it is probably advisable to consider number theory in some more detail.

§5. NUMBER THEORY AND STRONG CONJUNCTION.

5.5 A CONCRETE MODEL FOR THE SENTENTIAL LANGUAGE $\mathcal{L}$. Using the techniques of Troelstra [1979] one can show that there are primitive recursive terms ε, ρ, λ, primitive recursive predicates P, $\tilde{=}$ and for each sentence A of $\mathcal{L}$ a primitive recursive predicate P_A such that:

(.1) $PRA \vdash P_{B \supset C}(\underset{\sim}{n}) \equiv P(\underset{\sim}{n}, \ulcorner P_B(x) \supset P_C(\varepsilon(\underset{\sim}{n},x))\urcorner)$,

(.2) $PRA \vdash P_{B \& C}(\underset{\sim}{n}) \equiv P(\underset{\sim}{n}, \ulcorner P_B(\lambda\underset{\sim}{n}) \wedge P_C(\rho\underset{\sim}{n}) \wedge \lambda\underset{\sim}{n} \tilde{=} \rho\underset{\sim}{n}\urcorner)$,

(.3) $PRA \vdash \underset{\sim}{n} \tilde{=} \underset{\sim}{n} \wedge (\underset{\sim}{n} \tilde{=} \underset{\sim}{m} \supset \underset{\sim}{m} \tilde{=} \underset{\sim}{n}) \wedge (\underset{\sim}{n} \tilde{=} \underset{\sim}{m} \wedge \underset{\sim}{m} \tilde{=} \underset{\sim}{r} \supset \underset{\sim}{n} \tilde{=} \underset{\sim}{r})$,

(.4) $PRA \vdash P_A(\underset{\sim}{n}) \wedge \underset{\sim}{n} \tilde{=} \underset{\sim}{m} \supset P_A(\underset{\sim}{m})$,

(.5) if $A \in FVAL_{\mathcal{L}}$, then for some n, $PRA \vdash P_A(\underset{\sim}{n})$,

where PRA is primitive recursive arithmetic and $\ulcorner\;\urcorner$ gives the numeral corresponding to the Gödel number.

5.2 A FORMAL SYSTEM OF NUMBER THEORY FOR STRONG CONJUNCTION. Let $\underset{\sim}{HA}$ be intuitionistic number theory formulated in a sequent calculus. Then $\underset{\sim}{HA}(\&)$ is the extension of $\underset{\sim}{HA}$ obtained by

(1) Enlarging the class of formulae so as to include $\&$.

(2) Add the inference schemas corresponding to ($\& \Rightarrow$) and ($\Rightarrow \&$),

(3) Add the (binary) rule of repetition,

(4) Define pre-derivations and equivalent pre-derivations analogously to §3.4.7.

(5) Define derivations in $HA(\&)$ analogously to §3.4.8.

5.2.1 FORMAL REALIZABILITY FOR $HA(\&)$. Using an induction on the logical complexity of the formula A of $HA(\&)$ one can show that there is a formula $x\,r\,A$ of HA which formally expresses the recursive realizability of A. The only addition required to Troelstra [1973] is the clause:

$$x\,r\,(A \& B) \;=\; x\,r\,A \wedge x\,r\,B$$

For each formula A of $HA(\&)$, $x\,r\,A$ is an almost-negative formula of HA. (see page 193 of Troelstra [1973]).
The usual techniques then give us that:

(.1) if $HA(\&) \vdash A$, then $HA \vdash \exists x[x\,r\,A]$

Now let ECT_0 be the schema of the "extended Church's thesis". In Troelstra [1973] , page 196 it is shown that for formulae B of HA

(.2) $HA \vdash \exists x[x\,r\,B]$ iff $HA + ECT_0 \vdash B$.

Combining (.1) and (.2) we then obtain that $HA(\&)$ is conservative over $HA + ECT_0$ (with respect to the formulae of HA). However, of more interest would be to show that $HA(\&)$ is conservative over HA (which in our opinion is a minimum requirement for $\&$ to be considered as an "intuitionistic connective"). In any case, the above conservative extension result further supports the contention that the completeness of the axiomatization of §3.4 will involve some form of Church's thesis.

§6. CONCLUSION.

Although we feel confident that the results obtained in this note show that proof-functional connectives are viable concepts, we do not find that strong conjunction is, per se, of great interest. Of much more interest would be some kind of strong equivalence; two possible candidates would be given by:

(1) $A \Leftrightarrow B \;=\; (A \supset B) \;\&\; (B \supset A)$,

(2) $\pi_{A \sim B}(c) \;\equiv \pi(c, \ulcorner \pi_A(x) \equiv \pi_B(x) \urcorner)$.

But of course, it was strong conjunction that led us to the concept of a proof-functional connective.

It perhaps should be remarked that accepting proof-functional connectives, such as strong conjunction, requires rejecting the assumption that a construction proves a unique sentence and thus forces us to distinguish between a construction as an object and a construction as a method.

REFERENCES

Pottinger, G. [1980] A Type Assignment for Strongly Normalizable λ-Terms, in To H.B. Curry: Essays on Combinatory Logic, Lambda Calculus and Formalism. Edited by J.P. Seldin and J.R. Hindley, Academic Press, N.Y.

Troelstra, A.S. [1979] The interplay between logic and Mathematics: Intuitionism. Report 79-01 of the Mathematisch Instituut, University of Amsterdam.

Troelstra, A.S. [1973] Mathematical Investigation of Intuitionistic Arithmetic and Analysis. Lecture Notes in Mathematics, vol. 344, Springer-Verlag Publishing Co.

ULTRAPRODUCTS AND CATEGORICAL LOGIC

M. Makkai*
McGill University
Department of Mathematics and Statistics
805 Sherbrooke St. West
Montreal H3A 2K6 Quebec, Canada

Introduction

In categorical logic, there is a natural way of considering the idea of a logical operation in general. A logical operation can be construed as an operation acting on finite diagrams of a given type in a given category, yielding finite diagrams in the same category as values.

In symbolic logic, one starts with symbolic expressions having certain standard interpretations, and one proceeds to considering the symbolic expressions uninterpreted, governed by formal rules of manipulation. The parallel development in categorical logic starts by considering a standard category, most frequently Set, the category of sets, and certain selected operations, acting on finite diagrams in Set. The next step is to consider the 'same' operations in arbitrary categories; the original sets and functions of the diagrams become disembodied abstract objects and morphisms, just as sets and relations become formulas in symbolic logic.

In symbolic logic the main step of the transition from the real life situation to the symbolic one consists in choosing the rules of formal manipulation, including sentence formation, and inference. In this we are uniquely helped by experience in natural languages; one only has to codify in precise terms what is already given intuitively in a practically complete form.

In an attempt to clarify the general, or the "true", nature of logic, we may, however, be hampered rather than helped by our linguistic background. This background may have been shaped in an arbitrary manner by an evolutionary process. If one believes, as I do, that mathematical logic is essentially a part of the makeup of the physical universe, rather than being an addition to it contributed by intelligent life, then a theory of logic independent of concrete aspects of language is desirable.

A particular difficulty arising in symbolic logic is the one encountered when trying to coach in symbolic terms a general concept of

* The author's research is supported by NSERC Canada and FCAC Québec.

logical operation. In abstract model theory, in an attempt at arriving at such a general concept, one considers generalized quantifiers, operations on strings of variables, and strings of formulas. It is strikingly clear that the formal aspects of the concept are too closely modelled on previous linguistic patterns; the concept does not have an abstract integrity, a good chance of being comprehensive.

In categorical logic, the main step of the transition to the abstract situation is to decide what constitutes, in a more or less arbitrary category, the *same* operation as the one we start with in the standard category, say Set. The general shape of the procedure is a natural one. One considers those properties, as many as one can find, or maybe, only some of them, of the given concrete operation (or combinations of various given concrete operations) that can be expressed in the language of categories (i.e. in terms of composition of arrows), and one simply imposes these properties as requirements on the kind of categories; "theories", that serve to embody the abstract essence of the logic under consideration. The crucial point is that the "theory" is a structure of the same kind as the standard structure we start with: they are both categories. In the symbolic context, the formulas form a structure radically different from the entities making up the standard interpretation.

Although the choice of the properties may not be straightforward, and a priori it is not clear if the proposed procedure can be successful in capturing at least as much as symbolic logic does, the proposal at least has a clear enough outline to constitute a program for an investigation of logic in a general manner.

This is not the place to give a historical introduction to categorical logic. Nevertheless, we may make the blanket statement that categorical logic has been successfully developped along the lines indicated above to cover at least first order, and also, higher order logic, in both the so-called classical and intuitionistic versions. A particularly attractive feature of the resulting theory is its unified character. E.g., the definable concepts (roughly, formulas) of a first order theory on the one hand, and the models of the theory on the other hand, both form the same kind of structure: a category. Not only that but those two structures are connected, as a result of a general categorical construction, by a crucial functor, the so-called evaluation functor, mapping the theory-category into the category of functors from the model-category to Set. This, or other similar constructions, could, ultimately, be explained without category theory, but the explanations so gotten are clumsy, and they lack the coherence of a good theory.

In first order logic (our sole interest in this paper), categorical logic has revealed that, in order to have a good general theory of logic, one should consider certain kinds of 'definable concepts' associated with any given first order theory that do not fall under the usual definable sets or relations of symbolic logic. Those new definable sets arise either as a disjoint sum of definable sets, or as the set of equivalence classes of a definable equivalence relation on a definable set. Recently, the use of the more general definable sets has become important in stability theory; c.f. [Sh] and especially [CHL]. (These developments have been independent, so far, of categorical logic proper).

The central concept of first order categorical logic, replacing the notion of first order theory, is the notion of pretopos. It is the notion that underlies, implicitly, the generalized definable sets of stability theory mentioned above. The concept originally arose in abstract algebraic geometry (sheaf theory), without any connection to logic. The notion of pretopos has its central role, from the point of view of logic, since it enjoys completeness properties, each indicating that the concept cannot be made to comprise further logical operations without causing the notion of theory to lose essential properties.

The subject matter of this paper is one such completeness property. It is based on a very general (in fact, too general) concept of operation on diagrams, and it says that the operations in Set ('logical' operations) that commute with the ultraproduct functors on Set are exactly the composites of pretopos operations augmented with Boolean complement. (Thus, after all, in the given context, the definition of a logical operation is shown to be not too general). The result, of course, follows a familiar pattern: 'algebraic' characterizations of logical concepts via ultraproducts.

The main result is closely related to an unpublished theorem of Haim Gaifman. The main result of the present paper is, in essence, a stronger form of Gaifman's theorem. The proof of the theorem of the present paper will, again in essence, consist of a proof of Gaifman's original theorem followed by additional arguments. The comparison with Gaifman's result is made harder, although only in a superficial way, by the fact that Gaifman's terminology (not using categories) is different. Part 3 of the paper contains the details of the comparison.

The exposition of the theorem is used as an opportunity to give an introduction to categorical logic. The methods used here differ from those in [MR] inasmuch as I have tried to use, wherever I could, methods inherent in the categorical formulations. All unexplained category theoretical terminology can be found in [CWM].

I owe special thanks to Andrew Pitts, from whom I have learned much on conservative and quotient morphism. I also thank Michael Barr who pointed out an embarassing error in an earlier version of the paper.

Part 1. Logical Operations

1.1 Operations on Diagrams

Logical operations are seen, on the simplest level, as operations on sets: the operation, applied to one or more sets (depending on the arity of the operation), yields another set. On the second level, the operation on sets is abstracted, with retaining its essential properties, and one ends up with an algebraic operation, or a logical operation in the customary sense: an operation acting on formulas. The set operations union, intersection, etc. giving rise to abstract Boolean algebras, and giving rise to the logical connectives, are the obvious illustration of this state of affairs.

In fact, not only the connectives, but also the quantifiers can be so construed. To do so, one should, however, slightly refine the idea of an operation on sets. Instead of talking about operations acting on ordered finite tuples of sets, one has to consider ones operating on finite systems of a slightly more elaborate kind. It is a basic insight of category theory that, by talking about systems of sets *and* functions, each mapping one of the given sets into another one, complex ideas concerning sets can be elegantly expressed. A system of the kind we need to have is usually called a *diagram of sets and functions*. In this paper, we are interested mainly in finite diagrams. For a while, a diagram will always mean a diagram of sets and functions. A finite diagram is a system of finitely many sets and functions 'of a given shape'. An example would be the diagram

$$A \underset{g}{\overset{f}{\rightrightarrows}} B \xrightarrow{\;h\;} C$$

consisting of three sets and three functions, such that the domain of f is A, and f maps S into B, etc. The 'shape' of this diagram is this:

$$\cdot \rightrightarrows \cdot \longrightarrow \cdot \; ; \qquad (1)$$

such a thing is called a *graph*.

Before we make these concepts general and precise, let us represent the existential quantifier as an operation that yields, from a finite diagram, more sets and functions making up, together with the given one, a larger finite diagram.

One starts with a relation R whose first-place variable ranges over a set X, its second-place variable on Y (usually, Y = X). We can also write Rxy for R. We want to talk about ∃yRxy, a subset of X. Now, R itself is a subset of X×Y. So, we have the following five sets: X, Y, X×Y, R, ∃yRxy. These sets are connected by maps. First of all, R being a subset of X×Y means the presence of the inclusion map: $i: R \to X\times Y$. We will have another inclusion map: $j: \exists yRxy \to X$. The fact that X×Y is the Cartesian product of X and Y will be 'explained', by two projection maps: $\pi_1: X\times Y \to X$, $\pi_2: X\times Y \to Y$. Thus, the operation gives, from the diagram

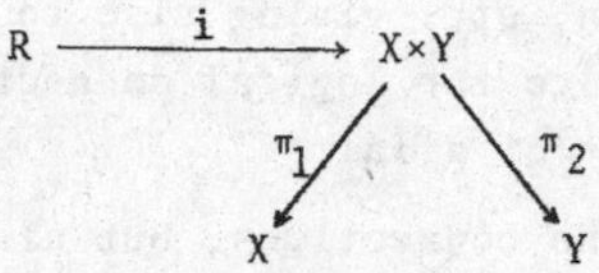

one further set and one further function, extending the given diagram to

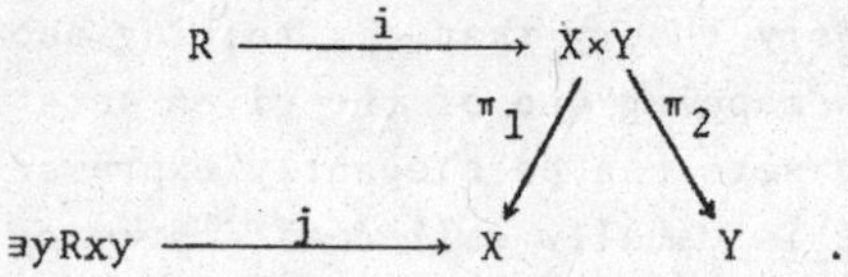

We have not mentioned yet how actually the operation is defined: we only wanted to make clear its 'shape', or arity. In fact, the operations yields from each of *certain* diagrams of the shape

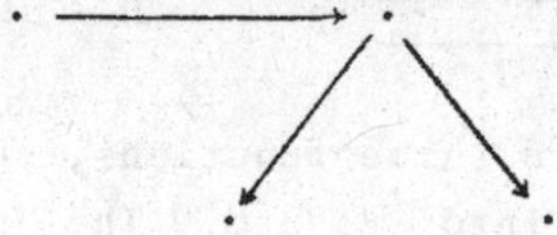

a diagram of the shape

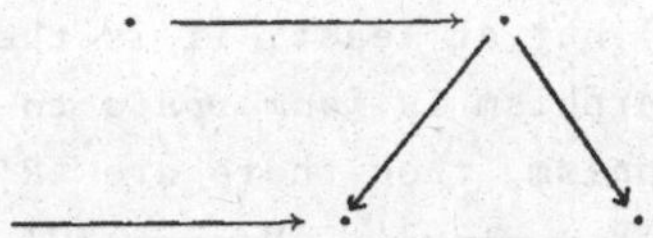

extending the first diagram in the obvious sense.

The above example can be streamlined, and generalized, in the following way. We start with a single function

$$R \xrightarrow{f} X ;$$

in the above situation, f is the composite of i and π_1. Then clearly, the subset $\exists y R x y$ of X is the same as the *range* (or *image*) of the function f. So, we consider the following operation: with any

$$R \xrightarrow{f} X$$

we associate

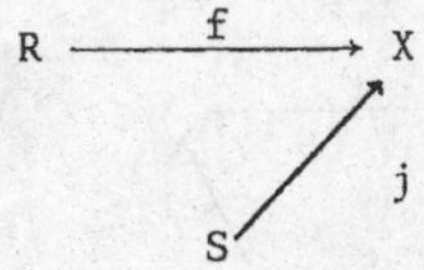

with S the range of f, and j the inclusion map of it into X. The latter operation has the 'shape'

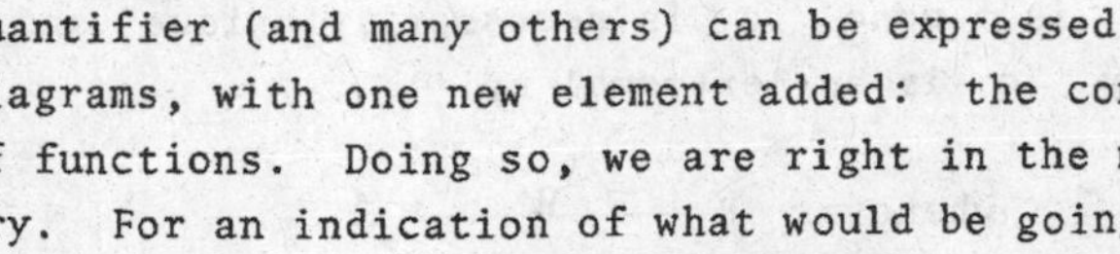

It is the next step in category theoretical insight to realize that not only the form, but also the content of the operation of the existential quantifier (and many others) can be expressed in the language of diagrams, with one new element added: the concept of *composition* of functions. Doing so, we are right in the middle of category theory. For an indication of what would be going on, let us look at the above simplified operation. The first thing would be to explicate the idea of 'inclusion'. This is done by the notion of *monomorphism*, whose definition I will not repeat here (c.f. [CWM]). To be sure, a monomorphism in Set is not necessarily an inclusion (the converse is true, of course) but at least, it is the case that (in the category of sets) any monomorphism is *isomorphic* to an inclusion: if $R \longrightarrow U$ is a monomorphism, then there are $R' \longrightarrow U'$, an inclusion, and *isomorphisms* $R \longrightarrow R'$, $U \longrightarrow U'$ such that the square

$$\begin{array}{ccc} R & \longrightarrow & U \\ \downarrow & & \downarrow \\ R' & \longrightarrow & U' \end{array}$$

commutes. In fact, U' can be chosen to be equal to U, with $U \longrightarrow U'$ the *identity* function.

Starting with the diagram

$$R \xrightarrow{\;f\;} X$$

we consider an extension diagram

$$\begin{array}{ccc} R & \xrightarrow{\;f\;} & X \\ & & \nearrow{\scriptstyle j} \\ & S & \end{array}$$

with the following properties: j is a mono; moreover, there is $q\colon R \longrightarrow S$ making

$$\begin{array}{ccc} R & \longrightarrow & X \\ {\scriptstyle q}\searrow & & \nearrow{\scriptstyle j} \\ & S & \end{array}$$

commute [since j is a mono, q is necessarily unique] and having the following property (q is an 'extremal epi'):

$$\text{whenever}\quad \begin{array}{ccc} R & \xrightarrow{\;q\;} & S \\ {\scriptstyle q'}\searrow & & \nearrow{\scriptstyle k} \\ & S' & \end{array}$$

commutes, and k is a monomorphism, then k must be an isomorphism.

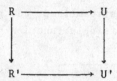

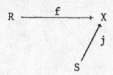

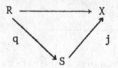

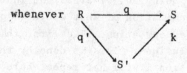

(We have described the extremal epi-mono factorization of the morphism f). It turns out that the property described defines j *uniquely up to a unique isomorphism*: whenever

$$S \xrightarrow{\;j\;} X$$

$$S' \xrightarrow{\;j'\;} X$$

(with the same X) both answer the description, then there is a *unique* isomorphism $S \longrightarrow S'$ making

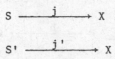

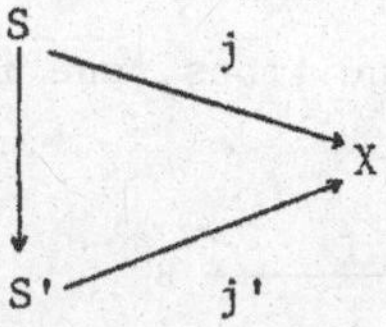

commute. In fact, the inclusion of the range of f into the set X does answer the description; thus, by the above 'categorical' definition, we have described, up to a unique isomorphism, the concept of 'range'.

What is important about the above 'definition' of range, and ultimately, that of the existential quantifier, is that it makes sense in a context where we talk about sets and functions abstractly, without mentioning *elements* of the sets. Now, it is clear that in logic we do precisely that: we talk about formulas representing abstract, "unspecified", sets; their being unspecified means in particular that their elements are not given. Therefore, it is natural to try to talk about 'formulas' as objects in a *category*. In fact, we will give in the 'category of formulas' the *same* definition of the operation of existential quantifier (or 'range') as we did above in the category of sets.

In the foregoing discussion, I already mentioned some technical terms of category theory: commute, category, monomorphism, isomorphism. We will have to continue to use such elementary categorical concepts; for explanations, I refer to [CWM].

A *graph* (see [CWM]), is like a category but with reference to composition and identity morphisms removed. The appropriate structure preserving maps between graphs are called *diagrams*. Since every category is a graph (has an underlying graph), we may speak of a diagram *of type* G *in* a category C: a diagram $D\colon G \longrightarrow C$.

Given a graph G and a category C, the diagrams of type G in C form the objects of a category (G,C) whose morphisms are *natural transformations* (see [CWM]). E.g., a morphism between two diagrams of type (1) is a system of arrows α, β, γ in C as in the picture:

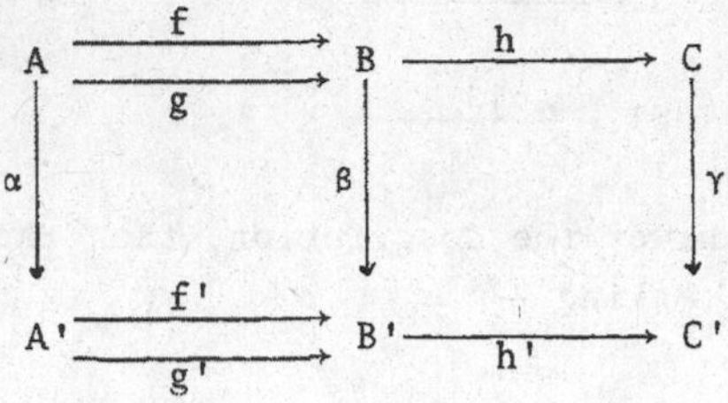

satisfying three commutativity conditions, one for each arrow in (1): the square

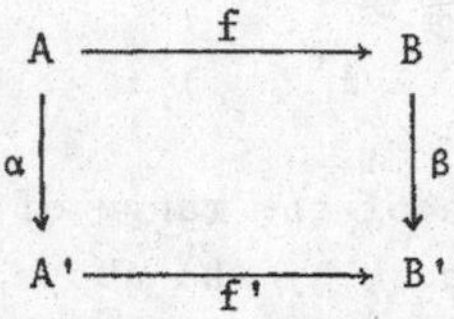

should be commutative; and two more similar conditions. In particular, an *isomorphism* of two diagrams is an invertible natural transformation; it is one in which every component (in the example, α, β or γ) is, individually, an isomorphism.

We are ready to give a very general definition of what an *operation* in a category should be. Upon reading the definition, the reader may feel that we should have talked about *partial operations* instead. However, it turns out that what one would call 'fully defined operations' are not sufficient in the important context of first order logic. Thus the shorter term 'operation' is preferable.

Let C be a category, G and G' graphs, G a subgraph of G'. An *operation in* C *of type* (G,G') is given by a class K' of diagrams of type G' in C $(K' \subset \mathrm{Ob}(G',C))$ satisfying the following conditions: (i) K' is closed under isomorphisms (if D_1' and D_2' are isomorphic diagrams in (G',C), and $D_1' \in K'$, then $D_2' \in K'$) and (ii) if $D_i' \in K'$ $(i = 1,2)$, $D_i = D_i'|G$ (restriction to G), and $\varphi\colon D_1 \longrightarrow D_2$ is an isomorphism, then there is a *unique* isomorphism $\varphi'\colon D_1' \longrightarrow D_2'$ that restricts to φ: $\varphi'|G = \varphi$.

K' may be called the *graph* of the given operation (in obvious analogy to ordinary algebraic operations).

An operation is *finitary* if G' (hence G too) is a finite graph (having finitely many objects and arrows).

The *domain* of the operation is the class K of those diagrams in (G,C) which are restrictions of ones in K': $K = \{D'|G: D': G' \longrightarrow C\}$. It is easy to see that K is closed under isomorphisms.

We are going to call the operation *fully defined* if $K = (G,C)$. We will soon see an important example of a non-full operation.

When we have two categories, with an operation specified in each so that the two operations are of the same type, we might want to consider the two operations to be the 'same', or "realizations of the same operation-symbol", as it is frequently done with ordinary operations in algebra. In particular, we have the important notion of a functor from one category to the other being "operation-preserving" (with respect to the given operations): this is the case if the image under the functor of any diagram in the graph of the operation in the domain category is in the graph of the operation in the codomain category.

Before we come to examples, let us mention that the most important operations satisfy a stronger condition, namely (ii) above with φ and φ' meaning arbitrary morphisms, not just isomorphisms (it is easy to see that, in fact, the modified condition is stronger than the original one). Let us call such operations *strong* operations. Nevertheless, there are important operations which do not satisfy the stronger condition.

Logical operations are operations in the category of sets, and, more generally, operations in other categories that 'behave like' operations in Set. Of course, the designation is imprecise; nevertheless, e.g. our description of pretoposes in the next section will be explicitly based on the idea that pretopos operations should be like certain operations in Set.

1.2. Pretopos operations.

The reader is now asked to recall (or look up, say in [CWM]) the concepts of limit and colimit in categories. Finite limits and colimits will be important for us as logical operations; also, infinite limits and colimits will be important in a different role, as 'semantic' operations (operations on models). Limits and colimits can be construed as operations in the above sense, in the obvious way.

With any given category, we obtain one (partial) operation of limit for each graph serving as the type of the diagram the limit of which we take; this graph is the type of the diagrams in the domain of the operation, denoted G above. E.g., in any category, the operation of product of two objects is given by the class K' of product-diagrams

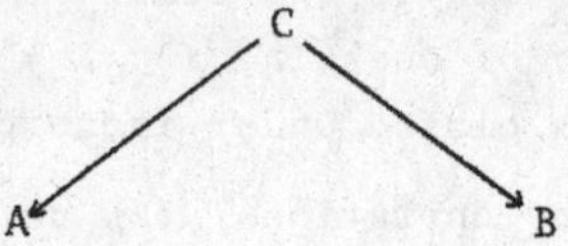

of type the graph G':

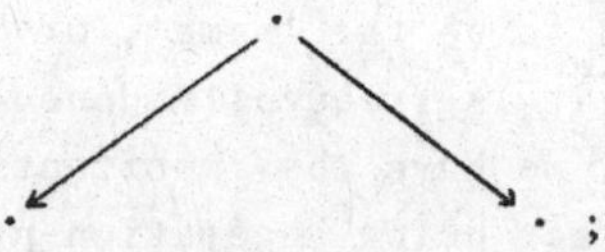

the domain of the operation consists of those pairs of objects A, B (diagrams of type G: • •) which have a product in C. The limit and colimit operations are, in fact, obviously, strong operations. A limit, or colimit, operation of a particular kind is fully defined in a given category just in case that category "has all limits (colimits) of the given kind", in the usual terminology.

In Set, the category of sets and functions, all small limits and colimits are fully defined, and they have meanings which are familiar from many contexts. E.g., any diagram of the form

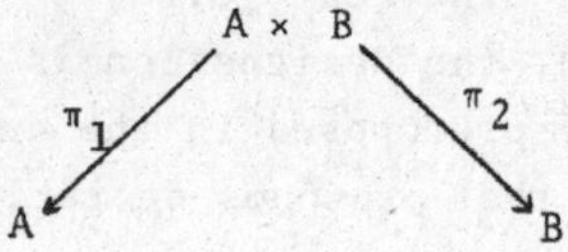

with $A \times B$ the Cartesian product: $A \times B = \{<a,b>: a \in A, b \in B\}$, and $\pi_1(<a,b>) = a$, $\pi_2(<a,b>) = b$, is a product diagram. The diagram in Set:

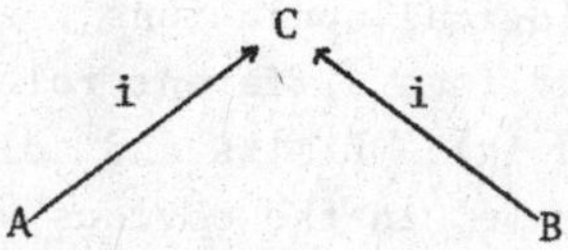

is a coproduct diagram if and only if both i, i' are one-to-one, and C is the disjoint union of range(i) and range(i'). (Therefore, in Set, coproducts are also called *disjoint sums*).

The limit of the empty diagram is called the *terminal object*; in Set, any one-element set serves as such. The colimit of the empty diagram is the *initial object*; it is the empty set in Set.

It is a well-known fact (see [CWM]) that if a category has a terminal object [this is a $\emptyset$-ary operation: G is now the empty graph], has binary products and has equalizers of parallel pairs of morphisms, then it has all finite limits. An exactly dual statement holds, as a consequence, for colimits. The notion of pretopos will be based on finite limits and finite colimits, the latter suitably restricted.

A diagram

$$A \underset{q'}{\overset{q}{\rightrightarrows}} B \qquad (1)$$

in Set will be called an *equivalence relation* if the map $a \longmapsto \langle q(a), q'(a)\rangle$ is a bijection of A onto an equivalence relation, in the usual sense, on B. In an arbitrary small category C, a diagram (2) is, by definition, an equivalence relation iff for all functors $F: C \longrightarrow \mathrm{Set}$ preserving finite limits, F takes (2) into an equivalance relation, in the previous sense, in Set:

$$F(A) \underset{F(q')}{\overset{F(q)}{\rightrightarrows}} F(B)$$

is an equivalence relation in Set.

It turns out (see the next section) that, in a category with finite limits, the notion of equivalence relation can be 'internally' defined by referring to finite limit diagrams; in particular, a functor between categories with finite limits, preserving finite limits, will take an equivalence relation in the domain category into one in the codomain category.

Let (1) be an equivalence relation in Set, let $R \subset B\times B$ be the (ordinary) equivalence induced by (1) on B as explained in the definition above. Let B/R be the set of equivalence classes of R, and let $B \longrightarrow B/R$ be the map that takes b into its equivalence class. Then the diagram

$$A \underset{q'}{\overset{q}{\rightrightarrows}} B \longrightarrow B/R$$

is a coequalizer diagram, as it is easily checked. For this reason, a coequalizer of a pair of morphisms that form an equivalence relation (in any category) is called a *quotient* of the equivalence relation.

Now, part of the definition of 'pretopos' can be put as follows. A *pretopos* is a category which has finite limits, an initial object (colimit of the empty diagram), coproduct of any two objects, and finally, coequalizers of equivalence relations; the definition is completed by imposing conditions on the mentioned operations. Thus we have as basic operations in any pretopos the following *pretopos operations*: the fully defined finite limit operations (or: the three particular ones mentioned above),the initial object, and the binary coproduct operations, both fully defined, and finally the non-full operation whose domain is the class of equivalence relations and whose graph consists of coequalizer diagrams (in this operation, G is $\cdot \rightrightarrows \cdot$, and G' is: $\cdot \rightrightarrows \cdot \longrightarrow \cdot$). A *morphism of pretoposes*, also called a logical functor in [MR], and an elementary functor in [M1], [M2], is a functor that preserves all the listed pretopos-operations (the meaning of a functor preserving an operation was explained above). A *pretopos embedding* of pretoposes is a pretopos morphism which is *conservative*: if it takes a morphism in the domain-pretopos into an isomorphism, than the original morphism is an isomorphism too.

The rest of the definition of 'pretopos' consists of some conditions put on the pretopos operations, all originating in properties of the pretopos operations in Set.

In the next section, the definition of the notion of pretopos will be completed and the following theorem will be proved.

Theorem 1.2.1. (Gödel-Deligne-Joyal representation theorem). *Every small pretopos has a pretopos embedding into a Cartesian power of Set.*

This theorem can be read as a definition: a pretopos is a category having the pretopos operations and having a pretopos embedding into a Cartesian power of Set. Gödel's name is in the name of the theorem because the theorem is closely related to, and in fact, essentially equivalent to, Gödel's completeness theorem. This relationship is explained in detail in [MR]. Deligne proved an essentially equivalent theorem in [SGA4], Exp. VI., formulated for coherent toposes instead of pretoposes. A suitably modified form of Deligne's proof will be given in the next section. The concept of pretopos appears in passing (as a set of exercises) in [SGA4, Exp. VI.]. It was André Joyal who first advocated the concept of pretopos as the basic notion of categorical first order logic.

We now consider the non-strong operation of Boolean complement. Let T be a pretopos. For A an object of T, let Sub(A) denote the partial ordering of subobjects of A (see V.7 in [CWM]). As we will see in the next section, Sub(A) is a distributive lattice. The Boolean complement of an element $\Phi \in$ Sub(A), if exists, is an element $\Phi' \in$ Sub(A) such that $\Phi \wedge \Phi' = 0_A$, $\Phi \vee \Phi' = 1_A$ (0_A, 1_A are the minimal and maximal elements, respectively, of Sub(A)). The Boolean complement is unique (if it exists); moreover, it is preserved by lattice-homomorphisms. As a consequence, if Φ' is the Boolean complement of Φ in T, and $F: T \longrightarrow T'$ is a pretopos morphism, then $F(\Phi')$ is the Boolean complement of $F(\Phi)$ in T'.

A *Boolean pretopos* is a pretopos in which Boolean complements always exist, i.e., in which Sub(A) is a Boolean algebra for every object A. It is important that there is no new notion of morphism of Boolean pretoposes: morphisms of pretoposes between Boolean pretoposes preserve the additional structure.

"Boolean complement" is an operation in our general setting: it assigns to any monomorphism $B \longrightarrow A$ a diagram $B \longrightarrow A \longleftarrow B'$ with appropriate properties; for suitable categories (such as pretoposes) the condition of uniqueness up to a unique isomorphism is clearly assured.

1.3. Semantical operations in Set.

Properties of interchangeability of certain limits and colimits in Set are fundamental to much of the theory of logical operations. Limits are freely interchangeable, in any category (c.f. IX. 2 in [CWM]). As an example, if

$$A_i \xrightarrow{e_i} B_i \overset{f_i}{\underset{g_i}{\rightrightarrows}} C_i \qquad (1)$$

in an equalizer diagram for i = 1 and 2, then

$$A_1 \times A_2 \xrightarrow{e_1 \times e_2} B_1 \times B_2 \overset{f_1 \times f_2}{\underset{g_1 \times g_2}{\rightrightarrows}} C_1 \times C_2$$

is again one.

The interchangeability of finite limits and filtered colimits (c.f. loc. cit.) is an important property of Set. Thus, if the equalizer diagrams (1) in Set are given for all $i \in I$, and I is, say, a directed partial order, and if, moreover, we have, for each $i \le j$, connecting morphisms $\alpha_{ij}: A_i \longrightarrow A_j$ with

$$\alpha_{jk}\circ\alpha_{ij} = \alpha_{ik} \quad (i \le j \le k)$$

and similar ones β_{ij}, γ_{ij}, for the B's and C's such that the diagram

$$\begin{array}{ccc} A_i & \xrightarrow{e_i} & B_i \\ \downarrow{\scriptstyle \alpha_{ij}} & & \downarrow{\scriptstyle \beta_{ij}} \\ A_j & \xrightarrow[e_j]{} & B_j \end{array}$$

and the similar ones with the f's and g's all commute, then

$$\operatorname*{colim}_i A_i \xrightarrow{\operatorname*{colim}_i e_i} \operatorname*{colim}_i B_i \underset{\operatorname*{colim}_i g_i}{\overset{\operatorname*{colim}_i f_i}{\rightrightarrows}} \operatorname*{colim}_i C_i$$

is again an equalizer diagram.

Finite colimits in Set offer less interchangeability features. Of course, they are interchangeable with colimits. Coequalizers (quotients) of equivalence relations are interchangeable with (finite or infinite) products but not necessarily with more general limits; even this much cannot be said about coequalizers in general. For binary coproducts, the only positive fact we can state is that they are interchangeable with certain combinations of products and directed colimits, the *ultraproducts*.

Let U be an ultrafilter on a set I, let A_i be a set for each $i \in I$. Let U^{op} be the partial ordering with underlying set U, in which $P \le Q$ iff $Q \subset P$, and consider U^{op} a category in the usual way. Consider the diagram

$$U^{op} \longrightarrow \text{Set}$$

$$P \longmapsto \prod_{i\in P} A_i$$

$$Q \supset P \longrightarrow \left(\prod_{i\in Q} A_i \xrightarrow{\text{canon.proj.}} \prod_{i\in P} A_i\right);$$

the colimit of this diagram is the ultraproduct of the sets A_i, $i \in I$. In short-hand:

$$\prod_{i\in I} A_i/U = \operatorname*{colim}_{P\in U^{op}} \prod_{i\in P} A_i .$$

When we choose the standard representation of products and colimits, we obtain the usual definition of ultraproduct of sets (except, maybe, for the fact that under the present definition, one has to consider vectors $\langle a_i \rangle_{i\in P}$ of elements $a_i \in A_i$ indexed by arbitrary $P \in U$, and not just $P = I$; but, when one thinks of the possibility of some of the A_i being empty, one sees that ours is the right definition!). Moreover, with the standard choices made, we may consider the ultraproduct (with given I and U) a functor

$$\mathrm{Set}^I \longrightarrow \mathrm{Set}$$

where the action of the functor on arrows is given by the universal properties of products and colimits: given $f_i : A_i \longrightarrow B_i$ for $i \in I$, we put

$$\prod_{i\in I} f_i/U = \operatorname*{colim}_{P\in U^{op}} \prod_{i\in P} f_i .$$

It is a basic fact that coproducts (disjoint sums) commute with ultraproducts: if

$$A_i \xrightarrow{\;f_i\;} C_i \xleftarrow{\;g_i\;} B_i$$

is a coproduct diagram in Set for all $i \in I$, then

$$\prod_i A_i/U \xrightarrow{\;\prod_i f_i/U\;} \prod_i C_i/U \xleftarrow{\;\prod_i g_i/U\;} \prod_i B_i/U$$

is again one. The reader is invited to check this directly. Also, the ultraproduct of copies of the empty set is again empty. In other words, ultraproducts commute with finite coproducts.

Since ultraproducts are combinations of products and directed colimits, what we said above implies that they commute with finite limits, and quotients of equivalence relations. We therefore conclude:

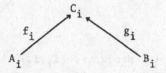

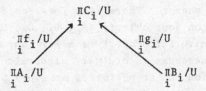

<u>Theorem</u> 1.3.1. (<u>Los's theorem</u>) *Any ultraproduct functor*

$$\prod_{i\in I}(-)/U\colon \mathrm{Set}^I \longrightarrow \mathrm{Set}$$

is a morphism of pretoposes.

The relation of the last theorem to the 'fundamental property of ultraproducts' (Los's theorem in the usual sense) will be explained in Part 2.

1.4. <u>Generating diagrams, and the statement of the main theorem.</u>

Composites of pretopos operations will commute with ultraproducts as a direct consequence of 1.3.1. Such a composite is, e.g., the operation on Set of taking $(A\times B)\amalg(C\times D)$, whose graph can be construed as the class of all diagrams

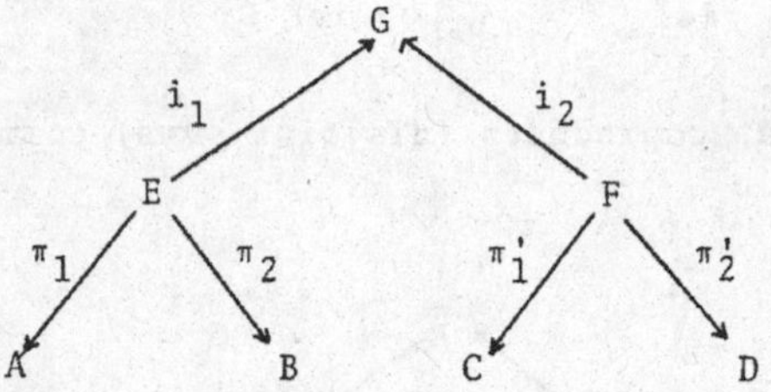

(1)

with (π_1,π_2), (π_1',π_2') being products, (i_1,i_2) being a coproduct; the domain of the operation is the collection of all tuples (A,B,C,D) of objects.

The general concept of a composite operation will be approached in a somewhat abstract manner. There is a *free* Boolean pretopos $F(G)$ on any graph G; e.g. on four objects, A_0, B_0, C_0, D_0, as generators. This is defined by a universal property similar (but not identical) to the one used for free algebras in ordinary algebraic situations. The above composite operation can be performed in every pretopos, hence in particular in the free Boolean pretopos on $\{A_0,B_0,C_0,D_0\}$, and we can consider its result on the particular argument $\langle A_0,B_0,C_0,D_0\rangle$ having say E_0,F_0,G_0 as additional objects and appropriate arrows. Then, the operation in question can be performed in any Boolean pretopos T in the following manner. Say, the arguments are A, B, C, D; one finds the essentially unique pretopos-morphism $\mathbb{F}\colon F(G) \longrightarrow T$ that takes A_0 to A, etc; $\mathbb{F}$ will pick out the desired diagram(1) as the values $F(E_0)$, $F(F_0)$, $F(G_0)$ and appropriate ones for arrows. In other words, the operation is entirely described by two diagrams, $D_0\colon G \longrightarrow F(G)$, and an extension of it, $D_0': G' \longrightarrow F(G)$; here G is the type of the domain, G' the type of the graph of the operation.

Moreover, conversely, every object and morphism in $F(G)$ is obtained as part of a "composite Boolean pretopos operation"; this could be made clear by explicitly constructing $F(G)$ as made up of an increasing union of larger and larger diagrams of such composites. Thus, every operation obtained from a D_0 and a D_0' as above *is* a "composite Boolean pretopos operation", if we have in mind some direct, syntactical meaning of the latter. Instead, we take the description using the free Boolean pretopos as the first approximation of a *definition* of composite operation.

However, the free pretopos construction is not enough. Consider the non-full pretopos operation of quotient of an equivalence relation. The domain of this operation consists of diagrams

$$A \rightrightarrows B \qquad (2)$$

satisfying additional conditions, amounting to (2) being an equivalence relation. The free Boolean pretopos, say F, on a fixed diagram of the form (2) will not 'contain' the operation in question, simply because (2) will not be an equivalence relation in it. But, if we 'impose' the further condition on F that (2) be an equivalence relation, the operation will be available. There is a process of forming a *quotient* of a pretopos T by imposing conditions on it in the form of inverting a class of morphisms in T. This is entirely analogous to the algebraic procedure of identifying elements in a (universal) algebra, which takes the form of passing from the algebra $\mathbb{M}$ to $\mathbb{M}/E$, with E a congruence relation of $\mathbb{M}$.

We will adopt the definition that a composite pretopos operation, with domain of type the finite graph G, means one that 'factors through' a pretopos which is a quotient of the free pretopos on G. For pretoposes T, T', Pretop(T,T') denotes the category of pretopos morphisms $T \longrightarrow T'$; it is the full subcategory of (T,T') the category of all functors from T to T', with objects the pretopos morphisms. Mod(T) = Pretop(T,Set).

Definition-Proposition 1.4.1.

Let G be a small graph.

(i) The free pretopos over G is a pretopos T (= $F(G) = F_{pt}(G)$) together with a diagram $\varphi: G \longrightarrow T$ ($\varphi = \varphi_G = \varphi_G^{pt}$) with the following universal property: for any pretopos T' the functor

$$()\circ\varphi: \text{Pretop}(T,T') \longrightarrow (G,T')$$

defined by composition:

$$F: T \longrightarrow T' \longmapsto F\circ\varphi: G \longrightarrow T'$$

is an equivalence of categories. The free pretopos over G exists, and is determined up to an equivalence over G.

(ii) The free Boolean pretopos over G is a Boolean pretopos T (= $F(G) = F_{Bpt}(G)$) together with a diagram $\varphi: G \longrightarrow T$ ($\varphi = \varphi_G = \varphi_G^{Bpt}$) with the following universal property: for any Boolean pretopos T', the functor

$$()\circ\varphi: \text{Pretop}(T,T') \longrightarrow (G,T')$$

is faithful, full on isomorphisms, and essentially surjective. The free Boolean pretopos over G exists, and is determined up to an equivalence over G.

(A functor $F: C \longrightarrow \mathcal{D}$ is *full on isomorphisms* if for every isomorphism $g: FC \longrightarrow FC'$ there is an isomorphism $f: C \longrightarrow C'$ such that $Ff = g$. F is *essentially surjective* if for all $D \in Ob(\mathcal{D})$ there is $C \in Ob(C)$ such that FC is isomorphic to D).

In the next section, we'll indicate the proof of the proposition; the construction of the free objects will use formulas. The reason for the weaker statement in part (ii) is the presence of a non-strong operation, Boolean complement.

These definitions of free objects are natural versions of the well-known diagrammatical definitions of free groups and other free algebraic structures. In either the case of pretopos or that of Boolean pretopos, we have the following form of the universal property as a consequence of the definition: given T' and any $F: G \longrightarrow T'$, there is M such that

$$\begin{array}{ccc} G & \xrightarrow{\quad\varphi\quad} & F(G) \\ & {\scriptstyle F}\searrow{\scriptstyle\cong} & \vdots\ {\scriptstyle M} \\ & & T' \end{array}$$

commutes up to isomorphism: there is an isomorphism

$$\mu\colon F \xrightarrow{\ \sim\ } M\circ\varphi;$$

and if

$$\mu\colon F \xrightarrow{\ \sim\ } M\circ\varphi$$

$$\mu'\colon F \xrightarrow{\ \sim\ } M'\circ\varphi$$

then there is a unique isomorphism $\nu\colon M \xrightarrow{\ \sim\ } M'$ such that

$$\begin{array}{ccc} & & M\circ\varphi \\ & \overset{\mu}{\nearrow}{\scriptstyle\sim} & \\ F & & \downarrow{\scriptstyle \nu\circ\varphi} \\ & \underset{\mu'}{\searrow}{\scriptstyle\sim} & \\ & & M'\circ\varphi \end{array}$$

commutes. The essentially unique M with the said property is said to be *induced by* F, and may be denoted by M_F.

For a functor $F\colon C \longrightarrow \mathcal{D}$, $\mathrm{Inv}(F)$ denotes the collection of all those morphisms f in C for which Ff is an isomorphism (invertible) in $\mathcal{D}$. E.g., F is conservative if $\mathrm{Inv}(F) = \mathrm{Inv}(\mathrm{Id}_C)$ = the class of all iso's in C.

Definition 1.4.2 (i) Let Σ be a collection of arrows in the pretopos T. A pretopos morphism $Q\colon T \longrightarrow T'$ is said to be *obtained by inverting the morphisms in* Σ if we have the following universal property: for any pretopos T'', the functor

$$(\,)\circ Q\colon \mathrm{Pretop}(T',T'') \longrightarrow \mathrm{Pretop}(T,T'')$$

induces an equivalence of $\mathrm{Pretop}(T',T'')$ onto the full subcategory of $\mathrm{Pretop}(T,T'')$ consisting of those $G\colon T \longrightarrow T''$ for which $\Sigma \subset \mathrm{Inv}(G)$.

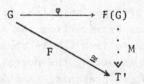

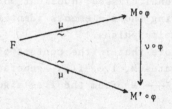

(ii) $Q: T \longrightarrow T'$, a pretopos morphism, is a *quotient morphism* (a *quotient*) if it is obtained by inverting the morphisms in $\mathrm{Inv}(Q)$. Q is a *finite quotient* if there is a finite subset Σ of $\mathrm{Inv}(Q)$ such that Q is obtained by inverting the morphisms in Σ.

(iii) Let T be a Boolean pretopos. A diagram $F: G \longrightarrow T$ (*finitely*) *generates* T (*as a Boolean pretopos*) if the pretopos-morphism $M_F: \mathcal{F}_{\mathcal{B}pt}(G) \longrightarrow T$ induced by F is a finite quotient morphism.

If M is a (universal) algebra, in a given equational class, say, and E is an arbitrary subset of $|M|\times|M|$, then there is a universal property similar to 1.4.2 (i) of a morphism $q: M \longrightarrow M'$ being "obtained by identifying a with b", for all pairs $\langle a,b\rangle \in E$: for any M'', the map $\mathrm{Hom}(M',M'') \longrightarrow \mathrm{Hom}(M,M'')$ defined by composition with q induces a bijection onto the set of those $M \longrightarrow M''$ which do identify all pairs in E. Of course, such M' is nothing but the quotient $M/[E]$ of M by the congruence relation $[E]$ generated by E, with q the map taking an element of $|M|$ into its equivalence class in $[E]$. (In equational classes, quotient morphism (those that are "obtained by identifying those elements identified by the morphisms") are the same as regular epimorphisms.)

Concerning (iii), note that in the context of an equational class, a map $f: X \longrightarrow M$ generates M, i.e. $f(X)$ generates M in the usual sense, iff the map $F(X) \longrightarrow M$ from the free algebra on X induced by f is a quotient.

Definition 1.4.3. *An abstract composite Boolean pretopos operation* (ACBPO) is given by a commutative triangle

$$\begin{array}{ccc} G & \xrightarrow{\;D_0\;} & T \\ & {\scriptstyle i=\text{inclusion}}\searrow & \nearrow{\scriptstyle D_0'} \\ & G' & \end{array}$$

where G, G' are finite graphs, T is a Boolean pretopos, and both D_0 and D_0' finitely generate T (see 1.4.2). In any pretopos S, e.g. $S = \mathrm{Set}$, an ACBPU as shown *defines the composite Boolean pretopos operation* of type (G,G') whose graph is

$$K' = \mathrm{Iso}\{M\circ D_0' : M \in \mathrm{Pretop}(T,S)\}.$$

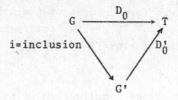

<u>Remark</u>. K' as defined in 1.4.3 from an ACBPU is indeed an operation in S of type (G,G'). To see this, let us write C^{iso}, with C any category, for the subcategory of C whose objects are those of C, and whose morphisms are the isomorphisms in C. With the notation of 1.4.3, let $(K')^{iso}$ be the full subcategory of $(G',S)^{iso}$ whose objects are the diagrams in K'. Let $K = \{D' \circ i \colon D' \in K'\}$, and define $K^{iso} \subset (G,S)$ similarly. The functor i induces by composition (restriction) the functor $i^* \colon (K')^{iso} \longrightarrow K^{iso}$. Note that to say that K' is the graph of an operation of type (G,G') is to say that i^* is full and faithful. With $F = F_{Bpt}(G)$, $F' = F_{Bpt}(G')$, $\hat{D} = M_{D_0}$, $\hat{D}' = M_{D_0'}$, $\hat{i} = M_{(\varphi^{Bpt}_{G'} \circ i)}$, we have that the given ACBPU induces the diagram

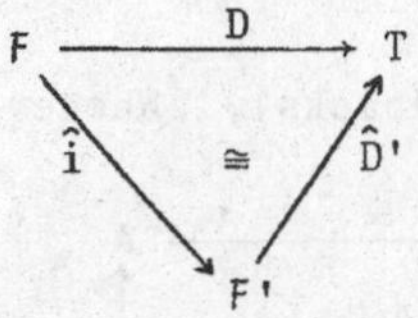

commuting up to an isomorphism. Passing to the induced diagram

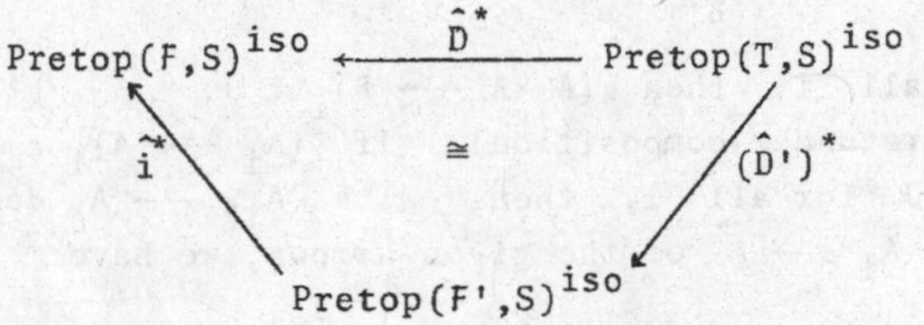

we note that $(K')^{iso}$ the full replete image of $(\hat{D}')^*$. Since both $\hat{D}^*$ and $(\hat{D}')^*$ are full and faithful as a direct consequence of $\hat{D}$, $\hat{D}'$ being quotients, it follows that $\hat{i}^* | (K')^{iso}$, the restriction of $\hat{i}^*$ to $(K')^{iso}$, is full and faithful. It follows that i^* is full and faithful.

<u>Definition</u> 1.4.3' (i) An operation is a *restriction* of another one of the same type if the graph of the first is contained in the graph of the second.

(ii) An operation in Set of type (G,G') having the graph K' *commutes with ultraproducts* if the following holds: when U is an ultrafilter on I, and $D' \colon G' \longrightarrow \mathrm{Set}^I$ has all its components $D_i \colon G' \xrightarrow{D'} \mathrm{Set}^I \xrightarrow{\mathrm{proj}_i} \mathrm{Set}$ $(i \in I)$ in K', then $[U] \circ D' \in K'$, where $[U]$ is the ultraproduct functor

$$\prod_{i \in I} (-)/U \colon \mathrm{Set}^I \longrightarrow \mathrm{Set}.$$

Theorem 1.4.4. Every finitary operation in Set commuting with ultraproducts is a restriction of a composite Boolean pretopos operation.

Part 2. Categories for the working logician

2.1. Sites

Let C be a category with finite limits. A (*Grothendieck*) *topology* on C, J, is a collection of families $\{A_i \longrightarrow A\colon i \in I\}$ (abbreviated as $(A_i \longrightarrow A)_i$) of morphisms in C with a fixed codomain (such a family is called a J-*covering* of A), with J required to satisfy the following closure conditions:

(i) Every isomorphism $A' \xrightarrow{\approx} A$ is a 1-element covering: $\{A' \xrightarrow{\approx} A\} \in J$.

(ii) (Stability under pullbacks). Whenever $(A_i \longrightarrow A)_i \in J$ and

$$\begin{array}{ccc} A_i & \longrightarrow & A \\ \uparrow & & \uparrow \\ A_i \times_B A & \longrightarrow & B \end{array}$$

is a pullback for all i, then $(A_i \times_B A \longrightarrow B)_i \in J$.

(iii) (Closure under composition). If $(A_i \longrightarrow A)_i \in J$ and $(A_{ik} \longrightarrow A_i)_{k \in K} \in J$ for all i, then, with $A_{ik} \longrightarrow A$ denoting the composite $A_{ik} \longrightarrow A_i \longrightarrow A$ of the given arrows, we have $(A_{ik} \longrightarrow A)_{i,k} \in J$.

(iv) (Monotonicity). If $(A_i \longrightarrow A)_i \in J$, and $(A'_j \longrightarrow A)_j$ is a family so that for all i there is j with $A_i \longrightarrow A$ factoring through $A'_j \longrightarrow A$, then $(A'_j \longrightarrow A)_j \in J$.

It is important that we may allow the empty family as a covering of one object, but not of another. Therefore, since in the empty set the codomain is not given, we should, more precisely, have J as a collection of pairs $\langle A, \mathcal{A}\rangle$ with $\mathcal{A}$ a (possibly empty) set of morphisms with codomain A. The reading of the closure conditions under the more precise definition should be clear.

Examples of topologies are the following.

With $C = \text{Set}$, let J be defined by: $(f_i\colon A \longrightarrow A)_{i \in I} \in J$ iff the family is jointly surjective, i.e. $\bigcup\{\text{Im}(f_i)\colon i \in I\} = A$. When Set is regarded a site, we always mean this *canonical* topology.

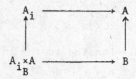

If X is a topological space, and $C = \mathcal{O}(X)$ is the partial ordering (under inclusion) of the open sets of X, regarded as a category in the usual way, we let J be the set of all $(U_i \leq U)_i$ such that $\bigcup_i U_i = U$.

The Grothendieck topology J obtained in this way may replace the ordinary topology on X in contexts like sheaf theory.

A *site* (C,J) is a category C with finite limits, together with a topology J on C. A *morphism of sites* $(C,J) \longrightarrow (\mathcal{D},J')$ is a functor $F: C \to \mathcal{D}$ preserving finite limits and taking J-covers into J'-covers: $(f_i: A_i \to A)_i \in J$ implies $(Ff_i: FA_i \to FA)_i \in J'$. We also talk about a (J,J')-continuous functor, or even a J-continuous one, if J' is understood.

Given any collection J_0 of families of morphisms in C, each with a fixed codomain [such J_0 may be called a *pre-topology*], one has a least topology J on C containing J_0. J, or (C,J) is said to be *generated* by J_0, or by (C,J_0). Also note that if, in addition, $(\mathcal{D},J')$ is a site, and the functor $C \to \mathcal{D}$ preserving finite limits takes J_0-covers into J'-covers [we may consider it a morphism $(C,J_0) \to (\mathcal{D},J')$], then the functor is a morphism of sites: $(C,J) \to (\mathcal{D},J')$.

A *finitary* topology, or site, is one which is generated by a *finitary* pre-topology, i.e. one in which every covering is finite. Note that a topology J is finitary just in case for all $(A_i \to A)_{i\in I} \in J$ there is a finite $I' \subset I$ such that $(A_i \to A)_{i\in I'} \in J$.

Theorem 2.1.1 (Deligne's completeness theorem). Let (C,J) be a small finitary site, and $A = (A_i \to A)_i$ any family *not* in J. Then there is a J-continuous functor $C \to$ Set taking A into a family in Set which is *not* a covering (in the canonical topology).

Remark. Suppose C is a small category with finite limits, and J_0 is a finitary pretopology on C. One may consider the collection J of all coverings in C which are carried into canonical coverings by all morphisms $(C,J_0) \to$ Set. The completeness theorem asserts that J is the same as the topology generated by J_0. The theorem will be proved using the following

Proposition 2.1.2. Suppose (C,J) is a small finitary site such that 1, the terminal object of C, is not empty, i.e. not covered by the empty family in J. Then there is at least one morphism $(C,J) \to$ Set.

Proof: Let $\mathrm{Lex}(\mathcal{C},\mathrm{Set})$ denote the category of all functors $\mathcal{C} \to \mathrm{Set}$ preserving finite limits; $\mathrm{Lex}(\mathcal{C},\mathrm{Set})$ is a full subcategory of $(\mathcal{C},\mathrm{Set})$, the category of all functors $\mathcal{C} \to \mathrm{Set}$. We have to construct $M \in \mathrm{Lex}(\mathcal{C},\mathrm{Set})$ which, in addition, carries J-coverings into canonical coverings. The construction will give M as a suitable directed colimit of representable functors $\mathcal{C}(C,-)$ $(C \in \mathcal{C})$. Since each representable functor is in $\mathrm{Lex}(\mathcal{C},\mathrm{Set})$, and directed colimits (in the sense of $(\mathcal{C},\mathrm{Set})$) of members of $\mathrm{Lex}(\mathcal{C},\mathrm{Set})$ are again in $\mathrm{Lex}(\mathcal{C},\mathrm{Set})$ [this is a consequence of directed colimits commuting with finite limits in Set], the construction will, at least, ensure that $M \in \mathrm{Lex}(\mathcal{C},\mathrm{Set})$. It will be a separate matter to ensure that M 'respect' the coverings as well.

Let $S = (S,\le)$ be a directed partial ordering, and let

$$\begin{array}{rcl} D\colon S & \longrightarrow & \mathcal{C}^{op} \\ s & \longmapsto & D_s \\ s \le t & \longmapsto & \delta_{ts}\colon D_t \longrightarrow D_s \end{array}$$

be a diagram of type S in $\mathcal{C}^{op}$. Let us consider the composite

$$S \xrightarrow{\ D\ } \mathcal{C}^{op} \xrightarrow{\ y\ } (\mathcal{C},\mathrm{Set})$$

(with y the Yoneda embedding: $C \longmapsto \mathcal{C}(C,-)$), and its colimit:

$$M(D) = \operatorname*{colim}_{s\in S} \mathcal{C}(D_s,-) \in \mathrm{Lex}(\mathcal{C},\mathrm{Set}).$$

Let $\mathcal{A} = (A_i \xrightarrow{f_i} A)_{i\in I} \in J$. For M to carry $\mathcal{A}$ into a canonical covering in Set, it is necessary and sufficient that for every $s \in S$ and every $g\colon D_s \longrightarrow A$, the following hold:

(*) there are $i \in I$, $t \ge s$ and an arrow g' making the diagram

$$\begin{array}{ccc} D_s & \xrightarrow{\ g\ } & A \\ {\scriptstyle \delta_{ts}}\uparrow & & \uparrow{\scriptstyle f_i} \\ D_t & \xrightarrow[\ g'\ ]{} & A_i \end{array}$$

commute. (This is a matter of inspection.) When condition (*) holds, we will say that g is *captured* for $\mathcal{A}$ in the diagram D by g'.

Let us call a diagram D of the above type *consistent* if none of the objects D_s is empty. A *continuation* of D is a directed diagram $D'\colon S' \longrightarrow \mathcal{C}$ with S a sub-poset of S', and D the restriction of D' to S. The main step in the proof is the following

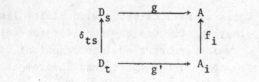

Lemma 2.1.3. Suppose D is a consistent diagram, $g\colon D \longrightarrow A$, and $A = (f_i\colon A_i \longrightarrow A)_{i\in I}$ is a finite J-covering of A. Then there is a consistent continuation of D in which g is captured for A.

Once the lemma is proved, the proof of the proposition is straightforward. Note that, once an arrow is captured for a covering in a diagram, it remains so captured in any continuation. If $(D_\alpha)_{\alpha<\beta}$ is a system of consistent diagrams, with β ordinal, such that for $\alpha < \alpha' < \beta$, D_α is a continuation of $D_{\alpha'}$, then we have an obvious colimit,

$$\bigcup_{\alpha<\beta} D_\alpha,$$

a consistent diagram, a continuation of each D_α. Note also that, to start with, we have the consistent diagram D^0, with $S = \{*\}$, $(D^0)(*) = 1 \in C$. Now, given any consistent diagram, by repeatedly using the lemma, and using the colimit construction of diagrams, we can construct a consistent continuation in which every arrow with domain in the original diagram is captured for every J-covering. An ω-type colimit of continuations so constructed will give us a continuation D of D^0 in which every arrow, with domain any object in D, is captured for all J-coverings. As we said above, this means that $M(D)$ is a morphism of sites $(C,J) \longrightarrow \mathrm{Set}$.

It remains to prove the lemma.

Let us choose, for any $t \geq s$, and $i \in I$, the object D_{ti} and morphisms δ_{ti}, g_{ti} to form a pullback diagram

$$\begin{array}{ccccc}
D_t & \xrightarrow{\ \delta_{ts}\ } & D_s & \xrightarrow{\ g\ } & A \\
\uparrow{\scriptstyle \delta_{ti}} & & & & \uparrow{\scriptstyle f_i} \\
D_{ti} & & \xrightarrow{\qquad g_{ti}\qquad} & & A_i
\end{array}$$

Then, for any t and u such that $s \leq t \leq u$ we have a unique arrow δ_{uti} making the following commute:

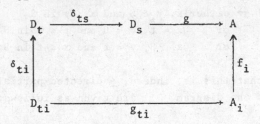

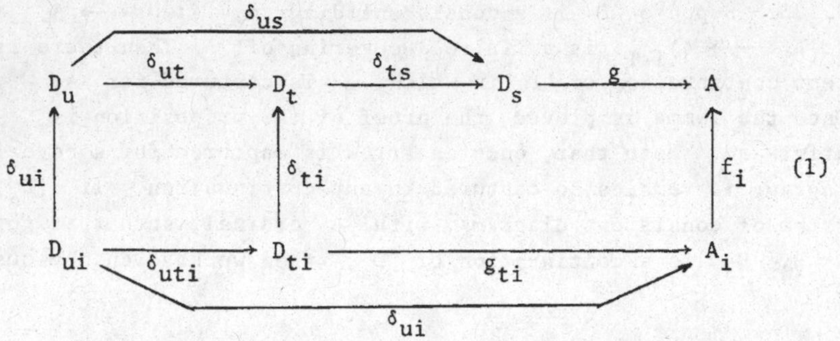

If D_{ti} is non-empty, and $u \geq t$, so is D_{ui}. We claim that there is $i \in I$ such that D_{ti} is non-empty for all $t \geq s$. Otherwise, for all $i \in I$ there is $t_i \geq s$ such that $D_{t_i i}$ is empty. Since I is finite, and since S is directed, there is $t \geq s$ such that $t \geq t_i$ for all $i \in I$. Thus, D_{ti} is empty for all $i \in I$. But, $(\delta_{ti}: D_{ti} \longrightarrow D_t)_{i \in i}$ is a covering, by axiom (ii) of topology. Hence, by axiom (iii), D_t is empty, contradicting the assumption that D is consistent. The claim is thus shown.

Let us fix i as in the claim.

We are ready to construct the desired continuation D'. The underlying poset will be $S' = S \amalg \{t: t \geq s\} = S \cup \{\hat{t}: t \geq s\}$. We put $D'_t = D_t$ for $t \in S$, and $D'_{\hat{t}} = D_{ti}$ for $t \geq s$. Thus, by the claim, we have ensured that D' is consistent. The partial ordering on S' is given as follows: for any $x, y \in S'$,

$$
\begin{aligned}
x \leq y \text{ in } S' \Longleftrightarrow\ & \text{either } x, y \in S \text{ and } x \leq y \text{ in } S \\
& \text{or} \quad x = \hat{t},\ y = \hat{u} \text{ and } t \leq u \text{ in } S \\
& \text{or} \quad x \in S,\ y = \hat{t} \text{ and } x \leq t \text{ in } S.
\end{aligned}
$$

One verifies easily that this is, indeed, a directed partial ordering. The remaining data of the diagram D' are given as follows:

$$\delta'_{\hat{u}\hat{t}} = \delta_{uti} \qquad (t \leq u)$$

$$\delta'_{\hat{u}t} = \delta_{ti} \circ \delta_{uti} \qquad (t \leq u)$$

(see diagram (1)). The commutativity of the left square in (1) is used to show that D' so constituted is, in fact, a functor.

Now, looking at (1), we see that g is captured for A in D' by g_{ti}, completing the proof of the lemma.

■ 2.1.3.
■ 2.1.2.

Theorem 2.1.1 will now be reduced to Proposition 2.1.2.

Lemma 2.1.3. Let J be a topology on C, and X an arbitrary collection of objects of C. Let J' be the topology generated by J together with the empty cover for each $A \in X$. Then J' can be described as follows. Let X' be the collection of all objects A' for which there is at least one arrow $A' \longrightarrow A$ into an object A in X. Then $A = (A_i \longrightarrow A)_{i \in I} \in J'$ if and only if there is $I' \supset I$ and an extension $A' = (A_i \longrightarrow A)_{i \in I}$ of A such that $A' \in J$ and $A_i \in X'$ for all $i \in I' - I$.

Proof: It is immediate that any A satisfying the condition is in J'. Conversely, it suffices to show that the collection of the families satisfying the condition is a topology; this is an easy verification.

■ 2.1.3.

Lemma 2.1.4. Suppose the family $(A_i \longrightarrow \mathbf{1})_{i \in I}$ with the joint codomain $\mathbf{1}$, the terminal object of C, is *not* a J-covering. Then in the topology J' generated by J and the empty coverings for the objects A_i, $\mathbf{1}$ is non-empty.

Proof: Suppose $\mathbf{1}$ is empty in J'. Then, by Lemma 2.1.3, there is a J-covering $(B_j \longrightarrow 1)_j$ such that for each j, there is an arrow $B_j \longrightarrow A_{i_j}$ to some $i_j \in I$. Since

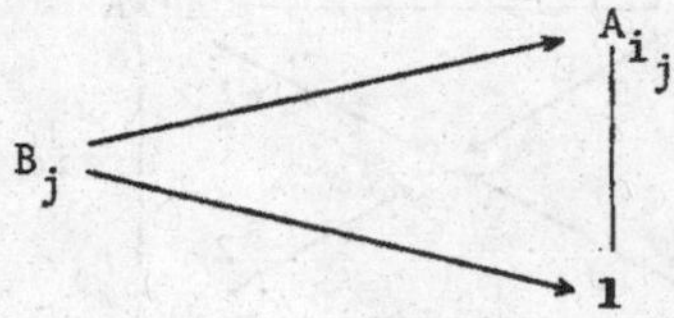

commutes ($\mathbf{1}$ being terminal), by axiom (iv) $(A_i \longrightarrow \mathbf{1})_i$ is a J-covering, contrary to the assumption.

■ 2.1.4.

Proof of Theorem 2.1.1.: We consider the comma category C/A (denoted $C{\downarrow}A$ in [CWM]). C/A has finite limits; the terminal object in C/A is the arrow $\mathrm{id}: A \longrightarrow A$. We have the functor

$$F\colon \mathcal{C} \longrightarrow \mathcal{C}/A \;:\quad C \longmapsto \left(\begin{array}{c} C\times A \\ \downarrow{\scriptstyle \pi_2} \\ A \end{array}\right)$$

$$\begin{array}{c} C \\ {\scriptstyle f}\downarrow \\ C' \end{array} \longmapsto f\times id_A \qquad \begin{array}{ccc} C\times A & & \\ \downarrow & \searrow{\scriptstyle \pi_2} & \\ & & A \\ & \nearrow{\scriptstyle \pi_2'} & \\ C'\times A & & \end{array}$$

embedding $\mathcal{C}$ into $\mathcal{C}/A$; F preserves finite limits. We define the topology J/A on $\mathcal{C}/A$ as follows:

$$\left(\begin{array}{ccc} B_k & \longrightarrow & B \\ & \searrow \quad \swarrow & \\ & A & \end{array}\right)_{k\in K} \in J/A \iff (B_k \longrightarrow B)_{k\in K} \in J$$

F is a morphism of sites $(\mathcal{C},J) \longrightarrow (\mathcal{C}/A,\ J/A)$. The following is always a pullback diagram in J/A:

$$\begin{array}{ccccc} A' & & \xrightarrow{\langle id,f\rangle} & & A'\times A \\ & \searrow{\scriptstyle f} & & \swarrow{\scriptstyle \pi_2'} & \\ {\scriptstyle f}\Big| & & A & & \Big|{\scriptstyle f\times id} \\ & \nearrow{\scriptstyle id} & & \nwarrow{\scriptstyle \pi_2} & \\ A & & \xrightarrow[\Delta=\langle id,id\rangle]{} & & A\times A \end{array} \qquad (2)$$

All the above facts can be verified easily, if not already known. Now, assume the hypotheses of the theorem. The family

$$A/A = \left(\begin{array}{ccc} A_i & \xrightarrow{f_i} & A \\ & {\scriptstyle f_i}\searrow \quad \nearrow{\scriptstyle id} & \\ & A & \end{array}\right)_{i\in I}$$

is not a covering on $\mathbf{1}_{\mathcal{C}/A}$ in J/A. By Lemma 2.1.4, in the topology J' on $\mathcal{C}/A$ generated by J/A and the empty covering of each $f_i\colon A_i \longrightarrow A \in \mathrm{Ob}(\mathcal{C}/A)$, the terminal object $\mathbf{1}_{\mathcal{C}/A}$ is not empty. By

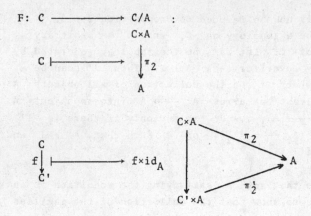

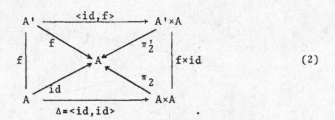

Proposition 2.1.2, there is a morphism of sites

$$M'\colon (C/A, J') \longrightarrow \mathrm{Set}.$$

By the definition of J', $M'(f_i) = \emptyset$, for all $i \in I$. Now, let $M = M' \circ F\colon C \longrightarrow \mathrm{Set}$. M is a morphism $(C,J) \longrightarrow \mathrm{Set}$. With any fixed $i \in I$, $A' = A_i$, $f = f_i$, the pullback diagram (2) is taken into a pullback in Set by M', i.e. the diagram

$$\begin{array}{ccc} M'(f_i) & \longrightarrow & M(A_i) \\ \downarrow & & \downarrow{\scriptstyle M(f_i)} \\ \mathbf{1} = M'(1_{C/A}) & \xrightarrow[M'(\Delta)]{} & M(A) \end{array}$$

is a pullback in Set. Since $M'(f_i) = \emptyset$, this means that the element of $M(A)$ picked out by $M'(\Delta)$ is not in the image of any $M(f_i)$, $i \in I$, i.e., the image of the family A under M is not a canonical covering.

□ 2.1.1.

2.2. Coherent categories.

Let C be a category with finite limits, A an object in C. Then the subobjects (see [CWM]) of A form a partially ordered set, denoted $\mathrm{Sub}(A)$. In fact, $\mathrm{Sub}(A)$ is a meet ($\wedge$-) semilattice, the meet of the subobjects represented by the monos $B \longrightarrow A$, $C \longrightarrow A$ being represented by $D \longrightarrow S$ from a pullback diagram

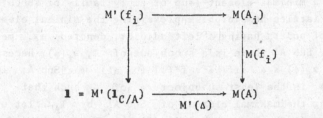

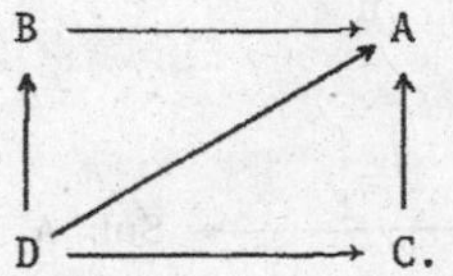

Given an arrow $f\colon A' \longrightarrow A$, the map

$$f^*\colon \mathrm{Sub}\,A \longrightarrow \mathrm{Sub}\,A' \qquad (1)$$

is defined by pullback: $f^*[B \longrightarrow A] = [B' \longrightarrow A']$ where

$$\begin{array}{ccc} B & \longrightarrow & A \\ \uparrow & & \uparrow f \\ B' & \longrightarrow & A' \end{array}$$

is a pullback. f^* is a homomorphism of meet semilattices.

By the condition that "C has stable finite sups of subobjects" we mean that each $\mathrm{Sub}\,A$ $(A \in \mathrm{Ob}(C))$ is a lattice (having a join operation, $\vee$) with a minimal element (sup of empty family of subobjects), and each f^* is a lattice homomorphism preserving the minimal element.

The map (1) of posets having a left adjoint, denoted $\exists_f$, means that for every $\phi \in \mathrm{Sub}\,A'$ there is a subobject of A, $\exists_f(\phi)$, necessarily unique, such that $\exists_f(\phi) \le \psi$ iff $\phi \le f^*(\psi)$ for all $\psi \in \mathrm{Sub}\,A$; in other words, $\exists_f(\phi)$ is the least subobject ψ of A such that $\phi \le f^*(\psi)$. Denoting the maximal element of $\mathrm{Sub}\,A'$ by $1_{A'}$, let us write $\mathrm{Im}(f)$ for $\exists_f(1_{A'})$. To say that $\mathrm{Im}(f) = 1_A$ means that f does not factor through any mono with codomain A unless the mono is an isomorphism; this circumstance is referred to by saying that f is an *extremal epimorphism*. The existence of $\exists_f$, for all f, is equivalent to saying that every morphism can be factored as the composition of an extremal epi followed by a mono. It is immediate that a morphism is an isomorphism just in case it is both a mono and an extremal epi.

The so-called *Beck-condition* for $\exists$ means that for any f as above, any $g: B \longrightarrow A$ and a pullback

$$\begin{array}{ccc} A' & \xrightarrow{f} & A \\ g' \uparrow & & \uparrow g \\ B' & \xrightarrow[h]{} & B \end{array} \qquad (2)$$

the diagram

$$\begin{array}{ccc} \mathrm{Sub}\,A' & \xrightarrow{\exists_f} & \mathrm{Sub}\,A \\ (g')^* \downarrow & & \downarrow g^* \\ \mathrm{Sub}\,B' & \xrightarrow[\exists_h]{} & \mathrm{Sub}\,B \end{array}$$

commutes. This is equivalent to saying that whenever in the pullback (2) f is an extremal epi, then so is h. If $\exists_f$ always exists and satisfies the Beck-condition, we say that C "has stable images".

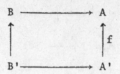

A *category suitable for finitary coherent (geometric) logic*, or simply: a *coherent category* (in [MR]: "logical category") is a category T with finite limits having stable finite sups of subobjects and stable images.

Let T be a coherent category. Define J ($= J_T$) to be the collection of all families $(f_i: A_i \longrightarrow A)_{i \in I}$ of morphisms of T such that some finite $I \subset I'$ we have

$$\bigvee_{i \in I} I_m(f_i) = 1_A.$$

J is a topology on T. Indeed, axiom (ii) follows by the stability conditions on sups and images. Axiom (iii) easily follows from the following two formulas: for morphisms

$$A'' \xrightarrow{\quad g \quad} A' \xrightarrow{\quad f \quad} A$$

we have

$$\exists_{fg}\ (\phi) = \exists_f(\exists_g(\phi)) \qquad (\phi \in \text{Sub } A'')$$

and for

$$A' \xrightarrow{\quad f \quad} A$$

we have

$$\exists_f(\bigvee_{i \in I} \phi_i) = \bigvee_{i \in I} \exists_f(\phi_i) \qquad (I \text{ finite}, \ \phi_i \in \text{Sub } A').$$

The two formulas are, in turn, easily checked. Axioms (i) and (iv) are immediate.

Note that a monomorphism is a covering in J (as a singleton) iff it is an isomorphism.

For a functor $F: T \longrightarrow T'$ between coherent T and T' preserving finite limits, saying that F is morphism of sites $(T, J_T) \longrightarrow (T', J_{T'})$ is equivalent to saying that F preserves finite sups and images (this latter condition means, naturally, that for

$$F_A: \text{Sub}_T\ A \longrightarrow \text{Sub}_{T'}(FA)$$

the order preserving map induced by F, F_A is a lattice homomorphism preserving minimal elements, and that whenever $f: A' \longrightarrow A$ is in T, then

$$F_A\,(\exists_f(\phi)) = \exists_{Ff}(F_A(\phi)) \qquad (\phi \in \text{Sub } A)).$$

The truth of this assertion is quite clear. Such an F may be called, naturally, a *coherent functor*, or *coherent morphism* ("logical functor", in [MR]). The category of all coherent $T \longrightarrow T'$ is denoted Coh(T,T'); it is a full subcategory of (T,T'). A coherent functor $T \longrightarrow$ Set is a *model* of T; Coh(T,Set) is denoted Mod(T).

A functor $F\colon T \longrightarrow T'$ *reflects isomorphisms* (or is *conservative*) if Ff being an isomorphism implies that f is an isomorphism. If T, T' have finite limits and F preserves them, then F being conservative implies that F is faithful.

One can apply Theorem 2.1.1 to conclude that for any small coherent category T, and any monomorphism in T which is not an isomorphism, there is a coherent $T \longrightarrow$ Set which takes the given monomorphism into a mono which is, again, not an isomorphism. Putting several such coherent functors together, one for each mono in T that is not an isomorphism, we obtain

Theorem 2.2.2. (Completeness for finitary coherent logic). Any small coherent category has a conservative coherent embedding into a small Cartesian power of Set.

■ 2.2.2.

One should note the easily seen facts that if $F\colon T \longrightarrow T'$ is coherent and conservative, then, for any finite diagram in T, if its F-image is a limit diagram in T', then the original diagram is a limit diagram in T as well; moreover, similar reflection properties hold for sups and images. This means that any 'diagrammatic property' of objects and morphisms involving commutation of diagrams, finite limits, sups of subobjects and images that holds throughout in Set, and hence in any Cartesian power of Set, will hold in any coherent category as well. This fact expresses the completeness of the defining axioms for 'coherent category' for the 'standard coherent logic', that of Set.

Definition 2.2.3. In a category with finite limits, a family $(f_i\colon A_i \longrightarrow A)_{i\in I}$ is said to be an *effective epimorphic family* if the following holds, with

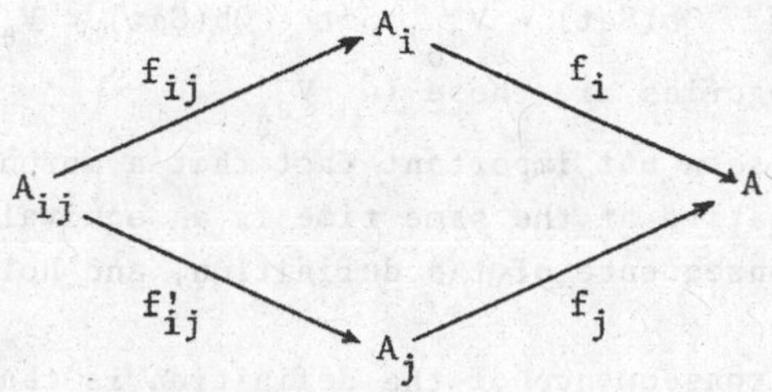

a pullback diagram for all $i,j \in I$:

for any object B and any system $(g_i: A_i \longrightarrow B)_{i\epsilon i}$ of morphisms, if $g_i f_{ij} = g_j f'_{ij}$ for all i and j in I, then there is a unique morphism $h: A \longrightarrow B$ such that $hf_i = g_i$ for all i in I.

Proposition 2.2.4. The topology J_T defined above in a coherent T coincides with the so-called precanonical topology, i.e. the one in which a family is a covering iff it contains a finite effective epimorphic family.

The fact that any family in J_T is effective epimorphic is seen, for $T = Set$, by a careful inspection. Then the same fact, for a general T, is essentially a consequence of completeness, 2.2.2. The detailed proof of 2.2.4 is postponed until Section 2.5.

2.3. The quotient-conservative factorization.

The definition of a quotient-morphism is given in Section 1.4 in the context of pretoposes. The definition has a general character; it can be repeated in other, similar, situations. Such a situation is given by a (concrete) 2-category (c.f. [CWM]): certain categories (the "objects" of the 2-category), certain functors between them (the "morphisms" of the 2-category), and certain natural transformations between those functors (the "2-cells" of the 2-category). E.g., Pretop is the 2-category of pretoposes, pretopos-morphisms, and all natural transformations between such. Similarly, we may talk about Lex, the 2-category of categories with finite limits, functors preserving them ("Lex-morphisms") and all natural transformations between such. The 2-category of coherent categories is Coh.

The 'standard' 2-category is Cat, the 2-category of categories, functors and natural transformations.

We want Set, the category of sets to be an object of Cat, and also of Lex, Pretop, etc. Therefore, we have in mind *three* set-theoretical universes: V_{θ_i} for $i = \emptyset, 1, 2$, with $\theta_0 < \theta_1 < \theta_2$

inaccessible cardinals. Ob(Set) = V_{θ_0}, Set $\in$ Ob(Cat) $\subset V_{\theta_1}$, Cat $\in V_{\theta_2}$. Small categories are those in V_{θ_0}.

Note the easily seen but important fact that a morphism which is a quotient and conservative at the same time is an equivalence; this is true as a direct consequence of the definition, and holds in any of our 2-categories.

Another obvious consequence of the definition is that if $I: T \longrightarrow T'$ is a quotient morphism in the 2-category $\mathbb{D}$, then, for any $T'' \in \mathbb{D}$, $(\,)\circ I: \mathbb{D}(T',T'') \longrightarrow \mathbb{D}(T,T'')$ is full and faithful. A result essentially contained in [MR] is that for $\mathbb{D}$ = Pretop, and T'' = Set, if $I: T \longrightarrow T'$ is such that $I^* = (\,)\circ I$: Mod $T' \longrightarrow$ Mod T is full and faithful, then I is a quotient (cf. § 3.1 below).

First, we will be concerned with Lex; as we said above, the concept of a quotient-morphism is defined as in 1.4.2, with Lex replacing Pretop.

<u>Proposition</u> 2.3.1. Any Lex-morphism $F: T \longrightarrow T'$ can be factorized into Lex-morphisms:

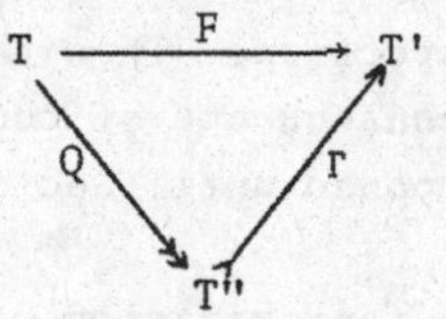

$F = \Gamma\circ Q$, so that Q is a quotient, and Γ is conservative.

Proof: Let $\Sigma = \mathrm{Inv}(F) \subset \mathrm{Morph}(T)$. We define T'' to have the same objects as T. The morphisms

$$A \cdots\cdots\cdots\to B$$

of T'' will be equivalence classes of pairs (f,s) as shown:

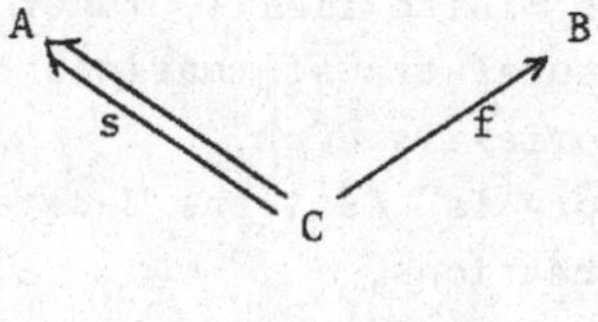

with varying C and with $s \in \Sigma$, under the following equivalence relation:

$$(f,s) \sim (f',s') \Leftrightarrow (Ff)(Fs)^{-1} = (Ff')(Fs')^{-1}$$

(remember that for $s \in \Sigma$, Fs is an isomorphism). To have a simpler notation, for the right hand side of the last defining equivalence we rather write

$$fs^{-1} \underset{T'}{=} f'(s')^{-1},$$

and read: "fs^{-1} equals $f's'^{-1}$ in T'". Thus, the definition is:

$$(f,s) \sim (f',s') \Leftrightarrow fs^{-1} \underset{T'}{=} f's'^{-1}.$$

This clearly defines an equivalence relation on the pairs (f,s) with $f \in T(C,A)$, $s \in T(C,B) \cap \Sigma$, with fixed A and B but varying C. For $(f,s)/\sim$, the morphism represented by (f,s), we'll write $\ulcorner fs^{-1} \urcorner$. The letters s, t, u always stand for morphism in Σ; in diagrams, arrows in Σ are indicated by double arrows.

To define composition, what we need to show is that, in the situation

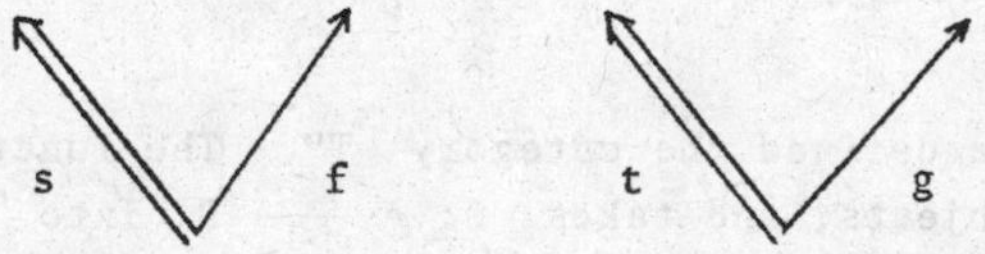

the morphism $gt^{-1}fs^{-1}$ is equal, *in* T', to one of the form hu^{-1}. Consider the extension

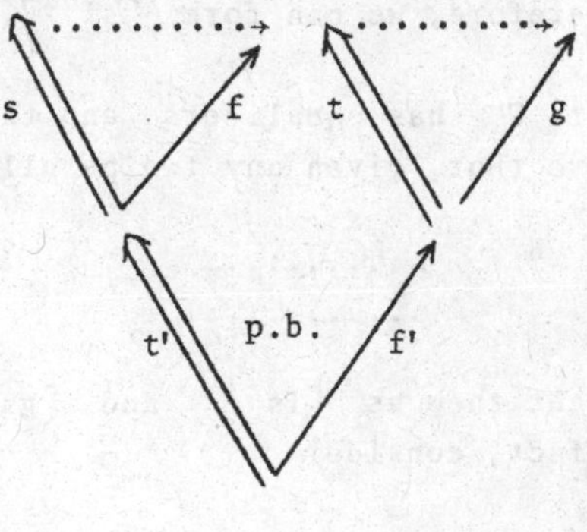

with the pullback indicated; this diagram is in T, except for the dotted arrows. Reading it in T', we also have the dotted arrows; it is commutative in T', and the pullback remains a pullback. It follows that t' is an iso in T', since t is; thus, $t' \in \Sigma$. Now, it is clear that

$$(gt^{-1})(fs^{-1}) \underset{T'}{=} (gf')(st')^{-1}.$$

Therefore, we *define*:

$$\ulcorner gt^{-1} \urcorner \underset{T''}{\circ} \ulcorner fs^{-1} \urcorner = \ulcorner (gf')(st')^{-1} \urcorner.$$

This definition is legitimate; if different representatives of the equivalence classes are chosen, the resulting right-hand-sides will again be equivalent, simply because they will be the same morphism *in* T'.

The identity morphisms in T' are given in the obvious way:

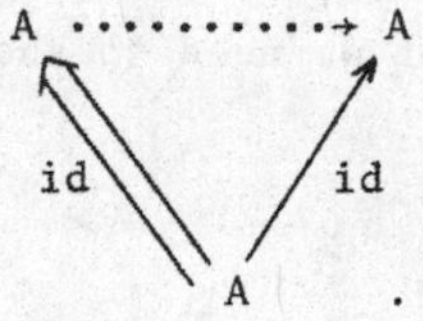

So far, we have defined the category T". The functor Q acts as the identity on objects, and takes $f: A \longrightarrow B$ into $\ulcorner f(id_A)^{-1} \urcorner$.

The functor Γ takes A into F(A) (so that $F = \Gamma \circ Q$ on objects), and $\ulcorner fs^{-1} \urcorner$ into fs^{-1} *in* T', i.e. $(Ff)(Fs)^{-1}$; clearly, Γ is well-defined, it is faithful, and $F = \Gamma \circ Q$. Q is also conservative: if fs^{-1} becomes an isomorphism under Q, i.e. fs^{-1} is an iso *in* T', then f is an iso *in* T' (since s is an iso *in* T'), hence $f \in \Sigma$; therefore, we can form $\ulcorner sf^{-1} \urcorner$, and it is clearly the inverse of $\ulcorner fs^{-1} \urcorner$.

Let us check that T" has equalizers, and that Γ preserves them. First of all, we observe that, given any two parallel morphisms

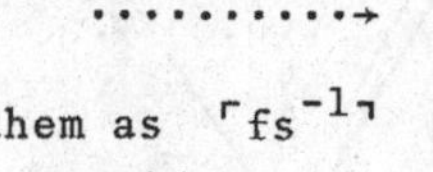

in T", we may represent them as $\ulcorner fs^{-1} \urcorner$ and $\ulcorner gs^{-1} \urcorner$ simultaneously with the same s. In fact, consider

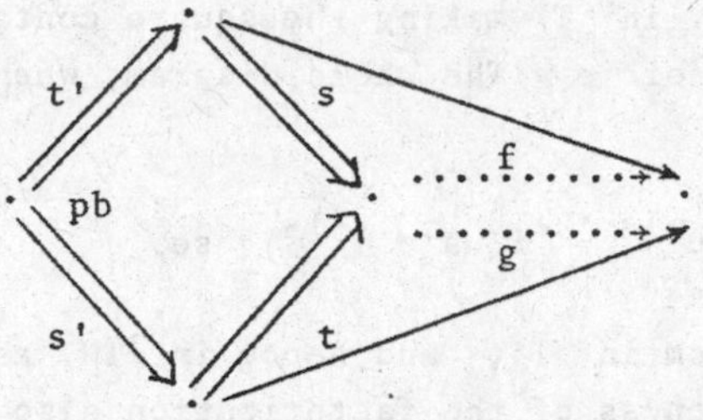

It is clear that, with $u = st' = ts'$, we have $\ulcorner fs^{-1} \urcorner = \ulcorner (ft')u^{-1} \urcorner$, $\ulcorner gt^{-1} \urcorner = \ulcorner (gs')u^{-1} \urcorner$.

Given two parallel morphisms $\alpha = \ulcorner fs^{-1} \urcorner$, $\beta = \ulcorner gs^{-1} \urcorner$ in T", let e be the equalizer of f and g (in T); since F preserves equalizers, and s is an iso *in* T', es is the equalizer of fs^{-1} and gs^{-1} *in* T'. Therefore, we are left with the task of showing that se $= \ulcorner se(id)^{-1} \urcorner$ is an equalizer of $\ulcorner fs^{-1} \urcorner$, $\ulcorner gs^{-1} \urcorner$ in T". To do this, we consider a morphism $\gamma = \ulcorner hu^{-1} \urcorner$ satisfying $\alpha\gamma = \beta\gamma$. Consider the following diagram in T:

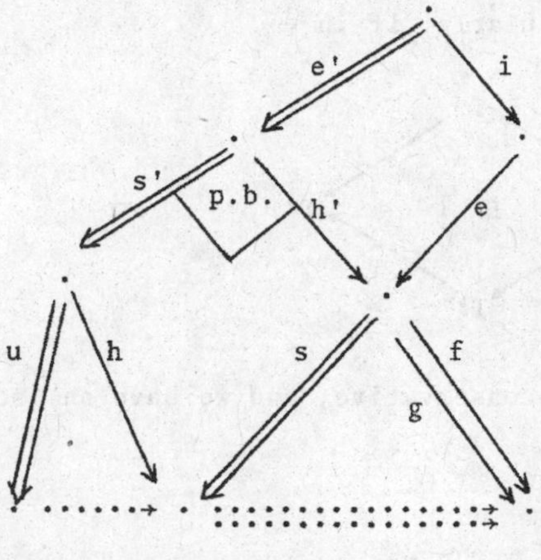

(the dotted arrows appear only in T' and T"). s' and h' form a pullback with s and h; e' is the equalizer of fh' and gh'. Now, the composite $\alpha\gamma$ is $\ulcorner (fh')(us')^{-1} \urcorner$, $\beta\gamma$ is $\ulcorner (gh')(us')^{-1} \urcorner$. To say that they are equal is to say that they are equal *in* T', i.e. that

$$fh' \underset{T'}{=} gh' ;$$

in other words, that e' is an isomorphism *in* T', i.e., that $e' \in \Sigma$.

There is a unique arrow i in T making the square containing it commute, by the definition of e. The whole diagram, when read *in* T', is commutative. Hence

$$\gamma = \ulcorner hu^{-1} \urcorner = \ulcorner i(us'e')^{-1} \urcorner \circ se.$$

Since se is a monomorphism in T', and hence in T'' as well (as it is easily seen), the uniqueness of the factorization also follows. This completes the verification of equalizers in T''.

The verification of the terminal object and products in T'', as well as the remaining parts of the facts that Q and Γ are Lex-morphisms are left to the reader.

It is clear that $\mathrm{Inv}(Q) = \mathrm{Inv}(F)$. Since T'' is constructed by doing the bare minimum to invert the morphisms in $\mathrm{Inv}(Q)$, it is intuitively clear that Q is a quotient; it is not hard to prove it rigorously either.

■ 2.3.1.

<u>Proposition</u> 2.3.1'. The quotient-conservative factorization in Lex is essentially unique. More particularly, if in

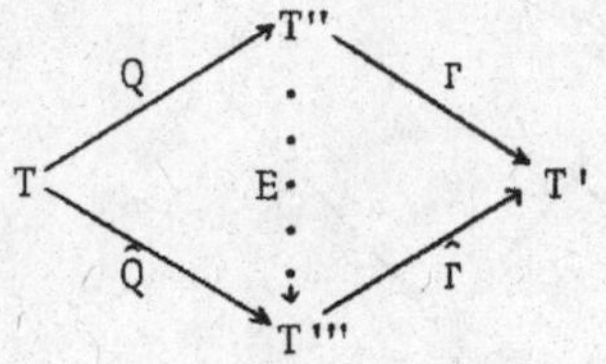

Q, $\hat{Q}$ are quotients, Γ, $\hat{\Gamma}$ are conservative, and we have an isomorphism

$$\Gamma\, Q \cong \hat{\Gamma}\, \hat{Q}, \tag{1}$$

then there are an equivalence functor E and isomorphisms

$$E\, Q \cong \hat{Q}$$
$$\Gamma \cong \hat{\Gamma}\, E$$

whose composite

$$\Gamma\, Q \cong \hat{\Gamma}\, E\, Q \cong \hat{\Gamma}\, \hat{Q}$$

is the one given in (1).

Proof: Straightforward.

■ 2.3.1'.

Let us work in a fixed category T with finite limits. The *graph* of a morphism $f: A \longrightarrow B$ is the subobject represented by the monomorphism

$$\langle id,f\rangle: A \longrightarrow A\times B.$$

A *functional subobject* ϕ of $A\times B$ is one that is represented by a monomorphism $X \longrightarrow A\times B$ for which the composite

$$X \longrightarrow A\times B \xrightarrow{\pi_1} A$$

is an isomorphism. (Verify that these concepts mean the expected things in Set.)

Mapping $T(A,B)$ into $Sub(A\times B)$ by taking the graph of the morphism, we obtain a bijection of $T(A,B)$ onto the set of functional subobjects of $A\times B$. In fact, given a functional subobject represented by $i: X \longrightarrow A\times B$, the morphism whose graph is i is $\pi_2 i(\pi_1 i)^{-1}$ ($\pi_2: A\times B \longrightarrow B$: the canonical projection).

Proposition 2.3.2. (Andrew Pitts) A Lex-morphism $F: T \longrightarrow T'$ is a quotient if and only if the following condition holds: for any morphism of the form $f: X \longrightarrow FA$ in T', there is a morphism $g: A' \longrightarrow A$ in T and an isomorphism $X \cong FA'$ such that

$$\begin{array}{ccc} X & \xrightarrow{f} & FA \\ & \searrow\!\!\!\!\!= \quad \nearrow_{Fg} & \\ & FA' & \end{array}$$

commutes.

Proof: Note that the particular construction in 2.3.1 of the quotient $Q: T \longrightarrow T''$, the first factor in the quotient-conservative factorization of any Lex-morphism, satisfies the condition. Now, by the uniqueness of this factorization (2.3.1'.), one easily deduces that the condition holds for any quotient-morphism.

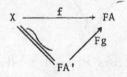

Conversely, assume that the condition holds, and consider the quotient-conservative factorization

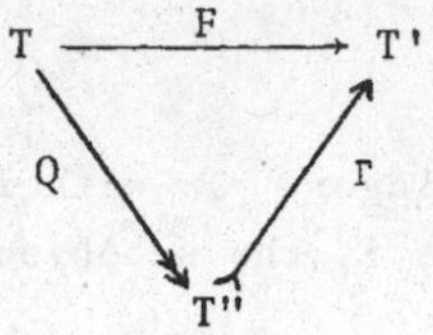

First, we may easily verify that Γ satisfies the same condition as the one assumed for F. Next, we use this fact to show that Γ is an equivalence. The only thing to prove is that Γ is full. Suppose we have a morphism $f\colon \Gamma A \longrightarrow \Gamma B$ in T'. Consider its graph $\Gamma A \longrightarrow \Gamma A \times \Gamma B \cong \Gamma(A \times B)$. By the condition, there is $g\colon C \longrightarrow A \times B$ in T'' such that Γg is a monomorphism, and Γg represents the graph of f. In particular, Γg represents a functional subobject, hence, since Γ is conservative, g represents a functional subobject $[g]$ of $A \times B$. Let $h\colon A \longrightarrow B$ be the morphism in T whose graph is $[g]$. But then Γh and f have the same graph, namely the one represented by Γg; hence $\Gamma h = f$, as required.

■ 2.3.2.

<u>Proposition</u> 2.3.3. Suppose T, T' are coherent categories, $F\colon T \longrightarrow T'$ a coherent functor. Then, taking the quotient-conservative factorization of F in Lex from 2.3.1:

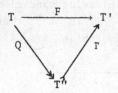

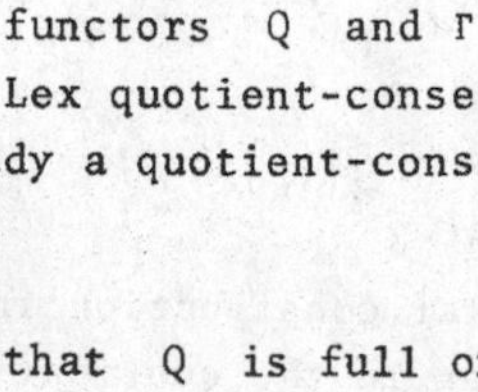

we find that T'' and the functors Q and Γ are, in fact, coherent.

It follows that the Lex quotient-conservative factorization of a coherent morphism is already a quotient-conservative factorization in Coh.

<u>Proof</u>: Let us note first that Q is full on subobjects: the function $Q_A\colon \mathrm{Sub}(A) \longrightarrow \mathrm{Sub}(QA)$ induced by Q is surjective. Indeed, by 2.3.2, every subobject ϕ of QA is represented by a mono of the form $Qf\colon QA' \longrightarrow QA$. Consider the image-mono factorization of f in

T: $A' \xrightarrow{q} A'' \xrightarrow{i} A$, $iq = f$. Since $Ff \cong \Gamma Qf$ is a mono, Fq is a mono as well. Since also, Fq is an extremal epi, Fq is an isomorphism. Since Γ is conservative, and $Fq \cong \Gamma Qq$, Qq is an isomorphism. But then Qi represents the same subobject of QA as Qf; thus $\phi = Q_A([i])$, as required.

Let $f: A \longrightarrow B$, and $\phi \in \mathrm{Sub}(A)$, be in T. To verify that Q preserves extremal epis, by the subobject-fullness of Q, it suffices to show the equivalence

$$Q(\exists_f(\phi)) \leq Q(\psi) \iff Q(\psi) \leq Q(f^*(\psi)) \qquad (\psi \in \mathrm{Sub}\ B).$$

But, applying Γ to both sides, the equivalence becomes true since F preserves extremal epis; since Γ is conservative, it follows that the equivalence itself holds.

Since Q is, also, essentially surjective (see 2.3.2) it follows that T" has images and that Γ preserves them. By the conservativeness of Γ, it easily follows that the property of stability under pullback of images is reflected from T' to T".

Similar arguments apply to sups of subobjects.

■ 2.3.3.

Corollary 2.3.4. If a Coh-morphism is a Lex-quotient, then it is a Coh-quotient.

■ 2.3.4.

Corollary 2.3.5. A Coh-morphism is a quotient in Coh iff it is full on subobjects and essentially surjective.

Proof: The 'only if' part follows from the proof of 2.3.3. For the 'if' part, assume that $F: T \longrightarrow T'$ satisfies the conditions. By 2.3.4. it suffices to show that F is a Lex-quotient, i.e., that it satisfies the condition of 2.3.2.

Let $f: X \longrightarrow FA$ be given. Find A' and an isomorphism $X \cong FA'$; consider the composite

$$f': FA' \cong X \xrightarrow{f} FA.$$

The graph of f', the subobject of $FA' \times FA$ represented by $\langle id, f' \rangle: FA' \longrightarrow FA' \times FA$ is represented by $Fi: FA'' \longrightarrow F(A' \times A) \cong FA' \times FA$, for some mono $i: A'' \longrightarrow A' \times A$. But then the diagram

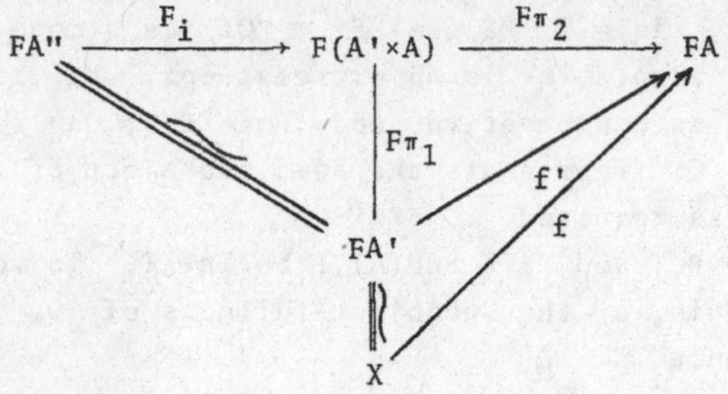

commutes, and so does

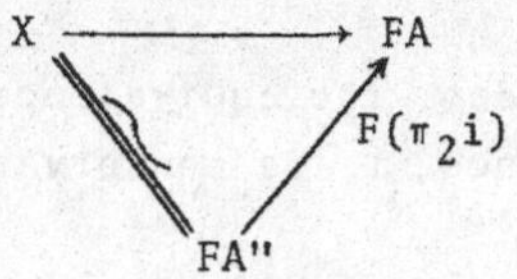

as required.

■ 2.3.5.

<u>Corollary</u> 2.3.6. If $F: T \longrightarrow T'$ is a Coh-quotient, and T is Boolean, then T' is Boolean as well.

Proof: Straightforward from 2.3.5.

■ 2.3.6.

2.4. <u>Pretoposes</u>.

Every category in this section has (at least) finite limits. If the context is not explicitly mentioned, we work in a fixed category with finite limits.

The ith projection $A_1 \times \cdots \times A_n \longrightarrow A_i$ is denoted by π_i, and for the morphism

$$<\pi_{i_1}, \cdots, \pi_{i_k}>: A_1 \times \cdots \times A_n \longrightarrow A_{i_1} \times \cdots \times A_{i_k}$$

we write $\pi_{i_1 \cdots i_k}$.

<u>Definition</u> 2.4.1. An *equivalence relation* is a pair of parallel morphisms

$$R \rightrightarrows A$$

such that $i = <p,q>: R \longrightarrow A \times A$ is a monomorphism, and such that if we also write R for the subobject of $A{\times}A$ represented by i, then we have the following:

$\Delta_A \le R$ (reflexivity)

$\pi_{21}{}^*(R) = R \qquad (\pi_{21}\colon A{\times}A \longrightarrow A;$ symmetry)

$\pi_{12}{}^*(R) \wedge \pi_{23}{}^*(R) \le \pi_{13}{}^*(R)$

$$(\pi_{12}, \pi_{23}, \pi_{13}\colon A{\times}A{\times}A \rightrightarrows A{\times}A; \text{ transitivity}).$$

One immediately notes that in Set, this is the definition given in 1.2. It is clear that an equivalence relation in the sense of the last definition is taken into another one by any functor preserving finite limits; hence the condition of 2.4.1 implies the defining condition in 1.2. Also, if a pair of parallel morphisms is taken by a conservative functor preserving finite limits into an equivalence relation according to 2.4.1, then the original pair is an equivalence relation too. It follows (by the 'completeness of the logic of finite limits', the special case of 2.1.1 for the trivial topology) that the two definitions are equivalent.

Next note that the kernel pair of any morphism is always an equivalence relation, since this is true in Set. Since in Set, any equivalence relation is the kernel pair of its own coequalizer (check!), equivalence relations could be defined by the condition that they are taken into a kernel pair by *some* conservative functor preserving finite limits.

It is easy to see that if a morphism is a coequalizer of some pair of morphisms, it is a coequalizer of its own kernel pair. Such a morphism is called an *effective epi*: clearly, a morphism f is an effective epi iff {f} is an effective epimorphic family according to 2.2.3. It is also easily seen that effective epis are extremal epis.

Definition 2.4.2. A pretopos is a category with finite limits having an initial object, coproducts of pairs of objects, and coequalizers of equivalence relations, and satisfying the following additional conditions:

(1) The initial object 0 is strict, i.e. every morphism with codomain 0 is an isomorphism;

(2) If

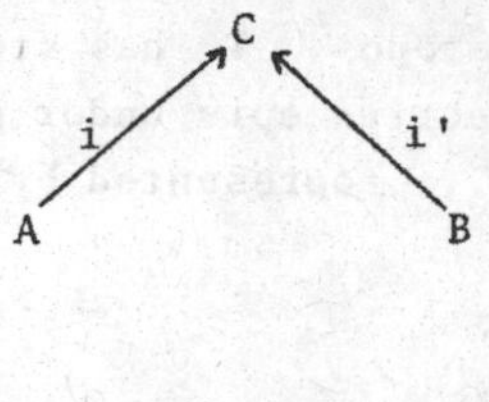

is a coproduct diagram, then i, i' are monos, and for the subobjects $[i]$, $[i']$ they represent, $[i] \wedge [i'] = O_C$ (O_C = subobject represented by $O \longrightarrow C$);

(3) If in

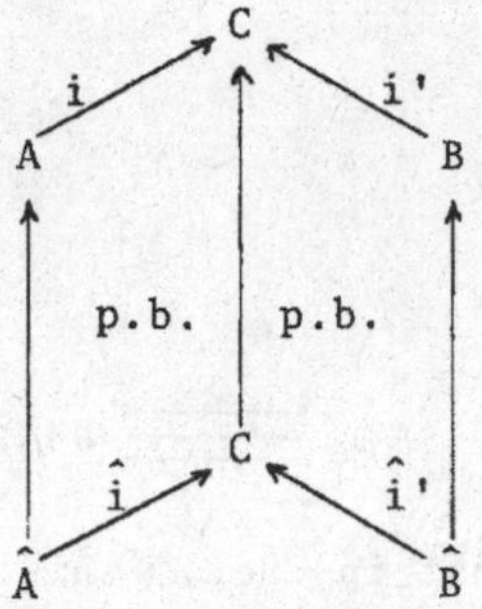

i, i' form a coproduct diagram, $\hat{A}$ and $\hat{B}$ are pullbacks as shown, then $\hat{i}, \hat{i}'$ also form a coproduct;

(4) Any pullback of an effective epi is again an effective epi,

(5) Any equivalence relation is a kernel pair of its coequalizer.

Lemma 2.4.3. In a pretopos, let a morphism $f: A \longrightarrow B$ be given. Let $s: A \longrightarrow C$ be the coequalizer of the kernel pair of f; by the universal property the coequalizer, we have a unique $i: C \longrightarrow B$ such that

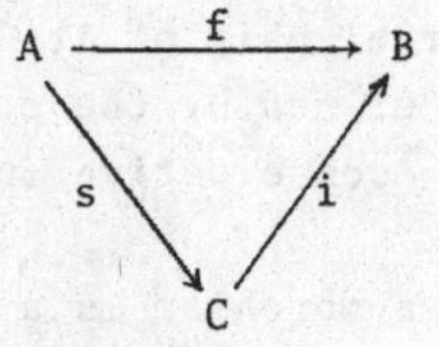

commutes. Then s and i give a factorization of f into an effective (hence extremal) epi followed by a mono.

We know that s is an effective epi. The proof that i is a mono is postponed until later.

Proposition 2.4.4. (i) Every pretopos is a coherent category.
(ii) A functor between pretoposes is a Coh-morphism iff it is a Pretop-morphism.

Proof: (i). Let T be a pretopos. T has stable images by 2.4.3 and the supposed stability of effective epis under pullback (2.4.2.(4)).

It is immediate that O_A, represented by $O \longrightarrow A$, is the minimal subobject of A; its pullback along any $A' \longrightarrow A$ is $0_{A'}$, by 2.4.2(1).

Here is the construction of joins of subobjects. Let two subobjects of A be represented by the monos $B \longrightarrow A$, $C \longrightarrow A$. Let us form the coproduct $B \amalg C$, with canonical injections $B \longrightarrow B \amalg C$, $C \longrightarrow B \amalg C$. We have a unique arrow $B \amalg C \longrightarrow A$ making the following commute:

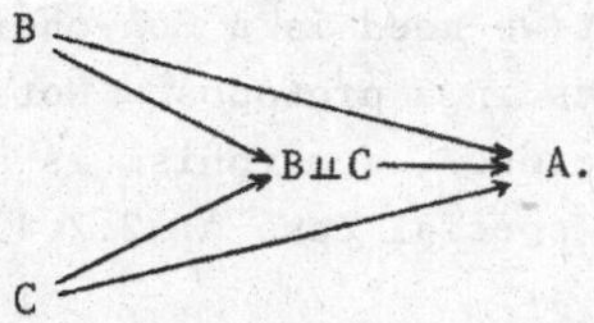

Now, let us form the effective epi - mono factorization of $B \amalg C \longrightarrow A$:

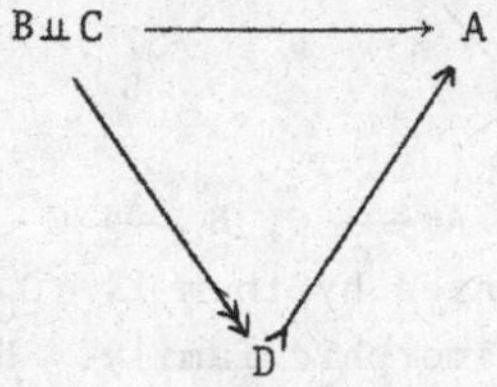

(1)

We claim that $D \longrightarrow A$ represents the desired join. Indeed, if $E \longrightarrow A$ is any mono such that both $B \longrightarrow A$, $C \longrightarrow A$ factor through it, then there is a unique $B \amalg C \longrightarrow E$ such that we have the commutative diagram

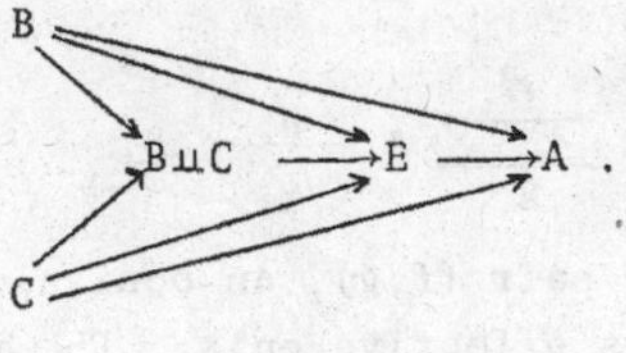

Now, if we form the effective epi - mono factorization of $B \amalg C \longrightarrow E$,

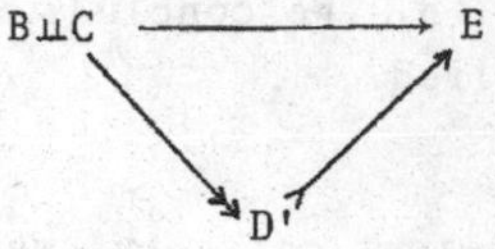

then, since $E \longrightarrow A$ is a mono, $B \amalg C \twoheadrightarrow D'$ and the composite $D' \longrightarrow E \longrightarrow A$ form an effective epi - mono factorization of $B \amalg C \longrightarrow A$, which has to be isomorphic to the one under (1). This means that $D \longrightarrow A$ factors through $E \longrightarrow A$, which was to be proved. The stability of sups under pullback is a consequence of 2.4.2(3) and (4).

(ii). The construction of images and joins in a pretopos makes it clear that any Pretop-morphism is a Coh-morphism.

For the converse, what we need is a Coh-characterization of effective epis and coproducts in a pretopos. Note that in a coherent category, and hence in a pretopos, a morphism is an effective epi iff it is an extremal epi, as a special case of 2.2.4. Also, by 2.4.2(2), in a pretopos T a diagram

$$A \longrightarrow C \longleftarrow B$$

is a coproduct diagram iff (a) $A \longrightarrow C$, $B \longrightarrow C$ are monos, (b) the meet of the subobjects represented by them is 0_C, and (c) the two morphisms form an effective epimorphic family. Hence, by 2.2.4, the same diagram is a coproduct diagram iff (a) and (b) hold, and $A \longrightarrow C$, $B \longrightarrow C$ form a covering in J_T. Thus, we have the required Coh-characterizations.

Given a coherent morphism $F: T \to T'$ between pretoposes T and T', we now see that F preserves finite limits and finite coproducts; it remains to see that F preserves coequalizers of equivalence relations. Let, in

$$R \underset{g}{\overset{f}{\rightrightarrows}} A \xrightarrow{\;h\;} B,$$

h be a coequalizer of the pair (f,g), an equivalence relation. Since, by the above, F preserves effective epis, Fh is an effective epi, hence a coequalizer of any of its kernel pairs. By the last clause(5) of the definition of pretopos, (f,g) is a kernel pair of h; hence (Ff, Fg) is a kernel pair of Fh. We conclude that Fh is a coequalizer of (Ff, Fg) as desired.

■ 2.4.4.

<u>Proposition</u> 2.4.5. For any coherent category, we have the *(free) pretopos completion of* T

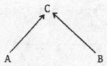

$$\gamma_T: T \longrightarrow P(T)$$

with $P(T)$ a pretopos, $\gamma = \gamma_T$ a Coh-morphism, and the following universal property: for any pretopos T', the functor

$$(\ ) \circ \gamma: \mathrm{Pretop}(P(T),T') \longrightarrow \mathrm{Coh}(T,T')$$

is an equivalence of categories.

Moreover, γ is conservative, full on subobjects (hence also full), and "every object of $P(T)$ is covered by γ", i.e. for every object X of $\gamma(T)$ there is a finite effective epimorphic family of the form $(\gamma(A_i) \longrightarrow X)_{i \in I}$.

Proof: The proof is quite straightforward, although long when written out fully. It resembles any one of a number of proofs in algebra of the type of the construction of the quotient-field of an integral domain.

It is convenient to break the construction into two steps. In the first, we adjoin disjoint sums; in the second, we adjoin quotients of equivalence relations to the result of the first step. It turns out that the final result is the desired pretopos completion.

To make this precise, we define *coherent categories with disjoint sums*, and their 2-category (Cohds for short), as follows. A Cohds-category is a coherent category having an initial object and coproducts of any two objects, satisfying conditions (1), (2) and (3) in 2.4.2; a Cohds-morphism is a Coh-morphism (automatically preserving the initial object and binary coproducts). A *coherent category with quotients of equivalence relations*, a Cohqe-category for short, is a coherent category in which every equivalence relation has a coequalizer, and condition (4) in 2.4.2 holds. We prove two propositions. One is our proposition, with Pretop replaced by Cohds, the other with Pretop replaced by Cohqe. The second proposition has the addendum that Cohqe(T) is also a Cohds-category, provided T is a Cohds-category. It is clear that the two propositions imply the one to be proved.

To carry out the first construction, let us make some preliminary observations. Let us put ourselves into a Cohds-category. Morphisms of the form

$$\coprod_i A_i \longrightarrow \coprod_j B_j$$

between finite disjoint sums are in one-to-one correspondence with tuples of the form

$$\langle A_i \longrightarrow \prod_j B_j \rangle_i .$$

Moreover, a morphism of the form $f\colon A \longrightarrow \coprod_j B_j$ gives rise to f_j, from the pullback

$$\begin{array}{ccc} A & \xrightarrow{\ f\ } & \coprod_j B_j \\ {\scriptstyle \ell_j}\Big| & & \Big|{\scriptstyle i_j} \\ A_j & \xrightarrow{\ f_j\ } & B_j \end{array}$$

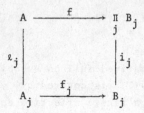

and in fact, the ℓ_j form a disjoint sum diagram. Conversely, given a disjoint sum representation $(A_j \longrightarrow A)_j$ of A, and morphisms $f_j\colon A_j \longrightarrow B_j$, one gets a unique f making the last diagram commute for all j.

This suggests the following definition for Cohds(T). Its objects are finite (possibly empty) tuples $(A_i)_i$ of objects A_i of T; such a tuple is denoted $\ulcorner \coprod_i A_i \urcorner$. A morphism

$$\ulcorner \coprod_i A_i \urcorner \longrightarrow \ulcorner \coprod_j B_j \urcorner$$

is given by: a disjoint sum representation $(A_{ij} \longrightarrow A_i)_j$ of each A_i, and a morphism $A_{ij} \longrightarrow B_j$ for each i and j. Two such morphisms, one as above, the other with primed items, are identified if and only if there is a system of isomorphisms $A_i \cong A'_i$, $A_{ij} \cong A'_{ij}$, such that all of the following commute:

$$\begin{array}{ccccc} & & A & & \\ & \nearrow & & \nwarrow & \\ A_i & & \cong & & A'_i \\ \uparrow & & & & \uparrow \\ A_{ij} & & \cong & & A'_{ij} \\ & \searrow & & \swarrow & \\ & & B & & \end{array}$$

To define composition, we return to a Cohds category, and in it, a pair of morphisms

$$\coprod_i A_i \longrightarrow \coprod_j B_j \longrightarrow \coprod_k C_k .$$

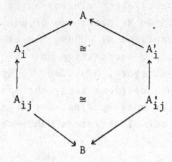

Consider the following

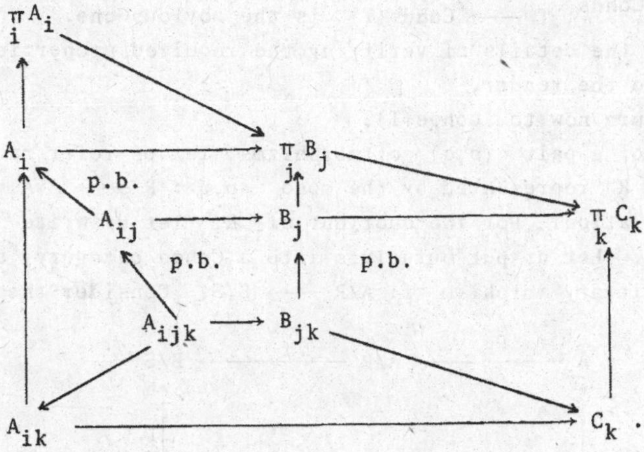

We are interested in deducing, for the sake of a later definition, the data

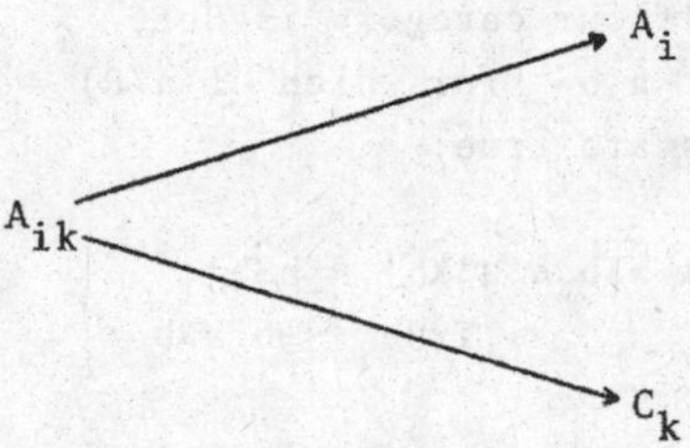

for the composite morphism, out of the ones

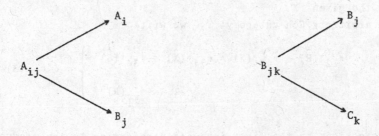

for the factors. We form the pullbacks A_{ijk}. It is clear that the arrows $(A_{ijk} \longrightarrow A_{ik})_j$ form a disjoint sum diagram. The desired morphism $A_{ik} \longrightarrow C_k$ is, therefore, given by the universal property of this last disjoint sum, from the components $A_{ijk} \longrightarrow B_{jk} \longrightarrow C_k$.

Now, it is clear how to formulate the definition of composition. The functor $\gamma_T^{Cohds}: T \longrightarrow Cohds(T)$ is the obvious one.

We leave the details of verifying the required properties of the construction to the reader.

Let us turn now to Cohqe(T).

Instead of a pair (p,q) of morphisms, let us refer rather to the subobject R represented by the mono $\langle p,q\rangle: R \longrightarrow A\times A$ as the equivalence relation. For the quotient of R, let us write $p_R: A \longrightarrow A/R$. Let us put ourselves into a Cohqe category, and let us analyze an arbitrary morphism $f: A/R \longrightarrow B/S$. Consider the pullback

$$\begin{array}{ccccc} A & \xrightarrow{p_R} & A/R & \xrightarrow{f} & B/S \\ \uparrow{\scriptstyle\alpha} & & & & \uparrow{\scriptstyle p_S} \\ X & & \xrightarrow[\beta]{} & & B \end{array}$$

and the subobject represented by $\langle\alpha,\beta\rangle: X \longrightarrow A\times B$; let us denote the latter subobject by $\hat{X}$. If our category is Set, $\hat{X}$ is the subset of $A\times B$ consisting of those $\langle a,b\rangle$ for which $f(a/R) = b/S$. Therefore, the following implications are true:

$$\left.\begin{array}{r} a\hat{R}a' \ \&\ a\hat{X}b \ \&\ a'\hat{X}b' \Rightarrow b\hat{S}b' \\ \text{True} \Rightarrow \exists b\ a\hat{R}b \end{array}\right\} \qquad (2)$$

with a, a' ranging over A, b, b' over B. Conversely, if X is a subset of $A\times A$ such that the implications (2) are true, then there is a unique $f: A/R \longrightarrow B/S$ such that $\hat{X}$ deduced from f as above is the same as the given $\hat{X}$.

In an arbitrary Coh-category T, we write

$$\left.\begin{array}{r} \pi_{12}{}^*(\hat{R}) \wedge \pi_{13}{}^*(\hat{X}) \wedge \pi_{24}{}^*(\hat{X}) \leq \pi_{34}{}^*(\hat{S}) \\ 1_A \leq \exists_{\pi_1}(\hat{R}) \end{array}\right\} \qquad (3)$$

with the projections in the first inequality referring to the product $A\times A\times B\times B$. Certainly, (2) and (3) mean the same in Set; and (3) holds in T iff for all Coh-morphisms $M: T \longrightarrow Set$, (2) holds with all of A, B, R, etc. meaning $M(A)$, $M(B)$, $M(R)$, etc.

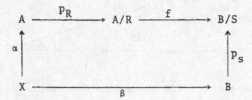

We make, therefore, the following definition. The objects of Cohqe(T) are given by pairs (A,R), with A an object of T, R an equivalence relation on A; we write ⌜A/R⌝ for (A,R). A morphism ⌜A/R⌝ $\longrightarrow$ ⌜B/S⌝ is given by an $\hat{X} \in \mathrm{Sub}(A\times B)$ (a 'relation from A to B') satisfying (3).

As for the composite of two morphisms, note that if we have a pair of morphisms

$$A/R \longrightarrow B/S \longrightarrow C/U$$

and $\hat{X}$, $\hat{Y}$ are the relations deduced from them, then the relation deduced from their composite is the relational product $\hat{X}\Box\hat{Y}$ of $\hat{X}$ and $\hat{Y}$, which, in Set, is defined as

$$a(\hat{X}\Box\hat{Y})c \iff \exists b\in B\ (a\hat{X}b\ \&\ b\hat{Y}c).$$

The official definition is

$$\hat{X}\Box\hat{Y} = \exists_{\pi_{13}}\ (\pi_{12}{}^*(\hat{X}) \wedge \pi_{23}{}^*(\hat{Y}))$$

with the projections referring to the product $A\times B\times C$.

Notice that we have to show that if X, Y satisfy (3) (with the appropriate data), then $X\Box Y$ satisfies (3) with R and U. For doing so, it suffices to verify the same fact in Set, and appeal to completeness.

We leave the remaining details to the reader.

■ 2.4.5.

<u>Corollary</u> 2.4.5'. If T is a Boolean category, then P(T) is a Boolean pretopos.

The *proof* is an instructive exercise, using the 'moreover' part of 2.4.5.

■ 2.4.5'.

Let us call a Coh-morphism *quotient-like* if it is full on subobjects, and every object in its codomain is covered by it (see the statement of the 'moreover' part of the last proposition). We will show that, for a Pretop-morphism, being quotient-like is equivalent to being a quotient. But first, we prove

Lemma 2.4.6. A quotient-like conservative Pretop-morphism is an equivalence.

Proof: Let $F: T \longrightarrow T'$ satisfy the conditions, and let $X \in Ob(T')$. We have an effective epi family $(F(A_i) \longrightarrow X)_i$ (one in J_T); with

$$A = \coprod_i A_i$$

we therefore have a single effective epi $f: F(A) \twoheadrightarrow X$. Let

$$Y \overset{p}{\underset{p'}{\rightrightarrows}} F(A)$$

be the kernel pair of f; let $\hat{Y}$ be the subobject of $A\times A$ represented by $\langle p,p'\rangle: Y \longrightarrow FA\times FA$. By assumption, we have a subobject, represented by $\langle g,g'\rangle: B \longrightarrow A\times A$, say, whose F-image is $\hat{Y}$. Since $\hat{Y}$ is an equivalence relation, and F is conservative,

$$B \overset{g}{\underset{g'}{\rightrightarrows}} A$$

is an equivalence relation. Let $h: A \longrightarrow C$ be its quotient. But then the two diagrams

$$Y \overset{p}{\underset{p'}{\rightrightarrows}} F(A) \xrightarrow{\;f\;} X$$

$$F(B) \overset{Fg}{\underset{Fg'}{\rightrightarrows}} F(A) \xrightarrow{\;Fh\;} F(C)$$

are isomorphic since each of f and Fh is a coequalizer in the first and second diagram, respectively: in particular $X \cong F(C)$. This shows that F is essentially surjective. F is full and faithful as a consequence of being conservative and subobject-full.

■ 2.4.6.

Proposition 2.4.7. (i) Every Pretop morphism can be factorized, up to an isomorphism, as the composition of a (Pretop-)quotient followed by a conservative (Pretop-)morphism.

(ii) A Pretop-morphism is a quotient if and only if it is full on subobjects, and every object in its codomain is covered by it (it is quotient-like).

(iii) If the domain of a Pretop quotient is Boolean, so is its codomain.

Proof: (i) Given the morphism $F: T \longrightarrow T'$ in Pretop, we consider it being in Coh, and factorize it in a quotient and a conservative one in Coh:

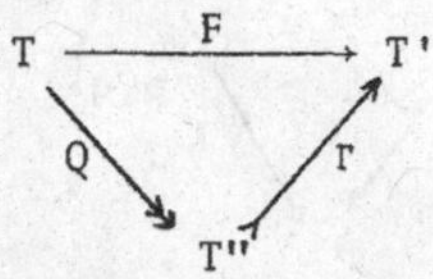

Now, form the pretopos completion $\gamma: T'' \longrightarrow P(T'')$ of T''. We clearly have a diagram

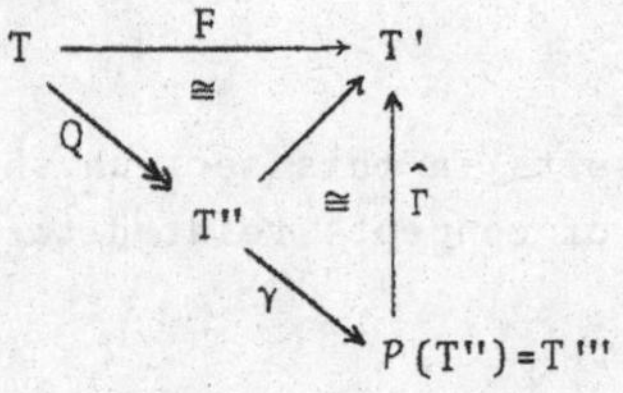

in which $\hat{\Gamma}\gamma \cong \Gamma$. The universal properties of Q and γ immediately imply that $\hat{Q} = Q\gamma$ is a quotient in Pretop. Moreover, it is quotient-like, by 2.3.5 applied to Q, and the "moreover" part of 2.4.5 applied to γ. Using the latter property of $\hat{Q}$, we now show that $\hat{\Gamma}$ is conservative.

Let $X \longrightarrow Y$ be a monomorphism in T''' taken by $\hat{\Gamma}$ into an isomorphism. Consider a finite covering family $(\hat{Q}A_i \longrightarrow Y)_i$, and the pullbacks, one for each i:

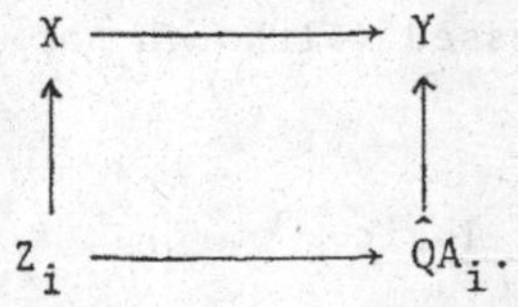

By $\hat{Q}$ being subobject-full, $Z_i \longrightarrow \hat{Q}A_i$ is isomorphic to some $\hat{Q}f_i: \hat{Q}B_i \longrightarrow \hat{Q}A_i$, with f_i a mono. $\hat{\Gamma}$ takes Qf_i into an isomorphism (as a pullback of an isomorphism). Since $F \cong \hat{\Gamma}\hat{Q}$, and F is conservative, Qf_i itself is an isomorphism, for every i. But then $(Z_i \longrightarrow Q\hat{A}_i \longrightarrow Y)_i$ is a covering, hence, since it 'factors through' $X \longrightarrow Y$, the monomorphism $X \longrightarrow Y$ must be an isomorphism.

It is a simple general fact that a Coh-morphism conservative with respect to monomorphisms in the domain category is conservative. This follows immediately when one reflects that a morphism $f: A \longrightarrow B$ is an isomorphism iff the mono representing the subobject $\exists_f(1_A)$ and the mono $A \longrightarrow C$, in a factorization

$$\begin{array}{ccc} A & \longrightarrow & C \\ & {\scriptstyle <id,id>} \searrow \quad \swarrow {\scriptstyle <p,q>} & \\ & A\times A & \end{array}$$

with (p,q) the kernel pair of f, are both isomorphisms.

(ii) By the proof of (i), and 2.4.6.

(iii) By 2.3.6. and 2.4.5'.

■ 2.4.7.

The final group of results in this section should, logically, be the first ones in the study of concepts related to the quotient-conservative factorization.

Proposition 2.4.8. In any one of the 2-categories Lex, Coh, Pretop, the following is true. If T is any object of the 2-category, Σ is any set of morphisms in T, then there is an arrow

$$Q: T \longrightarrow T[\Sigma^{-1}]$$

in the 2-category so that "Q is obtained by inverting the morphisms in Σ" in the sense of Definition 1.4.2.(i) (with Pretop replaced by the relevant 2-category). $T[\Sigma^{-1}]$ and Q are determined up to equivalence, resp. up to isomorphism.

The proof will be discussed (although not given) in the next section.

2.5. Relations with symbolic logic.

Let us start with a fundamental observation that goes a long way towards explaining the relation of model theory and categories. Let C be a small graph, or category, and consider a diagram (functor)

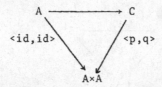

$M: C \longrightarrow Set$. Then M is the same as a *structure* for the many-sorted language given by C. M interprets each sort of the language, i.e. each object of C, say A, as a 'partial domain' $M(A)$; and it interprets each (sorted unary) operation symbol of the language, i.e., morphism of C, say $f: A \longrightarrow B$, as a corresponding kind of operation $M(f): M(A) \longrightarrow M(B)$. In fact, if C is a graph, then the diagrams $C \longrightarrow Set$ are precisely the structures for the *language* C, with possibly empty partial domains. If C is a category, the functors $C: \longrightarrow Set$ are those L-structures, for L the underlying graph of C that satisfy a certain set of identities: whenever

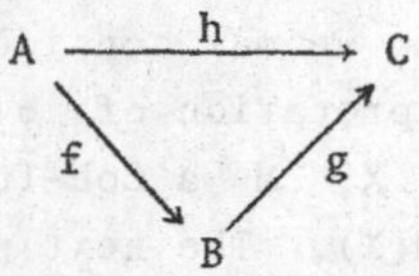

commutes, we have the identity

$$\left.\begin{array}{l} \forall a \in A \; g(f(a)) = h(a); \\ \text{also,} \\ \forall a \in A \; id_A(a) = a \end{array}\right\} \qquad (1)$$

for the identity morphism $id_A: A \longrightarrow A$.

Moreover, natural transformations between diagrams (functors)

$$C \; \substack{\longrightarrow \\ \longrightarrow} \; Set$$

are seen as the same as *homomorphisms* in the usual sense between structures.

In *first order logic* over the language L, we have variables each of which has a definite sort, an object of L. A quantifier $\forall x$, or $\exists x$ is meant to range over a fixed sort; to emphasize this, we write $(\forall x \in X)$, $(\exists x \in X)$ with X the sort of x. We have an equality sign that is sorted in the sense that only terms of the same sort are allowed to fill in the places of it (equivalently, there could be a separate equality sign for each sort). The only nonlogical symbols are sorted unary operation symbols (the arrows in L); their use is regulated in the natural way as illustrated under (1). In full first order logic over L, we have all the Boolean connectives, and the two quantifiers.

Although many-sorted languages with unary operation symbols may seem to lack expressive power, it turns out, as a biproduct of categorical logic, that in fact there is no loss of generality when we restrict attention to such languages.

A *coherent*, or positive existential, formula is one that is built up from atomic formulas by using (finite) conjunction, disjunction and existential quantification.

Let T be a coherent category, L its underlying graph, φ a formula, $\vec{x}$ a string of variables such that each free variable of φ occurs in $\vec{x}$. If $\vec{x} = \langle x_1,\ldots,x_n\rangle$, let X_i be the sort of x_i, and $X = [\vec{x}] = X_1\times\ldots\times X_n$. If M is any L-structure, in particular if M is a functor $M: T \longrightarrow \mathrm{Set}$, then we may consider $M_{\vec{x}}(\varphi) = \{a \in M(X): M \models \varphi[a]\}$, the interpretation of $\varphi(\vec{x})$ in M, a subset of $M(X)$. If ϕ is a subobject of X, M a Coh-functor, then $M(\phi)$ is a subobject, i.e. a subset, of $M(X)$. The next proposition tells us that the expressive power of coherent categorical logic is at least as great as that of coherent formulas.

<u>Proposition</u> 2.5.1. (i) For any coherent formula $\varphi(\vec{x})$ over (the underlying graph of) T, a coherent category, there is a unique subobject ϕ of $[\vec{x}] = X$ such that

$$M(\phi) = M_{\vec{x}}(\varphi) \quad (\in \mathrm{Sub}\, M(X))$$

for any $M \in \mathrm{Mod}(T)$. ϕ is called the *canonical interpretation* of $\varphi(\vec{x})$; it may be denoted by $[\vec{x}{:}\varphi]$.

(ii) For T a Boolean category, the same conclusion holds for any first order formula φ over T.

The *proof* of the existence of ϕ, proceeding by an induction on the complexity of φ, is an instructive exercise. In fact, in previous sections, there were implicit examples of constructions of canonical interpretations.

■ 2.5.1.

The last proposition can profitably be used in the proof of Proposition 2.2.4. Suppose $(f_i: A_i \longrightarrow A)_{i\in I} \in J_T$; we would like to show that it is an effective epimorphic family. Using the notation of 2.2.3, assume we have $g_i: A_i \longrightarrow B$ $(i \in I)$ such that $g_i f_{ij} = g_j f'_{ij}$ $(i,j \in I)$. Consider the following formula:

$$\bigvee_i (\exists x \in A_i)(f_i(x) = a \wedge g_i(x) = b)$$

with two free variables a,b of respective sorts A and B. The idea is that, if we are in Set, then this defines the graph of the desired function $A \longrightarrow B$, as immediately seen.

Now, let $\phi \in \mathrm{Sub}(A\times B)$ be the canonical interpretation of φ, $\phi = [ab:\varphi]$. If $M \in \mathrm{Mod}(T)$, then M takes the data to objects and morphisms in Set for which the same hypotheses hold, and as we know already there is a unique function $h_M: M(A) \longrightarrow M(B)$ satisfying the requirements; in fact, the graph of h_M is $M_{ab}(\varphi)$, i.e. $M(\phi)$. We infer, at least, that $M(\phi)$ is a functional subobject of $M(A)\times M(B)$ for all $M \in \mathrm{Mod}(T)$. By completeness, clearly, ϕ is a functional subobject of $A\times B$ in T. Let $h: A \longrightarrow B$ be the morphism whose graph is ϕ. The equalities $hf_i = g_i$ follow, since they are true in any $M \in \mathrm{Mod}(T)$: $M(h) = h_M$, and $h_M \circ M(f_i) = M(g_i)$.

The uniqueness of h can be deduced in the same manner.

■ 2.2.4.

Note that for any subobject $\phi \in \mathrm{Sub}(X)$, in any fixed T, there is a formula $\hat{\phi}(x)$ with one free variable x of sort X whose canonical interpretation is ϕ. If ϕ is represented by $m: A \longrightarrow X$, then $\hat{\phi}(x)$ can be taken to be $(\exists a \in A)(m(a) = x)$. Although $\hat{\phi}(x)$ is not uniquely determined by ϕ, we will call it *the formula representing* ϕ.

Once we accept that functors into Set are structures, we can directly interface with model theory. E.g., it makes sense to say that a natural transformation $h: M \longrightarrow N$ between two functors M, N from T to Set is an elementary embedding. Note the following immediate consequence of 2.5.1.: if T is Boolean, $M, N \in \mathrm{Mod}(T)$, then *any* $h: M \longrightarrow N$ is an elementary embedding. (Why is this not that paradoxical as it may sound first?)

Let us complete the proof of 2.4.3. In this proof, we have to remember that, at this stage, we have not yet established that a pretopos is a coherent category. With the notation of 2.4.3, we want to prove that i is a monomorphism. Let $M: T \longrightarrow \mathrm{Set}$ be a Lex-functor such that, in addition, M(s) is surjective. Then, in Set, M(f) and M(s) have the same kernel-pair, and by taking into account the meaning of pullbacks in Set, we see that

$$(\forall a, a'\in A)\ [f(a) = f(a') \Leftrightarrow s(a) = s(a')]$$

holds in M. This immediately implies that M(i) is a mono in Set.

In order to be able to apply completeness, consider the topology J generated by the single covering {s} of the object C. Since s is an effective epi, effective epis are stable under pullback in T (a condition on pretoposes), effective epis are extremal, extremal epis are closed under composition (*exercise*; or see 3.1.5 in [MR]), and since the pullback of a composite morphism can be formed as the composite of two pullbacks, it follows that every covering in J contains a morphism that is an extremal epi. It follows that a mono which is not an isomorphism is not a covering by itself in J. By the general completeness theorem 2.1.1, if i is not a mono, i.e. the mono $A \longrightarrow C$ in the factorization

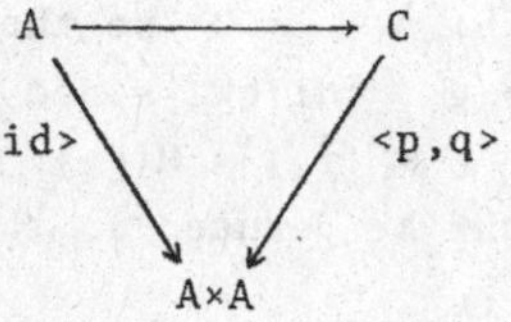

with (p,q) the kernel pair of i, is not an iso, then there is $M: T \longrightarrow \mathrm{Set}$, M J-continuous, such that M(i) is not a mono, a contradiction.

Although this proof is not even shorter than the direct proof (see the proof of Theorem 1.52 in [TT]), I find it preferable because it is more comprehensible.

Let us now look at the interaction of the symbolical and categorical formulations in the opposite direction: the expressibility of categorical conditions by symbolic means.

The first observation is that the graphs of all the Boolean pretopos operations (and of course, the Coh-operations) are strict (finitely axiomatizable) elementary classes. Note that each such graph is a class of structures over a finite similarity type; thus, our statement makes sense. The assertion is verified by writing out the well-known characterizations of the particular limit and colimit diagrams in Set. E.g., for the operation of binary products, we have this: a diagram

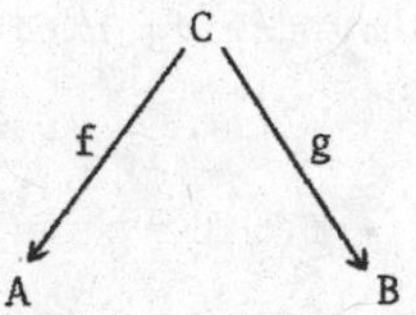

(2)

in Set is a product diagram iff the following is true:

$$(\forall a \in A)(\forall b \in B)(\exists c \in C)(fc=a \wedge fc=b) \wedge$$

$$(\forall c \in C)(\forall c' \in C)((fc=fc' \wedge gc=gc') \rightarrow c=c') . \qquad (3)$$

It is instructive to see that the class of all coequalizer diagrams (without restriction on the domain of the operation) is defined by an infinitary disjunction; in fact, it is not an elementary class.

Next, as a consequence, we see that, for any (small) pretopos T, the class of objects of Mod(T) is an elementary class (over the language the underlying graph of T). In fact, the axioms for this class are the following: for each pretopos operation and for each individual diagram in T which is in the graph of that operation, we write down the first-order sentence, with symbols appearing in that diagram, that expresses that the diagram would be the given kind were it in Set. E.g., if we find that the diagram (2) in T is a product diagram, we add the sentence (3) to our axiom system. We also add axioms ensuring that we have a functor; see (1) above. By the definition of what constitutes a Pretop-morphism T ⟶ Set, it is clear that the resulting axiom system will axiomatize exactly the class Ob(Mod T).

Let us discuss ultraproducts of structures, i.e. functors, from the categorical point of view.

With C a (small) graph, the category (C,Set) of all diagrams C ⟶ Set has limits and colimits that can be computed pointwise (c.f. V.3 in [CWM]). Therefore, if we define ultraproducts in (C,Set) by the same formula as in Set:

$$\prod_{i\in I} M_i/U = \operatorname{colim}_{P\in U^{op}} \prod_{i\in P} M_i ,$$

and similarly for morphisms in (C,Set), then, for $A \in Ob(T)$, we have

$$(\prod_{i\in I} M_i/U)(A) = \prod_{i\in I} M_i(A)/U \qquad (4)$$

and a similar formula for morphisms in T. This shows that the definition given here coincides with the usual one, considering the diagrams C ⟶ Set as structures.

Now, let T be a small pretopos, and let C be its underlying graph. Then, if in the above the M_i are in Mod(T) then $\prod M_i/U$ is again in Mod T. This is an immediate consequence of 1.3.1 (which we called Los's theorem) and the formulas (4). Of course, it also follows from the fact that Mod(T) is an elementary class in the sense of model theory, and the usual formulation of Los's theorem.

Let us turn to the construction of the free pretopos, and the free Boolean pretopos, needed for proving the assertions in 1.4.1. In fact, the constructions could be given by a direct application of the "method of words" used for constructing free groups and other free algebraic structures. The "words" would be literal composites of the pretopos (Boolean pretopos) operations, and they would take the form of labelled finite diagrams. It would also be possible to give the proof of the existence of the free objects by appropriate (2-categorical) versions of the adjoint functor theorem. However, in these two cases symbolic logic provides a simple and intuitive procedure.

First, we construct the free coherent category, and the free Boolean category on a graph G. They satisfy universal properties analogous to (i) and (ii), respectively, in 1.4.1.

For $F_{Coh}(G)$, we take as objects all the pairs $(\vec{x},\varphi)$ with $\vec{x}$ a finite tuple of distinct variables, φ a coherent formula with free variables all included in $\vec{x}$. We write $[\vec{x}:\varphi]$ instead of $(\vec{x},\varphi)$, since we mean, really, "the set of $\vec{x}$ such that φ". Given two objects

$$[\vec{x}:\varphi] \text{ and } [\vec{y}:\psi]$$

with $\vec{x}$ and $\vec{y}$ disjoint, we consider coherent formulas μ with free variables among $\vec{x}\vec{y}$, and satisfying

$$\vdash \forall\vec{x}\ \forall\vec{y}\ (\mu\ (\vec{x}\vec{y}) \longrightarrow (\varphi(\vec{x})\ \wedge\ \psi(\vec{y})))\ \wedge\ \forall\vec{x}(\varphi(\vec{x}) \longrightarrow \exists!\ \vec{y}\mu(\vec{x},\vec{y}))$$

with $\exists!$ meaning unique existence and "$\vdash$" referring to provability in first order logic. Such formulas μ are called *provably functional relations* from the one object to the other. Now, two provably functional relations between the same objects are identified if they are provably equivalent formulas. A morphism is an equivalence class of provably functional relations. The case when, in the given objects, $\vec{x}$ and $\vec{y}$ are not necessarily disjoint tuples, is left to the reader to formulate, as well as the definition of composition of morphisms, and that of the G-diagram

$$\varphi_G^{Coh}: G \longrightarrow F_{Coh}(G).$$

Although this defines the free Coh-category over G completely, of course more work is to be done to verify the required properties. This will not be done here; for details, see [MR].

The construction of the free Boolean category over G,

$\varphi_G^{Boole}: G \longrightarrow F_{Boole}(G)$ is similar, but uses all first order formulas over G as language, in defining both the objects and morphisms.

The free pretopos,

$$\varphi = \varphi_G^{Pt}: G \longrightarrow F_{Pt}(G)$$

is given as the composite

$$G \xrightarrow[\varphi]{} F_{coh}(G) \xrightarrow[\gamma]{} P(F_{coh}(G))$$

with the pretopos completion taken from 2.4.5. The verification of the required universal property is straightforward, on the basis of those of φ and γ. The free Boolean pretopos is obtained similarly, with φ_G^{Boole} in place of φ; one uses 2.4.5'.

Finally, let us turn to Proposition 2.4.8. Again, just as in the case of the free objects just discussed, the mere existence asserted in the proposition are consequences of general facts concerning "2-categories of category-based structures defined by essentially algebraic conditions". Since I am not quite ready to describe these general facts (see however a forthcoming paper by G.M. Kelly and myself), I point out some more concrete constructions of the desired quotients. Let me first of all mention that Professor Jean Benabou recently described, in lectures in Montreal, detailed constructions for the case of Lex and Coh in elegant category theoretical terms. These constructions are closely related to constructions in [GZ] ("calculus of fraction"), but unless I have missed it, the construction, even for the case of Lex, is not found in full generality in [GZ]. Let me add that, of course, the construction e.g. in 2.3.1 is a special case of what we want for the case of Lex, with $\Sigma = \mathrm{Inv}(F)$, with the notation of that proposition. (Note also that although the mere existence of the factorization in 2.3.1 for Lex, and the ones for Coh and Pretop are, of course, consequences of the general facts alluded to above, the special properties of the factorizations, crucial for our purposes, are not.)

The constructions for Coh and Pretop are given in [MR] in a different terminology. Given $R \in$ Coh, and $\Sigma \subset \mathrm{Morph}(R)$, consider T_R, the internal theory of R (see p. 128 in [MR]). Add the following axioms (Gentzen-sequents) expressing that each $\sigma \in \Sigma$ should be invertible:

$$\mathrm{True} \Rightarrow (\exists! a \in A)\sigma(a) = b \qquad (\sigma: A \to B \text{ in } \Sigma).$$

Let the theory so obtained be denoted by T'. Consider the categorization $R_{T'}$ of T' (see Theorem 8.1.3., p. 239 in [MR]), and call it $R[\Sigma^{-1}]$. The canonical interpretation of T_R in T', and that of T' in $R_{T'}$ induces the required Coh-functor

$$R \longrightarrow R[\Sigma^{-1}].$$

Given the properties of T_R and $R_{T'}$, it is easy to verify that we have the required universal property.

The case of Pretop can be handled as it was with respect to free objects: first, we carry out the construction on the given pretopos in Coh, then we take the pretopos completion.

Part 3. Conceptual completeness.

3.1. The statement and proof of conceptual completeness.

Consider the notion of the pretopos completion (see 2.4.5.): with any coherent category T, one associated a pretopos $P(T)$, and a Coh-morphism $T \to P(T)$ that induces, for any other pretopos S (e.g., for S = Set) an equivalence of categories $\mathrm{Coh}(P(T),S) \longrightarrow \mathrm{Coh}(T,S)$, hence, in particular, an equivalence of the categories of models: $\mathrm{Mod}\, P(T) \longrightarrow \mathrm{Mod}\, T$.

We interpret this situation by saying that the notion of coherent category is conceptually incomplete: when extending a coherent category, a 'theory', T by the concepts (abstract definable sets) inherent in its pretopos completion (finite disjoint sums of 'concepts' in T, quotients of concepts in T by definable equivalence relations), the resulting extension 'theory' "behaves semantically in the same way as T itself", by having the same category of models, or even, the same category of S-valued models, for any (pre-)topos S.

By contrast, the notion of pretopos is conceptually complete in the sense that it does not admit any extension conservative with respect to models. More precisely, the following theorem holds.

Theorem 3.1.1. (Conceptual completeness of pretoposes). For a morphism $I: T \longrightarrow T'$ of small pretoposes to be an equivalence of categories, it is sufficient that I induce an equivalence $I^*: \mathrm{Mod}\, T' \longrightarrow \mathrm{Mod}\, T$ on the categories of models.

This is Theorem 7.1.8 of [MR]. Recently, Andrew Pitts has found an essentially different proof yielding a stronger result, a constructive version of the theorem.

On the basis of this theorem, it seems natural to say that the "real" or "complete" form of the theory T is its pretopos completion $\mathcal{P}(T)$. E.g., if we ever try to recover T in an abstract way from Mod T, we should expect to get $\mathcal{P}(T)$ instead of T since there should not be any difference in principle between concepts in T and those in $\mathcal{P}(T)$ from the point of view of the category of models. This expectation is borne out by results in this area (see [M3], [M4], [L], [M1], [M2]).

The proof of Theorem 3.1.1 in [MR] gives a stronger result, not so stated in [MR]. Note that $I: T \to T'$ (in Pretop) being a quotient means two things: first, that $I^*: \text{Pretop}(T',S) \longrightarrow \text{Pretop}(T,S)$ is full and faithful, and second, that I^* is essentially surjective onto those $T \longrightarrow S$ that invert all morphisms inverted by I; both conditions being required to hold for all $S \in$ Pretop. The strong form of conceptual completeness says that it suffices to require the first condition, and even that for S = Set only.

Theorem 3.1.2. (Strong conceptual completeness for pretoposes). For a morphism $I: T \longrightarrow T'$ of small pretoposes to be a quotient, it is sufficient that the induced $I^*: \text{Mod } T' \longrightarrow \text{Mod } T$ be full and faithful.

(In fact, Theorem 7.1.4. in [MR] says that "I^* is full" implies that "I is full with respect to subobjects", Theorem 7.1.6 says that "I^* is faithful", implies that "every object in T' is subcovered by I". Therefore, 3.1.2 follows by 2.4.7(ii).)

Since I^* being essentially surjective on objects clearly implies that I is conservative, the strong version of conceptual completeness implies the original version.

It is interesting to note that in the "doctrine" Lex, (the obvious analog of) conceptual completeness holds, (a classical result, see [GU]) but its strong form does not (see [MP]).

Speaking about Boolean pretoposes, we have considered them as objects in Pretop, the 2-category of all pretoposes. Being interested in results specific for Boolean pretoposes (corresponding to theories in the common sense of model theory, in full first order logic), it is more natural to consider the category BPretop of Boolean pretoposes, with morphisms the Pretop morphisms. Furthermore, it is more natural to consider BPretop to be *Groupoid*-enriched rather than *Cat*-enriched;

this means that BPretop(T,T') should be the groupoid of all BPretop morphisms $T \rightarrow T'$, with only those natural transformations that are isomorphisms. Writing C^{iso} for the subcategory of C whose objects are those of C and whose morphisms are the isomorphisms of C, we are now saying that we should put $\text{BPretop}(T,T') = \text{Pretop}(T,T')^{iso}$. The reason is that BPretop construed in this way has good general algebraic properties (it is an 'essentially algebraic' Gpd-category), but not if we put in all natural transformations as 2-cells. Even without understanding the general concepts involved, the reader can see this plainly on the way the free Boolean pretopos had to be defined (1.4.1 (ii)). Obviously, the fact we are considering here is intimately related to the fact that not all operations defining Boolean pretoposes are strong (see Section 1.1).

Therefore, in the context of BPretop, the natural definition of a quotient-morphism is the following

Definition 3.1.3. A morphism $I: T \rightarrow T'$ in BPretop is a quotient if $I^{iso}: (\text{Mod } T')^{iso} \rightarrow (\text{Mod } T)^{iso}$ is full and faithful and essentially surjective onto those models of T which invert all morphisms inverted by I.

Note that, by 2.4.7 (iii), any quotient of a Boolean pretopos in Pretop is automatically Boolean. It follows easily that, for a morphism $I: T \rightarrow T'$ in BPretop to be a quotient in the sense of 3.1.3 is the same as to be a quotient in Pretop in the sense previously specified. Definition 3.1.3 is still useful since it suggests the question if an analog of conceptual completeness holds for Boolean pretoposes. The answer is "yes" with a curious qualification: the Boolean pretoposes involved should be countable (meaning that they have countably many objects and arrows).

Theorem 3.1.4. (Strong conceptual completeness for countable Boolean pretoposes) For a Pretop-morphism $I: T \rightarrow T'$ of countable Boolean pretoposes T, T' to be a quotient, it is sufficient that $I^{iso}: (\text{Mod } T')^{iso} \rightarrow (\text{Mod } T)^{iso}$ induced by I be full and faithful.

This result is essentially due to Haim Gaifman. In 1975, at the Abraham Robinson memorial conference, Gaifman announced a result which, although stated in a language not employing categories, is closely related to Theorem 3.1.4. A version closer to Gaifman's result would say that a Coh-morphism I between Boolean (Coh-)categories is quotient-like (see before 2.4.6) provided I^{iso} is full and faithful.

In this section, we will first prove Theorem 3.1.4, and then we will apply it to prove a stronger form of it, which will be directly applicable to the proof of the main result, Theorem 1.4.4.

The proofs in this section will use various methods of model theory.

The first lemma, Lemma 3.1.5., is the heart of Gaifman's theorem. It is analogous to Theorem 7.1.6 in [MR], but its proof is very different. As above, with $I: T \to T'$, $I^{iso}: (\text{Mod } T')^{iso} \to (\text{Mod } T)^{iso}$ denotes the functor obtained from $I^*: \text{Mod } T' \to \text{Mod } T$ restricted to $(\text{Mod } T')^{iso}$.

Lemma 3.1.5. *Assume that* $I: T \to T'$ *is a Coh-morphism between countable Boolean categories such that* I^{iso} *is full and faithful. Then* I *"subcovers" every object* X *in* T' *in the sense that there is a finite family of diagrams*

$$\begin{array}{ccc} X_i & \xrightarrow{\quad m_i \quad} & IA_i \\ {\scriptstyle p_i}\downarrow & & \qquad (i \in I) \\ X & & \end{array}$$

in T', *with the* m_i *being monomorphisms, such that the family* $(p_i)_{i \in I}$ *is effective epimorphic.*

Proof: Recall the omitting types theorem (OTT) (see [CK]). Given a consistent theory T (having at least one model) in a *countable* language L (Morph(L) is a countable set), a set $\Sigma(\vec{x})$ of L-formulas $\sigma(\vec{x})$ with a fixed finite tuple of free variables $\vec{x}$ is called simply a *type* (although "incomplete type" may be more appropriate). A type $\Sigma(\vec{x})$ is *isolated by* $\phi(\vec{x})$ if $\phi(\vec{x})$ is consistent with T (i.e. $T \cup \{\exists x \phi(\vec{x})\}$ is consistent) and $T \models \forall \vec{x}(\phi(\vec{x}) \to \sigma(\vec{x}))$ for all $\sigma(\vec{x}) \in \Sigma(\vec{x})$. OTT says that, given a countable family $\{\Sigma_i(\vec{x}_i): i \in I\}$ of types each of which is non-isolated, then T has a model M *omitting* each $\Sigma_i(\vec{x}_i)$. "M omits $\Sigma(\vec{x})$" means that there is no tuple $\vec{a}$ (matching $\vec{x}$ for sorts) in M that would satisfy all $\sigma(\vec{x})$ in $\Sigma(\vec{x})$ simultaneously; in the opposite case, we may say "M realizes $\Sigma(\vec{x})$:. If T is complete (for all L-sentence ϕ, $T \cup \{\phi\}$ consistent implies $T \models \phi$), then we have an obvious converse: any isolated type must be realized in all models of T.

Let us return to the notation of the Lemma. Let M be any *countable* model in Mod T' (each M(X), X ∈ Ob(T'), is countable);

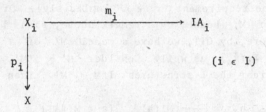

let us fix M for a while. Consider the language $\hat{L}$ obtained from L' (= the underlying graph of T') by adding a new individual constant $\langle a,A\rangle$ (written simply $\underline{a}$) for each $A \in Ob(L)$ (L = the underlying graph of T) and for each $a \in M(IA)$. Let $i: L' \to \hat{L}$ denote the inclusion. We have the structure $\hat{M} = (M,a)_{a\in IA, A\in Ob(L)}$ of the language $\hat{L}$ whose L'-reduct $i^*\hat{M}$ is M and for which $(\underline{a})^{\hat{M}} = a$. Let $\hat{T}$ be the theory of $\hat{M}$ in $\hat{L}$: the set of all sentences in $\hat{L}$ true in $\hat{M}$; $\hat{T}$ is clearly complete (see above).

Now, let us consider any finite tuple $\vec{b}$ of elements of the various MX, $X \in Ob(L')$ (using products in T', we could pass to a single element b of an appropriate product sort). We *claim* that *the (complete) type of* $\vec{b}$ *in* $\hat{M}$, i.e. the set $\Sigma_{\vec{b}}(\vec{x})$ of all $\hat{L}$-formulas $\sigma(\vec{x})$ such that $\hat{M} \models \sigma[\vec{b}]$, is isolated. Suppose $\Sigma_{\vec{b}}(\vec{x})$ is non-isolated. We now also consider, for each $A \in Ob(L)$, the type $\Sigma_A(x_A) = \{\neg x_A = \underline{a}: a \in M(A)\}$ (x_A is a variable of sort IA: hence, $\Sigma_X(x_A)$ contains the requirement "$x_A \in A$" implicitly). Of course, each $\Sigma_A(x_A)$ omitted in $\hat{M}$; hence, by a remark above, each $\Sigma_A(x_A)$ is non-isolated. Therefore, by OTT, we have a model $\hat{M}'$ of $\hat{T}$ omitting $\Sigma_{\vec{b}}(\vec{x})$ and every $\Sigma_A(\vec{x}_A)$ as well. Consider $M' = i^*\hat{M}'$, the L'-reduct of $\hat{M}'$, and consider the L-structures I^*M, I^*M'. Then the maps

$$h_A: a \longmapsto \hat{M}'(\underline{a}) \qquad (a \in M(IA))$$

make up an *isomorphism* $h: I^*M \to I^*M'$: indeed, by the fact that $\hat{M}'$ omits each Σ_A, each h_A is surjective onto $(I^*M')(A)$; since M' is a model of $\hat{T}$, h_A is "$\hat{L}$'-elementary on elements of I^*M" implying that h is elementary, in particular, h is structure preserving and reflecting. However, h cannot be extended to an isomorphism $h': M \to M'$ with $I^*h' = h$, since with any such h', $h'(\vec{b})$ would clearly have the same type in $\hat{M}'$ as $\vec{b}$ does in $\hat{M}$, but, since $\hat{M}'$ omits $\Sigma_{\vec{b}}(\vec{x})$, there is nothing in $\hat{M}'$ having the same type as $\vec{b}$.

We have constructed a counter-example to the fullness of I^{iso}; the *claim* is, therefore, proved by contradiction.

We conclude that, still with the fixed countable $\hat{M}$ of $\hat{T}$, each finite tuple $\vec{b}$ of elements has an isolated type. In model theory, we say that $\hat{M}$ is *atomic*. It is well-known (c.f. [CK]) (and not hard to show) that any countable atomic model is *homogeneous* in the sense that if $\vec{b}$ and $\vec{c}$ have the same type in it, then there is an automorphism of the model mapping $\vec{b}$ to $\vec{c}$. In our case, this implies that if $\vec{b}$ and $\vec{c}$ have the same type in $\hat{M}$, they are identical: $\vec{b} = \vec{c}$. The reason is that an automorphism of $\hat{M}$ is the same as an automorphism h

of M such that $I^*h = Id_{I^*M}$, and by the faithfulness of I^{iso}, therefore, the only automorphism of $\hat{M}$ is the identity.

We now see that for any $X \in Ob(L')$ and $b \in M(X)$, there is an L'-formula $\phi(x,\vec{y})$ with x of sort X, $\vec{y}$ a tuple of variables of various sorts IA with $A \in Ob(L)$, and there are elements $\vec{a}$ of M matching $\vec{y}$ such that $M \models \phi[b,\vec{a}]$ and $M = \forall xx'(\phi(x,\vec{a}) \wedge \phi(x',\vec{a}) \to x=x')$. Namely, if $\phi(x,\vec{a})$ isolates the type of b in $\hat{M}$, then of course $M \models \phi[b,\vec{a}]$, and if $M \models \phi[b',\vec{a}]$, then b and b' have the same type in $\hat{M}$, hence $b = b'$.

We can, in fact, make $\vec{y}$ and $\vec{a}$ into singletons by passing to an appropriate product sort in L.

We can re-write our last conclusion this way: for the fixed M and X,

$$M \models \forall x \bigvee_{\phi} \exists y(\phi(x,y) \wedge \forall xx'(\phi(x,y) \wedge \phi(x',y) \to x = x')). \quad (1)$$

Here in the *infinite* disjunction ϕ ranges over all formulas of the form $\phi(x,y)$ with x a free variable of X, y a variable of a sort of the form IA, with $A \in Ob(L)$ depending on ϕ. Since the displayed formula holds in all countable models M of $\mathbf{T}$, with $\mathbf{T}$ the L'-theory whose models are exactly the objects of Mod T', by the downward Löwenheim-Skolem and the compactness theorems, it is easy to conclude that there are finitely many formulas $\phi_o(x,y_o),\ldots,\phi_{n-1}(x,y_{n-1})$ as the ϕ above such that

$$\mathbf{T} \vdash \forall x \bigvee_{i<n} \hat{\phi}_i(x) \quad (2)$$

with $\hat{\phi}_i(x)$ formed from ϕ_i as the formula after "$\bigvee_{\phi}$" is formed from ϕ in (1).

It should now be clear that (2) implies that X is subcovered by I. Namely, let $i < n$ and consider the canonical interpretation in T' of the formula

$$\psi(y_i) \underset{df}{=} \forall xx'(\phi_i(x,y_i) \wedge \phi_i(x',y_i) \to x = x')$$

(remember, T' is Boolean). $[y_i: \psi(y_i)]$ is a subobject of IA_i (IA_i the sort of y_i); let $m_i: X_i \to IA_i$ represent this subobject. Let p_i be the morphism $X_i \to X$ in T' whose graph is $[x_i x: \phi(x,m_i(x_i))]$ (it is clear from the definitions that the last subobject of $X_i \times X$ is indeed functional). Then (2) clearly translates into saying that the p_i form an effective epimorphic family in T' (see 2.2.4) (here we use the relation of $\mathbf{T}$ to T', ensuring, by completeness, that sentences provable from $\mathbf{T}$ 'become true' in T').

■ 3.1.5.

Lemma 3.1.6. Suppose $I: T \to T'$ is a Coh-morphism between Boolean categories such that I^{iso} is full. Then I is full on subobjects.

Proof: This is essentially a direct consequence of Beth's definability theorem. Let us state Beth's theorem in a notation convenient for our purposes.

Assume L and L' are graphs (languages), and $I: L \longrightarrow L'$ is a graph-map (diagram). Let $\mathbf{T}$ be a theory over L', let $A \in Ob(L)$, and let $\varphi(a)$ be a formula of L' with a single free variable a of sort $I(A)$. Assume that "$\varphi(a)$ is preserved by all L-isomorphisms of models $\mathbf{T}$", i.e. that for any $M, N \in \mathrm{Mod}\,\mathbf{T}$, and any isomorphism $h: I^*(M) \longrightarrow I^*(N)$, we have a commutative diagram

$$\begin{array}{cccc} (\{a \in M(IA): M \models \varphi[a]\} =) & M(\varphi) & \xrightarrow{\text{incl.}} & M(IA) & (= I^*(M)(A)) \\ & \downarrow f & & \downarrow h_A & \\ & N(\varphi) & \xrightarrow{\text{incl.}} & N(IA) & (= I^*(N)(A)). \end{array}$$

Then there is an L-formula $\psi(a)$ such that $M(\varphi) = I^*(M)(\psi)$ for all $M \in \mathrm{Mod}\,\mathbf{T}$.

Assume the hypotheses of the lemma. To apply Beth's theorem, let L and L' be the underlying graphs of T and T', respectively, and let $\mathbf{T}$ be the theory axiomatizing Mod T'. Let ϕ be an arbitrary subobject of $I(A)$, and let $\varphi(a)$ be the L'-formula representing ϕ. We claim that $\varphi(a)$ is preserved by all L-isomorphisms of models of $\mathbf{T}$. Indeed, given M,N and h as above, by I^{iso} being full, there is $\hat{h}: M \to N$ such that $I^*\hat{h} = h$. We can now take $f = \hat{h}_\Phi$ where $\Phi = [a: \varphi(a)]_{T'}$, the canonical iterpretation of φ in T'. Thus, the claim is shown.

By Beth's theorem, we have ψ as stated; it is clear that $I([a: \psi(a)]_T) = \phi$ as required.

■ 3.1.6.

Proof of 3.1.4: The theorem follows directly from 3.1.5, 3.1.6, and 2.4.7 (ii).

■ 3.1.4.

With T a pretopos, $\mathbb{M}$ a (not necessarily full) subcategory of Mod(T), we call $\mathbb{M}$ *dense* if the evaluation functor

$$T \longrightarrow (\mathbb{M}, \mathrm{Set})$$

(defined by

$$A \longmapsto \left\{ \begin{array}{ccc} M & \longmapsto & M(A) \\ M & & M(A) \\ h \big\downarrow & \longmapsto & \big\downarrow h_A \\ N & & N(A) \end{array} \right.$$

$$A \xrightarrow{f} B \quad \longmapsto \quad [M \longmapsto M(A) \xrightarrow[M(f)]{} M(B)] \qquad (M \in \mathbb{M})$$

is conservative. Let's say that $\mathbb{M}$ is *ultraclosed* if the class of its objects is closed under taking ultraproducts in Mod(T). Both 'dense' and 'ultraclosed' depend only on Ob($\mathbb{M}$).

Theorem 3.1.7. Let $I: T \to T'$ be a pretopos morphism between countable Boolean pretoposes. Assume that $\mathbb{M}$ is a dense full ultraclosed subcategory of $(\mathrm{Mod}\ T')^{\mathrm{iso}}$ such that $I^{\mathrm{iso}}\restriction\mathbb{M}: \mathbb{M} \to (\mathrm{Mod}\ T)^{\mathrm{iso}}$ is full and faithful. Then I is a quotient morphism.

The proof will obviously be accomplished by 3.1.4 and the following

Lemma 3.1.8. *Assume the hypotheses of* 3.1.7. *Then* I^{iso} *is full and faithful.*

Proof: Recall that every $M \in \mathrm{Mod}\ T'$ is a structure over L', the underlying graph of T', and $I^*M \in \mathrm{Mod}\ T$ is a structure over L, the underlying graph of T.

The fact that the ultraclosed subclass $\mathbb{M}$ of Mod T' is dense implies that for every $M \in \mathrm{Mod}\ T'$ there is $N \in \mathbb{M}$ such that $M \equiv N$ (elementary equivalence). Indeed, consider any L'-sentence (formula without free variables) ϕ; its canonical interpretation (cf. 2.5.1(ii)), $[\phi]$, is a subobject of 1, the terminal object of T'. The density of $\mathbb{M}$ implies that if $[\phi] \neq 0_1$ (0_1 = the minimal subobject of 1), then there is $N \in \mathbb{M}$ such that $N([\phi]) = 1_{(1_S)}$ i.e. $N \models \phi$. Now, let $M \in \mathrm{Mod}\ T'$ be given; by the last sentence and by the "ultraproduct version of the compactness theorem", 4.1.11 in [CK], since $\mathbb{M}$ is ultraclosed, we have $N \in \mathbb{M}$ satisfying all ϕ satisfied by M, i.e. $N \equiv M$.

Next, recall the Keisler-Shelah isomorphism theorem. Given two (small) families $\langle M_i \rangle_{i\in I}$, $\langle N_i \rangle_{i\in I}$ of structures such that $M_i \equiv N_i$ for all $i \in I$, then there is an ultrafilter U (over some set J) such that $(M_i)^U \cong (N_i)^U$ for all $i \in I$ (M^U, the ultraproduct of M with exponent U, is the ultraproduct $\Pi\langle M: j\in J\rangle/U$ of the constant family. The isomorphism theorem is 6.1.15 in [CK] stated for the case when I is a singleton. The general case follows from the special case by considering the composite structure $\langle \ldots M_i \ldots \rangle_{i\in I}$ over the disjoint sum of the languages of the M_i. Another remark is that the isomorphism theorem is proved, in [CK], only for one-sorted logic; however, the general case has the same proof, and it can also be deduced from the one-sorted case.)

By passing to the full subcategory of $(\mathrm{Mod}\ T')^{\mathrm{iso}}$ whose objects are all those that are isomorphic to a member of $\mathbb{M}$, we may assume that $\mathbb{M}$ is closed under taking isomorphic copies in $(\mathrm{Mod}\ T')^{\mathrm{iso}}$ ($\mathbb{M}$ is 'replete' in $(\mathrm{Mod}\ T')^{\mathrm{iso}}$).

As a consequence of the preceding, we have the following. Whenever $\langle M_i \rangle_{i\in I}$ is a small family of models in $\mathrm{Mod}\ T'$, there is an ultrafilter U (on some set) such that M_i^U belongs to $\mathbb{M}$ for all $i \in I$. Namely, choose $N_i \in \mathbb{M}$ such that $M_i \equiv N_i$; then choose U such that $M_i^U \approx N_i^U$ $(i \in I)$; since $\mathbb{M}$ is replete and ultraclosed, $M_i^U \in \mathbb{M}$ for all $i \in I$.

Next, consider the language (graph) L'', the similarity type of a composite structure consisting of two L'-structures M_1, M_2, and a homomorphism $I^*M_1 \to I^*M_2$. In detail, let L'' be the disjoint sum of two copies of L', together with a new edge $\chi_A: (IA)_1 \to (IA)_2$ for each $A \in \mathrm{Ob}(L)$; here X_1, X_2 denote the two copies of $X \in \mathrm{Ob}(L')$ in L'' under the canonical injections $I_1, I_2: L' \to L''$. It is easy to write down a set $\mathbf{T}''$ of sentences in the language L'' such that an L''-structure N is a model of $\mathbf{T}''$ if and only if $M_1 \underset{df}{=} I_1^*N$ and $M_2 \underset{df}{=} I_2^*N$ are in $\mathrm{Mod}\ T'$, and the $N(\chi_A)$ are the components h_A of an isomorphism $h: I^*M_1 \to I^*M_2$.

Let us consider the extension L''' of L'' obtained by adding an edge $\chi_X: X_1 \to X_2$ to L'' for each $X \in S \underset{df}{=} \mathrm{Ob}(L') - \{IA: A \in \mathrm{Ob}(L)\}$. Let $i: L'' \to L'''$ denote the inclusion. Furthermore, let $\mathbf{T}'''$ be the theory in L''' whose models are exactly those L'''-structures P for which $M_1 \underset{df}{=} I_1^*i^*P$, $M_2 \underset{df}{=} I_2^*i^*P$ are in $\mathrm{Mod}\ T'$, and for which there is an isomorphism $h: M_1 \to M_2$ such that $h_X = P(\chi_X)$ for $X \in S$, and $h_{IA} = P(\chi_A)$ for $A \in \mathrm{Ob}(L)$ (Of course, $\mathbf{T}'''$ includes, among others, axioms saying that $x_A = x_{A'}$ whenever $IA = IA'$).

Now we see that the conclusion of the lemma is equivalent to the following statement:

(*) For every model N of $\mathbb{T}''$ there is *exactly one* model P of $\mathbb{T}'''$ such that $i^*P = N$.

Consider the class $\hat{\mathbb{M}}$ of L''-structures N such that N is a model of $\mathbb{T}''$ and I_1^*N, I_2^*N belong to $\mathbb{M}$. Notice that the hypotheses of the lemma imply the following statement:

(**) For every N in $\hat{\mathbb{M}}$, there is exactly one model P of $\mathbb{T}''$ such that $i^*P = N$.

We are left with the task of deducing (*) from (**). To this end, we will make use of Beth's definability theorem (cf [CK]).

The first remark is that we have the weakening of (*), called (***), obtained by replacing 'exactly one' by 'at most one'. To show this, let N be a model of $\mathbb{T}''$, and let P_1, P_2 be expansions of N ($i^*P_1 = i^*P_2 = N$) which are models of $\mathbb{T}'''$. Find an ultrafilter U such that $(I_1^*N)^U$, $(I_2^*N)^U$ both belong to $\mathbb{M}$. Hence $N^U \in \hat{\mathbb{M}}$. Then P_1^U, P_2^U are both expansions of $N^U \in \hat{\mathbb{M}}$ which are models of $\mathbb{T}'''$. Hence, by the uniqueness part of (**), $P_1^U = P_2^U$. But then it follows that $P_1 = P_2$: were the two maps $P_1(\chi_X)$, $P_2(\chi_X)$: $N(X_1) \to N(X_2)$ different, so would their U-ultrapowers be (if you like because two parallel maps in Set being distinct is a first order property).

It is at this point that we can invoke Beth's theorem. We use the theorem in the following form. Suppose L'' and L''' are two languages (graphs) such that L'' is contained in L''' but L''' does not contain any new sort with respect to L''. Suppose also that $\mathbb{T}'''$ is a theory in L''' and we have that for any L''-structure N, there is at most one expansion P of N that is a model of $\mathbb{T}'''$. In this case, there is an assignment of an L''-formula $\phi_f(x,y)$ to each $f: X \to Y$ in L'''-L'' such that if P is a model of $\mathbb{T}''$, then $P(f)$ is the interpretation of ϕ_f in P (i.e. the interpretation of ϕ_f in N, the L''-reduct of P); in other words, $P(f)(a) = b \iff N \models \phi_f[a,b]$ for $a \in NX$, $b \in NY$. (Usually, Beth's theorem is stated for the situation when L'''-L'' contains only a single new symbol. However, the general case is well-known too, and can be proved in much the same way as the special case.)

Since in our situation, the reduct $N = i^*P$ of any model P of $\mathbb{T}'''$ is automatically a model of $\mathbb{T}''$, (***) expresses that the hypotheses of Beth's theorem are satisfied. Thus for each $X \in S$, we have a

$\phi_X(x_1,x_2)$, an L''-formula, such that for any model P of $\mathbb{T}'''$, the interpretation of ϕ_X in P is $P(\chi_X)$.

We can now prove (*) as follows. Suppose N is a model of $\mathbb{T}''$. Remember that each $\phi_X(x_1,x_2)$ is an L''-formula, hence we can consider its interpretation in N; it is a subset $\hat{\chi}_X$ of $N(X_1) \times N(X_2)$. We claim that each $\hat{\chi}_X$ is the graph of a map $N(X_1) \to N(X_2)$, also denoted by $\hat{\chi}_X$, and for the L'''-expansion P of N obtained by letting $P(\chi_X) = \hat{\chi}_X$, we have that P is a model of $\mathbb{T}'''$. (This will, of course, complete the proof.) As before, there is an ultrafilter U such that $N^U \in \hat{M}$. By (**), there is an expansion P' of N^U which is a model of $\mathbb{T}'''$. By the properties of the ϕ_X, $P'(\chi_X)$ is the same as the interpretation of ϕ_X in N^U. This means that P' is nothing but the ultrapower P^U of a structure P which is an expansion of N, and for which $P(\chi_X)$ $(X \in S)$ is the interpretation of ϕ_X in N. Since P' is a model of $\mathbb{T}'''$, and $P \equiv P^U = P'$, P is a model of $\mathbb{T}'''$ as well. This completes the proof. ■ 3.1.8.

■ 3.1.7.

Theorem 3.1.4. throws some light on a situation considered in [MP] (for this discussion, we use the terminology and notation of [MP]). Let $\mathcal{B}$ be a locally finitely presentable (l.f.p.) category, and $F: \mathcal{A} \longrightarrow \mathcal{B}$ a faithful functor, full on isomorphisms, which creates limits and filtered colimits in $\mathcal{A}$. Proposition 2.3. in [MP] says that in this case $\mathcal{A}$ is l.f.p. itself. If $\mathbb{C} = \mathcal{B}_{f.p.}$, the category of finitely presentable objects in $\mathcal{B}$ (hence, $\mathcal{B} \cong \mathrm{LEX}(\mathbb{C}^{op},\mathrm{Set})$) and $\mathbb{D} = \mathcal{A}_{f.p.}$, then F induces a Lex-morphism $I: \mathbb{C} \to \mathbb{D}$ (so that $F \cong I^*$). It is pointed out in [MP] that I is not necessarily a Lex-quotient even if F is full. In the general situation, we can still say that $\mathbb{C}$ "generates $\mathbb{D}$ via I on the level of Boolean pretoposes".

To formulate this statement precisely, we introduce some analogs of definitions in Section 1.4. With C in Lex, the free Boolean pretopos $T = F(C)$ over C, with $\varphi: C \longrightarrow F(C)$ a canonical Lex-morphism, is defined by the universal property that for any Boolean pretopos T', $()\circ\varphi$: Pretop(T,T') $\longrightarrow$ Lex(C,T') is faithful, full on isomorphisms, and essentially surjective. The existence of $F(C)$ can be proved by considerations like those in Section 2.5 used to prove the assertions in 1.4.1. For a Lex-morphism $F: C \longrightarrow T$, with T a Boolean pretopos, we may say that F *generates* T if the Pretop-morphism $F(C) \longrightarrow T$ induced by F is a quotient in Pretop.

Returning to our situation above, the precise formulation of the assertion is that the composite

$$\mathcal{D} \longrightarrow \mathcal{C} \xrightarrow[\varphi]{} F(\mathcal{C})$$

generates $F(\mathcal{C})$ provided that $\mathbb{C}$ is essentially countable. This claim is easily verified by applying 3.1.4.

Let me point out that the Craig interpolation theorem (of which Beth's definability theorem, a result that played a key role in this section, is a consequence) has recently been given a purely categorical treatment, in fact, in the more general context of intuitionistic logic, by A.M. Pitts in [P1] and [P2]. It is possible that Pitts' methods, and the methods in [JT] he uses, will eventually lead to a (more) "categorical" proof of all results of this paper.

3.2. <u>The proof of the main result.</u>

In preparation for the proof of 1.4.4, I remind the reader of a fundamental, and 'trivial', 2-categorical adjunction. Let $\mathcal{S}$ = Set; at first, $\mathcal{S}$ in fact could be any category. There is an adjunction between the 2-functors

$$\mathrm{Cat}^{\mathrm{op}} \underset{\mathbb{F}=(-,\mathcal{S})}{\overset{\mathbb{G}=(-,\mathcal{S})}{\rightleftarrows}} \mathrm{Cat}$$

making $\mathbb{F}$ a left adjoint of $\mathbb{G}$. In detail: the effect of $\mathbb{G}$ is this:

$$\mathcal{C} \longmapsto (\mathcal{C},\mathcal{S}) = \mathrm{Cat}(\mathcal{C},\mathcal{S})$$

$$\mathcal{C} \xrightarrow{F} \mathcal{C}' \longmapsto (-,F) = (\,)\circ F\colon (\mathcal{C}',\mathcal{S}) \longrightarrow (\mathcal{C},\mathcal{S})$$

$$\mathcal{C} \underset{F'}{\overset{F}{\Downarrow f}} \mathcal{C}' \longmapsto (H \longmapsto f\circ \mathrm{id}_H \qquad (H\in \mathrm{Ob}(\mathcal{C}',\mathcal{S}))).$$

All arrows are meant in Cat, and not in $\mathrm{Cat}^{\mathrm{op}}$; we have a contravariant functor Cat $\longrightarrow$ Cat. $\mathbb{F}$, as a contravariant functor Cat $\longrightarrow$ Cat, is *identical* to $\mathbb{G}$.

The adjunction consists of an *isomorphism* between two categories:

$$\theta_{\mathcal{C},\mathcal{D}}\colon (\mathcal{C},(\mathcal{D},\mathcal{S})) \xrightarrow{\cong} (\mathcal{D},(\mathcal{C},\mathcal{S}))$$

which is natural in both $\mathcal{C}$ and $\mathcal{D}$. In detail: $\theta_{\mathcal{C},\mathcal{D}}$ maps X into the following Y:

$$\left(\begin{array}{c} \mathcal{C} \xrightarrow{\ X\ } (\mathcal{D},\mathcal{S}) \\ \hline \mathcal{D} \xrightarrow[\ Y\ ]{} (\mathcal{C},\mathcal{S}) \end{array} \right)$$

$$Y: \begin{cases} D \longmapsto \left[\begin{array}{l} C \longmapsto X(C)(D) \\ \\ C \xrightarrow{\ f\ } C' \longmapsto X(f)_D \end{array} \right] \\ \\ D \xrightarrow{\ g\ } D' \longmapsto [\, C \longmapsto X(C)(g) \,]. \end{cases}$$

With a morphism $\xi: X \longrightarrow X'$ in $(\mathcal{C},(\mathcal{D},\mathcal{S}))$, $\theta_{\mathcal{C},\mathcal{D}}$ associates the natural transformation

$$D \longmapsto (C \longmapsto \xi_C)$$

(of course, D ranges over $\mathrm{Ob}(\mathcal{D})$, C over $\mathrm{Ob}(\mathcal{C})$, etc). Naturality in $\mathcal{C}$, $\mathcal{D}$ means that given

$$\mathcal{C}' \underset{F'}{\overset{F}{\rightrightarrows}} \!\!{\scriptstyle \downarrow f}\ \ \mathcal{C} \underset{X'}{\overset{X}{\rightrightarrows}} \!\!{\scriptstyle \downarrow \xi}\ \ (\mathcal{D},\mathcal{S}) \underset{(H',\mathcal{S})}{\overset{(H,\mathcal{S})}{\rightrightarrows}} \!\!{\scriptstyle \downarrow (h,\mathcal{S})}\ \ (\mathcal{D}',\mathcal{S})$$

$$\left(\mathcal{D}' \underset{H'}{\overset{H}{\rightrightarrows}} \!\!{\scriptstyle \downarrow h}\ \ \mathcal{D} \right),$$

under the isomorphism $\theta_{\mathcal{C}',\mathcal{D}'}$ the appropriate composites go over to the corresponding composites in

$$\mathcal{D}' \underset{H'}{\overset{H}{\rightrightarrows}} \!\!{\scriptstyle \downarrow h}\ \ \mathcal{D} \underset{\theta_{\mathcal{C},\mathcal{D}}(X')}{\overset{\theta_{\mathcal{C},\mathcal{D}}(X)}{\rightrightarrows}} \!\!{\scriptstyle \downarrow \theta_{\mathcal{C},\mathcal{D}}(\xi)}\ \ (\mathcal{C},\mathcal{S}) \underset{(F',\mathcal{S})}{\overset{(F,\mathcal{S})}{\rightrightarrows}} \!\!{\scriptstyle \downarrow (f,\mathcal{S})}\ \ (\mathcal{C}',\mathcal{S}).$$

Having the above adjunction, we can construct another, this time between 2-functors

$$\mathrm{Pretop}^{\mathrm{op}} \underset{\mathbb{F}}{\overset{\mathbb{G}}{\rightleftarrows}} \mathrm{Cat}\ .$$

Now, $\mathbb{G}$ (as a contravariant functor $\mathrm{Pretop} \longrightarrow \mathrm{Cat}$) is given by:

$$G(T) = \mathrm{Mod}(T) \qquad \text{(a full subcategory of } (T,\mathcal{S})\text{)}$$

and otherwise similar formulas as before. Also,

$$\mathbb{F}(\mathcal{C}) = (\mathcal{C},\mathcal{S}).$$

It is essential, of course, that the functor-category $(\mathcal{C},\mathcal{S})$ is a pretopos, for any category $\mathcal{C}$, with operations 'inherited' from $\mathcal{S}$. This is a special case of the fact that limits and colimits in functor categories are computed pointwise, provided the codomain category has them. We have, for the rest of the effect of $\mathbb{F}$, similar formulas as before; the 'pointwise' character of the pretopos operations in $(\mathcal{C},\mathcal{S})$, $(\mathcal{C}',\mathcal{S})$ will ensure that $\mathbb{F}(F)$, for arbitrary $F\colon \mathcal{C} \longrightarrow \mathcal{C}'$, as defined above, will indeed be a Pretop-morphism.

The adjunction-isomorphisms are defined as before.

Below, referring to the last adjunction, we will write F^* for certain restrictions of $\mathbb{F}(F)$, or $\mathbb{G}(F)$, for a morphism F in the appropriate 2-category, and $\llcorner H \lrcorner$, $\ulcorner H \urcorner$ for $\theta(H)$, $\theta^{-1}(H)$, respectively ($\theta = \theta_{\mathcal{C},\mathcal{D}}$ for some appropriate $\mathcal{C},\mathcal{D}$).

We turn to the proof of 1.4.4.

Let $\mathcal{O}$ be an operation in Set commuting with ultraproducts; let $\mathcal{O}$ be of type (G,G'), with graph K', and domain K. Below, K' will denote the full subcategory of $(G',\mathcal{S})^{\mathrm{iso}}$ with the objects in K'; similarly for $K \subset (G,\mathcal{S})^{\mathrm{iso}}$.

With $i\colon G \longrightarrow G'$ the inclusion, $F(G)$ denoting the Boolean pretopos completion, $\varphi = \varphi^{\mathrm{Bpt}}$, and similarly for the primed items, we have F making

$$\begin{array}{ccc} G & \xrightarrow{\ \varphi\ } & F(G) \\ {\scriptstyle i}\downarrow & & \downarrow{\scriptstyle F} \\ G' & \xrightarrow{\ \varphi'\ } & F(G') \end{array} \qquad (1)$$

commute up to an isomorphism. We can now construct the following diagram:

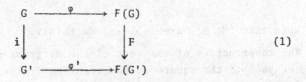

$$\begin{array}{ccccc}
K & \xrightarrow{\ \text{incl}\ } & (G,S)^{\text{iso}} & \underset{\varphi^*}{\xleftarrow{\ \sim\ }} & (\text{Mod } F(G))^{\text{iso}} \\
{\scriptstyle p}\uparrow & & \uparrow{\scriptstyle i^*} \quad \cong & & \uparrow{\scriptstyle F^*} \\
K' & \xrightarrow{\ \text{incl}\ } & (G',S)^{\text{iso}} & \underset{\varphi'^*}{\xleftarrow{\ \sim\ }} & (\text{Mod } F(G'))^{\text{iso}}
\end{array} \qquad (2)$$

Here, all the starred functors are defined by composition with φ; F^* is in fact $\mathbf{F}(F)$, properly restricted. The right-hand-side square commutes up to an isomorphism inherited from (1). Moreover, by the universal properties of $F(G)$, $F(G')$, φ^* and φ'^* are equivalences of categories. The left-hand-side square commutes when p is defined as the restriction of i^* to K'; in fact, $Ob(K)$ is the image of $Ob(K')$ under i^*. By the definition of "operation in a category", p is full and faithful.

Taking quasi-inverses of φ^*, φ'^*, and composing them with the inclusion, we obtain the upper square in the left-hand-side one of the following two diagrams:

$$\begin{array}{ccc}
K & \xrightarrow{\ H\ } & (\text{Mod } FG)^{\text{iso}} \\
{\scriptstyle p}\uparrow & \cong & \uparrow{\scriptstyle F^*} \\
K' & \xrightarrow{\ H'\ } & (\text{Mod } FG')^{\text{iso}} \\
{\scriptstyle \ulcorner\Gamma\urcorner}\searrow & & \nearrow{\scriptstyle Q^*}\ \cong \\
 & (\text{Mod } T)^{\text{iso}} &
\end{array}
\qquad\qquad
\begin{array}{ccc}
(K,S) & \xleftarrow{\ \llcorner H\lrcorner\ } & F(G) \\
{\scriptstyle p^*}\downarrow & \cong & \downarrow{\scriptstyle F} \\
(K',S) & \xleftarrow{\ \llcorner H'\lrcorner\ } & F(G') \\
\cong\ \nwarrow{\scriptstyle \Gamma} & & \swarrow{\scriptstyle Q} \\
 & T &
\end{array}$$

Note that H, H' are full and faithful.

The construction of the rest of the diagrams proceeds as follows. First, we pass to the square on the right-hand-side, by the help of our adjunction. Second, Q and Γ form the quotient-conservative factorization of $\llcorner H'\lrcorner$. Finally, again by the adjunction, we return to the triangle on the left (and take appropriate restrictions.)

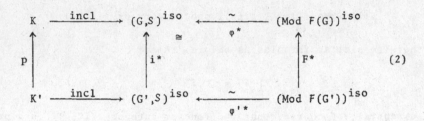

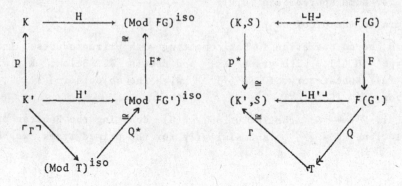

Claim 3.2.1. $F(G) \xrightarrow{QF} T$ is a quotient.

To deduce this from 3.1.8, let's consider

$$K' \xrightarrow[\ulcorner\Gamma\urcorner]{} (\text{Mod } T)^{\text{iso}} \xrightarrow[(QF)^*]{} (\text{Mod } F(G))^{\text{iso}}.$$

$\ulcorner\Gamma\urcorner$ is full and faithful since H', Q^* are (Q^* is because Q is a quotient), and $Q^*\ulcorner\Gamma\urcorner \cong H'$. The composite $(QF)^*\ulcorner\Gamma\urcorner$ is full and faithful, since it is isomorphic to Hp, and both H and p are full and faithful.

Let $\mathbb{M}$ be the *image* of $\ulcorner\Gamma\urcorner$; the subcategory of $(\text{Mod } T)^{\text{iso}}$ whose objects are of the form $\ulcorner\Gamma\urcorner(D')$, with $D' \in \text{Ob}(K')$, and morphisms $\ulcorner\Gamma\urcorner(\delta)$, with $\delta\colon D' \longrightarrow D''$ in K'. Since $\ulcorner\Gamma\urcorner$ if full, the image is well-defined, and we have an equivalence $K' \longrightarrow \mathbb{M}$ such that

$$\begin{array}{ccc} K' & \xrightarrow{\sim} & \mathbb{M} \\ & {\scriptstyle\ulcorner\Gamma\urcorner}\searrow \quad \swarrow{\scriptstyle\text{incl}} & \\ & (\text{Mod } T)^{\text{iso}} & \end{array}$$

commutes. Under the adjunction, the functor $T \longrightarrow (\mathbb{M}, \mathbb{S})$ corresponding to the inclusion is the evaluation functor appearing in the definition of 'dense' above. Since it is isomorphic to Γ, and Γ is conservative, we conclude that $\mathbb{M}$ is dense in Mod T. By the facts that $\ulcorner\Gamma\urcorner$ and $(QF)^*\ulcorner\Gamma\urcorner$ are full and faithful, and by the last factorization, we have that $\mathbb{M}$ is full in $(\text{Mod } T)^{\text{iso}}$ and that $(QF)^*\restriction\mathbb{M}$ is full and faithful. By the constructions of the free Boolean pretopos and quotients, it is clear that $T(G)$ and T are both countable, since G and G' are finite.

It remains to show that $\mathbb{M}$ is ultraclosed in Mod T.

A *u-functor* $X\colon K_1 \longrightarrow K_2$ ('pre-ultrafunctor' in [M1]) with categories K_1, K_2 in which ultraproducts have been defined (e.g., K' or Mod T above) is a functor that preserves ultraproducts up to specified isomorphisms. Thus, X consists of a functor, denoted by X itself, $X\colon K_1 \longrightarrow K_2$, together with a *transition-structure* consisting of an isomorphism $[X,U]$ for every ultrafilter U on any set I as follows:

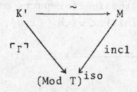

$$\begin{array}{ccc} (K_1)^I & \xrightarrow{[U]} & K_1 \\ {\scriptstyle X^I}\downarrow & \overset{[X,U]}{\swarrow} & \downarrow{\scriptstyle X} \\ (K_2)^I & \xrightarrow[{[U]}]{} & K_2 \end{array}$$

$$[X,U]: X \circ [U]_{K_1} \longrightarrow [U]_{K_2} \circ X^I$$

(here the [U] are the ultraproduct functors on K_1, K_2). A *strict* u-functor is one in which the [X,U]'s are all identities.

The u-functors $K_1 \longrightarrow K_2$ form a category, $u(K_1,K_2)$, with the u-*transformations* as morphisms. A u-transformation $\sigma: X \longrightarrow Y$ is a natural transformation between the functor-parts of X and Y with the additional condition of compatibility with [X,U], [Y,U]: the diagram

$$\begin{array}{ccc} (K_1)^I & \xrightarrow{[U]} & K_1 \\ X^I \Big\downarrow \xrightarrow{\sigma^I} \Big\downarrow Y^I & & X \Big\downarrow \xrightarrow{\sigma} \Big\downarrow Y \\ (K_2)^I & \xrightarrow[{[U]}]{} & K_2 \end{array}$$

gives a commutative diagram of 1-cells and 2-cells:

$$\begin{array}{ccc} X[U] & \longrightarrow & Y[U] \\ {\scriptstyle [X,U]}\Big\downarrow & & \Big\downarrow{\scriptstyle [Y,U]} \\ [U]X^I & \longrightarrow & [U]Y^I. \end{array}$$

If, in particular, $K_1 = K$, $K_2 = S$, then $u(K,S)$ is a pretopos, and in fact, the forgetful functor

$$u(K,S) \xrightarrow{q_K} (K,S)$$

(forgetting the [X,U]) is a conservative Pretop-morphism; this is easy to see (using Los's theorem).

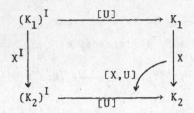

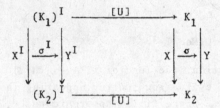

If $I: T_1 \longrightarrow T_2$ is a Pretop-morphism, then $I^*: \text{Mod } T_2 \to \text{Mod } T_1$ is a strict u-functor. Similarly, for a diagram $D_0: G \longrightarrow T$, $(D_0)^*: \text{Mod } T \longrightarrow (G, \mathcal{S})$ is a strict u-functor.

Now, let us turn to the diagram under (2) and (3). As we said, $(\varphi')^*$ is a strict u-functor; also, the inclusion of K' in $(G', \mathcal{S})^{\text{iso}}$ is one. It is easy to see that the quasi-inverse of the functor-part of a u-functor whose functor-part is an equivalence can be made into a u-functor by endowing it with a transition structure (such that, in fact it will be a quasi-inverse with isomorphisms at head and tail that are u-transformations, but we don't need this additional fact). Thus, H' in (3) can be made into a u-functor, also denoted by H'. It is then readily seen that $\llcorner H' \lrcorner$ factors (exactly) in the form

$$F(G') \xrightarrow{\ \hat{H}\ } u(K', \mathcal{S}) \xrightarrow{\ q_{K'}\ } (K', \mathcal{S}).$$

Let

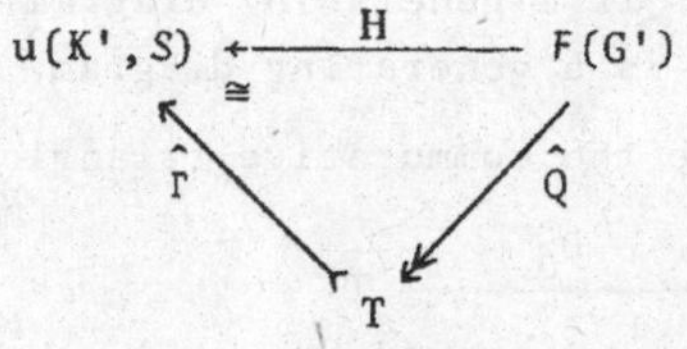

be the quotient-conservative factorization of $\hat{H}$. Since $q = q_{K'}$ is conservative, and hence $q\hat{\Gamma}$ is conservative, by the uniqueness of the q-e factorization, we have

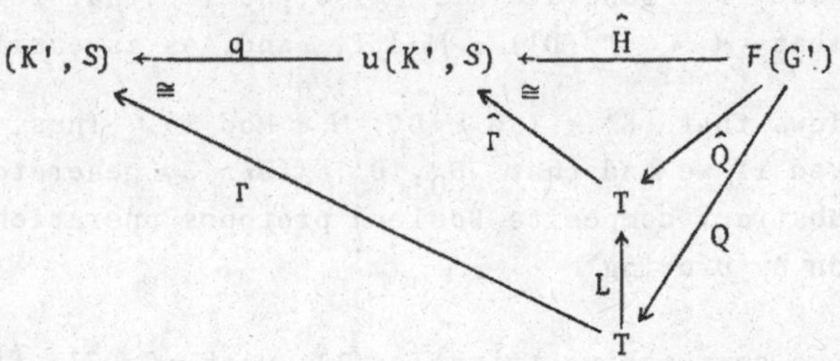

with L an equivalence of categories. Now, we verify directly that $\ulcorner q\hat{\Gamma} \urcorner$ is a u-functor. In fact, we may put

$$([q\hat{\Gamma}]_{(D_i')_{i\in I}})_A = [\hat{\Gamma}(A), U]_{(D_i')_{i\in I}}: \ ((q\hat{\Gamma})(A))(\prod_i D_i'/U) \to \prod_i (q\hat{\Gamma}(A))(D_i')/U.$$

Remember that each $\hat{\Gamma}(A) \in \text{Ob}(u(K', \mathcal{S}))$ is a u-functor, hence $[\hat{\Gamma}(A), U]$ is defined.

We have that $\ulcorner\Gamma\urcorner \cong L^*\ulcorner q\hat{\Gamma}\urcorner$. Thus, up to isomorphisms, the image of $\ulcorner\Gamma\urcorner$ is the same as that of $\ulcorner q\hat{\Gamma}\urcorner$. The image of $\ulcorner q\hat{\Gamma}\urcorner$ is clearly ultraclosed since K' is closed under ultraproducts and $\ulcorner q\hat{\Gamma}\urcorner$ preserves them. Thus $\mathbb{M}$ is ultraclosed.

This argument could have been subsumed in the basic adjunction, by replacing Cat by a 2-category of categories with ultraproducts; c.f. [M1].

We have shown that 3.1.8. is applicable to conclude the Claim.

■ 3.2.1.

Let us define

$$D_0' = Q\,\varphi':\ G' \longrightarrow T$$

$$D_0 = D_0' \circ i := Q \circ \varphi' \circ i:\ G \longrightarrow T.$$

Since Q is a quotient, D_0' is a generating diagram. Since $D_0 \cong QF\varphi$, and QF is a quotient, D_0 is a generating diagram.

In other words, we have the commutative triangle

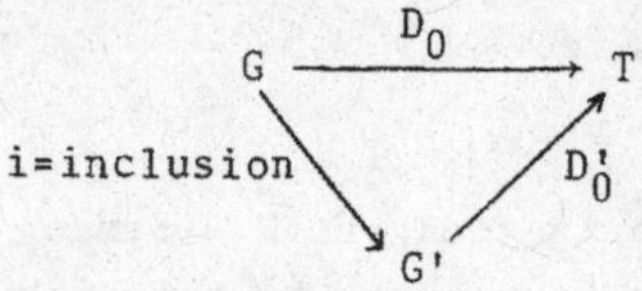

(1)

in which both D_0 and D_0' generate T. Also, notice that for every $D' \in K'$, we have that $M \underset{df}{=} \ulcorner\Gamma\urcorner(D') \in \text{Mod}\ T$, and, as is easily seen, $M \circ D_0' \cong D'$. It follows that $K' \subseteq \text{Iso}\{M \circ D_0' : M \in \text{Mod}\ T\}$. Thus, Theorem 1.4.4 would be proved if we had that D_0, D_0' *finitely* generate T (and (1) gives an abstract composite Boolean pretopos operation). We repair the situation by proving

<u>Lemma</u> 3.2.2. Given a commutative triangle (1), with G, G' finite graphs, and D_0, D_0' both generating T, we have a factorization

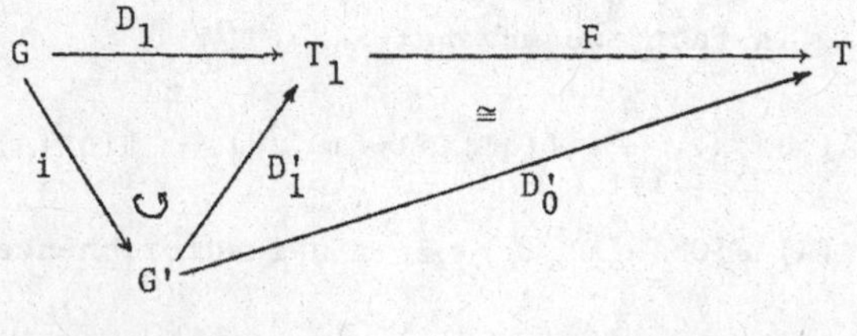

in which each of D_1, D_1' finitely generates T_1.

Note that, if the assertion of the lemma is true, we get that for $M \in \mathrm{Mod}\ T$, we have $M \circ D_0' \cong (M \circ F) \circ D_1'$; hence $K' \subset \mathrm{Iso}\{N \circ D_1' : N \in \mathrm{Mod}\ T_1\}$, i.e. K' is a restriction of the abstract Boolean pretopos operation given by D_1 and D_1', proving the Theorem.

Proof of 3.2.2: The lemma certainly has the flavor of being 'abstract nonsense'. The proof that follows is straightforward, but somewhat tedious. We start with some preliminary remarks.

The process of (universally) adding new axioms to a theory can be mimicked by (universally) inverting morphisms, i.e. constructing $T \longrightarrow T[\Sigma^{-1}]$ out of T and some $\Sigma \subset \mathrm{Morph}(T)$. E.g., if R and S are subobjects of A in T, then the "axiom" $R \le S$ can be mimicked by inverting a single arrow. The reason is that there is an arrow σ in T, explicitly constructed from representing monos r and s for R and S, respectively, such that for any Pretop morphism $F: T \longrightarrow T'$ with domain T, $FR \le FS$ iff $F\sigma$ is invertible. In fact, σ is the monomorphism in the pullback

$$\begin{array}{ccc} R & \xrightarrow{\ r\ } & A \\ {\scriptstyle\sigma}\uparrow & & \uparrow{\scriptstyle s} \\ R \cap S & \longrightarrow & S \end{array}$$

as shown; the assertion is clear. To give another example, if $\Phi \in \mathrm{Sub}(A \times B)$ in T, then the axiom "Φ is a functional subobject" is (universally) imposed on T by (universally) inverting the composite $\Phi \hookrightarrow A \times B \xrightarrow{\pi_1} A$ (see before 2.3.2).

To start the proof of the lemma, let us consider the triangle

$$\begin{array}{ccc} F & \xrightarrow{\ \hat{D}\ } & T \\ & {\scriptstyle\hat{i}}\searrow \quad \cong \quad \nearrow{\scriptstyle \hat{D}'} & \\ & F' & \end{array}$$

induced by (1) (here we use the same notation as after Definition 1.4.3). We pretend that $\hat{i}$ is an inclusion (i.e., we also write simply R for $\hat{i}R$ with R in F), and similarly for the canonical $\varphi: G \longrightarrow F$, $\varphi': G' \longrightarrow F'$. We will write $\hat{R}$ for $\hat{D}(R)$ (R in F) and for $\hat{D}(R)$ (R in F'). Remember that $\hat{D}$ and $\hat{D}'$ are quotients in Pretop.

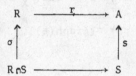

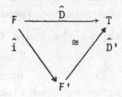

This means that they are full on subobjects, and for every X in $\mathcal{T}$ there are A in $\mathcal{F}$ (in $\mathcal{F}'$) and an effective epi $p\colon \hat{A} \twoheadrightarrow X$ (by 2.4.7 (ii) and also by taking the sum of the finitely many objects mentioned by 2.4.7 (ii)).

To each X, an object in $\mathcal{G}'$, we assign an object A_X of $\mathcal{F}$ and an effective epi

$$p_X\colon \hat{A}_X \twoheadrightarrow \hat{X} \quad (\text{in } \mathcal{T}).$$

Let $\Phi_X \in \mathrm{Sub}(A_X{\times}X)$ in $\mathcal{F}'$ be such that

$$\hat{\Phi}_X = \text{graph of } p_X \quad (\text{in } \mathcal{T}).$$

Let $R_X \in \mathrm{Sub}(A_X{\times}A_X)$ in $\mathcal{F}$ be such that

$$\begin{aligned}\hat{R}_X = {} & (\text{the subobject of } \hat{A}_X{\times}\hat{A}_X \text{ induced by})\\ & \text{the kernel pair of } p_X \quad (\text{in } \mathcal{T}).\end{aligned}$$

Let, for any $\xi\colon X \longrightarrow Y$ in $\mathcal{G}'$, $S_\xi \in \mathrm{Sub}(A_X{\times}A_Y)$ in $\mathcal{F}$ be such that

$$\hat{S}_\xi = (p_X{\times}p_Y)^{-1}(\mathrm{graph}(\xi)),$$

i.e., there is a pullback

$$\begin{array}{ccc} \hat{A}_X{\times}\hat{A}_Y & \xrightarrow{\;p_X\times p_Y\;} & \hat{X}{\times}\hat{Y} \\ \uparrow & & \uparrow \\ \hat{S}_\xi & \longrightarrow & \mathrm{graph}(\xi). \end{array}$$

In addition, let us make sure that in case X is in $\mathcal{G}$, then $A_X = X$ and R_X is the trivial equivalence relation (equality).

Since $\mathcal{G}'$ is a finite graph, in the above finitely many items were introduced.

We are going to define $\mathcal{T}_1$ as a finite quotient of $\mathcal{F}'$. We universally impose on $\mathcal{F}'$ the following conditions:

A1. "Φ_X is functional". $(X \in \mathrm{Ob}(\mathcal{G}'))$.

A2. "R_X is the kernel pair of the morphism $q_X\colon A_X \longrightarrow X$ whose graph is Φ_X." $(X \in \mathrm{Ob}(\mathcal{G}'))$.

A3. "S_ξ is the pullback $(q_X{\times}q_Y)^{-1}(\mathrm{graph}(\xi))$ with q_X, q_Y as in A2." $(\xi\colon X \to Y$ in $\mathcal{G}')$.

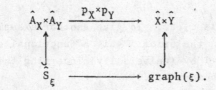

Some explanations are in order. First of all, we intend here the specification of certain definite finite sets Σ_1, Σ_2, Σ_3 of arrows in F' whose inversion is equivalent to the respective "axioms" A1, A2, A3. Σ_1 is clear from remarks above. Concerning A2 and Σ_2, note that q_X does not exist as an arrow in F' (it will exist only when Φ_X has become functional). The specification of Σ_2 is done as follows. Fix $X \in G'$. Let $A = A_X$. Consider the subobjects $\Phi = \Phi_X$ of $A \times X$, $R = R_X$ of $A \times A$ specified above, and construct the subobject

$$R' :=_{df} [a \in A, a' \in A: (\exists x \in X)(\Phi(a,x) \wedge \Phi(a',x)]$$

of $A \times A$ in F'. The axioms under A2 are

$$R' = R \qquad (\text{i.e.,} \quad R' \leq R, \quad R' \leq R)$$

one for each $X \in \mathrm{Ob}(G')$. Note that for any $F: T \to T'$ in Pretop satisfying A1, FR is the kernel pair of the morphism $q: A \to X$ whose graph is $F\Phi$ if and only if $FR' = FR$. The set Σ_2 is obtained by remarks above concerning how to impose an inequality of subobjects.

The description of Σ_3 corresponding to A3 is left to the reader.

As promised, we define

$$T_1 = F'[(\Sigma_1 \cup \Sigma_2 \cup \Sigma_3)^{-1}]$$

with $Q_1: T \to T_1$ the canonical quotient. Since the conditions A1, A2, A3 are all satisfied by $\hat{D}'$, clearly, F as in the diagram in the statement of the lemma exists. It remains to show that the composite

$$F \xrightarrow{\hat{i}} F' \xrightarrow{Q_1} T_1$$

is a finite quotient as well.

Let T_2 be the finite quotient of F obtained by imposing on F the following conditions:

A2'. "R_X is an equivalence relation" $(X \in \mathrm{Ob}(G'))$.

A3'. "The subobject S_ξ of $A_X \times A_Y$ gives rise to a morphism $A_X/R_X \to A_Y/R_Y$" $(\xi: X \to Y$ in $G')$.

We intend finite sets $\Sigma_{2'}$, $\Sigma_{3'}$ of morphisms of F. $\Sigma_{2'}$ is pretty clear (see 2.4.1). Concerning $\Sigma_{3'}$, the inequalities (2) (or (3)) in the proof of 2.4.5 show what we need here. We let

$$T_2 = F[(\Sigma_{2'} \cup \Sigma_{3'})^{-1}]$$

with $Q_2\colon F \to T_2$ the canonical quotient.

Next, note that we have a triangle

$$\begin{array}{ccc} F & \xrightarrow{\hat{i}} & F' \\ & {\scriptstyle Q_2}\searrow \cong & \downarrow {\scriptstyle Q_2'} \\ & & T_2 \end{array}$$

commuting up to isomorphism, with Q_2' induced by

$$G' \longrightarrow T_2$$

given as

$$\begin{array}{ccl} X & & Q_2A_X/Q_2R_X \\ \xi\downarrow & \longmapsto & \downarrow \text{arrow induced by } Q_2S_\xi \\ Y & & Q_2A_Y/Q_2R_Y\ . \end{array}$$

Then, there is an essentially unique $G\colon T_2 \to T$ in Pretop such that in

$$\begin{array}{ccccc} F & \xrightarrow{\hat{i}} & F' & \xrightarrow{Q_1} & T_1 \\ & {\scriptstyle Q_2}\searrow\cong & \downarrow {\scriptstyle Q_2'} & \cong\nearrow {\scriptstyle G} & \\ & & T_2 & & \end{array} \qquad (2)$$

the right-hand-side triangle commutes up to an isomorphism. To obtain G, note that $Q_1\hat{i}$ satisfies the conditions A2', A3' whose universal solution is Q_2. This gives an essentially unique G with an isomorphism

$$GQ_2 \cong Q_1\hat{i}. \qquad (3)$$

The concrete definition of Q_2' gives an isomorphism

$$GQ_2'\varphi' \cong Q_1\varphi' \qquad (4)$$

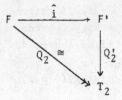

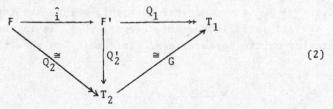

for φ' the canonical diagram $G' \to F'$ such that the pasting of the isomorphism $GQ_2' \cong Q_1$ induced by (4) and the isomorphism in the left-hand-side triangle in (2) is the same as the isomorphism (3).

We claim that G is a finite quotient, in fact that it is obtained by imposing on T_2 the conditions

A1'. "Φ_X is the graph of the canonical arrow $A_X \to A_X/R_X$" ($X \in Ob(G')$).

To show the claim, we have to verify that G satisfies the corresponding universal property. Let T_3 be an arbitrary pretopos. The fact that composition with G induces a full and faithful functor $\text{Pretop}(T_1,T_3) \to \text{Pretop}(T_2,T_3)$ is a consequence of the facts that in (2), Q_1 and Q_2 are quotients. G certainly satisfies A1'. Finally, if $H: T_2 \to T_3$ in Pretop satisfies A1', then (as inspection shows) HQ_2' satisfies A1, A2, A3, hence there is $I: T_1 \to T_3$ such that $IQ_1 \cong HQ_2'$. Then $IGQ_2 \cong HQ_2$, hence $IG \cong H$, as required. This shows the claim.

Q_2 and G being finite quotients, their composite is easily seen to be a finite quotient as well. But then $\hat{i}Q_1$, being isomorphic to GQ_2, is a finite quotient as well.

This completes the proof of the lemma.

■ 3.2.2.

■ 1.4.4.

Let us note that the same proof, using 3.1.2 instead of 3.1.4, establishes the following version of 1.4.4.

<u>Theorem</u> 3.2.3. Any strong finitary operation in Set commuting with ultraproduct functors is the restriction of a composite pretopos operation.

"Composite pretopos operation" is defined in the same way as "composite Boolean pretopos operation" with the free Boolean pretopos replaced by the free pretopos (over a graph).

Let me point out that Theorems 1.4.4 and 3.2.3 express a "completeness property of Set in Pretop". In fact, ultraproduct functors are particular Pretop morphisms of the form $\text{Set}^I \to \text{Set}$, I a set. The versions of the theorem talking about *all* Pretop morphisms of the form $\text{Set}^I \to \text{Set}$, instead of just ultraproducts, although weaker than the original versions, still seem to be non-trivial. The formulation of the general phenomenon of completeness two instances of which are 1.4.4 and 3.2.3 requires the context of "essentially algebraic 2-categories".

References

[MR] M. Makkai and G.E. Reyes, First order categorical logic. Lecture Notes in Mathematics no. 611, Springer Verlag, 1977.

[MP] M. Makkai and A. Pitts, Some results on finitely presentable categories.

[GZ] P. Gabriel and M. Zisman, Calculus of Fractions and Homotopy Theory, Springer-Verlag, 1967.

[CK] C.C. Chang and H.J. Keisler,, Model Theory, North-Holland, 1973.

[SGA4] Theorie des Topos et Cohomologie Etale des Schemas, ed. M. Artin, A. Grothendieck and J.L. Verdier. Lecture Notes in Mathematics no.'s 269 and 270. Springer-Verlag, 1972.

[CHL] G. Cherlin, L. Harrington and A.H. Lachlan, $\aleph_0$-categorical $\aleph_0$-stable structures.

[Sh] S. Shelah, Classification Theory, North-Holland, 1978.

[M1] M. Makkai, Stone duality for first order logic, Proceedings of the Herbrand Symposium, ed. J. Stern, North-Holland, 1982, pp. 217-232.

[M2] M. Makkai, Stone duality for first order logic, to appear in Advances in Mathematics.

[TT] P.T. Johnstone, Topos Theory, Academic Press, 1977.

[MR1] M. Makkai and G.E. Reyes, Model theoretic methods in the theory of topoi and related categories, Bull. Acad. Polon. Sci. 24, pp. 379-392.

[CWM] S. MacLane, Categories for the Working Mathematician. Springer-Verlag, 1971.

[P1] A.M. Pitts, Amalgamation and interpolation in the category of Heyting algebras. J. Pure and Applied Algebra 29(1983), 155-165.

[P2] A.M. Pitts, An application of open maps to categorical logic. J. Pure and Applied Algebra 29(1983), 313-326.

[JT] A. Joyal and M. Tierney, An extension of the Galois theory of Grothendieck, Memoirs A.M.S.

[M3] M. Makkai, The topos of types, Lecture Notes in Math., no. 859, Springer-Verlag, pp. 157-201.

[M4] M. Makkai, Full continuous embeddings of toposes, Trans. A.M.S. 269(1982), pp. 167-196.

[L] D. Lascar, On the category of models of a complete theory, J. Symbolic Logic 47(1982), pp. 249-266.

[GU] P. Gabriel and F. Ulmer, Lokal präsentierbare Kategorien. Lecture Notes in Math., no. 221, Springer-Verlag.

"Added in proof.
A.M. Pitts has recently proved an important result which is a strong conceptual completeness theorem for Heyting pretoposes (conceptualizing full first order intuitionistic logic). His proof uses results of [P1], [P2] and [JT]. Theorem 3.1.2 is essentially a special case of Pitts' theorem, in fact with the countability condition removed. After the fact, it is not hard to see that one can deduce the stronger version of 3.1.2, i.e. strong conceptual completeness for small Boolean pretoposes, from the countable case, i.e. 3.1.2 itself. One uses a forcing argument and an absoluteness argument, after suitable preparation."

PROBLEMS IN TAXONOMY, A FLOATING LOG

Jerome I. Malitz
University of Colorado, Boulder
Department of Mathematics
Boulder, Colorado 80309
U.S.A.

ABSTRACT.
Given a region R in an n-dimensional normed linear space, is there a partition of R into k pieces $R_1, R_j, \ldots, R_k$ with centers of mass $c_1, c_2, \ldots, c_k$ respectively such that for each $i \leq k$ and each $x \in R_i$ we have $|x-c_i| \leq |x-c_j|$ for all $j \leq k$. The problem is open even for $n = 2$ and $k = 2$ but some partial results are known.

The problem is highly relevant to the foundations of taxonomy, medical diagnosis and classification in general.

§1. INTRODUCTION. The problem of classifying individuals according to certain characteristics arises in several fields such as taxonomy and medical diagnosis. This problem has been treated by different authors (see [4],[5],[6],[7]). If the characteristics can be measured by some metric then the individuals are represented by points in some product space. A classification of the set X of these points is a partition $\{X_1, X_2, \ldots, X_k\}$ of X. Of course, the partition should reflect relationships between the points. A point $X \in X_i$ should be as close to the other members of X_i as it is to the members of X_j for $j \neq i$. We call such a partition 'good'.

We consider two notions of closeness. The first arose in a seminar led by A. Ehrenfeucht of the computer science department and L. Gold of the biology department at the University of Colorado at Boulder. This definition measures closeness by average distance and is discussed in §3. The second definition suggested by Matt Foreman and the author, measures closeness by distance from the centers of mass of the X_i's. This is the paradigm method of classification and is discussed in §4. For both definitions the main problem is the existence of good partitions. For the definition based on average distances good partitions need not exist. For the definition based on distance from paradigms we have only partial results and the existence problem is open.

The notation and basic definitions are presented in §2.

In §5 a heuristic argument is given which suggests that good partitions exist.

§6 presents some open problems.

§2. DEFINITIONS AND NOTATIONS. We will be working in Euclidian n space with the usual norm. Two notions of distance of a point p to a set Y will be discussed and for each there is a discrete version and a continuous version. In the discrete case each point in Y is given unit mass and in the continuous case Y has uniform density.

i) $D(p,Y) = |p-c|$ where c is the center of mass of Y.

ii) $D^{\#}(p,Y) = \frac{1}{n}\Sigma\{|p-y| : y \in Y\}$ for Y finite with n the cardinality of $Y - \{p\}$.

iii) $D^{\#}(p,Y) = [\int_y |p-y|dy]/\int_y dy$ for Y bounded and measurable.

Let $P(X) = \{X_1,X_2,\ldots,X_k\}$ be a partition of X with each $X_i \neq 0$ and let $x \in X_i$. Then $C(x)$, the contentment of x, is $\min_{j\neq i} D(x,X_j) - D(x,X_i)$. The goodness of the partition P, $G(P)$ is $\min_{x\in X} C(x)$. P is good if $G(P) \geq 0$.

If we use the $D^{\#}$ distance instead of the D distance, we get $C^{\#}$ and $G^{\#}$.

§3. GOODNESS IN THE SENSE OF $D^{\#}$. A. Ehrenfeucht proved (unpublished) that for $n = 1$ and k arbitrary there is always a #-good P that can be obtained in polynomial time, and raised the question for $n > 1$. For $n = 2$ and $k = 2$ the answer is no.

3.1. EXAMPLE. Let $X = \{(0,-1), (0,0), (0,1), (1.1,0)\}$. It is an easy matter to check each P and show $G(P) < 0$. For example, if $P = \{\{(0,-1), (0,0)\}, \{(0,1), (1.1,0)\}\}$, then $C^{\#}((1.1,0)) < 0$.

We do not have a general method for constructing counterexamples with ℓ points for arbitrary ℓ, or for partitions in $k > 2$ pieces. The following theorem and its proof indicate the difficulty of generalizing the above example.

3.2. THEOREM. For every $\varepsilon > 0$ and every k there is an N such that for any X contained in the unit disc and having at least N elements there is a partition P into k parts with $G(P) > -\varepsilon$.

PROOF. Let $\varepsilon > \delta > 0$. Let f be a 1-1 function on a subset A of X into X satisfying

a) $|x-f(x)| < \delta$ for all $x \in A$,

and b) $|x-y| \geq \varepsilon$ for all $x,y \notin A \cup F[A]$.

Let $B = X - A$. $B = f[A] \cup K$ where $|x-y| \geq \delta$ for all $x,y \in K$. Let n be the number of points in A and k the number of elements in K. Notice that k is bounded, in fact $k(\frac{\delta}{2})^2\Pi < \Pi(1+\frac{\delta}{2})^2$ since the disk of radius $\frac{\delta}{2}$ centered at the members of $x \in K$ are pairwise disjoint and contained in the disk of radius $1+\frac{\delta}{2}$

For $x \in B - K$,

$$C^{\#}(x) = \frac{1}{n}\sum_{y\in A}|x-y| - \frac{1}{n+k-1}\sum_{y\in B}|x-y|$$

$$= \frac{1}{n}\sum_{y\in A}|x-y| - \frac{1}{n+k-1}\sum_{y\in A}|x-f(y)| - \frac{1}{n+k-1}\sum_{y\in K}|x-y|$$

$$> \frac{1}{n+k-1}\sum_{y\in A}[|x-y| - |x-f(y)|] - \frac{1}{n+k-1}\sum_{y\in K}|x-y|$$

$$> \frac{-n\delta}{n+k-1} - \frac{k}{n+k-1} \geq -\delta - \frac{k}{n} .$$

Hence for n large enough, $C^{\#}(x) > -\varepsilon$ for $x \in B - K$ and K non-empty. For K empty, or $x \in K$, or $x \in A$ a similar and easier argument shows $C^{\#}(x) > -\varepsilon$ for large enough n.

Hence for N large enough (since k is bounded) we have $G^{\#}(A,B)\ \ -\varepsilon$.

Note that the proof gives an approximately good partition by separating points that are close to each other, an undesirable approach from the standpoint of taxonomy.

If the set X of points is not required to be bounded, then example 3.1 can be modified so that for each k we get a four point example with $G(P) < k$ for all P.

In the proof of 3.2 points that are close together are placed in different cells of the partition. From the point of view of taxonomy or medical diagnosis this makes little sense. This shortcoming is brought into focus by the following example.

3.3 EXAMPLE.

Let $X_1 = \{(d,\varepsilon_i) : i \le n\} \cup \{(1,\varepsilon_i) : i \le n\}$ and let $X_2 = \{(d,1+\varepsilon_i) : i \le n\} \cup \{(1,1+\varepsilon_i) : i \le n\}$ where $\varepsilon_i > 0$ for each i and $\Sigma\varepsilon_i < \varepsilon$. For $x \in X_i$,

$$C(x) > \frac{1-\varepsilon + \sqrt{1+(d-\varepsilon)^2}}{2} - \frac{nd+\varepsilon}{2n-1}.$$

Hence for any d and large enough n, $P = \{X_1,X_2\}$ is a # - good partition of $X = X_1 \cup X_2$. In fact, for $n = 1$, $\varepsilon_1 = 0$ and $1 < d < \frac{4}{3}$, P is # - good.

§4. GOODNESS IN THE SENSE OF D. Here we consider only bounded measurable sets X. A cut C in n space is a plane of dimension n - 1, and so for n = 2 a cut is a line.

For k = 2, a good partition has a very simple geometric characterization.

4.1 THEOREM. Let $P = \{X_1,X_2\}$ with c_i the center of mass of X_i and $c_1 \ne c_2$. Let $\overline{c_1c_2}$ be the line segment joining c_1 and c_2. Let C be the cut that is the perpendicular bisector of $\overline{c_1c_2}$. Let P_i be the closed half space determined by C that contains c_i. Then the following statements are equivalent:

i) P is good.

ii) $X_i \subseteq P_i$ for $i = 1,2$.

The restriction that $c_1 \ne c_2$ is needed. For example, if X is the unit disk centered at the origin and X_1 the disk of radius 1/2 centered at the origin, then $P = \{X_1, X - X_1\}$ is good. From the point of view of taxonomy or classification theory, good partitions with $c_1 = c_2$ are of no interest.

Let C be a cut that partitions X into $\{X_1,X_2\}$ with c_i the center of mass of X_i. Say that L has property:

I if C bisects $\overline{c_1c_2}$

II if C perpendicular to $\overline{c_1c_2}$.

So $\{X_1,X_2\}$ is good if C has properties I and II.

Property I is easily satisfied.

4.2. THEOREM. For every line L there is a cut C normal to L that satisfies I.

PROOF. Consider L to be the x-axis and let C(x) be the cut at x. If C(x) partitions X into X_1 and X_2 with centers of mass c_1 and c_2 we let $u_1(x)$ be the distance between c_1 and C(x) and $u_2(x)$ be the distance between c_2 and C(x). Let $f(x) = u_1(x)/u_2(x)$. f is continuous and as x increases over the domain of f, f(x) goes from 0 to ∞. Hence for some x, f(x) = 1 as needed.

In general, there may be many cuts normal to a given line which have property I. We conjectured that if X is convex, then there is only one cut perpendicular to a given line that satisfies I. This was recently proved by M. Elgueta (unpublished). I do not have an example of a bounded X and a line L with infinitely many cuts perpendicular to L that satisfies I.

The next result giving the existence of some cut satisfying II was obtained by Matt Foreman. His beautiful proof, which we give below, works only in odd dimensional Euclidean space since it depends on the fact that a continuous function f on the surface of the sphere S^{2n-1} has a fixed point $x = f(x)$ or an antipodal point $-x = f(x)$ (here -x is the reflection of x through the origin).

4.3. THEOREM (Matt Foreman). Let X be a bounded measurable region in an odd dimensional Euclidean space. Then X has a cut with property II.

PROOF. We may suppose that the center of mass c of X is the center of a sphere $S^{2n-1} \supseteq X$. For each $s \in S^{2n-1}$ let C(s) be the cut through c normal to the vector $\overline{c,s}$. Let C divide X into regions X_1, X_2 with centers of mass $c_1(s)$, $c_2(s)$ respectively. One of the c's, say $c_1(s)$ is on the same side of C as s. The line from c through $c_1(s)$ intersects S^{2n-1} at the point we call f(s).
So $f : S^{2n-1} \to S^{2n-1}$ and f is continuous. Since s and f(s) are on the same side of C(s), f cannot map s onto -s. Hence f has a fixed point $s^* = f(s^*)$. But this means that $C(s^*)$ is perpendicular to the line extending $c, c_1(s)$ and so is perpendicular to $\overline{c_1(s), c_2(s)}$ since $c = m_1c_1 + m_2c_2$ where m_i is the mass of X_i.

We have a different argument for 2-space giving a cut with property II but for n even and greater than two we do not know how to get such a cut.

Of course, for a cut to be good it has to have both properties I and II.

As yet we do not see how to prove the existence of such cuts even for regions X in 2-space.

§5. A FLOATING LOG, HEURISTICS. Here is a physical model which strongly suggests to us that good cuts exist. We will consider a region X in the plane, although the argument can be given for 3-space and is a bit simpler.

Cylindrify X in 3-space and truncate the cylinder at two different levels. This results in a 3-dimensional object like a section cut from a tree limb, a wooden log. Float the log in liquid with its faces (each a copy of X) perpendicular to the level of the liquid. Rotate the log until stable position is reached. At such a position, with C the surface of the water, X_1 the region above C and X_2 the region below C with centers of mass c_1 and c_2, we have $\overline{c_1 c_2}$ perpendicular to C. By increasing or decreasing the density of the liquid while rotating the log so as to maintain stability ($\overline{c_1 c_2} \perp C$) a position will be attained where C satisfies both conditions I and II.

We have not been able to convert this heuristic argument into a mathematical proof.

§6. PROBLEMS. Of course, the main problem is to show that a bounded measurable region in Euclidean n-space has at least one good cut. In fact, we believe that there must be several.

At the moment, we do not even see that regions in 2-space have at least one good cut. Nor do we see that regions in even dimensions $2n > 2$ have cuts with property II.

For discrete point masses the questions are also open even if we allow C to cut through some points partitioning the mass of these points as convenient.

The proof of 3.2 suggests the following problem. Let X be bounded and of measure > 0. Is there a partition $\{X_1, X_2\}$ of X such that X_1 and X_2 have the same measure and

i) $\int_{X_1} |p-q|\,dq \le \int_{X_2} |p-r|\,dr$ for all $p \in X_1$.

ii) $\int_{X_1} |p-q|\,dq \le \int_{X_2} |p-q|\,dq$ for all $p \in X_2$.

Postcript. Shortly after the conference we obtained a proof for $k = 2$ and all odd n. Fixed point theorems are used in this proof. Recently, A. Ehrenfeucht settled the question positively for all n and all k using methods that are not only simpler but more constructive. However, related problems in taxonomy that are settled by the fixed point methods do not yield to his methods.

REFERENCES

[1] Ehrenfeucht, A. Classification Theory (unpublished).

[2] Ehrenfeucht, A. & Malitz, J. Problems in mathematical taxonomy (in preparation).

[3] Elgueta, M. Unicity of cuts equalizing distances to centers of mass in convex regions (unpublished).

[4] Chen, C. Statistical pattern recognition. Hayden Books, (1973)(See Chapter VIII in particular).

[5] Murtagh, A. Survey of recent advances in hierarchical clustering algorithms, Comp.J., 26,nº 4 (1983) pp.354-359.

[6] Mauldin, R.D. ed., The Scottish Book. Birkhauser (1981)(See problem 19, p. 90 in particular).

[7] Steinhaus, H. Mathematical Snapshots. Oxford University Press (1969)(See p. 50).

COUNTING PROBLEMS IN BOUNDED ARITHMETIC

J.Paris and A.Wilkie
Department of Mathematics
Manchester University
Manchester M13 9PL, England

§ INTRODUCTION.

In this paper we shall consider the following problems.

PROBLEM 1. Let A be a Δ_0 subset of $\mathbb{N}$. Then does

$$\{\langle n,m\rangle \mid m = |A \cap n|\}$$

have a Δ_0 definition?.

PROBLEM 2. Let A be a Δ_0 subset of $\mathbb{N}$. Then for $k \in \mathbb{N}$ does

$$\{\langle n,i\rangle \mid i < k \wedge |A \cap n| = i \bmod k\}$$

have a Δ_0 definition?.

PROBLEM 3. Does $I\Delta_0 \vdash \Delta_0 PHP$ where $\Delta_0 PHP$ (Δ_0 pigeon hole principle) is the schema

$$\neg \exists x \,[\, \forall y \le x \;\; \exists z < x \;\; \theta(y,z) \wedge \forall y_1, y_2 \le x \forall z < x(\theta(y_2,z) \wedge \theta(y_2,z) \to y_1 = y_2)$$

for $\theta\Delta_0$?

There are several reasons for studying these problems. One is that it is possible, even tempting, to view the Δ_0 subsets of $\mathbb{N}$ as the meaningful, feasible collections of natural numbers and to view $I\Delta_0$ (Δ_0-induction) as the self evident fragment of Peano's axioms. From this standpoint it is very natural to enquire whether we can enumerate sets in increasing order and whether the size of sets is unique.

And apart from this there are practical reasons for wanting to know how extensive the Δ_0 sets are and in how weak a fragment of Peano's Axioms we can still prove the PHP. For example Ritchie [9] has shown that every Δ_0 subset of $\mathbb{N}$ is in Linear Space. The general feeling however is that these classes are not equal and since

$$A \in \text{Lin. Space} \Rightarrow \{\langle n,m\rangle \mid m = |A \cap n|\} \in \text{Lin.Space}$$

a way to confirm this belief would be to give a negative answer to problem 1.

Problems 1,2,3 are all open and our intention in this paper will be to survey the area (as we know it) and to prove some new results of our own.

NOTATION AND DISCUSSION. Most of our notation is standard, as in [7], [8]. When we employ expressions like $x^y = z$ we shall be using the Δ_0 definition of exponentiation given in [3]. All logarithms will be to base 2 and in expressions like $\log(x)$, x^ε it will always be assumed that we mean the integer part of these quantities.

Throughout this paper M will stand for a countable model of $I\Delta_0$ and we identify $a \in M$ with $\{x | M \models x < a\}$. Δ_0^M is the collection of subsets of M defined by Δ_0 formulae, with parameters from M. For $A \in \Delta_0^{\mathbb{N}}$ let

$$A^{(\infty)} = \{<n, m> | m = |A \cap n|\},$$
$$A^{(k)} = \{n | \ |A \cap n| = 0 \bmod k\}.$$

It is easy to see that $A^{(k)} \in \Delta_0^{\mathbb{N}}$ just if

$$\{<n,i> | i < k \wedge |A \cap n| = i \bmod k\} \in \Delta_0^{\mathbb{N}}.$$

Thus problem 1 (2) is equivalent to :

$$\text{If } A \in \Delta_0^{\mathbb{N}} \text{ is } A^{(\infty)} \in \Delta_0^{\mathbb{N}} ? \quad (A^{(k)} \in \Delta_0^{\mathbb{N}} ?)$$

Another way of expressing problem 2 is as follows.

Let $C_k x < t$ (where t is a term of LA) be the quantifier

$$C_k x < y \quad \theta(x) \Leftrightarrow |\{x < t | \theta(x)\}| = 0 \bmod k,$$

and let $C_k\Delta_0$ be the formulae formed as for Δ_0 formulae but with this additional quantifier adjoined. Let $C_k\Delta_0^{\mathbb{N}}$ be the subsets of $\mathbb{N}$ defined by such formulae. Then problem 2 is equivalent to:

$$\text{Is } \Delta_0^{\mathbb{N}} = C_k\Delta_0^{\mathbb{N}} ?$$

We shall also be interested in problems 1, 2 when we replace the standard model $\mathbb{N}$ by M. However since the expression $|A \cap n| = m$ may have no meaning in M we must generalize these problems to:

PROBLEM 1'. Let $A \in \Delta_0^M$. Is there $B \in \Delta_0^M$ such that, in M

$$\forall x,y[<x, y> \in B \leftrightarrow \{(x = 0 \wedge y = 0) \vee (x > 0 \wedge x\text{-}1 \in A \wedge y > 0 \wedge <x\text{-}1,y\text{-}1> \in B)$$
$$\vee (x > 0 \wedge x\text{-}1 \in' A \wedge <x\text{-}1, y> \in B)\}]? \ (*)$$

i.e. is Δ_0^M closed under counting, or, to put it another way, can we count in M?

PROBLEM 2'. Let $A \in \Delta_0^M$. Is there $B \in \Delta_0^M$ such that, in M,

$$\forall x,y[\langle x, y\rangle \in B \leftrightarrow \{(x = 0 \wedge y = 0)$$
$$\vee(x > 0 \wedge x\text{-}1 \in A \wedge 0 < y < k \wedge \langle x\text{-}1, y\text{-}1\rangle \in B)$$
$$\vee(x > 0 \wedge x\text{-}1 \in A \wedge y = 0 \wedge \langle x\text{-}1, k\text{-}1\rangle \in B)$$
$$\vee(x > 0 \wedge x\text{-}1 \in' A \wedge \langle x\text{-}1,y\rangle \in B)\}]?$$

i.e. is Δ_0^M closed under counting mod k, or, can we count mod k in M? Notice that by the compactness theorem problem 1' (say) has an affirmative answer for all such M just if given any Δ_0 formula $\theta(x)$ we can exhibit a Δ_0 formula $\psi(x,y)$ and prove (*) in $I\Delta_0$ (with $x\text{-}1 \in A$ replaced by $\theta(x\text{-}1)$ etc).

We remark here that the general feeling is that problems 1,2,3 all have negative solutions. Indeed we suspect that problems 1, 2 fail for quite simple $A \in \Delta_0^{\mathbb{N}}$, for example for the set of primes.

Finally, in connection with problem 3, notice that by the standard proof

$$I\Delta_0 + \forall x,\ 2^x \text{ exist} \vdash \Delta_0 PHP,$$

so the problem amounts to whether we can remove exponentiation in this proof.

SECTION 1. In this section we mention some simple connections between these problems. Clearly if we can count mod 2 then we can count mod 4 since given $A \in \Delta_0^M$, $0 \le i \le 3, |A \cap n| = i \bmod 4 \Leftrightarrow$

$$(i = 0 \wedge n \in A^{(2)} \cap (A \cap A^{(2)})^{(2)}) \vee (i = 1 \wedge \exists x < n(x \in A \wedge x \in A^{(2)} \cap (A \cap A^{(2)})^{(2)}$$
$$\wedge \neg \exists z < n(x < z \wedge z \in A)) \vee (i = 2 \wedge \exists x < y < n(x,y \in A \wedge x \in A^{(2)} \cap (A \cap A^{(2)})^{(2)}$$
$$\wedge \neg \exists z < n(x < z \wedge z \neq y \wedge z \in A)) \vee$$
$$(i = 3 \wedge \exists x < y < w < n(x,y,w \in A \wedge x \in A^{(2)} \cap (A \cap A^{(2)})^{(2)} \wedge \neg \exists z < n(x < z \wedge z \neq y \wedge z \neq w \wedge z \in A))$$

In general we can count mod k just if we can count mod $p_1,\ldots,p_j$ where $p_1,\ldots,p_j$ are all the prime divisors of k. However the following problem is open.

PROBLEM. Let p, q be distinct (standard) primes. If we can count mod p in M can we count mod q?

In this connection we do have the following result.

<u>THEOREM 1</u>. If in M we can uniformly count mod p for all primes p then in M we can count.

<u>PROOF.</u> By the hypothesis we mean that given $A \in \Delta_0^M$ there is a $B \in \Delta_0^M$ such that whenever $n,m,p \in M$, p prime (in M) and $m < p$ then

$$\begin{aligned} \langle n,m,p\rangle \in B \iff & [n = 0 \wedge m = 0] \\ & \vee[n > 0 \wedge n-1 \in A \wedge m > 0 \wedge \langle n-1, m-1, p\rangle \in B] \\ & \vee[n > 0 \wedge n-1 \in A \wedge m = 0 \wedge \langle n-1, p-1, p\rangle \in B] \\ & \vee[n > 0 \wedge n-1 \in' A \wedge \langle n-1, m, p\rangle \in B] \end{aligned}$$

i.e. "$\langle n, m, p\rangle \in B \iff |A \cap n| = m \bmod p$".

Then we can 'define'

$$|A \cap n| = j \iff j \text{ is minimal such that for all primes } p \le n$$
$$|A \cap n| = j \bmod p.$$

To show that such a j exists we show that by induction on n, for fixed m,

$$\forall n\, \exists j \le n\ \forall \text{ primes } p \le m,\ |A \cap n| = j \bmod p.$$

It is now possible to show that as defined above $|A \cap n| = j$ satisfies the required counting conditions. □

A close connection between counting and the Δ_0PHP was discovered by Woods [11]. Woods' result will be discussed in section 4 (Theorem 20) but for the moment we content ourselves with the following special case.

<u>THEOREM 2.</u> Woods [11]. If in M we can count then M satisfies the Δ_0PHP. □

Woods' proof (of Theorem 20) is quite long. Perhaps the simplest way to prove the theorem is to show by induction on y that if $F \in \Delta_0^{\mathbb{N}}$ and $F : a \rightarrowtail a-1$ (i.e. F maps a 1-1 into a-1) then

$$|F \cap (y\times(a-1))| = y \quad \text{for } y \le a$$
$$|F \cap (a\times y)| \le y \quad \text{for } y \le a-1$$

which gives the required contradiction

$$a = |F \cap (a\times(a-1))| \le a-1.$$

Notice that this only requires a single counting, of $F \cap (y_0 \times y_1)$.

Of course this result is not at all surprising, it simply says that if a set has a reasonable definition of its size then there cannot be another such definition of its size i.e. its size is unique.

Further results connecting these problems can be proved similarly. For example, assuming counting mod 2, if $F : a \rightarrowtail\!\!\twoheadrightarrow (a-1)$ (i.e. F maps a 1-1 onto a-1) then by induction on y

$$|F \cap (y\times(a-1))| = y \bmod 2 \quad \text{for } y \leq a$$

$$|F \cap (a\times y)| = y \bmod 2 \quad \text{for } y \leq a-1$$

which gives a contradiction as above. Similarly

THEOREM 3. Let $0 < k \in M$ and suppose that in M we can count mod k. Then for $F \in \Delta_0^M$, $\neg \exists x$, $F : X \rightarrowtail\!\!\twoheadrightarrow x-p$, for all $0 < p < k^{\mathbb{N}}$. □

Returning for a moment to Theorem 1 we would of course have had a much easier proof of this result if we could have shown that M had unboundedly many primes. In fact by Theorems 1, 2 and the following result due to Woods this is the case although to prove it we seem to need Theorem 1.

THEOREM 4. [Woods, [11].] If M has the Δ_0PHP then M has arbitrarily large primes. □

SECTION 2. In this section we shall give the best results (to our knowledge) concerning the problems in section 0.

Of course the obvious way in arithmetic of stating that $|A \cap n| = m$ is simply

$$\exists f, \ f \text{ codes a 1-1 map from } m \text{ onto } A \cap n.$$

Unfortunately this will give a best upper bound for f of around n^m so f cannot in general be bounded by a term in LA. However if we put suitable conditions on A we can do this. We can now refine this basic idea (a trick due to Nepomnyascii, see [6]) by disecting f so that

$$|A \cap n| = m \Leftrightarrow \exists k \exists g \in {}^{(k+1)}(n+1),\ f \in {}^{(k+1)}(m+1) \text{ such that } g, f \text{ are increasing, } f(0) = 0,\ f(k) = m,\ g(0) = 0,\ g(k)=n \text{ and } \forall i < k \exists h_i : [f(i), f(i+1) \rightarrowtail\!\!\twoheadrightarrow A \cap [g(i), g(i+1)).$$

Iterating this device gives

THEOREM 5. Let $k \in \mathbb{N}$, $0 < \alpha < 1$ and $A \in \Delta_0^{\mathbb{N}}$. For $n \in \mathbb{N}$ let

$$A = \{m \mid m < n \ \ <n, m> \in A\},$$

and assume $A_n \leq 2^{\log(n)^\alpha}$ for all $n \in \mathbb{N}$. Then

$$B^k = \{\langle i,\ j,\ n,\ m\rangle \mid j \le n \wedge m = \min(|A_n \cap [i,\ j)|,\ (\log(n)^{\frac{1-\alpha}{2}})^k + 1)\} \in \Delta_0^{\mathbb{N}}.$$

PROOF. By induction on k. Clearly we may assume n is large.

For $k = 1$ we have

$$m = \min(|A_n \in [i,\ j)|,\ \log(n)^{\frac{1-\alpha}{2}} + 1) \Leftrightarrow$$

$$\Leftrightarrow \exists f(f : m \twoheadrightarrow A_n \cap [i,\ j) \wedge m \le \log(n)^{\frac{1-\alpha}{2}})$$

$$\vee \text{ no such } f \text{ exists and } m = \log(n)^{\frac{1-\alpha}{2}} + 1.$$

Then with the given condition on the size of elements of A_n and the usual coding

$$f \le (2^{\log(n)^\alpha})^{\log(n)^{\frac{1-\alpha}{2}}} < n.$$

Now assume the result for k. Then

$$m = \min(|A_n \cap [i,\ j)|,\ (\log(n)^{\frac{1-\alpha}{2}})^{(k+1)} + 1) \Leftrightarrow$$

$$\Leftrightarrow \{\exists h : (\log(n)^{\frac{1-\alpha}{2}} + 1) \to [i,j]\ \exists g : (\log(n)^{\frac{1-\alpha}{2}} + 1) \to (m+1),$$

$$\text{such that } m \le (\log(n)^{\frac{1-\alpha}{2}})^{(k+1)},\ g,\ h \text{ increasing}$$

$$h(0) = i,\ h(\log(n)^{\frac{1-\alpha}{2}}) = j,\ g(0) = 0,\ g(\log(n)^{\frac{1-\alpha}{2}}) = m$$

$$\text{and } \forall s < \log(n)^{\frac{1-\alpha}{2}},\ g(s+1) - g(s) \le (\log(n)^{\frac{1-\alpha}{2}})^k \text{ and}$$

$$\langle h(s),\ h(s+1),\ n,\ g(s+1) - g(s)\rangle \in B^k\} \text{ or}$$

$$\text{or } \{\text{no such } h,\ g \text{ exist and } m = (\log(n)^{\frac{1-\alpha}{2}})^{(k+1)} + 1\}. \square$$

This immediately gives:-

THEOREM 6. Let $k \in \mathbb{N}$, $0 < \alpha < 1$, $A \in \Delta_0^{\mathbb{N}}$ and suppose that

$$A_n = \{m \mid \langle n,\ m\rangle \in A \wedge m < n\} \subseteq 2^{\log(n)^\alpha} \quad \text{for all } n \in \mathbb{N}.$$

Then there is a function $F \in \Delta_0^{\mathbb{N}}$ such taht for all $n \in \mathbb{N}$

$$F(n) = \min(|A_n|,\ \log(n)^k). \qquad \square$$

Having got this amount of counting we can drop the condition

$A_n \subseteq 2^{\log(n)^\alpha}$. Precisely:

THEOREM 6'. Let $k \in \mathbb{N}$, $A \in \Delta_0^{\mathbb{N}}$ and suppose that

$$A_n = \{m \mid \langle n, m\rangle \in A \wedge m < n\}.$$

Then there is a function $F \in \Delta_0^{\mathbb{N}}$ such that for all $n \in \mathbb{N}$,

$$F(n) = \min(|A_n|, \log(n)^k). \qquad \square$$

This theorem follows from Theorem 6 and the following theorem which will be proved in a forthcoming paper by the authors:

THEOREM 6". Let A_n be as in Theorem 6' and let $\varepsilon > 0$. Then there is a function $F \in \Delta_0^{\mathbb{N}}$ such that for all $n \in \mathbb{N}$,

$$F(n) : A_n \twoheadrightarrow |A_n|^{1+\varepsilon}. \qquad \square$$

We remark here that Theorems 5 and 6 hold with M in place of $\mathbb{N}$ (but k, α, ε still standard) provided we interpret "$|A_n| \geq \log(n)^k$" by

$$\exists F,\ F : \log(n)^k \twoheadrightarrow A_n \quad \text{for a}$$

suitable Δ_0^M function F etc.

Furthermore combining this limited amount of counting with the methods of the previous section we can show

THEOREM 7. For $k \in \mathbb{N}$ and $F \in \Delta_0^M$

$$M \models \neg \exists x,\ F : \log(x)^k \twoheadrightarrow \log(x)^k - 1. \qquad \square$$

As an aside we mention here that this can be combined with Woods' proof of Theorem 4 to give

THEOREM 8. For $k \in \mathbb{N}$, $M \models \forall x \exists$ prime p, $p > \log(x)^k$. $\square$

It is not clear at present that Theorem 6' can be similarly generalized. However we can show the following weak theorem:

THEOREM 9. Let $A \in \Delta_0^M$ and suppose that for all $\alpha, \beta \in A$,

$\alpha < \beta \Rightarrow \alpha(1 + \log(\alpha)^{-k}) < \beta \quad (k \in \mathbb{N})$.

Then $A^{(\infty)} \in \Delta_0^M$.

PROOF. Let $N \in M$. Since $(1 + p^{-1})^p \geq 2$ for $p > 0$,

$$(1 + \log(n)^{-k})^{\alpha} \geq n \quad \text{for some} \quad \alpha \leq \log(n)^{k+2}.$$

Hence we can define $F : A \cap n \longrightarrow \log(n)^{k+2}$ by

$$F(x) = \text{least } \beta,\ (1 + \log(n)^{-k})^{\beta} \leq x < (1 + \log(n)^{-k})^{\beta+1}$$

and count $A \cap n$ by counting $F'' A \cap n$. □

SECTION 3. In this section we give an illuminating characterization of counting mod 2. This characterization is a natural variation of a machine based description of the classes ε_{f*} (the Grzegorczyk class of relations corresponding to the function f) and the class $\Delta_o^{\mathbb{N}}$ given by Bel'tyukov in [2]. We shall show that these machines can be used to describe $C_2\Delta_0^{\mathbb{N}}$ and hence highlight what it means for $\Delta_0^{\mathbb{N}}$ to be closed under counting mod 2.

We first describe Bel'tyukov's Stack Register Machines (SRMSs).
Such a machine M consists of a finite number of input registers $x_1, x_2, \ldots, x_m$, stack registers $t_0, t_1, \ldots, t_k$ and a work register r. Initially the input goes into $x_0, \ldots, x_m$ and the other registers are all zero. The program for M is a sequence of instructions $L_1, L_2 \ldots L_p$ where each L_q has one of the forms

(i) $t_i := t_i + 1$ & $\forall j < i,\ t_j := 0$ & goto L_{q+1},

(ii) $r := z$ & goto L_{q+1},

(iii) if $z_1 + z_2 = z_3$ goto L_i, else goto L_j,

(iv) if $z_1 \cdot z_2 = z_3$ goto L_i, else goto L_j,

where z, z_1, z_2, z_3 are chosen from $0, \vec{x}, t, r$. In this program for each i at most one of $L_1, L_2 \ldots L_p$ contains the instruction $t_i := t_i + 1$ and M halts just if he is told to go to L_{p+1}. M accepts an input $\vec{x}$ if M eventually halts with $t_k = 0$. [Our definition disagrees inessentially with Bel'tyukov's in order to make the later work run more smoothly].

For functions $f, g : \mathbb{N} \to \mathbb{N}$ and $A \subseteq \mathbb{N}^m$ we say that $A \in$ SRM Space (f,g) if there is a SRM M with the following properties:-

(i) M halts on all inputs $\vec{x}$ and accepts iff $\bar{x} \in A$.

(ii) $\exists i, j \in \mathbb{N}$ such that for any input $\vec{x}$, during the computation of M on $\vec{x}$

$$r \le f^i(\max(\vec{x})) \quad \text{and}$$

$$t_o, t_1, \ldots, t_k \le g^j(\max(\vec{x})).$$

Let $\mathbb{0}(x) = 0$, $\mathbb{1}(x) = 1$, etc., $p_2(x) = x^2 + 2$ for all $x \in \mathbb{N}$. Then

THEOREM 10. (a) [Bel'tyukov, [2].] SRM Space (p_2, p_2) = Lin.Space,

(b) [Bel'tyukov, [2].] SRM Space $(\mathbb{0}, p_2) = \Delta_0^{\mathbb{N}}$,

(c) SRM Space $(\mathbb{1}, p_2) = C_2\Delta_0^{\mathbb{N}}$,

(d) SRM Space $(\mathbb{2}, p_2) = C_6\Delta_0^{\mathbb{N}}$,

(e) SRM Space $(\mathbb{3}, p_2) = C_6\Delta_0^{\mathbb{N}}$.

[Actually Bel'tyukov shows that for reasonable functions f, SRM Space$(f,f) = \varepsilon_{f*}$.
Result (a) follows immediately since by the result of Ritchie [9], Lin.Space $= \varepsilon_{p_2*}$.]

PROOF. Result (a) is proved in [2] and (b) is stated there. We outline a proof of (c) borrowing heavily on ideas of Bel'styukov in [2]. The proofs of (d) and (e) are along similar lines.

We first show that $C_2\Delta_0^{\mathbb{N}} \subseteq$ SRM Space $(\mathbb{1}, p_2)$. Clearly we can find suitable machines to decide if $x_1 = 0$, $x_1 + x_2 = x_3$, $x_1 . x_2 = x_3$ so it only remains to show that SRM Space $(\mathbb{1}, p_2)$ is closed under negation, conjunction, bounded quantification and counting mod 2. To show this last property suppose that $A \subseteq \mathbb{N}^{m+1}$, $A \in$ SRM Space $(\mathbb{1}, p_2)$ and that M is a suitable machine (with the above notation) to accept A. We may assume that at the end of any computation of M, $t_k = 0$ or $t_k = 1$. Let

$$\langle y, \vec{u}\rangle \in B \iff C_2 x < y, \langle x, \vec{u}\rangle \in A.$$

We amend M to form a suitable machine M' to accept B. Phisically M' looks like M but has further stack registers t_{-1}, t_{k+1}, t_{k+2}, t_{k+3}. Suppose that on input $\langle y, \vec{u}\rangle$ at some stage M' has $t_{k+2} = z$, $t_{k+3} = 0$ and $r = 0$ if $C_2 x < z$, $\langle x, \vec{u}\rangle \in A$, else $r = 1$.

If $z = y$ then apply $t_{k+3} := t_{k+3} + 1$ (etc) if $r \neq 0$ and halt.

If $z \neq y$ (so $z < y$ in fact) set $t_{k+2} := t_{k+2} + 1$(etc). Then if

$r \neq 0$ set $t_{k+1} := t_{k+1} + 1$ (etc) and $r = 0$ and run M on the input $<z+1, \vec{u}>$ i.e. with x_0 replaced everywhere by t_{k+2}. If at the end of this computation we have $t_k = 0$ put $r = 0$ if $t_k \neq t_{k+1}$, $r = 1$ if $t_k = t_{k+1}$ (i.e. by setting $t_{-1} := t_{-1} + 1$, (etc) and $r = t_{-1}$).

Clearly this can be written out as a suitable program and the accepted class for M' is B. The proofs that SRM Space ($\mathbb{1}$, p_2) is closed under negation, conjunction and bounded quantification are similar.

It remains to show that SRM Space ($\mathbf{1}$, p_2) $\subseteq C_2 \Delta_0^{\mathbb{N}}$. The method of proof is due to Bel'tyukov [2]. Let M be an SRM running in this space and halting on all inputs. Let M have stack registers $t_0, t_1, \ldots, t_k$ (bounded by $p_2^d(\max(x))$), input registers $x_0, \ldots, x_m$ and program $L_1, L_2, \ldots, L_p$. Let I_i be the instruction amongst $L_1, L_2, \ldots, L_p$ (if any) which starts

$$t_i := t_i + 1 \;\&\; \forall j < i,\ t_j := 0.$$

For convenience we make some unimportant assumptions about M, namely that $L_1 = I_0$ and that at the end of the computation the answer 0 (accept) or 1 (reject) appears in r rather than t_k.

Then for $0 \le i, q \le k$ we can find $\theta_{iq}(\vec{t}, \vec{x}, r) \in \Delta_0$ and Δ_0 functions $F_{iq}(\vec{t}, \vec{x}, r)$ (i.e. having Δ_0 graph and value bounded by a polynomial) such that if in any computation of M the stacks contain $\vec{t}$, $\vec{x}$, r and we have just applied instruction I_i then

$\theta_{iq}(\vec{t}, \vec{x}, r) \iff$ [the next instruction of the form I_p to be executed is I_q] or [$q = k+1$ and we will halt without applying another instruction of the form I_p].

Furthermore if $\theta_{iq}(\vec{t}, \vec{x}, r)$ then $F_{iq}(\vec{t}, \vec{x}, r)$ is the value of r when we are just about to execute I_q or halt if $q = k+1$. That such θ_{iq}'s exist follows from the fact that no computation of M can go on for more than, say, $p(m + k + 3)$ moves without applying an instruction of the form I_i, otherwise it would contain a loop.

So M is equivalent to a machine K whose instructions are:-

$(i)_q$ $(q \le k)$ In state L if $\psi_q(\vec{t},\vec{x},r)$ set $r := G_q(\vec{t},\vec{x},r)$, $t_q := t_q + 1$,

$t_j := 0$ for $j < q$ and return to L.

(ii) In state L if $\psi_{k+1}(\vec{t},\vec{x},r)$ set $r := G_{k+1}(\vec{t}, \vec{x}, r)$ and halt, where

$$\psi_q = \bigvee_i (\theta_{iq}(\vec{t}, \vec{x}, r) \wedge t_i \neq 0 \wedge \bigwedge_{j<i} t_j = 0) \vee (q = 0 \wedge r = 0 \wedge \bigwedge_j t_j = 0)$$

and

$$G_q(\vec{t},\vec{x}, r) = z \Leftrightarrow \bigvee_i (\theta_{iq}(\vec{t},\vec{x}, r) \wedge t_i \neq 0 \wedge \bigwedge_{j<i} t_j = 0 \wedge F_{iq}(\vec{t},\vec{x}, r) = z)$$
$$\vee (q = 0 \wedge \bigwedge_j t_j = 0 \wedge r = 0 \wedge z = 0).$$

The idea now is to find a machine K' equivalent to K with one less instruction and with ψ'_q, $G'_q \in C_2\Delta_0$. Repeating this process until we are left with just the final "halt" instruction gives the answer.

We shall remove the loop corresponding to $(i)_0$. Let $\tilde{t} = t_1, t_2, \ldots, t_k$ and consider a computation of K starting with $0, \tilde{t}, \vec{x}, r$ and in state L. It will continue to execute $(i)_0$, producing $r_0, r_1, r_2, \ldots$ until we arrive at a value s_0 of t_0 such that $\neg\psi_0(s_0, \tilde{t}, \vec{x}, r)$. Now if we knew s_0, r_{s_0} beforehand we could bypass $(i)_0$ by replacing t_0 and r by s_0 and r_{s_0} in the remaining instructions of K. (Notice that whenever we apply $(i)_q$ for $0 < q \le k$ we reset t_0 to 0).

Since r_s can only take values 0, 1 we can define r_s by

$r_s = 0 \Leftrightarrow \exists b < s$ [b is maximal such that $G_0(b, \tilde{t}, \vec{x}, 0)$ $G_0(b, \tilde{t}, \vec{x}, 1)$

and either $(G_0(b, \tilde{t}, \vec{x}, 0) = 0 \wedge C_2x < s(b < x \wedge G_0(x, \tilde{t}, \vec{x}, 0) = 1))$

or $(G_0(b, \tilde{t}, \vec{x}, 0) = 1 \wedge \neg\ C_2x < s(b < x \wedge G_0(x, \tilde{t}, \vec{x}, 0) = 1))$]

or [no such b exists and either

$(r = 0 \wedge C_2x < s,\ G_0(x, \tilde{t}, \vec{x}, 0) = 1)$

or$(r = 1 \wedge \neg\ C_2x < s,\ G_0(x, \tilde{t}, \vec{x}, 0) = 1)$].

Hence

s_0 = least $s \le p_2^d(\max(\vec{x}))$ such that $\neg\ \psi_0(s, \tilde{t}, \vec{x}, r_s)$

and K is equivalent to K' with instructions:-

(i)$'_q$ $(1 \le q \le k)$ In state L if $\psi_q(s_0, \tilde{t}, \vec{x}, r_{s_0})$ set

$$r := G_q(s_0, \tilde{t}, \vec{x}, r_{s_0}),\ t_q := t_q + 1,\ t_j := 0 \text{ for } 0 < j < q$$

and return to L. □

(ii)' In state L if $\psi_{k+1}(s_0, \tilde{t}, \vec{x}, r_{s_0})$ set $r := G_{k+1}(s_0, \tilde{t}, \vec{x}, r_{s_0})$

and halt.

The theorem now follows,

The fact that SRM Space $(\mathbf{2}, p_2)$ = SRM Space $(\mathbf{3}, p_2)$ is rather surprising and suggests that perhaps the same result holds for $\mathbf{1}$ and $\mathbf{2}$, i.e. that if you can count mod 2 then you can count mod 3. Of course a natural conjecture at this stage is

<u>Conjecture.</u> SRM Space $(\mathbf{k}, p_2) = C_{(k+1)!}\Delta_0^{\mathbb{N}}$.

For other investigations along these lines see Hicks [5] and a forthcoming paper by Handley-Paris-Wilkie.

By directly applying the last theorem we can obtain a further characterization of $C_2\,\Delta_0^{\mathbb{N}}$ due to Gandy (by a different proof).

<u>THEOREM 11</u>. [Gandy, unpublished.] Let G be the class of functions formed from the characteristic functions of the graphs of $+, \cdot$ by substitution and (bounded) primitive recursion. Then G is precisely the characteristic functions of the sets in $C_2\Delta_0^{\mathbb{N}}$. □

<u>SECTION 4.</u> In situations such as we have with these counting problems a currently popular reaction is to prove some oracle independence results, thus confirming (what we already knew!) that these problems are not entirely trivial. In this section we give some such results.

The relativized versions of $\Delta_0^{\mathbb{N}}$ are formed in the obvious way. Namely add to LA a new unary relation symbol X, define bounded formulae, $\Delta_0(X)$, of LA(X) as before and for $A, B \subseteq \mathbb{N}$ say that $B \in \Delta_0^A$ iff there is some $\phi(\vec{x}, X) \in \Delta_0(X)$ such that $B = \{\vec{n} \in \mathbb{N} \mid \phi(\vec{n}, A)\}$.

Notice that by induction on the length of $\phi(\vec{x}, X)$ we can show that there is $m_\phi \in \mathbb{N}$ such that if $n > 1$ and

$$A \cap n^{m_\phi} = D \cap n^{m_\phi} \text{ then}$$
$$\{\vec{x} \le n \mid \phi(\vec{x}, A)\} = \{\vec{x} \le n \mid \phi(\vec{x}, D)\}.$$

Our first theorem is very straightforward.

THEOREM 12. $\exists A \subseteq \mathbb{N}$ such that Δ_0^A is closed under counting.

PROOF. Suppose that $A \cap p$ has so far been defined and $A-p$ is completely unspecified. If p has the form

$$\langle n, n^{m_\phi}, \ulcorner \phi(x_0, \vec{x}, X) \urcorner, \vec{a}, k \rangle$$

for some $\vec{a} \leq n$ and $\phi(x_0, \vec{x}, X) \in \Delta_0(X)$ set

$$p \in A \iff |\{x < n \mid \phi(x, \vec{a}, A \cap p\}| = k.$$

Otherwise put $p \notin A$. The result follows by the above remark. □

Whilst this tells us that any proof that $\Delta_0^{\mathbb{N}}$ is not closed under counting cannot relativize it really gives no new insights. Our next result however seems more promising.

THEOREM 13.(*) $\exists A \subseteq \mathbb{N}$ such that Δ_0^A is not closed under counting mod 2.

PROOF. The proof is an easy application of a very beautiful theorem due to Ajtai [1] characterizing sets of the form

$$\{A \subseteq n \mid \phi(n, A)\}$$

for $\phi(x, X) \in \Delta_0(X)$. [Actually Ajtai's result is much more general than the special case we shall need.] First we introduce some notation,

Let be the language of arithmetic but with +,. treated as 3-ary relations and for $n \in \mathbb{N}$ let $\underset{\sim}{n}$ be the obvious structure for L with domain n. It is straightforward to show that for any $\phi(x,X) \in \Delta_0(X)$ there is a formula $\overline{\Phi}(x) \in L(X)$ such that whenever $A \subseteq n$, $n \in \mathbb{N}$,

$$\phi(n, A) \iff \langle \underset{\sim}{n}, A \rangle \models \overline{\Phi}(A).$$

This result, for L, is proved in [7]. Since we assume $A-n = \emptyset$ the result for $L(X)$ is a simple generalization

For $f \in {}^n 2$ let $X_f = \{m \mid m < n \wedge f(m) = 0\}$. Let ${}^{\subseteq n}2 = \{f \mid f$ maps a subset of n into $2\}$ and for $f \in {}^{\subseteq n}2$ let $\overline{f} = \{h \in {}^n 2 \mid h \supseteq f\}$. (This notation will only be used when n is implicit.) Then the version of Ajtai's Theorem which we need is : -

THEOREM 14. [Ajtai, [1].] Let $\varepsilon > 0$ and $\theta(X) \in L(X)$. Then for all n eventually $\exists S \subseteq T \subseteq {}^{\subseteq n}2$ such that

(*) There is a striking resamblance between this theorem and results of [12] which were unknown to the authors at the moment of writing this paper.

(i) $\{\bar{f} \mid f \in T\}$ form a partition of ${}^{n}2$,

(ii) $f \in T \Rightarrow |\mathrm{dom}(f)| = n - n^{1-\varepsilon}$,

(iii) $|\{h \in {}^{n}2 \mid \langle \underset{\sim}{n}, X_h\rangle \models \theta(X_h)\} \Delta \bigcup_{f \in S} \bar{f}| \le 2^{n-n^{1-\varepsilon}}$ □

Here Δ stands for symmetric difference. The result says then that $\{h \in {}^{n}2 \mid \langle \underset{\sim}{n}, X_h\rangle \models \theta(X_h)\}$ can be approximated very closely by the simple set $\bigcup_{f \in S} \bar{f}$. To see just how good this approximation is divided both sides of (iii) by 2^n. This shows that the probability that $h \in {}^{n}2$ chosen at random is in one of $\{h \in {}^{n}2 \mid \langle \underset{\sim}{n}, X_h\rangle \models \theta(X_h)\}$, $\bigcup_{f \in S} \bar{f}$ but not in the other is at most $1/2^{n^{1-\varepsilon}}$

As a simple corollary to this we have.

THEOREM 15. Let $g \in {}^{t}2$, $\theta(X) \in L(X)$. Then for all n eventually we can find $q \in {}^{\subseteq n}2$ such that $q \supseteq g$, $|\mathrm{dom}(q)| \le n - \sqrt{n} + 2^{t+1}$ and either

$$\bar{q} \subseteq \{h \in {}^{n}2 \mid \langle \underset{\sim}{n}, X_h\rangle \models \theta(X_h)\}$$

$$\text{or} \quad \bar{q} \subseteq \{h \in {}^{n}2 \mid \langle \underset{\sim}{n}, X_h\rangle \models \neg\theta(X_h)\}.$$

PROOF. Let n be large and let T etc be as in theorem with $\varepsilon = \frac{1}{2}$. Let $H = \{f \in T \mid \bar{f} \cap \bar{g} \neq \phi\}$ and suppose that whenever $f \in H$, $|\bar{f} \cap \{h \in {}^{n}2 \mid \langle \underset{\sim}{n}, X_h\rangle \models \theta(X_h)\}|$, $|\bar{f} \cap \{h \in {}^{n}2 \mid \langle \underset{\sim}{n}, X_h\rangle \models \neg\theta(X_h)\}| > 2^t$. Then

$$2^{n-\sqrt{n}} \ge |\{h \in {}^{n}2 \mid \langle \underset{\sim}{n}, X_h\rangle \models \theta(X_h)\} \Delta \bigcup_{f \in S} \bar{f}| > 2^t . |H|.$$

But $\bigcup_{f \in H} \bar{f} \cap \bar{g} = \bigcup_{f \in H} \overline{(f \cup g)} = \bar{g}$ so by counting elements,

$$|H| . 2^{\sqrt{n}} \ge |\bar{g}| = 2^{n-t}$$

which gives a contradiction. So, without loss of generality we can find $f \in H$ such that

$$|\bar{f} \cap \{h \in {}^{n}2 \mid \langle \underset{\sim}{n}, X_h\rangle \models \theta(X_h)\}| \le 2^t$$

$$\text{and} \quad f \cup g \in {}^{\subseteq n}2, \ |\mathrm{dom}(f \cup g)| \le n - \sqrt{n} + t.$$

Hence, since n is large we can find $q \in {}^{\subseteq n}2$ such that $q \supseteq f \cup g$, $|\mathrm{dom}(q)| \le n - \sqrt{n} + t + 2^t \le n - \sqrt{n} + 2^{t+1}$ and

$$\overline{q} \cap \{h \in {}^{n}2 \mid \langle \underset{\sim}{n}, X_h\rangle \models \theta(X_h)\} = \emptyset$$

as required.

PROOF OF THEOREM 13. We construct A in stages. Suppose we have found $A \cap t$ and $\psi(x,X) \in \Delta_0(X)$ and we wish to arrange that

$$\{n \mid \psi(n,A)\} \neq \{n \mid |A \cap n| = 0 \bmod 2\} = A^{(2)}$$

Fix n large. We arrange that

$$\neg\, \psi(n,A) \iff n \in A^{(2)}$$

We shall set $A \cap [n, n^{m_\psi}) = \emptyset$ so that by earlier remarks it will be enough to find $A \cap [t,n)$ to satisfy

$$\langle \underset{\sim}{n}, A \cap n\rangle \models \neg\, \overline{\psi}(A \cap n) \iff n \in A^{(2)}.$$

By Theorem 15 we can find $q \in {}^{\subseteq n}2$ such that $q \restriction t \in {}^{t}2, |\mathrm{dom}(q)| < n$, $X_{q \restriction t} = A \cap t$ and

$$\text{either} \quad \overline{q} \subseteq \{h \in {}^{n}2 \mid \langle \underset{\sim}{n}, X_h\rangle \models \overline{\psi}(X_h)\},$$
$$\text{or} \quad \overline{q} \subseteq \{h \in {}^{n}2 \mid \langle \underset{\sim}{n}, X_h\rangle = \neg\, \overline{\psi}(X_h)\}.$$

Without loss of generality assume the former. Then since $|\mathrm{dom}(q)| < n$ we can find $h \in {}^{n}2$ such that $h \supseteq q$ and $|X_h|$ is odd. Set $A \cap n = X_h$. Then $n \in' A^{(2)}$ whilst

$$\langle \underset{\sim}{n}, A \cap n\rangle \models \overline{\psi}(A \cap n). \qquad \square$$

In view of our present inability to count even sets with fairly few elements it would be nice if we could arrange $|A|$ here to be very small. However Ajtai's Theorem does not seem to give this. Another similar open problem is whether or not we can find A such that Δ_0^A is closed under counting mod 2 but not, say, closed under counting.

Unlike some oracle results this one gives rather more than just that any proof that $\Delta_0^{\mathbb{N}}$ is closed under counting mod 2 will not relativize. For example an obvious way to attempt to prove that whenever $\theta(x) \in \Delta_0$ then

$$\{x \mid \theta(x)\}^{(2)} \in \Delta_0^{\mathbb{N}}$$

would be by induction on the complexity of θ. But the Theorem puts a block on doing this in most natural ways. For example with A as in the theorem if

$A_1 = \{2n \mid n \in \mathbb{N}\}$, $A_2 = \{2n \mid n \in A\} \cup \{2n+1 \mid n \in' A\}$

then we do have $A_1^{(2)}$, $A_2^{(2)} \in \Delta_0^A$ but we do not have $(A_1 \cap A_2)^{(2)} \in \Delta_0^A$.

In [1] Ajtai also uses his Theorem 14 to show that distinguishing between sets of odd or even size is difficult. Precisely:-

THEOREM 18.[Ajtai, [1].] Let K be a countable non-standard model of Peano Arithmetic, $a \in K$, a non-standard. Then $\exists P, Q \subseteq a$, P, Q coded in K such that in K, $|P|$ is even, $|Q|$ is odd and $\langle \underset{\sim}{a}, P\rangle \cong \langle \underset{\sim}{a}, Q\rangle$. □

THEOREM 19.[Ajtai, [1] .] Let K be a countable non-standard model of Peano Arithmetic, $a \in K$, a non-standard and $R \subseteq a^k$ coded in K. Then $\exists P \subseteq a^2$, P coded in K such that for all $i < a$, $|\{j \mid \langle i, j\rangle \in P\}|$ is even and for all $A_1, \ldots, A_m \subseteq a$ coded in $K\ \exists R', Q, A_1', \ldots, A_m'$ coded in K such that

$$\langle \underset{\sim}{a}, R, P, A_1, \ldots, A_m\rangle \cong \langle \underset{\sim}{a}, R', Q, A_1', \ldots, A_m'\rangle$$

and for some $i < a$, $|\{j \mid \langle i, j\rangle \in Q\}|$ is odd. □

We now turn to look at oracle versions of the PHP.

PROBLEM. Let F be a new unary function symbol. Then is

$$I\Delta_0(F) + \exists x,\ F : x \twoheadrightarrow x\text{-}1 \text{ consistent?}$$

[It does not matter if we allow F to appear in the bounds on the quantifiers in $I\Delta_0$ (F) since we can always replace F by the identity above "x".]

At present we know of no full solution to this problem although there are several partial results. Firstly Woods has shown:-

THEOREM 20. [Woods, [11].] Let Def(G, F) be the axiom

$$\forall x,y\ [G(0, y) = 0 \wedge G(x+1, y) = \begin{cases} G(x,y) + 1 & \text{if } F(x) \le y \\ G(x, y) & \text{otherwise} \end{cases}].$$

Then $I\Delta_0(G, F) + \text{Def}(G, F) \vdash \neg \exists x,\ F : x \twoheadrightarrow x\text{-}1$. □

Notice that Theorem 2 is a special case of this.

Our next result shows that we can give a positive answer to our problem for the fragment $I\ \exists_1(f)$.

THEOREM 21. $I\ \exists_1(F) + \exists x,\ F : x \twoheadrightarrow x\text{-}1$ is consistent, where $I\ \exists_1(F)$ is the induction schema for formulae of LA(F) of the form

$$\exists x_1\ \exists x_2,\ldots,\exists x_n\theta,$$

θ quantifier free.

<u>PROOF.</u> Let K be a countable non-standard model of Peano's axioms, $\alpha \in K$ and α non-standard. We produce $F : \alpha+1 \twoheadrightarrow \alpha$ by a simple forcing argument. Forcing conditions are finite sets of the form

$$F(x_1) = y_1,\ F(x_2) = y_2,\ldots,\ F(x_n) = y_n,$$

with $\vec{x} \le \alpha$, $\vec{y} < \alpha$ and F 1-1. For σ a forcing condition and $\psi(x)$ a formula from $LA(F)$ and $\vec{a} \in M$,

$$\sigma \Vdash \psi(\vec{a})$$

is defined in the obvious way, e.g.

$$\sigma \Vdash F(a) = b \iff (f(a) = b) \in \sigma,$$

$$\sigma \Vdash \lambda \iff M \models \lambda \text{ for } \lambda \text{ not involving } F, \text{ etc.}$$

Now suppose that $\lambda(x) = \exists\vec{y}\theta(x,\vec{y}) \in \exists_1(F)$. Then there is a fixed $n \in \mathbb{N}$, dependent only on θ such that for any condition σ and $b \in M$ either $\sigma \Vdash \neg\lambda(b)$ or $\exists\tau \supseteq \sigma$, $\tau \Vdash \lambda(b)$ and $|\tau - \sigma| \le n$. To see this notice that to decide $\theta(x, y_1,\ldots,y_m)$ we only need to know the values of $F^i(x), F^i(y_1),\ldots, F^i(y_m), F^i(e_1),\ldots,F^i(e_k)$ for $i \le j$, some fixed j, where $e_1,\ldots,e_k$ are the constants appearing in θ. Suppose there are n such values. Then either $\sigma \Vdash \neg\lambda(b)$ or $\exists\nu \supseteq \sigma$, $\nu \Vdash \theta(b,\vec{a})$ for some $\vec{a}$. Hence we can pick $\tau \supseteq \sigma$ such that τ specifies $F^i(b), F^i(a_1),\ldots,F^i(a_m), F^i(e_1)$, $F^i(e_k)$ for $i \le j$ and is compatibile with ν. This is possible since α is non-standard and σ is finite. Then $\tau \cup \nu \Vdash \theta(b,\vec{a})$ so since τ decides $\theta(b,\vec{a})$, $\tau \Vdash \theta(b,\vec{a})$ and hence $\tau \Vdash \lambda(b)$.

We can now pick a complete sequence of forcing conditions to ensure that $\langle K, F\rangle \models I\exists_1(F) + F : \alpha+1 \twoheadrightarrow \alpha$. For it is enough to force that every non-empty $\exists_1(F)$ set has a least element. So suppose we have a forcing condition σ and $\sigma \Vdash \lambda(b)$ for $\lambda \in \exists_1(F)$. Then for n as above

$$\{a \le b \mid \exists \text{ forcing condition } \tau \supseteq \sigma, |\tau - \sigma| \le n \ \&\ \tau \Vdash \lambda(a)\}$$

is definable in K and hence has a least element, e say. Let $\tau \supseteq \sigma$. $\tau \Vdash \lambda(e)$.

Then by the above remarks $\sigma \Vdash \neg\lambda(a)$ for all $a < e$ so τ forces that e

is the least element satisfying $\lambda(x)$ in $<K,F>$. □

Along similar lines we have

THEOREM 22. [Goad, [4].] Let L_0 be the language with just < and ′ (successor) and let T_0 be the L_0 theory of $\mathbb{N}$. Then

T_0 + full induction for $L_0(F) + \exists x, F : x' \mapsto x$

is consistent. □

Goad proves this by a quantifier elimination argument. His argument can be somewhat simplified by working with models as follows. Let J be a countable non-standard model of Peano Arithmetic, let $\mathbb{Z}_J$ be the "integers" of J and let K be the structure for L_0 with domain $J + \mathbb{Z}_J$, in that order. Then $K \models T_0$. Let $d \in \mathbb{Z}_J$ and $\eta \in J$, η non-standard. Define $F : K \to K$ by

$$F(x) = \begin{cases} x + \eta & \text{if } x \in J, \\ x + \eta - 1 & \text{if } J < x < d, \\ x - d & \text{if } d \le x < d + \eta, \\ x & \text{if } d + \eta \le x, \end{cases}$$

so $F : d + \eta \mapsto d + \eta - 1$. Also $<K,F>$ satisfies full induction for, if not, since $<K,F>$ is definable in J, J must be definable in $<K, F>$ from parameters. But we can now construct an isomorphism σ of $<K,F>$ which fixes these parameters but does not fix J, hence giving the required contradiction.

SECTION 5. In this short final section we indicate how our previous results might be improved by making various assumptions about $\Delta_0^{\mathbb{N}}$ and $I\Delta_0$. It must be admitted however that in all cases except for Theorem 26 the general opinion is that the assumption is false.

Recall that we know that Lin.Space is closed under counting and so of course is $\Delta_0^{\mathbb{N}}$ if $\Delta_0^{\mathbb{N}}$ = Lin.Space. By directly amending this proof we obtain, for $f : \mathbb{N} \to \mathbb{N}$,

THEOREM 23. Assume that $\Delta_0^{\mathbb{N}} \supseteq \text{Space}(f(n))$ and let $A \in \Delta_0$, $A \subseteq \mathbb{N}^2$ and such that for all $n \in \mathbb{N}$,

$$A_n = \{m \mid <n,m> \in A\} \subseteq 2^{f(\log(n))}.$$

Then $\{<n,k> \mid k = |A_n|\} \in \Delta_0^{\mathbb{N}}$. □

The assumption here concerns the amount of computational space available in $\Delta_0^{\mathbb{N}}$. Assumptions about available time also yield more counting. Precisely let $f : \mathbb{N} \to \mathbb{N}$ have $\Delta_0^{\mathbb{N}}$ graph, f strictly increasing and $f(n) \geq n^{\log(n)}$ for all n. Define $\Delta^{\mathbb{N}}(f)$ as for $\Delta_0^{\mathbb{N}}$ but with the new function f added to $+, .$ (allowing f to appear in the bounding terms in the quantifier). In [8] an equivalence between the assumption $\Delta^{\mathbb{N}}(f) = \Delta_0$ and the computational time available in $\Delta_0^{\mathbb{N}}$ is proved.

<u>THEOREM 24.</u> Assume $\Delta_0^{\mathbb{N}} = \Delta^{\mathbb{N}}(f)$ and let $A \in \Delta_0^{\mathbb{N}}$, $A \subseteq \mathbb{N}^2$ and such that for all $n \in \mathbb{N}$,

$$|A_n| = |\{m \mid m < n \;\&\; \langle n, m\rangle \in A\}| \leq \log(f^j(n)).$$

Then $\{\langle n, k\rangle \mid k = |A_n|\} \in \Delta_0^{\mathbb{N}}$. □

This theorem follows easily from Theorem 6.

Turning now to the status of the Δ_0PHP we mention the following result.

<u>THEOREM 25.</u> Assume that

$$I\Delta_0 + \forall x,\ x^{\log(x)} \text{ exists} \vdash \text{Matijasevic-Rovinson-Davis-Putnam Thm.}$$

Then $I\Delta_0 + \forall x,\ x^{\log(x)}$ exists $+ \mathrm{Con}(I\Delta_0) \vdash \Delta_0$PHP. □

A proof of a rather more general result will appear in [10]. The assumption $\mathrm{Con}(I\Delta_0)$ can be weakened but our proof still needs some consistency. It would indeed be surprising however if there really was a formal connection between the consistency of $I\Delta_0$ and the Δ_0PHP.

For our next result we show that the problem of section 4 on the relativized Δ_0PHP has an affirmative answer assuming the Cook-Reckhov conjecture viz.

$\forall k\, \exists n$ such that every proof in the propositional calculus of

$$\bigwedge_{i \leq n} \bigvee_{j<n} p_{ij} \to \bigvee_{i<e\leq n} \bigvee_{j<n} (p_{ij} \wedge p_{ej})$$

uses more than n^k symbols.

<u>THEOREM 26.</u> Assuming the Cook-Reckhov Conjecture,

$$I\Delta_0(F) + \exists x,\ F : x \rightarrowtail\!\!\!\!\to x-1$$

is consistent.

<u>PROOF.</u> Let K be a countable proper elementary extension of $\mathbb{N}$. Pick

non-standard $c, b \in K$ such that, in K, every proof of

$$\bigwedge_{i\leq c} \bigvee_{j<c} p_{ij} \rightarrow \bigvee_{i<e\leq c} \bigvee_{j<c} (p_{ij} \wedge p_{ej})$$

uses at least c^b symbols.

For $\theta(x)$ a formula from $L(R)$, where R is a new binary relation symbol, and $\vec{a} \leq c$ we define a formula $\theta(\vec{a})^*$ of the propositional calculus as follows.

$\underline{\theta}$	$\underline{\theta^*}$	
$R(a_1, a_2)$	p_{a_1,a_2}	
$a_1 + a_2 = a_3$	s_{a_1,a_2,a_3}	new propositional variables
$a_1 \cdot a_2 = a_3$	t_{a_1,a_2,a_3}	
$a_1 = a_2$	e_{a_1,a_2}	
$\theta_1 \wedge \theta_2$	$\theta_1^* \wedge \theta_1^*$	
$\neg \theta_1$	$\neg \theta_1^*$	
$\exists x\theta(x)$	$\bigvee_{a\leq c} \theta(a)^*$	
$\forall x\theta(x)$	$\bigwedge_{a\leq c} \theta(a)^*$ etc.	

Not let T_0 consist of

$$\bigwedge_{i\leq c} \bigvee_{j<c} p_{ij} \wedge \bigwedge_{i<e\leq c} \bigwedge_{j<c} \neg (p_{ij} \wedge p_{ej}) \ldots \qquad \dagger$$

together with

$\{\theta^* \mid \theta$ is an atomic or negation of atomic sentence of L and $\underset{\sim}{c+1} \models \theta\}$.

Then there is still no proof of an inconsistency from T_0 using less than c^b symbols. Now we can form increasing sets of sentences $T_0 \subseteq T_1 \subseteq T_2 \subseteq \ldots$ such that each T_n is coded in K and

(i) There is no proof of an inconsistency (in K) from T_n using less than $c^{b/2^n}$ symbols.

(ii) For every formula $\theta(x)$ of $L(R)$ and $\bar{a} \leq c$ there is an $n \in \mathbb{N}$ such that $\theta(\vec{a})^* \in T_n$ or $\neg \theta(\vec{a})^* \in T_n$.

(iii) Whenever $\bigvee_{a\leq c} \theta(a)^* \in T_n$ then $\exists n < m \in \mathbb{N}$ and $d \leq c$ such that $\theta(d)^* \wedge \bigwedge_{a<d} \neg\theta(a)^* \in T_m$.

Define $\overline{R} \subseteq (c+1) \times (c+1)$ by

$$\overline{R}(a_1,a_2) \iff p_{a_1,a_2} \in \bigcup_{n\in\mathbb{N}} T_n.$$

Then it is easy to see that

$$<\underset{\sim}{c+1},\ \overline{R}> \models \text{Induction} + \{\theta \mid \theta^* \in \bigcup_{n\in\mathbb{N}} T_n\}.$$

Define for $a \leq c$, $F(a)$ = the least d such that $\overline{R}(a,d)$. Then because the formula † is in T_0, $F : c+1 \rightarrowtail\!\!\!\rightarrow c$, and (as in [7]),

$$<c^{\mathbb{N}}, F> \models I\Delta_0(F) + F : c+1 \rightarrowtail\!\!\!\rightarrow c$$

where $c^{\mathbb{N}}$ denotes the substructure of K with universe $\{d \mid d < c^i$ some $i \in \mathbb{N}\}$. □

We now develop a rather surprising connection between the Δ_0-PHP and $\neg\Delta_0 H$, that is the collapse of the Δ_0 hierarchy [see[8].] First we prove a rather special result which we later generalise.

THEOREM 27. Assume that $F \in \Delta_0^M$, $\mathbb{N} < a, \nu$ and that $F : a^\nu \rightarrowtail\!\!\!\rightarrow a$. Assume also that the only parameters appearing in this Δ_0 definition of F are a^ν, a. Then for each $r \in \mathbb{N}$ there is a Δ_0 formula $\theta(x,y,z)$ with all quantifiers restricted to x such that $\{b < a^\nu \mid M \models \theta(a^\nu, b, a)\}$ is not equal to any set of the form

$$\{b < a^\nu\ M \models \exists\vec{x}_1 < a^\nu\ \forall\vec{x}_2 < a^\nu\ \ldots\ Q\vec{x}_r < a^\nu\ [g(\vec{x}_1,\ldots,\vec{x}_r, b, a) = h(\vec{x}_1,\ldots,\vec{x}_r, b, a)]\}$$

with g, h polynomials over $\mathbb{N}$.

PROOF. It is shown in [7] that any set of this form can be written as

$$*\ldots \{b < a^\nu \mid M \models \exists\vec{x}_0 < a^\nu\ \forall\vec{x}_1 < a^\nu\ \ldots\ Q\vec{x}_r < a^\nu \bigwedge_{m=1}^{s} z_{m_1} \cdot z_{m_2} + z_{m_3} = z_{m_4}$$

where the z_m's are from $\{\vec{x}_0, \vec{x}_1,\ldots,\vec{x}_r, 1, a, b, 0\}$.

The idea now is to replace a^ν by a and $z_1 \cdot z_2 + z_3 = z_4$ by $F^{-1}(z_1)\cdot F^{-1}(z_2) + F^{-1}(z_3) = F^{-1}(z_4)$ etc. Precisely we can write the formula in * as $\exists y_0 < a^{\nu k} \quad \forall y_1 < a^{\nu k} \; \ldots \; Q y_r < a^{\nu k}$ [such that for all $\langle p_1, p_2, p_3, p_4 \rangle \in T$, $x_1 . x_2 + x_3 = x_4$] where for $1 \le i \le 4$,

$$x_i = \begin{cases} y_q(n) & \text{if} \quad p_i = \langle q, n\rangle \;\&\; q \le r,\ n < k \\ 0 & \text{if} \quad p_i = \langle r+1, 0\rangle \\ 1 & \text{if} \quad p_i = \langle r+1, 1\rangle \\ a & \text{if} \quad p_i = \langle r+1, 2\rangle \\ b & \text{otherwise} \end{cases}$$

for suitable k and finite $T \subseteq \mathbb{N}^4$.

So using F this can be written as

$\exists y_0 < a^k$, $y_0''k \subseteq \mathrm{Rg}(F)$ $\quad \forall y_1 < a^k$, $\quad y_1''k \subseteq \mathrm{Rg}(F) \ldots \quad Q y_r < a^k, y_r''k \subseteq \mathrm{Rg}(F)$ [such that for all $\langle p_1, p_2, p_3, p_4\rangle \in T \; \exists \alpha_1, \alpha_2, \alpha_3, \alpha_4 < a^\nu$ such that $\alpha_1 \cdot \alpha_2 + \alpha_3 = \alpha_4$] where for $1 \le i \le 4$

$$F(\alpha_i) = \begin{cases} y_q(n) & \text{if} \quad p_i = \langle q, n\rangle \;\&\; q \le r,\ \ n < k \\ 0 & \text{if} \quad p_i = \langle r+1, 0\rangle \\ 1 & \text{if} \quad p_i = \langle r+1, 1\rangle \\ F(a) & \text{if} \quad p_i = \langle r+1, 2\rangle \\ F(b) & \text{otherwise} \end{cases}$$

and hence as $\Gamma(a^\nu, a, b, \langle k,T\rangle)$ where $k, T \in \mathbb{N}$. Γ is Δ_0, and all quantifiers in Γ are restricted to a^ν. Furthermore the <u>formula</u> Γ only depends on r. It follows that $\{b < a^\nu \mid M \models \neg\, \Gamma(a^\nu, a, b, b)\}$ is not of the form given in the theorem. □

<u>COROLLARY 28.</u> Suppose that $a^\nu \in M$ some $\nu > \mathbb{N}$, $F \in \Delta_0^M$ and $F: a^2 \mapsto\!\!\!\to a$. Then $M \models \Delta_0 H$.

<u>PROOF.</u> By $M \models \Delta_0 H$ we mean that there is no fixed $r \in \mathbb{N}$ such that every Δ_0^M set (without parameters) is of the form

$$\{b \in M \mid M \models \exists \vec{x}_1 < b \quad \forall \vec{x}_2 < b \; \ldots \; Q \vec{x}_r < b,\ \ g(\vec{x}_1, \ldots, x_r, b) = h(\vec{x}_1, \ldots, \vec{x}_r, b)\}$$

with g, h polynomials over $\mathbb{N}$. To show this it is enough to show that the hypotheses of Theorem 27 hold since if any Δ_0^M set can be written in the above form then, with the notation of Theorem 27, the set

$$\{a^{\nu}, a, b> | b < a^{\nu} \ \& \ M \models \theta(a^{\nu}, b, a)\},$$

this set could be written in that form and hence (for a possibly larger, but fixed, value of r) in the form disallowed by Theorem 27. So let c be the parameter, if any, appearing in the Δ_0 definition of F. We may assume $c < a^{\nu}$ and $\nu = 2^{\delta}$.

Now define $\tilde{F} : M \to M$ by $\tilde{F}(\sum_i e_i . a^{2i}) = \sum_i F(e_i) a^i$ where $e_i < a^2$, and define $\overline{F} : a^{2^{\delta}} \leftrightarrow a$ by $\overline{F}(z) = \tilde{F}^{\delta}(z)$. $\overline{F}$ has parameters $a^{2^{\delta}}$, a, c but c can now be removed by replacing it by the least d such that $\theta(z, a^{2^{\delta}}, a, d)$ defines a 1-1 map from $a^{2^{\delta}}$ into a where $\theta(z, a^{2^{\delta}}, a, c)$ defines $\overline{F}$. □

REFERENCES

[1] Ajtai, M. Σ_1^1-formulae on finite structures . Annals of Pure and Applied Logic, 24(1),(1983).

[2] Bel'tyukov,A. A computer description and a hierarchy of initial Grzegorczyk classes. Zap.Navcu.Sem.Leningrad Otdel Mat. Inst.Steklov (LOMI) 88, (1979).

[3] Gaifman, H. & Dimitracopoulos, C. Fragments of Peano's Arithmetic and the MRDP theorem, Logic and Algorithmic, Monographie Nº 30 de L'Enseignement Mathematique

[4] Goad, C. Duplicated Notes on the pigeon hole principle.

[5] Hicks, J. A machine characterization of quantification and primitive recursion with applications to low level complexity classes. Ph.D. Thesis, Oxford University (Submitted 1983).

[6] Nepomnjascii, V. Rudimentary predicates and Turing calculations. Soviet Math.Dokl. 11, Nº 6 (1970).

[7] Paris, J. & Dimitracopoulos, C. Truth definitions for Δ_0 formulae. Logic and Algorithmic, Monographie Nº 30 de L'Enseignement Mathematique.

[8] Paris, J. & Wilkie, A. Δ_0 sets and induction. Open Days in Model Theory and Set Theory, Proceedings of the 1981 Jadwisin (Poland) Logic Conference, Ed. Guzicki, Marek, Pelc, Rauszer.

[9] Ritchie, R. Classes of predictably computable functions. TAMS 106 (1963).

[10] Wilkie, A. & Paris, J. On the schema of induction for bounded arithmetic formulas. (To appear).

[11] Woods, A. Some problems in logic and number theory and their connections. Ph.D. Thesis, Manchester University (1981)

[12] Furst, M., Saxe, J.B. & Sipser, M. Parity, circuits and the polynomial-time hierarchy. IEEE Foundations of Computer Science (22nd Symposium). (1981), pp. 260-270.

DEFINABLE ULTRAFILTERS AND ELEMENTARY END EXTENSIONS

Ramon Pino* and Jean-Pierre Ressayre
Université de Paris VII
U.E.R. de Mathématiques et Informatique
Tour 45-44 5éme étage- 2 Place Jussieu
75251 Paris Cedex 05, France

The purpose of this paper is to show that if $M \models Ad + V = L$. The existence of Δ_n-based Π_n-ultrafilters on $(On^M)^k$ implies the existence of $M_1, M_2, \ldots, M_k$ such that $M \precneqq^e_{n+1} M_1 \precneqq^e_{n+1} M_2 \precneqq^e_{n+1} \cdots \precneqq^e_{n+1} M_k$.

At the end of this paper we will assert without proof that if $M \not\models$ foundation schema, then the existence of Δ_n-based Π_n-ultrafilters on $(On^M)^k$ with a pseudonormality property is equivalent to the fact that M is (n,K)-extendible.

To begin let us recall some definitions and set some notations. Ad(non standard Admissible Set Theory) is defined by the following axioms and schema of axioms : pair, union, extensionality, foundation (axiom), Δ_o-separation, Δ_o-collection and $(\Sigma_1 \cup \Pi_1)$-foundation (see [B]).

The classes of $\Delta_n, \Pi_n, \Sigma_n$ formula for $n \geq 0$, are defined as usual.

On^M denotes the ordinals of the model M. Often we write On instead of On^M. On^k is the cartesian product of On k-times. The elements of On^k are denoted $\vec{\alpha}$ i.e. $\vec{\alpha} = (\alpha_1, \alpha_2, \ldots, \alpha_k)$.

If ϕ is a formula with parameters $a_1, \ldots, a_l$ we write it $\phi(\vec{a})$.

The notation $\vec{a} \in M$ means that $\vec{a}$ is $(a_1, \ldots, a_n)$ and every $a_i \in M$, for $i \in \{1, \ldots, n\}$. The notation $\vec{f}(A) \subseteq L_\alpha$ means that $\vec{f}$ is $(f_1, \ldots, f_m)$ where the f_i's are functions whose domains contain A and $\bigcup_{i=1}^{m} f_i''(A) \subseteq L_\alpha$.

---

*Current address: Universidad Simón Bolívar
Departamento de Matemáticas y Ciencia de la Computación
Apartado 80659
Caracas 1080A, Venezuela.

Often we manipulate $\vec{f}$ as if it were a single function.

$M \subset^e M_i$ means that M_i is an end extension of M, i.e. if $M_i \models a \in b$ and $b \in M$ then $a \in M$ and $M \models a \in b$.

$M \prec_n M_i$ means that M_i is a Σ_n-elementary extension of M, i.e. if ϕ is Σ_n and $\vec{a} \in M$ then $M \models \phi[\vec{a}] \Leftrightarrow M_i \models \phi[\vec{a}]$.

$M \precneqq^e_n M_i$ means that M_i is a proper Σ_n-elementary end extension of M.

A class X is Δ_n (Σ_n or Π_n) in M if there exists a Δ_n (Σ_n or Π_n) formula ϕ with parameters such that $X = \{a \in M : M \models \phi[a]\}$. Often we will only write X is a Δ_n-class, then the set M in which X is Δ_n will be clear from the context.

A Π_n-ultrafilter over On is a subset $\mathcal{U}$ of $P(On)$ which satisfies the following conditions:

(i) $A \in \mathcal{U}$ and $B \in \mathcal{U} \Rightarrow A \cap B \in \mathcal{U}$

(ii) $A \in \mathcal{U}$, $B \in P(On)$ and $A \subseteq B \Rightarrow B \in \mathcal{U}$

(iii) $A \in P(On)$ and A is $\Pi_n \Rightarrow A \in \mathcal{U}$ or $On \setminus A \in \mathcal{U}$

(iv) $A \in \mathcal{U} \Rightarrow A$ is unbounded

(v) $\mathcal{U}$ is Π_n-complete i.e. if $X \subseteq M \times On$ is Π_n and $\forall m \in a\ X_m \in \mathcal{U}$, where $X_m = \{\alpha \in On : (m,\alpha) \in X\}$ and $a \in M$, then $\bigcap_{m \in a} X_m \in \mathcal{U}$

M is resolvable if there exists a Δ_1 function $f : On \to M$ such that $M \models \forall x \exists \alpha\ x \in f(\alpha))$.

M has Δ_n skolem functions iff for every $k \geq 1$ and any Δ_n-class $A \subseteq M^{k+1}$ such that $M \models \forall a_1, \forall a_2, \ldots, \forall a_k \exists a\ (a_1, \ldots, a_k, a) \in A$ there exists a Δ_n-function $f : M^k \to M$ such that $M \models \forall a_1, \forall a_2, \ldots, \forall a_k\ (a_1, \ldots, a_k, f(\vec{a})) \in A$

Now we state a theorem of which our result is a partial generalization:

THEOREM 1. (Kaufmann, Kranakis) (see [Ka], [Kr]). Let M be a countable model of $Ad + V = L$. For every integer $n \geq 2$ the following conditions are equivalent:

(i) There exists M' such that $M \precneqq^e_n M'$

(ii) $M \models \Pi_{n-1}$-collection

(iii) There exists a Π_{n-1} ultrafilter over On^M

In fact if $M \models Ad$ then :

(a) (ii) $\Rightarrow$ (i) if M is countable

(b) (i) $\Rightarrow$ (ii) if M is resolvable

(c) (iii) $\Rightarrow$ (i) if M has Δ_{n-1} Skolem functions

(d) Always (i) $\Rightarrow$ (iii)

From now on we will work in a fixed model M of Ad.

The following two definitions extend the concepts of unbounded and Π_n-ultrafilter.

DEFINITION 2. Let A be contained in On^2. A is unbounded iff $\{\alpha : \{\beta : (\alpha,\beta) \in A\}$ is unbounded $\}$ is unbounded. More generaly, if $k > 2$ and $A \subseteq On^k$, we define inductively A is unbounded iff $\{\alpha_1 : \{(\alpha_2,\ldots,\alpha_k) : (\alpha_1,\alpha_2,\ldots,\alpha_k) \in A\}$ is unbounded $\}$ is unbounded.

DEFINITION 3. Let U be contained in $P(On^k)$, $k \geq 1$. U is a Π_n-ultrafilter on On^k iff the following conditions hold:

i) $A \in U$ and $B \in U \Rightarrow A \cap B \in U$

ii) $A \in U$, $B \in P(On^k)$ and $A \subseteq B \Rightarrow B \in U$

iii) $A \in P(On^k)$ and A is $\Pi_n \Rightarrow A \in U$ or $On^k \setminus A \in U$

iv) $A \in U \Rightarrow$ A is unbounded

v) U is Π_n-complete i.e. if $X \subseteq M \times On^k$ is Πn and $\forall m \in a\ X_m \in U$, where $X_m = \{(\alpha_1,\ldots,\alpha_k) : (m,\alpha_1,\ldots,\alpha_k) \in X\}$. Then $\bigcap_{m \in a} X_m \in U$.

A Π_n-ultrafilter U is Δ_n-based if for every $A \in U$ there exists $A' \in U$ such that A' is Δ_n and $A' \subseteq A$.

The following theorem extends the implication c) of Theorem 1.

THEOREM 4. Let M and n be such that $M \models Ad + V = L$ and $n \geq 1$. If there exists a Δ_n-based Π_n-ultrafilter on On^k then there exists $M_1, M_2, \ldots, M_k$ such that $M \precneqq^e_{n+1} M_1 \precneqq^e_{n+1} M_2 \precneqq^e_{n+1} \cdots \precneqq^e_{n+1} M_k$

PROOF. We use the following notation : if $\overline{x} = (x_1,\dots,x_i)$ $A \subseteq On^k$ and $i \in \{1,\dots,k-1\}$ then we put

$$A_{\overline{x}} = \{(\overline{x},x_{i+1},\dots,x_k) : (\overline{x},x_{i+1},\dots,x_k) \in A\}$$

$$\overline{\mathbb{P}}_i(A) = \{(x_1,\dots,x_i) : (\exists(x_{i+1},\dots,x_k))(x_1,\dots,x_i,x_{i+1},\dots,x_k) \in A\}$$

$D_i = \{f : f{:}On^k \to M,$ f is Δ_n and there exists $A \in \mathcal{U}$ such that for every $\overline{x} \in \overline{\mathbb{P}}_i(A)$ f is bounded over $A_{\overline{x}}\}$

where f is bounded over B means that there is β such that $f''(B) \subseteq L_\beta$

We say that A is a witness for the fact $f \in D_i$ if $A \in \mathcal{U}$, A is Δ_n and for every $\overline{x} \in \overline{\mathbb{P}}_i(A)$ f is bounded over $A_{\overline{x}}$.

We put $D_k = \{f : f{:}On^k \to M$ and f is $\Delta_n\}$

We define $M_i = {}^{D_i}/\mathcal{U}$ for $i \in \{1,\dots,k\}$. Recall if $f,g \in M_i$ then:

$$M_i \models f = g \Leftrightarrow \{\vec{\alpha} \in On^k : M \models f(\vec{\alpha}) = g(\vec{\alpha})\} \in \mathcal{U}, \text{ and}$$

$$M_i \models f \in g \Longleftrightarrow \{\vec{\alpha} \in On^k : M \models f(\vec{\alpha}) \in g(\vec{\alpha})\} \in \mathcal{U}$$

Remark that, if $i < k$ and $f{:}On^k \to M$ is Δ_n to ensure $f \in M_i$ it is enough to find a witness for $f \in D_i$.

Now, we set a series of lemmas that will prove the theorem.

LEMMA 5. $M \subseteq^e M_k$.

PROOF. If $m \in M$, m is identified with the constant function $C_m{:}On^k \to M$ such that $C_m(\vec{\alpha}) = m$ for every $\vec{\alpha} \in On^k$. Then $M \subseteq M_1$.

Let f and m be such that $M_k \models f \in m$, where $m \in M$. By definition of M_k, $\{\vec{\alpha} \in On^k : M \models f(\vec{\alpha}) \in m\} \in \mathcal{U}$. Suppose that for every $m' \in m$ $\{\vec{\alpha} \in On^k : M \models f(\vec{\alpha}) = m'\} \notin \mathcal{U}$ i.e. $f^{-1}(m') \notin \mathcal{U}$.

Then, by definiton of $\mathcal{U}$, $On^k \setminus f^{-1}(m') \in \mathcal{U}$ for every $m' \in m$, because $f^{-1}(m')$ is Δ_n. Put $B_{m'} = On^k \setminus f^{-1}(m')$, for every $m' \in m$.

By Π_n-completeness $\bigcap_{m' \in m} Bm' \in \mathcal{U}$, but $\bigcap_{m' \in m} Bm' = \emptyset$ and $\emptyset \notin \mathcal{U}$ because $\emptyset$ is bounded. This contradiction shows that $f^{-1}(m') \in \mathcal{U}$ for some $m' \in m$ i.e. $M_k \models f = m'$.

LEMMA 6. $M \subsetneq M_1$ and $M_i \subsetneq M_{i+1}$ for $i\in\{1,\ldots,k-1\}$

PROOF. We have seen that $M \subseteq M_1$. To see that $M_i \subseteq M_{i+1}$ it suffices to take $f\in D_i$ and show that $f\in D_{i+1}$. Let f be an element of D_i, then there is a witness A for $f\in D_i$. So, $f''(A_{\overline{x}})$ is bounded for every $\overline{x}\in\overline{\mathbb{P}}_i(A)$ and therefore $f''(A_{\overline{y}})$ is bounded for every $\overline{y}\in\overline{\mathbb{P}}_{i+1}(A)$. That is A is a witness for $f\in D_{i+1}$.

Let us define $p_i : On^k \to M$ by $p_i(\vec{\alpha}) = \alpha_i$, $i\in\{1,\ldots,k\}$. It is clear that $p_i\in M_i$ because p_i is bounded over $(On^k)_{\overline{x}}$ for each $\overline{x}\in\overline{\mathbb{P}}_i(On^k)$. Moreover it is easy to see that for every unbounded $A \subseteq On^k$ and for every $i\in\{1,\ldots,k-1\}$ there is $\overline{x}\in\overline{\mathbb{P}}_i(A)$ such that p_{i+1} is unbounded over $A_{\overline{x}}$. Thus $p_{i+1}\notin M_i$ because the elements of $\mathcal{U}$ are unbounded. That is $M_i \subsetneq M_{i+1}$.

We show that $p_1 \notin M$. Let α be an ordinal in M. Note that $\{\vec{x}\in On^k : \alpha\in x_1\}$ has bounded complement, thus $\{\vec{x}\in On^k : \alpha\in x_1\}\in\mathcal{U}$. And it is clear that $p_1\in On^{M_1}$. So, $M_1 \models \alpha\in p_1$ for every $\alpha\in On$ and therefore $p_1 \notin M$.

LEMMA 7. $M_i \subset^e M_{i+1}$ for $i\in\{1,\ldots,k-1\}$.

PROOF. Let f and g be such that $g\in M_i$ and $M_{i+1}\models f\in g$. By definition of M_{i+1}, $A = \{\vec{\alpha}\in On^k : M\models f(\vec{\alpha})\in g(\vec{\alpha})\}\in\mathcal{U}$. Let B be a witness for $g\in M_i$. Then $A\cap B\in\mathcal{U}$ and $A\cap B$ witness $f\in M_i$ because g is bounded over $(A\cap B)_{\overline{x}}$ for every $\overline{x}\in\overline{\mathbb{P}}_i(A\cap B)$ and f is bounded by g.

LEMMA 8. Łoś lemma for $(\Sigma_n\cup\Pi_n)$ formulas holds for each M_i, $i\in\{1,\ldots,k\}$. That is for every Σ_n or Π_n formula ϕ

$$M_i \models \phi[\vec{f}] \Leftrightarrow \{\vec{\alpha}\in On^k : M\models \phi[\vec{f}(\vec{\alpha})]\}\in\mathcal{U}. \quad (*)$$

Before the proof of lemma 8 we have to remark that $M\models\Pi_n$-collection if the hypothesis of theorem 4 holds; this is so because if there exists a Δ_n-based Πn-ultrafilter $\mathcal{U}$ on On^k then there exists a Πn-ultrafilter $\mathcal{U}'$ on On. By induction, since $M\models V = L$, M has Δ_n-Skolem functions. Then (theorem 1c) there exists M' such that $M \not\prec^e_{n+1} M'$. But, as $M\models V = L$, M is resolvable and therefore (theorem 1b) $M\models\Pi_n$-collection.

PROOF OF LEMMA 8. By induction on the complexity of ϕ. If ϕ is atomic the lemma is true by definition of M_i.

Note that if (*) holds for the formulas of some class $\Gamma \subseteq \Sigma_n$, then (*) holds for the boolean combinations of formulas of Γ.

The proof of (*) for Δ_o-formulas is essentially the same as we give for Σ_n-formulas assuming (*) for Σ_{n-1}-formulas. So, we will omit the proof of (*) for Δ_o-formula.

Assume (*) for Σ_{n-1}-formulas. Suppose that $M_i \models \exists x\phi(x,\vec{f})$ where ϕ is Π_{n-1}. Then there exists $g\epsilon M_i$ such that $M_i \models \phi(g,\vec{f})$. But, by the above remark, (*) holds for Π_{n-1} formulas because (*) holds for Σ_{n-1}-formulas. Then $\{\vec{\alpha}\epsilon On^k : M \models \phi(g(\vec{\alpha}), \vec{f}(\vec{\alpha}))\}\epsilon\mathcal{U}$, and therefore $\{\vec{\alpha}\epsilon On^k : M \models \exists x\phi(x,\vec{f}(\vec{\alpha}))\epsilon\mathcal{U}$.

Conversely, suppose $B = \{\vec{\alpha}\epsilon On^k : M \models \exists x\phi(x,f(\vec{\alpha}))\}\epsilon\mathcal{U}$. There exists $A \subseteq B$ such that A is Δ_n and $A\epsilon\mathcal{U}$, because $\mathcal{U}$ is Δ_n-based.

Define $g\colon On^k \to M$ by

$$g(\vec{\alpha}) = y \leftrightarrow [\vec{\alpha}\epsilon A \wedge \phi(y,\vec{f}(\vec{\alpha})) \wedge \forall z<_L y \neg \phi(z,\vec{f}(\vec{\alpha}))] \vee [\vec{\alpha}\notin A \wedge y = 0].$$

g is well define because $M \models \Delta_{n+1}$-foundation (this is a standard fact about M, see [P]). As A and $\vec{f}$ are Δ_n and $\forall z<_L y \neg \phi(z,\vec{f}(\vec{\alpha}))$ is Δ_n, since $M \models \Pi_n$-collection (c.f. the remark before this proof), g is also Δ_n. Clearly $\{\vec{\alpha}\epsilon On^k : M \models \phi(g(\vec{\alpha}), \vec{f}(\vec{\alpha})\}\epsilon\mathcal{U}$ and if $g\epsilon M_i$, by inductive hypothesis, $M_i \models \phi(g,\vec{f})$, so $M_i \models \exists x\phi(x,\vec{f})$.

Therefore it remains to prove that $g\epsilon M_i$. If $i = k$, $g \epsilon M_k$ because g is Δ_n. If $i<k$ we must find a witness for $g\epsilon M_i$.

Let C be a witness for $\vec{f}\epsilon M_i$. Put $D = A\cap C$. D is a witness for $\vec{f}\epsilon M_i$. So if $\bar{x} \epsilon \mathbb{P}_i(D)$ there exists α such that $\vec{f}''(D_{\bar{x}}) \subseteq L_\alpha$. But $M \models \Sigma_n$-separation because $M \models Ad + \Pi_n$-collection (this fact is standard, see [P] for a proof). So $\vec{f}''(D_{\bar{x}})\cap L_\alpha$ is a set. Put $a = \vec{f}''(D_{\bar{x}})\cap L_\alpha$. Thus $\forall z\epsilon a\ \exists y\phi(y,z)$. So, by Π_n-collection $\forall z\epsilon a\ \exists y\epsilon L_\beta\ \phi(y,z)$ for some β. Therefore $g''(D_{\bar{x}}) \subseteq L_\beta$ that is D is a witness for $g\epsilon M_i$.

LEMMA 9. $M \prec_n M_i$ and $M_i \prec_n M_{i+1}$ for every $i\,\{1,\ldots,k-1\}$.

PROOF. Let ϕ be a Σ_n-formula. Then, by Lemma 8

$M \models \phi(\vec{a}) \Leftrightarrow \{\vec{\alpha}\in On^k : M \models \phi(\vec{a})\} = On^k \in U \Leftrightarrow M_i \models \phi(\vec{a})$

i.e. $M \prec_n M_i$.

Let $\vec{f}$ be such that $\vec{f}\in M_i$. By Lemma 8 (ϕ is Σ_n).

$$M_i \models \phi(\vec{f}) \Leftrightarrow \{\vec{\alpha}\in On^k : M \models \phi(\vec{f}(\vec{\alpha}))\}\in U \Leftrightarrow M_{i+1} \models \phi(\vec{f}).$$

i.e. $M_i \prec_n M_{i+1}$.

LEMMA 10. $M \prec_{n+1} M_k$

PROOF. Let ϕ be a Σ_{n+1}-formula. If $M \models \phi[\vec{a}]$, then $M_k \models \phi[\vec{a}]$ because $M \prec_n M_k$ (Lemma 9) and Σ_n-elementary extensions preserve Σ_{n+1}-formulas.

To show that $M \prec_{n+1} M_k$ it suffices to show that Π_{n+1}-formulas satisfied in M are satisfied in M_k. So, let ϕ b a Π_{n+1}-formula, ϕ of the form $\forall x \exists y \Psi(x,y,z)$ where Ψ is Π_{n-1}. Suppose $M \models \phi(\vec{a})$. Define $f:M \to M$ by $f(x)=y \longleftrightarrow \Psi(x,y,\vec{a}) \wedge \forall z<_L y \neg \Psi(x,z,\vec{a})$. Note that f is Δ_n because $M \models \Pi_n$-collection. But it is clear that for every $g\in M_k$, $M \models \Psi(g(\vec{\alpha}),f(g(\vec{\alpha})),\vec{a})$ for any $\vec{\alpha}\in On^k$. Thus $\{\vec{\alpha} : M \models \Psi(g(\vec{\alpha}), f\circ g(\vec{\alpha}),\vec{a})\}\in U$, and as $f\circ g$ is Δ_n, we have, by Lemma 8, $M_k \models \Psi(g,f\circ g,\vec{a})$ for any $g\in M_k$. So, for any $g\in M_k$, $M_k \models \exists y\, \Psi(g,y,\vec{a})$, and therefore $M_k \models \forall x \exists y\, \Psi(x,y,\vec{a})$.

LEMMA 11. $M \prec_{n+1} M_i$ for every $i\in\{1,\ldots,k-1\}$.

PROOF. If $M_i \models \exists x \phi(x,f)$ where ϕ is Π_n, then $M_i \models \phi(h,f)$ for some h. So, by lemma 9, $M_k \models \phi(h,f)$ and hence $M_k \models \exists\, \phi(x,f)$. So by lemma 10, $M \models \exists\, x\phi(x,f)$.

The converse is a consequence of lemma 9.

LEMMA 12. $M_i \prec_{n+1} M_k$ for every $i\in\{1,\ldots,k\}$

PROOF. Let ϕ be a Σ_{n+1} formula. If $M_i \models \phi[\vec{f}]$ then $M_k \models \phi[\vec{f}]$ because $M_i \prec_n M_k$ (Lemma 9) and Σ_{n+1}-formulas are preserved by Σ_n-elementary extensions.

Conversely, assume $M_k \models \phi[\vec{f}]$. Since ϕ is Σ_{n+1}, ϕ is of the form $\exists x \forall y \Psi(x,y,\vec{v})$, where Ψ is Σ_{n-1}.

Note that $M_k \models$ "$\forall_\alpha L_\alpha$ exists" because $n \geq 1$, $M \prec_{n+1} M_k$, $M \models$ "$\forall_\alpha L_\alpha$ exists" and "$\forall_\alpha L_\alpha$ exists" is Π_2. So, $M_k \models L_{p_k}$ exists.

$M_k \models \exists x \forall y \in L_{p_k} \Psi(x,y,\vec{f})$ because $M_k \models \exists x \forall y \Psi(x,y,\vec{f})$. Take $h \in M_k$ such that $M_k \models \forall y \in L_{p_k} \Psi(h,y,\vec{f})$. Since $\forall y \in L_{p_k} \Psi(h,y,\vec{f})$ is Π_n, by the Lemma 8, $\{\vec{\alpha} \in On^k : M \models \forall y \in L_{\alpha_k} \Psi(h(\vec{\alpha}),y,\vec{f}(\vec{\alpha}))\} \in \mathcal{U}$. It follows that $B = \{\vec{\alpha} \in On^k : M \models \exists x \forall y \in L_{\alpha_k} \Psi(x,y,\vec{f}(\vec{\alpha}))\} \in \mathcal{U}$.

Let A be such that $A \subseteq B$, A is Δ_n and $A \in \mathcal{U}$. Define $g : On^k \to M$ by:

$$g(\vec{\alpha}) = x \longleftrightarrow (\vec{\alpha} \in A \wedge \forall y \in L_{\alpha_k} \Psi(x,y,\vec{f}(\vec{\alpha})) \wedge \forall z <_L x \neg \forall y \in L_{\alpha_k} \Psi(z,y,\vec{f}(\vec{\alpha})))$$
$$\vee (\vec{\alpha} \notin A \wedge x = 0)$$

g is well defined because $M \models \Delta_{n+1}$-foundation.

The formulas $\forall y \in L_{\alpha_k} \Psi(x,y,\vec{v})$ and $\forall z <_L x \neg \forall y \in L_{\alpha_k} \Psi(z,y,\vec{v})$ are Δ_n because $M \models \Pi_n$-collection. A is Δ_n and $\vec{f}$ is Δ_n. These three facts imply that g is Δ_n.

By definition of g, $\{\vec{\alpha} \in On^k : M \models \forall y \in L_{\alpha_k} \Psi(g(\vec{\alpha}),y,\vec{f}(\vec{\alpha}))\} \in \mathcal{U}$.

Then, by Lemma 8, $M_k \models \forall y \in L_{p_k} \Psi(g,y,\vec{f})$. Now, remark that $M_i \subseteq L_{p_k}$ since $M_i \models V = L$, $M_k \models V = L$ ($V = L$ is Π_2, $M \models V = L$ and $M \prec_{n+1} M_i$, $M \prec_{n+1} M_k$ and $n \geq 1$) and $On^{M_i} \subseteq P_k$ (Lemmas 6 and 7).

Assume $g \in M_i$, then for every $y \in M_i \subseteq L_{p_k}$, $M_k \models \Psi(g,y,\vec{f})$ but $M_i \prec_n M_k$, so $M_i \models \psi(g,y,\vec{f})$ for any $y \in M_i$, that is $M_i \models \forall y \Psi(g,y,\vec{f})$ and therefore $M_i \models \exists x \forall y \Psi(g,y,\vec{f})$.

It only remains to prove the assumption $g \in M_i$. We have $M_k \models \exists x \forall y \Psi(x,y,\vec{f})$. Take h' such that $M_k \models \forall y\ \Psi(h',y,\vec{f})$. Then, by Lemma 8, $E = \{\vec{\alpha} \in On^k : M \models \forall y\ \Psi(h'(\vec{\alpha}),y,\vec{f}(\vec{\alpha}))\} \in \mathcal{U}$. Let C_1 be such that $C_1 \subseteq E$, C_1 is Δ_n and $C_1 \in \mathcal{U}$. Let C_2 be a witness for $\vec{f} \in M_i$. Put

$D = A \cap C_1 \cap C_2$. D is Δ_n and $D \in \mathcal{U}$. We are going to prove that D is a witness for $g \in M_i$.

Take $\bar{x} \in \mathbb{P}_i(D)$. As D is a witness for $\vec{f} \in M_i$, there exists β such $\vec{f}''(D_{\bar{x}}) \subseteq L_\beta$. Recall that $M \models \Sigma_n$-separation (see [P]) so $\vec{f}''(D_{\bar{x}}) \cap L_\beta$ is a set. Put $b = \vec{f}''(D_{\bar{x}}) \cap L_\beta$. It is easy to see, by definition of D, $M \models \forall Z \in b\ \exists x \forall y\ \Psi(x,y,Z)$. But now, by Π_n-collection, $M \models \exists \gamma \forall\ Z \in b\ \exists x \in L_\gamma \forall y\ \Psi(x,y,Z)$. This last fact implies that g is bounded by L_γ on $D_{\bar{x}}$.

Theorem 4 now follows from lemmas 5,6,7,10 and 12.

We don't know if the converse of Theorem 4 is true. But this question has carried us to the equivalence of two concepts that we give immediately.

DEFINITION 14. Let $\mathcal{U}$ be a Δ_n-based Π_n-ultrafilter on On^k and $k \geq 2$. $\mathcal{U}$ is said to be (n,k)-pseudonormal if for any $i \in \{1,\ldots,k-1\}$, for every X such that X is Π_n, $X \subseteq On^i$ and $On^{k-i} \times X \in \mathcal{U}$ and for every $f: X \to M$ such that f is Δ_n, then there exists $a \in M$ such that $On^{k-i} \times f^{-1}(a) \in \mathcal{U}$ or

$\{(\alpha_1,\ldots,\alpha_k) \in On^k : \alpha_{k-i} < rk \circ f\ (\alpha_{k-i+1},\ldots,\alpha_k)\} \in \mathcal{U}$.

DEFINITION 15. Let n and k be such that $n \geq 1$ and $k \geq 2$ M is said to be (n,k)-extendible iff there exist $M_1, M_2, \ldots, M_k$ such that the following conditions hold:

i) $M \prec^e_{n+1} M_i$ for every $i \in \{1,\ldots,k\}$

ii) $M_i \prec_n M_{i+1}$ for every $i \in \{1,\ldots,k-1\}$

iii) $M_i \models D^1_n$-foundation for every $i \in \{1,\ldots,k\}$

where $D^1_n = \{\phi : \phi$ has the form $\exists x \in a \theta$ with θ Δ_n-formula

iv) There exists $\alpha_i \in On^{M_i}$ such that $On^M < \alpha_i < On^{M_{i-1}} \setminus On$

for every $i \in \{2,\ldots,k\}$.

The proof of the following theorem is given in [P].

THEOREM 16. If $M \models Ad + v = L$ and $M \not\models$ foundation schema then, for every $n \geq 1$ and $k \geq 2$, M is (n,k)-extendible iff there exists a (n,k)-pseudonormal ultrafilter.

We end with a proposition relating extendibility with collection. The proof is in [P].

PROPOSITION 17. If $M \models Ad + v = L$, $M \not\models$ foundation schema and M is (n,k)-extendible then $M \models \Sigma_{n+k}$-collection.

REFERENCES

[B] Barwise, J. Admissible sets and structures. Springer-Verlag, Berlin (1975).

[Ka] Kaufmann, M. On existence of Σ_n end extensions, in Logic Year 1979-1980. The University of Connecticut. Springer-Verlag, LNM, vol. 859

[Kr] Kranakis, E. Definable ultrafilters and end extensions of constructible sets. Z.Math.Logik Grundlagen Math. 28(1982) p. 4.

[P] Pino, R. Π_n-collection, indicatrices et ultrafiltres définissables. Thèse de 3ème cycle. Université Paris VII. (1983).

ON THE AXIOMATIZATION OF PRC-FIELDS

Alexander Prestel
Fakultät für Mathematik
Universität Konstanz
7750 Konstanz, West-Germany

1. Introduction

In [5] we introduced and investigated the notion of a PRC-field. A field K is called <u>pseudo real closed</u> (PRC) if every absolutely irreducible (affine) variety V, defined over K, which admits a simple point in each real closure $\overline{(K,P)}$ of K, has a K-rational point. In case the space X_K of orderings of K is empty, this notion coincides with that of a pseudo algebraically closed (PAC) field. In case X_K is finite, it was shown in [5] that if suffices to consider only plane curves in the definition. Meanwhile, Ershov proved this for arbitrary X_K (cf.[2]). The aim of this paper is to give an alternative proof which closely follows that of the finite case in [5].

Let us point out that, following the notation of [3], a PRC-field K should be called <u>regularly closed</u> with respect to X_K. For convenience we assume for the case $X_K = \emptyset$ that K should have <u>characteristic zero</u>.

2. The curve condition

Throughout this note we use the basic notations and results of [5]. We say that a field K satisfies the <u>curve condition</u> if in K the following holds:

(CC) <u>For every absolutely irreducible</u> $f(X,Y) \in K[X,Y]$ <u>which has a simple zero in each real closure</u> $(\overline{K,P})$ <u>of</u> K <u>there are</u> $x,y \in K$ <u>such that</u> $f(x,y) = 0$.

Clearly, every PRC-field satisfies (CC). The following theorem states the converse.

THEOREM <u>Every field</u> K <u>satisfying</u> (CC) <u>is pseudo real closed</u>.

Before we can prove this theorem in Section 3, we have to draw some <u>consequences</u> from (CC):

(1) <u>Every sum of squares</u> a <u>equals a sum of two squares in</u> K .

This follows immediately from (CC) using the absolutely irreducible polynomial (cf.[5], Prop. 1.5)

$$X^2 + Y^2 - a .$$

The next consequence deals with the space X_K of orderings of K . Thus let us recall some basic facts about X_K . A subset $S \subset K$ is called a <u>preordering</u> of K if

$$S + S \subset S, \ S \cdot S \subset S, \ K^2 \subset S, \ -1 \notin S .$$

(Note that this includes $0 \in S$, contrary to the convention in [5].) S is called an <u>ordering</u> of K if, in addition, $S \cup -S = K$. Orderings are exactly the maximal preorderings of K . If we set

$$a \leq_S b \ \text{ if and only if } \ b - a \in S ,$$

this defines a partial ordering on K which is linear in case S is an ordering. Let us always denote orderings by P . As it is well-known from the work of Artin-Schreier,

$$S = \bigcap \{P \mid S \subset P\} ,$$

and K admits a (pre-)ordering if and only if K is formally real which means that -1 is not a sum of squares in K . Now we consider the set

$$X_K = \{P \mid P \text{ an ordering of } K\}$$

which is called the <u>order space</u> of K . It is non-empty exactly if K is formally real. The sets

$$H(a) = \{P \in X_K \mid a \in P\}$$

with $a \in K^\times$ form a subbase of a totally disconnected and compact topology on X_K , i.e. with respect to this topology X_K is a boolean space (cf.[4]). Thus the clopen (=closed and open) sets of X_K form a boolean algebra with inclusion as partial ordering. Clearly, $H(a)$ is a clopen set. Indeed, $X_K \setminus H(a) = H(-a)$. In general, the sets $H(a)$ do not form a boolean algebra with respect to the inclusion. However, they form a boolean algebra if and only if the following consequence of (CC) holds:

(2) <u>For all</u> $a,b \in K^\times$ <u>there exists</u> $c \in K^\times$ <u>such that</u> $H(a) \cap H(b) = H(c)$.

This follows easily applying (CC) to the absolutely irreducible polynomial (see [5],Prop.1.3)

$$abX^2Y^2 + aX^2 + bY^2 - 1 .$$

Therefore, in a field K satisfying (CC) the clopen subsets of X_K are exactly the sets $H(a)$. In [5], Prop.1.4 , we used (2) to show that a field K with (CC) also satisfies

(3) K <u>is dense in every real closure</u> $(\overline{K,P})$.

From (3) it follows easily that the orderings $P \in X_K$ induce different topologies on K . Thus the Approximation Theorem (see [6], Theorem 4.1) holds for any finite number of orderings $P_1,\dots,P_m$ of K.

But even more can be proved:

LEMMA (Block Approximation)

<u>Assume that</u> K <u>satisfies</u> (CC) <u>and</u> $X = X_1 \dot{\cup} \dots \dot{\cup} X_m$ <u>is a partition of</u> X <u>into clopen subsets</u> X_i $(1 \le i \le m)$. <u>To each pair of sequences</u> $x_1,\dots,x_m \in K$, $\varepsilon_1,\dots,\varepsilon_m \in K^\times$ <u>there exists</u> $x \in K$ <u>such that for all</u> $1 \le i \le m$:

$$P \in X_i \Rightarrow |x-x_i|_P \le \varepsilon_i^2 .$$

The last expression means $\varepsilon_i^2+(x-x_i) \in P$ and $\varepsilon_i^2-(x-x_i) \in P$. Hence in the i-th 'block', x approximates x_i simultaneously for all orderings $P \in X_i$. In case X_K is finite this is just the above mentioned Approximation Theorem. The Block Approximation was introduced by Ershov [1] as one of the axioms for a certain class of fields which turned out to be exactly the 'maximal' PRC-fields.

<u>Proof</u>: By (2), we may assume $X_i = H(a_i)$ for some $a_i \in K^\times$. Replacing a_i by $a_i(1+a_i^2)^{-1}$ if necessary, we may assume that $|a_i|_P \le 1$ for all $P \in X_i$. Now consider the polynomial

$$f(X,Y) = Y^2 + \prod_{i=1}^{m} ((X-x_i)^2 - a_i\varepsilon_i^4).$$

Since $a_i \neq a_j \mod K^2$ for $i \neq j$, the polynomial

$h(X) = \prod_{i=1}^{m} ((X-x_i)^2 - a_i\varepsilon_i^4)$ is separable. Hence $f(X,Y)$ is absolutely irreducible. Clearly, $(x_j, \sqrt{-h(x_j)})$ is a simple zero of f in a real closure $(\overline{K,P})$ with $P \in X_j$. Indeed, $a_j \in P$ and $-a_i \in P$ for all $i \neq j$ implies $-h(x_j) \in P$. Now by (CC) there are $x,y \in K$ such that

$$y^2 + \prod_{i=1}^{m} ((x-x_i)^2 - a_i\varepsilon_i^4) = 0 .$$

Considering again $P \in X_j$ we see (as above) that

$$a_j\varepsilon_j^4 - (x-x_j)^2 \in P .$$

Thus we obtain $(x-x_j)^2 \leq_P a_j\varepsilon_j^4 \leq_P \varepsilon_j^4$ and hence $|x-x_j|_P \leq \varepsilon_j^2$.

q.e.d.

Let us make one more observation before we come to the proof of the theorem: From (CC) we easily obtain the following 'curve condition' which states the existence of infinitely many K-rational points on certain curves.

(C) <u>For every absolutely irreducible</u> $f(X,Y) \in K[X,Y]$, <u>monic in</u> Y , <u>such that there are</u> a,b <u>in each real closure</u> $(\overline{K,P})$ <u>satisfying</u> $f(a,b) = 0$ <u>and</u> $\frac{\partial f}{\partial Y}(a,b) \neq 0$, <u>and every</u> $h(X) \in K[X] \setminus \{0\}$ <u>there are</u> $x,y \in K$ <u>such that</u> $f(x,y) = 0$ <u>and</u> $h(x) \neq 0$.

Indeed, let $\alpha_1,\ldots,\alpha_n \in \tilde{K}$ be all zeros of h . Choose some $r \in \mathbb{N}$ such that $f(\alpha_i,r) \neq 0$ for all $1 \leq i \leq n$. Now let $Y_1 = (Y-r)h(X)^{-1}$ and $f_1(X,Y_1) \in K[X,Y_1]$ be such that

$$f_1(X,Y_1) = f(X,Y_1 \cdot h(X) + r) .$$

Clearly, f and f_1 have the same function field. Thus f_1 is absolutely irreducible and has a simple zero in each real closure $(\overline{K,P})$ of K . By (CC), f_1 has a zero, say $x,y_1 \in K$. Now x, $y = y_1h(x) + r$ is a zero of f in K . From the choice of r it follows that $h(x) \neq 0$.

3. Proof of the theorem

Let K satisfy the curve condition (CC). Assuming that V is an absolutely irreducible K-variety which has a simple point in each real closure $(\overline{K,P})$ of K , we have to show that V admits a K-rational point. Equivalently, we may show that the function field $L = K(V)$ embeds into K^* over K , where

(4) K^* <u>is a $|K|^+$-saturated elementary extension of</u> K .

From the assumption on V we know that (see e.g.[5], Preliminaries)

(5) L/K <u>is finitely generated, regular, and totally real.</u>

An extension L/K is called <u>totally real</u>, if each ordering of K extends to some ordering of L . Since (CC) can be expressed in the first order language of fields (see Theorem 4.1 and its proof in [5]; note also that (1) is used here), we also know that K^* satisfies (CC). Thus, in particular we have

(6) K^* <u>is dense in all its real closures, and satisfies</u> (C) <u>and the Block Approximation.</u>

From (4), (5), (6) we now deduce the embeddability of L into K^* over K .

Since L/K is regular and finitely generated, we conclude from Theorem 3D in [7],Ch.V, that there are generators $x_1,\dots,x_n,x,y$ of L over K such that

(i) $x_1,\dots,x_n,x$ are algebraically independent over K

(ii) y is algebraic over $K(x_1,\dots,x_n,x)$ and there is $f \in K(x_1,\dots,x_n)[X,Y]$, monic in Y , such that $f(x,y) = 0$ and f is irreducible over the algebraic closure of $K(x_1,\dots,x_n)$.

As we will prove below, there exist elements $x_1^*,\dots,x_n^*$ in K^* , algebraically independent over K , satisfying the following formula

$$\varphi(v_1,\dots,v_n) :\equiv \exists v,w(f(\bar{v},v,w) = 0 \wedge \frac{\partial f}{\partial Y}(\bar{v},v,w) \neq 0)$$

in every real closure $(\overline{K^*,P})$ of K . Then clearly $f(\bar{x}^*,X,Y)$ defines an absolutely irreducible curve over K^* which has a suitable simple point in each $(\overline{K^*,P})$. Thus by (6)(i.e. by (C) for K^*) there are $x',y' \in K^*$ satisfying $f(\bar{x}^*,x',y') = 0$ and $h(x') \neq 0$, where

$h \in K[\bar{x}^*,X] \setminus \{0\}$ is given. Since $K[\bar{x}^*,X]$ has cardinality $|K|$ and K^* is $|K|^+$-saturated, we even find $x^*,y^* \in K^*$ satisfying $f(\bar{x}^*,x^*,y^*) = 0$ and $h(x^*) \neq 0$ for all $h \in K[\bar{x}^*,X] \setminus \{0\}$. Thus $x_1^*,\ldots,x_n^*,x^*$ are algebraically independent over K and, by sending $x_1,\ldots,x_n,x,y$ onto $x_1^*,\ldots,x_n^*,x^*,y^*$ resp., the field L gets embedded into K^*, as contented. It remains to find the desired elements $x_1^*,\ldots,x_n^*$ in K^*.

Using Elimination of Quantifiers, over every real closed field R extending K, the formula $\varphi(\bar{v})$ is equivalent to some finite disjunction

$$\bigvee_i (p_i(\bar{v}) = 0 \wedge \bigwedge_j q_{ij}(\bar{v}) > 0)$$

where the conjunction $\bigwedge_j$ is also finite and the polynomials p_i and q_{ij} have coefficients in K.

Let P be any ordering of K^*, i.e. $P \in X^* = X_{K^*}$. Let Q be an ordering of L extending $P \cap K$ (such an ordering exists by (5)). Then, depending on Q, there exists some i such that

$$(L,Q) \models (p_i(\bar{x}) = 0 \wedge \bigwedge_j q_{ij}(\bar{x}) > 0).$$

Since $x_1,\ldots,x_n$ are algebraically independent over K, $p_i(\bar{x}) = 0$ must be trivially true, i.e. the coefficients are all equal to zero. Thus we may forget about p_i for a while. From

$$(\overline{L,Q}) \models \exists \bar{v} \quad \bigwedge_j q_{ij}(\bar{v}) > 0$$

we conclude

$$(\overline{K^*,P}) \models \exists \bar{v} \quad \bigwedge_j q_{ij}(\bar{v}) > 0 .$$

Since K^* is dense in $(\overline{K^*,P})$ by (6), we find elements $\bar{a}_P$ of K^* such that

$$(\overline{K^*,P}) \models \bigwedge_j q_{ij}(\bar{a}_P) > 0 .$$

As $p_i(\bar{a}_P) = 0$, this implies $(\overline{K^*,P}) \models \varphi(\bar{a}_P)$.

By continuity we can find some $\varepsilon_P \in K^*$, $\varepsilon_P \neq 0$, satisfying

$$(\overline{K^*,P}) \models \forall \bar{v}(|\bar{v} - \bar{a}_P| \leq \varepsilon_P^2 \rightarrow \bigwedge_j q_{ij}(\bar{v}) > 0).$$

Expressing this formula in quantifier-free form we find polynomials $s, r_\nu \in K[X_1,\ldots,X_n,Z]$ such that

$$(K^*,P) \models (s(\bar{a}_P,\varepsilon_P) = 0 \wedge \bigwedge_\nu r_\nu(\bar{a}_P,\varepsilon_P) > 0)$$

and, whenever (K^*,P') also satisfies this (finite) conjunction, then

$$(7) \qquad (\overline{K^*,P'}) \models \forall \bar{v}(|\bar{v} - \bar{a}_P| \leq \varepsilon_P^2 \rightarrow \bigwedge_j q_{ij}(\bar{v}) > 0) .$$

The set of $P' \in X^*$ satisfying (7) thus contains the clopen set

$$Y_P = \bigcap_\nu H(r_\nu(\bar{a}_P,\varepsilon_P)) .$$

Clearly, the sets Y_P cover X^* . Since X^* is compact, a finite number of clopen sets Y_P already covers X^*. Let $X^* = X_1 \dot{\cup} \ldots \dot{\cup} X_l$ be a partition of X^* such that each X_μ is contained in some Y_P . For each X_μ we thus can find some elements $\bar{a}_\mu$ and $\varepsilon_\mu \neq 0$ in K^* and some i_μ such that (7) holds for all $P' \in X_\mu$ with $\bar{a}_P, \varepsilon_P, i$ replaced by $\bar{a}_\mu, \varepsilon_\mu, i_\mu$.

Now we apply the Block Approximation of X^* to this situation in order to find elements $\bar{a}$ of K such that

$$|\bar{a} - \bar{a}_\mu|_{P'} \leq \varepsilon_\mu^2 \quad \text{for all } P' \in X_\mu .$$

Since (7) holds for all $P' \in X_\mu$, we obtain $(K^*,P') \models \bigwedge_j q_{i_\mu j}(\bar{a})$ for all $P' \in X_\mu$. Thus finally we get

$$(K^*,P) \models \bigvee_i \bigwedge_j q_{ij}(\bar{a}) > 0$$

for all $P \in X^*$. Since for every i which was actually used in the course of the proof, p_i was the zero polynomial, we even get

$$(K^*,P) \models \bigvee_i (p_i(\bar{a}) = 0 \wedge \bigwedge_j q_{ij}(\bar{a}) > 0).$$

Moreover, given a non-zero polynomial $g \in K[X_1,\ldots,X_n]$ we may assume that g^2 is among the q_{ij} for each i. Indeed, since $x_1,\ldots,x_n$ are algebraically independent over K, clearly $(L,Q) \models g^2(\bar{x}) > 0$ holds for all orderings Q of L. By this assumption, we also find $g(\bar{a}) \neq 0$. Here the choice of the elements $\bar{a}$ dependson the choice of $g \in K[\bar{X}]$. Using once more the $|K|^+$-saturatedness of K^*, we can even find elements $\bar{x}^*$ in K^* satisfying

$$(K^*,P) \models \bigvee_i (p_i(\bar{x}^*) = 0 \wedge \bigwedge_j q_{ij}(\bar{x}^*) > 0)$$

and $g(\bar{x}^*) \neq 0$ for all non-zero $g \in K[\bar{X}]$. Thus the elements $\bar{x}^*$ are algebraically independent over K and satisfy $(\overline{K^*,P}) \models \varphi(\bar{x}^*)$ for all $P \in X^*$. This finishes the proof of the theorem.

References

[1] ERSHOV,Yu.L.: Totally real field extensions. Soviet Math.Dokl. 25,No.2, 477-480 (1982)

[2] ERSHOV,Yu.L.: Two theorems on regularly r-closed fields. J. reine angew. Math. (to appear)

[3] HEINEMANN,B., PRESTEL,A.: Fields regularly closed with respect to finitely many valuations and orderings (to appear)

[4] PRESTEL,A.: Lectures on formally real fields. Monografias de Matemática, Vol.22.IMPA, Rio de Janeiro 1975

[5] PRESTEL,A.: Pseudo real closed fields. In: Set theory and model theory. Lecture Notes in Math.Vol.872, Berlin-Heidelberg-New York: Springer 1981

[6] PRESTEL,A., ZIEGLER,M.: Model theoretic methods in the theory of topological fields. J.reine angew.Math.299/300,318-341 (1978)

[7] SCHMIDT,W.M.: Equations over finite fields. An elementary approach. Lecture Notes in Math., Vol. 536, Berlin-Heidelberg-New York: Springer 1976

FORMALIZATIONS OF CERTAIN INTERMEDIATE LOGICS
Part I

Cecylia Rauszer
Uniwersytet Warszawaski - Instytyt Matematyki
Palac Kultury i Nauki
00-901 Warszawa, Polska

It was Godel who first observed the existence of a continuum of logics between the intuitionistic predicate logic LI and the classical predicate logic LK. These logics were named by Umezawa "intermediate logics".

There are many interesting results connected with intermediate logics. One of them asserts that only seven propositional intermediate logics have the interpolation property. This result has been proved by Maksimova [7]. She showed that the interpolation property is equivalent to the amalgamation property and then she proved the existence of seven classes of Heyting variaties with the amalgamation property.

For predicate intermediate logics the problem of the interpolation property was first examined by Gabbay [2]. He used the theorem that Craig's interpolation lemma is equivalent to a weaker version of Robinson's consistency theorem to prove Craig's interpolation lemma for the intermediate logics: LI, LM, LMH, where

(LM) $LI + (\neg\alpha \cup \neg\neg\alpha)$
(LMH) $LI + \forall x \neg\neg\alpha(x) \equiv \neg\neg\forall x\alpha(x)$.

Eight years later in [3] Gabbay extended his model theoretic methods to the so-called logic of constant domains LD, i.e. for the intermediate logic

(LD) $LI + \forall x(\alpha(x) \cup \beta) \Rightarrow (\forall x\alpha(x) \cup \beta)$,

where x does not appear as a free object variable in β and he also proved Craig's interpolation lemma for LD.

In 1981 López Escobar [4] proved Craig's interpolation lemma for LD using proof-theoretic methods. The same formalization for LD was earlier used by Umezawa [10] to show some syntactical properties of LD.

It turned out that both papers [3] and [4] contain gaps. A certain Kripke model for an LD theory constructed in [3] has no constant domains and the Gentzen type formalization of [4] is not complete.

It should be emphasized that both authors know about the errors in their papers. López-Escobar wrote two papers [5], [6] partially connected with the Gentzen formalization of LD and the interpolation property for LD.

Let me cite the following two sentences from the paper Stationary Logic by J.Barwise, M. Kaufmann and M. Makkai: "Alan Anderson often argued that every reasonable formal system has both a Hilbert-style notion of proof and a Gentzen-style notion of proof. While this may overstate the case a bit, it is certainly true that a Gentzen-style approach, with its emphasis on rules, rather than on axioms, lays bare the laws of thought inherent in any given logic in a way not done by a Hilbert-style system".

But if we want to examine the problem of the existence of a cut-free Gentzen-style formalization of an intermediate logic L we should know what is to be a cut-free Gentzen-style formalization of L. A possible definition was given by López-Escobar in [5] and he proved that in the sense of his definition there is no cut-free and complete Gentzen-style formalization for LD.

In the present paper the basic concept of a formal system S is different from the ones that were considered by López-Escobar in [5] and in [6].

Every formal system S consists of two parts : the first part called basic system (BS) is roughly speaking, a modification of the system for LI given by Schütte [8]. Intuitionistic logic LI is complete with respect to BS i.e. $\alpha \in LI$ iff α has a normal derivation in BS, i.e. there is a cut-free proof for α. The second part of the defined system S consists of a certain set of logical rules which are specific for the intermediate logic considered.

For the intermediate logics LM and LMH the appropriate formal system LS is complete, and cut-free.

As an application of LS we show, among others, the Craig interpolation lemma for L, where L = LM, LMH.

The cut-free calculus BS introduced in this paper can be extended to a formal system for Dummett's logic LC. This is done in a separate paper.

§1. BASIC SYSTEM.

We define a formal language with connectives $\bot$ (falsum), $\cup$ (disjunction), $\cap$ (conjuction), $\Rightarrow$(implication), $\exists$ (existensial quantifier), $\forall$ (general quantifier).

Assume we are given countably infinite sets of free and bound object variables, sentential variables and predicate symbols of each number of arguments ≥ 1.

By an atomic formula we mean every sentential variable, falsum and if p is an n-ary predicate symbol $(n \geq 1)$ and $a_1,\ldots,a_n$ are free object variables then $p(a_1,\ldots,a_n)$ is an atomic formula.

The set FORM of formulas is defined as usual. Denote by $\Gamma,\Delta,\ldots$ (with indeces, if necessary) finite (possibly empty) sequences of formulas. If $\Gamma = \alpha_1,\ldots,\alpha_n$, $\Delta = \beta_1,\ldots,\beta_m$, $\alpha,\beta \in$ FORM, then for brevity we write

1) $\Gamma \Rightarrow (\beta \cup (\Delta \Rightarrow \alpha))$ and $\Gamma \Rightarrow \alpha$

for

$$\alpha_1 \Rightarrow (\alpha_2 \Rightarrow (\ldots \Rightarrow (\alpha_n \Rightarrow (\beta \cup (\beta_1 \Rightarrow (\beta_2 \Rightarrow (\ldots = (\beta_n \Rightarrow \alpha)\ldots)))))\ldots)))$$

and

$$\alpha_1 \Rightarrow (\alpha_2 \Rightarrow (\ldots(\alpha_n \Rightarrow \alpha)\ldots))), \quad \text{respectively.}$$

If Γ is empty, then $\Gamma \Rightarrow (\beta \cup (\Delta \Rightarrow \alpha))$ is the formula $\beta \cup (\Delta \Rightarrow \alpha)$, if Δ empty, then $\Gamma \Rightarrow (\beta \cup (\Delta \Rightarrow \alpha))$ is the formula $\Gamma \Rightarrow (\beta \cup \alpha)$ and if Γ and Δ are empty, then $\Gamma \Rightarrow (\beta \cup (\Delta \Rightarrow \alpha))$ is the formula of the form $(\beta \cup \alpha)$.

Denote by R^β the set of inference rules listed below:

$$(str_1^\beta) \quad \frac{\Gamma \Rightarrow (\beta \cup (\alpha \Rightarrow (\alpha \Rightarrow \gamma)))}{\Gamma \Rightarrow (\beta \cup (\alpha \Rightarrow \gamma))}$$

$$(str_2^\beta) \quad \frac{\Gamma \Rightarrow (\beta \cup \gamma)}{\Gamma \Rightarrow (\beta \cup (\alpha \Rightarrow \gamma))}$$

$$(str_3^\beta) \quad \frac{\Gamma \Rightarrow (\beta \cup (\Delta \Rightarrow (\alpha \Rightarrow (\delta \Rightarrow \gamma))))}{\Gamma \Rightarrow (\beta \cup (\Delta \Rightarrow (\delta \Rightarrow (\alpha \Rightarrow \gamma))))}$$

$$(\cup r^\beta) \quad \frac{\Gamma \Rightarrow (\beta \cup (\Delta \Rightarrow \gamma_i))}{\Gamma \Rightarrow (\beta \cup (\Delta \Rightarrow (\gamma_1 \cup \gamma_2)))} \qquad i = 1,2$$

$$(\cap l^\beta) \quad \frac{\Gamma \Rightarrow (\beta \cup (\alpha_i \Rightarrow \gamma))}{\Gamma \Rightarrow (\beta \cup ((\alpha_1 \cap \alpha_2) \Rightarrow \gamma))} \qquad i = 1,2$$

$$(\cup l^\beta) \quad \frac{\Gamma \Rightarrow (\beta \cup (\alpha_1 \Rightarrow \gamma)) \quad \Gamma \Rightarrow (\beta \cup (\alpha_2 \Rightarrow \gamma))}{\Gamma \Rightarrow (\beta \cup ((\alpha_1 \cup \alpha_2) \Rightarrow \gamma))}$$

$$(\cap r^\beta) \quad \frac{\Gamma \Rightarrow (\beta \cup (\Delta \Rightarrow \gamma_1)) \quad \Gamma \Rightarrow (\beta \cup (\Delta \Rightarrow \gamma_2))}{\Gamma \Rightarrow (\beta \cup (\Delta \Rightarrow (\gamma_1 \cap \gamma_2)))}$$

$(\Rightarrow 1^{\beta})$ $$\frac{\Gamma \Rightarrow (\beta \cup (\Delta \Rightarrow \alpha)) \qquad \beta \cup (\eta \Rightarrow \gamma)}{\Gamma \Rightarrow (\beta \cup (\Delta \Rightarrow ((\alpha \Rightarrow \eta) \Rightarrow \gamma)))}$$

$(\bot r^{\beta})$ $$\frac{\Gamma \Rightarrow (\beta \cup (\Delta \Rightarrow \bot))}{\Gamma \Rightarrow (\beta \cup (\Delta \Rightarrow \gamma))}$$

$(\exists r^{\beta})$ $$\frac{\Gamma \Rightarrow (\beta \cup (\Delta \Rightarrow \alpha(a)))}{\Gamma \Rightarrow (\beta \cup (\Delta \Rightarrow \exists x \alpha(x)))}$$

$(\forall 1^{\beta})$ $$\frac{\Gamma \Rightarrow (\beta \cup (\alpha(a) \Rightarrow \gamma))}{\Gamma \Rightarrow (\beta \cup (\forall x \alpha(x) \Rightarrow \gamma))}$$

where $\alpha, \beta, \gamma, \gamma_i \in$ FORM and Γ, Δ are any sequence of formulas.

Let $r^{\beta} \in R^{\beta}$, then r is rule of inference such that the formula β does not appear, i.e. let us take for example as r^{β} the rule (str_2^{β}) then (str_2) is the following rule of inference:

(str_2) $$\frac{\Gamma \Rightarrow \gamma}{\Gamma \Rightarrow (\alpha \Rightarrow \gamma)}$$

Denote by R the set of all rules r such that $r^{\beta} \in R^{\beta}$, i.e.

$$R = \{r : r^{\beta} \in R^{\beta}\}.$$

By the basic system (BS) we mean the set of all rules of inference from $R \cup R^{\beta}$ and the following formulas and rules:

(ax_1) $(\bot \Rightarrow \alpha)$ (ax_2) $(\alpha \Rightarrow \alpha)$

(cut) $$\frac{\Gamma \Rightarrow \delta \quad \delta \Rightarrow \gamma}{\Gamma \Rightarrow \gamma}$$

$(\exists 1)$ $$\frac{\alpha(a) \Rightarrow \gamma}{\exists x \alpha(x)}$$ $(\forall r)$ $$\frac{\Gamma \Rightarrow \alpha(a)}{\Gamma \Rightarrow \forall x \alpha(x)}$$

where $\alpha, \delta, \gamma, \alpha(a) \in$ FORM and a does not occur as a free object variable in any formula in the conclusion of the rules $(\exists 1)$ and $(\forall r)$.

The formula δ in the cut rule will be called the cut formula and denoted cfl.

Let us mention that the rules $(\bot r^{\beta})$ and $(\bot r)$ are added only to simplify some rules that will be introduced later.

It is not difficult to verify that:

1.1 (ax_1) and (ax_2) can be restricted to $(\bot \Rightarrow a)$ and $(a \Rightarrow a)$, where a is an atomic formula. □

1.2 For each rule of inference in BS, if all premises are intuitionistically valid, so is the conclusion. □

A formula α is said to be provable in the BS (henceforth abbreviated as "$BS \vdash \alpha$") iff α is obtained from the axioms by means of the rules of inference.

Notice that the subsystem of BS consisting of (ax_1), (ax_2), the set R, the cut rule, $(\exists\, l)$ and $(\forall r)$ is the formal system for intuitionistic predicate calculus described by Schutte in [8] and [9]. Let us denote this system by IS. Then in the standard way we obtain

1.3 For any formula α

$$BS \vdash \alpha \quad \text{iff} \quad IS \vdash \alpha. \qquad \square$$

1.4 (Cut-elimination for BS). For any formula α, $BS \vdash \alpha$ iff α has a cut-free derivation in BS.

PROOF. It follows immediately from the Haptsatz theorem for IS [8] and 1.3. □

A derivation in which the cut rule does not appear is called normal.

COROLLARY. For any formula α , the following conditions are equivalent

1. $\alpha \in LI$
2. α has a normal derivation in IS
3. $BS \vdash \alpha$
4. α has a normal derivation in BS. □

We say that a formal system S is separable or the separation theorem holds for S iff the provable formulas of S have derivations using only the rules of inference containing those logical connectives appearing in the formula.

It is not difficult to show that:

1.5 The formal system BS without the cut rule is separable. □

1.6 The separation theorem holds for the system IS without the cut rule. □

1.7 In a normal derivation of α (in BS) only subformulas of α occur. □

Let $\xi \in FORM$. Recall that we can consider ξ as a formula of the form $\Gamma \Rightarrow \gamma$, where $\Gamma = \alpha_1,\dots,\alpha_n$ and $\alpha_i, \gamma \in FORM$. In the sequel $\alpha_1,\dots,\alpha_n$ will be called the antecedents of ξ and γ the succedent of ξ. If $\gamma = \gamma_1 \Rightarrow \gamma_2$ and γ_2 is not of the form $\gamma_2' \Rightarrow \gamma_2''$, then γ_1 is also called antecedent of ξ and γ_2 the end part of ξ.

Let Σ be the sequence of antecedents of ξ. A pair $[\Phi,\xi^*]$ is said to be a <u>partition</u> of ξ if Φ is a subsequence of Σ and ξ^* is obtained from ξ by omitting all antecedents occurring in Φ.

Let $\langle\alpha\rangle$ be the set of all free object variables, sentential variables and predicate letters which occur in $\alpha \in FORM$. In the same way we define $\langle\Gamma\rangle$, where Γ is a sequence of formulas.

Craig's Interpolation Lemma. Let L be an intermediate logic and let $\xi \in L$. Let $[\Phi,\xi^*]$ be any partition of ξ.

A formula γ such that

1. $\langle\gamma\rangle \subset \langle\Phi\rangle \cap \langle\xi^*\rangle$
2. $\Phi \Rightarrow \gamma \in L$ and $\gamma \Rightarrow \xi^* \in L$,

is called an interpolant for $[\Phi,\xi^*]$. Sometimes we say that $[\Phi,\xi^*]$ has an interpolant if there exists a γ such that 1 and 2 hold. We say that a formula ξ has an interpolant if every partition of ξ has an interpolant.

It is well known that

1.8 (Craig's interpolation lemma for LI). If $\xi \in LI$, then ξ has an interpolant. □

1.9 (Craig's interpolation lemma for BS). Let ξ have a normal derivation in BS. Then ξ has an interpolant.

<u>PROOF.</u> Let us assume that $BS \vdash \xi$. By the corollary $\xi \in LI$. According to the 1.8 ξ has an interpolant. □

§2. <u>LOGIC LM.</u>

The intermediate logic LM is obtained by adding to LI the following schema

(M) $$(\neg\alpha \cup \neg\neg\alpha),$$

where $\neg\alpha$ is $\alpha \Rightarrow \bot$.

It is well known that the logic LM is characterized by Kripke structures with the directed set of "worlds", i.e. if $\langle T,\leq\rangle$ is a Kripke structure for LM, then

$$\bigwedge t,s \bigvee u(t \leq u \ \ \& \ \ s \leq u).$$

The next two lemmas show some syntactical properties of LM.

2.1* The following formulas are equivalent to (M)

(1) $(((\neg\alpha \Leftrightarrow \beta) \cap (\alpha \Leftrightarrow \neg \beta)) \Rightarrow (\alpha \cup \beta))$,

(2) $((\neg \alpha \Leftrightarrow \neg \beta) \cup (\neg \alpha \Leftrightarrow \neg \neg \beta))$,

(3) $(\neg (\alpha \cap \beta) \Rightarrow (\neg \alpha \cup \neg \beta))$,

(4) $((\neg \neg \alpha \Rightarrow \alpha) \Rightarrow (\alpha \cup \neg \alpha))$,

(5) $((\neg \alpha \Rightarrow (\neg \beta \cup \neg \gamma)) \Rightarrow ((\neg \alpha \Rightarrow \neg \beta) \cup (\neg \alpha \Rightarrow \neg \gamma)))$,

(6) $((\neg \alpha \Rightarrow \beta) \cup (\beta \Rightarrow \neg \alpha))$,

(7) $((\alpha \Rightarrow \neg \beta) \Rightarrow (\neg \alpha \cup \neg \beta))$,

(8) $(\neg \neg (\neg \alpha \cup \neg \beta) \Rightarrow (\neg \alpha \cup \neg \beta))$,

where α, β, $\gamma \in$ FORM, and as usual $(\xi \Leftrightarrow \eta)$, $\neg \eta$ represent $((\xi \Rightarrow \eta) \cap (\eta \Rightarrow \xi))$, $(\eta \Rightarrow \bot)$, respectively. □

2.2 If $A, B \in \{(\alpha \cap \beta), (\neg \alpha \cup \neg \beta), (\neg \neg \alpha \Rightarrow \neg \beta), (\neg \neg \beta \Rightarrow \neg \alpha), (\beta \Rightarrow \neg \alpha), (\alpha \Rightarrow \neg \beta), \neg \neg (\alpha \Rightarrow \neg \beta), \neg \neg (\beta \Rightarrow \neg \alpha), \neg \neg (\neg \alpha \cup \neg \beta)\}$ where α, $\beta \in$ FORM, then $(A \Leftrightarrow B) \in$ LM. □

The main difficulty in constructing the formal system MS for LM is the choice of suitable rules inference.

It is known that the positive part of LM is equivalent to the positive part of LI. Thus the system MS should contain all inferences LI and some new rules characteristic for nonpositive tautologies of LM.

Let us consider the following set of rules:

$(m_1{}^{\beta})$ $$\frac{\Gamma \Rightarrow (\beta \cup (\Delta \Rightarrow (\alpha \Rightarrow \neg \delta)))}{\Gamma \Rightarrow (\beta \cup (\Delta \Rightarrow (\neg \alpha \cup \neg \delta)))}$$

$(m_2{}^{\beta})$ $$\frac{\Gamma \Rightarrow (\beta \cup (\neg \neg \alpha \Rightarrow \gamma)) \quad \Gamma \Rightarrow (\beta \cup (\delta \Rightarrow \gamma))}{\Gamma \Rightarrow (\beta \cup ((\neg \alpha \Rightarrow \delta) \Rightarrow \gamma))}$$

* Some of the mentioned equivalences were indicated to me by T. Prucnal.

$(m_3{}^{\beta})$ $$\frac{\Gamma \Rightarrow (\beta \cup (\neg\alpha \Rightarrow \gamma)) \quad \Gamma \Rightarrow (\beta \cup (\neg\delta \Rightarrow \gamma))}{\Gamma \Rightarrow (\beta \cup ((\alpha \Rightarrow \neg\delta) \Rightarrow \gamma))}$$

$(m_4{}^{\beta})$ $$\frac{\Gamma \Rightarrow (\beta \cup \Delta \Rightarrow \neg\neg\alpha) \quad \Gamma \Rightarrow (\beta \cup \Delta \Rightarrow \neg\delta)}{\Gamma \Rightarrow (\beta \cup \Delta \Rightarrow \neg(\alpha \Rightarrow \delta))}$$

2.3 If the premises of the rule $(m_i{}^{\beta})$, $i = 1,2,3,4$ belong to LM, so does the conclusion.

PROOF. Notice that $(\alpha \Rightarrow \neg\delta) \Leftrightarrow (\neg\alpha \cup \neg\delta) \in LM$. Thus the lemma follows immediately for $(m_1{}^{\beta})$ and $(m_3{}^{\beta})$. The rule $(m_4{}^{\beta})$ is intuitionistically valid.

PROOF of $(m_2{}^{\beta})$. It is sufficient to observe that if $(\neg\neg\alpha \Rightarrow \gamma) \in LM$ and $(\delta \Rightarrow \gamma) \in LM$, then $((\neg\alpha \Rightarrow \delta) \Rightarrow \gamma) \in LM$. To prove this let $(\neg\neg\alpha \Rightarrow \gamma) \in LM$ and $(\delta \Rightarrow \gamma) \in LM$.

Notice that

$$(((\neg\neg\alpha \Rightarrow \gamma) \cap (\neg\alpha \Rightarrow \delta) \cap (\delta \Rightarrow \gamma)) \Rightarrow ((\neg\neg\alpha \Rightarrow \gamma) \cap (\neg\alpha \Rightarrow \gamma))) \in LI.$$

Hence

$$(((\neg\neg\alpha \cup \neg\alpha) \Rightarrow \gamma) \Rightarrow \gamma) \Rightarrow ((\neg\neg\alpha \Rightarrow \gamma) \cap (\neg\alpha \Rightarrow \delta) \cap (\delta \Rightarrow \gamma)) \in LI$$

which together with our assumptions proves that $((\neg\alpha \Rightarrow \delta) \Rightarrow \gamma) \in LM$. □

We call MS the system based on the propositional part of the basic system BS, and the rules $(m_i{}^{\beta})$, (m_i), $i = 1,2,3,4$ where (m_i) is obtained from $(m_i{}^{\beta})$ in the same way as the rule r from r^{β} (cf. § 1). Now we will treat LM as propositional logic.

THEOREM 2.3. For any formula α, $\alpha \in LM$ iff $MS \vdash \alpha$.

PROOF. The standard proof is omitted. □

THEOREM 2.4. (Cut elimination for MS). Every theorem of LM has a normal proof in MS.

PROOF. To prove the theorem we carry out two complete inductions, one on the grade, the othter on the rank of the derivation.

Recall that the grade of a formula α, $g(\alpha)$, is the number of occurrences of logical connectives in α. The left rank R_1 of the derivation π is defined as the largest length of any thread of formulas in π that ends with the left hand side premise of the rule (cut) and in which the cut formula occurs in the succedents. The right rank R_r is defined analogously. The rank R is $R_1 + R_r$.

According to 1.4 the only new cases arise when one (or both) premise (or premises) of the rule cut is (are) of the conclusion of the rule (m_i^{β}), or (m_i) $i = 1,2,3,4$.

I. Assume that $R = 2$, i.e. $R_l = R_r = 1$, and the theorem is proved for the grade n.

Case of $(m_1\ )$. Suppose that $g(\beta\cup(\Delta\Rightarrow(\neg\alpha\cup\neg\delta))) = n+1$. Consider the following derivation π.

$$\pi\left\{\dfrac{\dfrac{\vdots}{\Gamma\Rightarrow(\beta\cup(\Delta\Rightarrow(\alpha\Rightarrow\neg\delta)))}{}\! \atop (m_1)\,\dfrac{\Gamma\Rightarrow(\beta\cup(\Delta\Rightarrow(\alpha\Rightarrow\neg\delta)))}{\Gamma\Rightarrow(\beta\cup(\Delta\Rightarrow(\neg\alpha\cup\neg\delta)))} \qquad \dfrac{\dfrac{\vdots}{\beta\Rightarrow\gamma} \quad \dfrac{\vdots}{(\Delta\Rightarrow(\neg\alpha\cup\neg\delta))\Rightarrow\gamma}}{(\beta\cup(\Delta\Rightarrow(\neg\alpha\cup\neg\delta)))\Rightarrow\gamma}\,(\cup 1)}{\Gamma\Rightarrow\gamma}\,(\text{cut})\right.$$

Then π is reduced to π_1 as follows

$$\pi_1\left\{\dfrac{\dfrac{\vdots}{\Gamma\Rightarrow(\beta\cup(\Delta\Rightarrow(\alpha\Rightarrow\neg\delta)))} \qquad \dfrac{\dfrac{\vdots}{\beta\Rightarrow\gamma} \quad \dfrac{(m_1)\,\dfrac{(\Delta\Rightarrow(\alpha\Rightarrow\neg\delta))\Rightarrow(\Delta\Rightarrow(\alpha\Rightarrow\neg\delta))}{(\Delta\Rightarrow(\alpha\Rightarrow\neg\delta))\Rightarrow(\Delta\Rightarrow(\neg\alpha\cup\neg\delta))} \quad \dfrac{\vdots}{(\Delta\Rightarrow(\neg\alpha\cup\neg\delta))\Rightarrow\gamma}}{(\Delta\Rightarrow(\alpha\Rightarrow\neg\delta))\Rightarrow\gamma}\,(\text{cut}1)}{(\beta\cup(\Delta\Rightarrow(\alpha\Rightarrow\neg\delta)))\Rightarrow\gamma}\,(\cup 1)}{\Gamma\Rightarrow\gamma}\,(\text{cut }2)\right.$$

Notice that $g(\Delta\Rightarrow(\neg\alpha\cup\neg\delta)) < n$ and $g(\beta\cup(\Delta\Rightarrow(\alpha\Rightarrow\neg\delta))) = n$. Hence by the induction hypothesis (cut_1) and (cut_2) are eliminable from π_1, that proves that the proof π, given above, can be transformed into a proof without the cut rule.

The case when the right premise of the cut rule in π is the result of the rule (str_2) is trivial.

Case of (m_1). Suppose that $g(cf2) = g(\neg\alpha\cup\neg\delta) = n + 1$ and consider the following derivation

$$\dfrac{\dfrac{\dfrac{\vdots}{\Gamma\Rightarrow(\alpha\Rightarrow\neg\delta)}}{\Gamma\Rightarrow(\neg\alpha\cup\neg\delta)} \qquad \dfrac{\dfrac{\vdots}{\neg\alpha\Rightarrow\gamma} \quad \dfrac{\vdots}{\neg\delta\Rightarrow\gamma}}{(\neg\alpha\cup\neg\delta)\Rightarrow\gamma}\,(\cup 1)}{\Gamma\Rightarrow\gamma}$$

It is reduced to

$$\dfrac{\dfrac{\vdots}{\Gamma \Rightarrow (\alpha \Rightarrow \neg\delta)} \qquad \dfrac{\dfrac{\vdots}{\neg\alpha \Rightarrow \gamma} \quad \dfrac{\vdots}{\neg\delta \Rightarrow \gamma}}{(\alpha \Rightarrow \neg\delta) \Rightarrow \gamma}\,(m_3)}{\Gamma \Rightarrow \gamma}\,(\mathrm{cut})$$

By the induction hypothesis on the grade $g(\alpha \Rightarrow \neg\delta) = n$, the cut rule is eliminable from the above proof.

Case of (m_2). Recall that $R = 2$ and theorem holds for the grade n. Notice that according to the assumption on the rank the cut formula can not be η. So $cfl = (\neg\alpha \Rightarrow \delta) = \eta$ and let $g(cfl) = n + 1$. Let the derivation π run as follows

$$\pi \left\{ \dfrac{(m_2)\dfrac{\dfrac{\vdots}{\Gamma \Rightarrow (\neg\neg\alpha \Rightarrow \eta)} \quad \dfrac{\vdots}{\Gamma \Rightarrow (\delta \Rightarrow \eta)}}{\Gamma \Rightarrow (\neg\alpha \Rightarrow \delta) \Rightarrow \eta)} \qquad \dfrac{\vdots}{((\neg\alpha \Rightarrow \delta) \Rightarrow \eta) \Rightarrow \gamma}}{\Gamma \Rightarrow \gamma}\,(\mathrm{cut}) \right.$$

Let π' be a cut free MS-proof of $((\neg\alpha \Rightarrow \delta) \Rightarrow \eta) \Rightarrow \gamma$. The only interesting cases are when the end-formula of π' is the result of the rule $(\Rightarrow 1)$ or of the rule (m_3). In the latter η must be of the form $\neg\eta_1$.

CASE 1. π' runs as follows

$$\pi' \left\{ \dfrac{\dfrac{\vdots}{\neg\alpha \Rightarrow \delta} \quad \dfrac{\vdots}{\eta \Rightarrow \gamma}}{((\neg\alpha \Rightarrow \delta) \Rightarrow \eta) \Rightarrow \gamma}\,(\Rightarrow 1) \right.$$

Then π is transofrmed into

$$\dfrac{\dfrac{\dfrac{\vdots}{\neg\alpha \Rightarrow \delta} \qquad \dfrac{\dfrac{\vdots}{\Gamma \Rightarrow ((\neg\alpha \Rightarrow \delta) \Rightarrow \eta)}}{(\neg\alpha \Rightarrow \delta) \Rightarrow (\Gamma \Rightarrow \eta)}}{\Gamma \Rightarrow \eta}\,(\mathrm{cut}_1) \qquad \dfrac{\vdots}{\eta \Rightarrow \gamma}}{\Gamma \Rightarrow \gamma}\,(\mathrm{cut}_2)$$

(cut_1) and (cut_2) are eliminable by the induction hypothesis on the grade.

CASE 2. π' runs as follows: $(\eta = \neg\eta_1)$

$$\dfrac{\dfrac{\vdots}{\neg(\neg\alpha\Rightarrow\delta)\Rightarrow\gamma} \qquad \dfrac{\vdots}{\neg\eta_1\Rightarrow\gamma}}{((\neg\alpha\Rightarrow\delta)\Rightarrow\neg\eta_1)\Rightarrow\gamma}\,(m_3)$$

Using the fact that $SI \vdash (\neg\neg\neg\alpha\Rightarrow(\neg\delta\Rightarrow\neg(\neg\alpha\Rightarrow\delta)))$ the proof π' is reduced to

$$\dfrac{\dfrac{\vdots}{\Gamma\Rightarrow(\delta\Rightarrow\neg\eta_1)} \qquad \dfrac{\dfrac{\Gamma\Rightarrow(\neg\neg\alpha\Rightarrow\neg\eta_1) \qquad \dfrac{(\mathrm{cut}_1)\,\dfrac{\dfrac{\text{SI-proof}}{\neg\neg\neg\alpha\Rightarrow(\neg\delta\Rightarrow\neg(\neg\alpha\Rightarrow\delta))} \qquad \dfrac{}{\neg(\neg\alpha\Rightarrow\delta)\Rightarrow\gamma}}{\neg\neg\neg\alpha\Rightarrow(\neg\delta\Rightarrow\gamma)} \qquad \dfrac{\dfrac{\vdots}{\neg\eta_1\Rightarrow\gamma}}{\neg\eta_1\Rightarrow(\neg\delta\Rightarrow\gamma)}}{(\neg\neg\alpha\Rightarrow\neg\eta_1)\Rightarrow(\neg\delta\Rightarrow\gamma)}\,(m_2)}{\Gamma\Rightarrow(\neg\delta\Rightarrow\gamma)}\,(\mathrm{cut}_2) \qquad \dfrac{\dfrac{\vdots}{\neg\eta_1\Rightarrow\gamma}}{\Gamma\Rightarrow(\neg\eta_1\Rightarrow\gamma)}}{(\delta\Rightarrow\neg\eta_1)\Rightarrow(\Gamma\Rightarrow\gamma)}\,(m_3)}{(\mathrm{cut}_3)\qquad \Gamma\Rightarrow\gamma}$$

(cut_1), (cut_2) and (cut_3) are eliminable by the induction hypothesis on the grade.

Now let the right premise of the cut rule be the result of (m_2) and let $\mathrm{cfl} = (\neg\alpha\Rightarrow\delta)$, i.e. the derivation π run as follows:

$$\pi\left\{\dfrac{\pi'\left\{\dfrac{\vdots}{\Gamma\Rightarrow(\neg\alpha\Rightarrow\delta)}(r)\right. \qquad \dfrac{\dfrac{\vdots}{\neg\neg\alpha\Rightarrow\gamma} \qquad \dfrac{\vdots}{\delta\Rightarrow\gamma}}{(\neg\alpha\Rightarrow\delta)\Rightarrow\gamma}\,(m_2)}{\Gamma\Rightarrow\gamma}\right.$$

The premise of the rule (r) can not contain $(\neg\alpha\Rightarrow\delta)$ as the left rank is 1. So the last step in π' must be one of the following cases:

CASE 1.

$$\dfrac{\dfrac{\vdots}{\Gamma\Rightarrow\alpha} \qquad \bot\Rightarrow\delta}{\Gamma\Rightarrow(\neg\alpha\Rightarrow\delta)}\,(\Rightarrow 1)$$

CASE 2.* $\delta = \delta_1 \cup \delta_2$ and

$$\pi' \left\{ \dfrac{\dfrac{\vdots}{\Gamma \Rightarrow (\neg\alpha \Rightarrow \delta_1)}}{\Gamma \Rightarrow (\neg\alpha \Rightarrow \delta)} \, (\cup r) \right. \quad \text{or} \quad \pi' \left\{ \dfrac{\dfrac{\vdots}{\Gamma \Rightarrow (\neg\alpha \Rightarrow \delta_2)}}{\Gamma \Rightarrow (\neg\alpha \Rightarrow \delta)} \, (\cup r) \right.$$

CASE 3. $\delta = \delta_1 \cap \delta_3$ and

$$\pi' \left\{ \dfrac{\dfrac{\vdots}{\Gamma \Rightarrow (\neg\alpha \Rightarrow \delta_1)} \qquad \dfrac{\vdots}{\Gamma \Rightarrow (\neg\alpha \Rightarrow \delta_2)}}{\Gamma \Rightarrow (\neg\alpha \Rightarrow \delta)} \right.$$

CASE 4. $\delta = (\delta_1 \Rightarrow \delta_2) \Rightarrow \delta_3$ and

$$\dfrac{\dfrac{\vdots}{\Gamma \Rightarrow (\neg\alpha \Rightarrow \delta_1)} \qquad \dfrac{\vdots}{\delta_2 \Rightarrow \delta_3}}{\Gamma \Rightarrow (\neg\alpha \Rightarrow ((\delta_1 \Rightarrow \delta_2) \Rightarrow \delta_3)} \, (\Rightarrow l)$$

CASE 5. $\delta = \neg\delta_1$ and

$$\begin{array}{c} \vdots \\ \hline \Gamma \Rightarrow (\delta_1 \Rightarrow \neg\neg\alpha) \\ \hline\hline \Gamma \Rightarrow (\neg\alpha \Rightarrow \neg\delta_1) \end{array}$$

---

* In the case when instead of $(\cup r)$ we apply a rule of the form (r^{β}) we proceed in an analogous way as in the case (1)-(7).
Note that the subformula $(\alpha \cup \beta)$ of the formula $(\alpha \cup \beta) \Rightarrow \gamma$ is decomposable only by the rule $(\cup l)$.

CASE 6. $\delta = \neg(\delta_1 \Rightarrow \eta)$ and

$$\dfrac{\dfrac{\vdots}{\Gamma \Rightarrow (\neg\alpha \Rightarrow \neg\neg\delta_1)} \qquad \dfrac{\vdots}{\Gamma \Rightarrow (\neg\alpha \Rightarrow \neg\eta)}}{\Gamma \Rightarrow (\neg\alpha \Rightarrow \neg(\delta_1 \Rightarrow \eta))}\ (m_4)$$

CASE 7. $\delta = \neg\delta_1 \cup \neg\delta_2$ and

$$\dfrac{\Gamma \Rightarrow (\neg\alpha \Rightarrow (\delta_1 \Rightarrow \neg\delta_2))}{\Gamma \Rightarrow (\neg\alpha \Rightarrow (\neg\delta_1 \cup \neg\delta_2)}\ (m_1)$$

We check only the Case 4. Then the derivation π'' of the right premise is as follows:

$$\pi'' \left\{ \dfrac{\neg\neg\alpha \Rightarrow \gamma \qquad \dfrac{\dfrac{\vdots}{A} \quad \text{or} \quad \dfrac{\vdots}{B} \quad \text{or} \quad \dfrac{\vdots}{C}}{((\delta_1 \Rightarrow \delta_2) \Rightarrow \delta_3) \Rightarrow \gamma}}{(\neg\alpha \Rightarrow ((\delta_1 \Rightarrow \delta_2) \Rightarrow \delta_3)) \Rightarrow \gamma}\ (m_2) \right.$$

Thus π can be transformed as follows:

$$\dfrac{\dfrac{\vdots}{\Gamma \Rightarrow (\neg\alpha \Rightarrow \delta_1)} \qquad \dfrac{\dfrac{\vdots}{\neg\neg\alpha \Rightarrow \gamma} \qquad \dfrac{(\Rightarrow 1)\ \dfrac{\delta_1 \Rightarrow \delta_1 \qquad \dfrac{\vdots}{\delta_2 \Rightarrow \delta_3}}{\delta_1 \Rightarrow ((\delta_1 \Rightarrow \delta_2) \Rightarrow \delta_3)} \qquad \dfrac{\vdots}{((\delta_1 \Rightarrow \delta_2) \Rightarrow \delta_3) \Rightarrow \gamma}}{\delta_1 \Rightarrow \gamma}\ (\text{cut } 1)}{(\neg\alpha \Rightarrow \delta_1) \Rightarrow \gamma}\ (m_2)}{\Gamma \Rightarrow \gamma}\ (\text{cut}_2)$$

By induction on the grade all cuts are eliminable.

Case of (m_3). Consider only the case when $\text{cf1} = (\alpha \Rightarrow \neg\delta) \Rightarrow \eta$ and let η be the form $\neg\eta_1$. Let π' be the proof of $((\alpha\Rightarrow\neg\delta)\Rightarrow\neg\eta_1)\Rightarrow\gamma$ such that the end formula of π' is obtained by application of (m_3) to $\neg(\alpha\Rightarrow\neg\delta)\Rightarrow\gamma$ and $\neg\eta_1\Rightarrow\gamma$.

More exactly, let the derivation π runs as follows:

$$\dfrac{(m_3)\,\dfrac{\dfrac{\vdots}{\Gamma\Rightarrow(\neg\alpha\Rightarrow\neg\eta_1)}\quad\dfrac{\vdots}{\Gamma\Rightarrow(\neg\delta\Rightarrow\neg\eta_1)}}{\Gamma\Rightarrow((\alpha\Rightarrow\neg\delta)\Rightarrow\neg\eta_1)}\qquad\dfrac{\dfrac{\vdots}{\neg(\alpha\Rightarrow\neg\delta)\Rightarrow\gamma}\quad\dfrac{\vdots}{\neg\eta_1\Rightarrow\gamma}}{((\alpha\Rightarrow\neg\delta)\Rightarrow\neg\eta_1)\Rightarrow\gamma}}{\Gamma\Rightarrow\gamma}\,(m_3)$$

As before, we use that $SI\vdash\neg\neg\alpha\Rightarrow(\neg\neg\delta\Rightarrow\neg(\alpha\Rightarrow\neg\delta))$

Thus π can be transformed into:

$$\dfrac{(cut_2)\,\dfrac{\dfrac{\vdots}{\Gamma\Rightarrow(\neg\alpha\Rightarrow\neg\eta_1)}\qquad\dfrac{(cut_1)\,\dfrac{\dfrac{SI\text{ - proof}}{\neg\neg\alpha\Rightarrow(\neg\neg\delta\Rightarrow\neg(\alpha\Rightarrow\neg\delta))}\quad\dfrac{\vdots}{\neg(\alpha\Rightarrow\neg\delta)\Rightarrow\gamma}}{\neg\neg\alpha\Rightarrow(\neg\neg\delta\Rightarrow\gamma)}\qquad\dfrac{\dfrac{\vdots}{\neg\eta_1\Rightarrow\gamma}}{\neg\eta_1\Rightarrow(\neg\neg\delta\Rightarrow\gamma)}}{(\neg\alpha\Rightarrow\neg\eta_1)\Rightarrow(\neg\neg\alpha\Rightarrow\gamma)}\,(m_2)}{\Gamma\Rightarrow(\neg\neg\delta\Rightarrow\gamma)}\qquad\dfrac{\vdots}{\Gamma\Rightarrow(\neg\eta_1\Rightarrow\gamma}}{\dfrac{\dfrac{\vdots}{\Gamma\Rightarrow(\neg\delta\Rightarrow\eta_1)}\qquad(\neg\delta\Rightarrow\neg\eta_1)\Rightarrow(\Gamma\Rightarrow\gamma)}{\dfrac{(\Gamma\Rightarrow\gamma)}{\Gamma\Rightarrow\gamma}}\,(cut_3)}\,(m_2)$$

Note that all cuts are eliminable by the induction on the grade.

We omit the case when the right premise of the cut rule is of the form $(\alpha\Rightarrow\neg\delta)\Rightarrow\gamma$ and $cfl=(\alpha\Rightarrow\neg\delta)$ as the proof is similar to the analogous case for (m_2).

Case of (m_4). For simplicity assume that $\Delta=\phi$ and let π run as follows:

$$\pi \left\{ \dfrac{\dfrac{\dfrac{\vdots}{\Gamma \Rightarrow (\beta \cup \neg\neg\alpha)} \quad \dfrac{\vdots}{\Gamma \Rightarrow (\beta \cup \neg\delta)}}{\Gamma \Rightarrow (\beta \cup \neg(\alpha \Rightarrow \delta))}(m_4\beta) \qquad \dfrac{\dfrac{\vdots}{\beta \Rightarrow \gamma} \quad \dfrac{\vdots}{\neg(\alpha \Rightarrow \delta) \Rightarrow \gamma}}{(\beta \cup \neg(\alpha \Rightarrow \delta)) \Rightarrow \gamma}(\cup l)}{\Gamma \Rightarrow \gamma}(cut) \right.$$

It is reduced to π' :

$$\dfrac{\dfrac{\dfrac{\Gamma \Rightarrow (\beta \neg\neg\alpha) \qquad \dfrac{\dfrac{\beta \Rightarrow \gamma}{\beta \Rightarrow (\neg\delta \Rightarrow \gamma)} \qquad \dfrac{\dfrac{\text{SI-proof}}{\neg\neg\alpha \Rightarrow (\neg\delta \Rightarrow \neg(\alpha \Rightarrow \delta))} \quad \dfrac{\vdots}{\neg(\alpha \Rightarrow \delta) \Rightarrow \gamma}}{\neg\neg\alpha \Rightarrow (\neg\delta \Rightarrow \gamma)}(cut_1)}{(\beta \cup \neg\neg\alpha) \Rightarrow (\neg\delta \Rightarrow \gamma)}(\cup l)}{\Gamma \Rightarrow (\neg\delta \Rightarrow \gamma)}(cut_2)}{\Gamma \Rightarrow (\beta \cup \neg\delta) \qquad \dfrac{\Gamma \Rightarrow (\neg\delta \Rightarrow \gamma) \quad \dfrac{\beta \Rightarrow \gamma}{\Gamma \Rightarrow (\beta \Rightarrow \gamma)}}{(\beta \cup \neg\delta) \Rightarrow (\Gamma \Rightarrow \gamma)}(\cup l)} \quad \dfrac{\dfrac{}{\Gamma \Rightarrow (\Gamma \Rightarrow \gamma)}(cut_3)}{\Gamma \Rightarrow \gamma}$$

Arguing as usual we can eliminate all cuts from the above derivation.

Analogously for the remaining rules:

It is not difficult to conclude that if $R_1 > 1$ and $R_r > 1$ then using the induction hypothesis on the rank we transform any derivation into a normal proof.

COROLLARY. For any formula α the following conditions are equivalent:

1) $\alpha \in LM$,
2) $MS \vdash \alpha$,
3) there exists a cut-free proof for α. □

Problem. Is it possible to eliminate from MS the rules (m_3) and $(m_4{}^\beta)$?.

THEOREM 2.5. (Interpolation lemma for LM). Craig's interpolation lemma holds for LM.

PROOF. Let $\xi \in LM$. To prove the theorem it is sufficient to show that for any partition $[\phi, \xi^*]$ of ξ there exists a formula γ such that

1) $(\Phi \Rightarrow \gamma) \in LM$ and $(\gamma \Rightarrow \xi^*) \in LM$

2) $\langle\gamma\rangle \subset \langle\Phi\rangle \cap \langle\xi^*\rangle$,

where for any formula α, $\langle\alpha\rangle$ is the set of all sentential variables ocurring in α.

The method which we use is due to Maehara and its significance lies in the fact that an interpolant of the formula of the form $(\sigma \Rightarrow \eta)$ can be constructively obtained from a proof of $(\sigma \Rightarrow \eta)$.

By the corollary $MS \vdash \xi$. Now we will construct an interpolant γ by induction on the number of inference rules, in a normal derivation π of ξ. At each stage there are several cases to consider. According to 1.9 the only new cases arise when ξ is a conclusion of the rule (m_i^{β}) or (m_i), $i = 1,2,3,4$.

We deal only with (m_i), $i = 1,2,3,4$ and suppose that for every $\xi' \in LM$ such that $\pi_{\xi'} = k$ the theorem holds, where now π_{ξ} is the length of derivation π of ξ.

Let $\pi_{\xi} = k + 1$ and the last inference be (m_1), i.e. $\xi = \Gamma \Rightarrow (\neg\alpha \cup \neg\delta)$. Let us consider the following partition $\Phi = \Gamma'$, $\xi^* = \Gamma^* \Rightarrow (\neg\alpha \cup \neg\delta)$. By the induction assumption we know that for $\xi_1 = \Gamma \Rightarrow (\alpha \Rightarrow \neg\delta)$ there is an interpolant γ. Hence for the partition $[\Phi_1 \xi_1^*]$, where $\Phi_1 = \Gamma'$, $\xi_1^* = \Gamma^* \Rightarrow (\alpha \Rightarrow \neg\delta)$ there exists a γ such that $MS \vdash \Gamma' \Rightarrow \gamma$ and

$MS \vdash (\gamma \Rightarrow (\Gamma^* \Rightarrow (\alpha \Rightarrow \neg\delta)))$.

Moreover $\langle\gamma\rangle \subset \langle\Gamma'\rangle \cap \langle\Gamma^* \Rightarrow (\alpha \Rightarrow \neg\delta)\rangle$.

Thus and by (m_1) we infer that

$$MS \vdash \Gamma' \Rightarrow \gamma \qquad \text{and} \qquad MS \vdash \Gamma^* \Rightarrow (\neg\alpha \cup \neg\delta).$$

It is obvious that $\langle\gamma\rangle \subset \langle\Gamma'\rangle \cap \langle\Gamma^* \Rightarrow (\neg\alpha \cup \neg\delta)\rangle$ which proves that γ is an interpolant for $\Gamma \Rightarrow (\neg\alpha \cup \neg\delta)$.

Let the last inference be (m_2), i.e. $\xi = \Gamma \Rightarrow ((\neg\alpha \Rightarrow \delta) \Rightarrow \eta)$. Moreover for the simplicity we assume that η is not of the form $\eta_1 \Rightarrow \eta_2$.

Then we have the following partitions

1) $\Phi = \Gamma'$; $\xi^* = \Gamma^* \Rightarrow ((\neg\alpha \Rightarrow \delta) \Rightarrow \eta)$

2) $\Phi = \Gamma'$; $(\neg\alpha \Rightarrow \delta)$; $\xi^* = \Gamma^* \Rightarrow \eta$.

We omit 1) as it is trivial. Let us take the latter and consider $\xi_1 = \Gamma \Rightarrow (\neg\neg\alpha \Rightarrow \eta)$, $\xi_2 = \Gamma \Rightarrow (\delta \Rightarrow \eta)$. By the induction assumption, since $\pi_{\xi_1}<k+1$, $\pi_{\xi_2}<k+1$, we infer that for the partition $[\Phi_1, \xi_1^*]$ and $[\Phi_2, \xi_2^*]$, where $\Phi_1 = \Gamma', \neg\neg\alpha$; $\Phi_2 = \Gamma', \delta$; $\xi_1^* = \xi_2^* = \Gamma^* \Rightarrow \eta$ there are interpolants γ_1 and γ_2, i.e.

$$MS \vdash \Gamma' \Rightarrow (\neg\neg\alpha \Rightarrow \gamma_1), \quad MS \vdash \Gamma' \Rightarrow (\delta \Rightarrow \gamma_2), \quad MS \vdash \gamma_1 \Rightarrow (\Gamma^* \Rightarrow \eta)$$
$$MS \vdash \gamma_2 \Rightarrow (\Gamma^* \Rightarrow \eta)$$

and

$$<\gamma_1> \subset <\Phi_1> \cap <\xi_1^*> \quad , \quad <\gamma_2> \subset <\Phi_2> \cap <\xi_2^*>$$

Using the rules $(\cup r)$, $(\cup l)$ we obtain

$$MS \vdash \Gamma' \Rightarrow (\neg\neg\alpha \Rightarrow (\gamma_1 \cup \gamma_2)), \quad MS \vdash \Gamma' \Rightarrow \delta \Rightarrow (\gamma_1 \cup \gamma_2)), \quad MS \vdash (\gamma_1 \cup \gamma_2) \Rightarrow (\Gamma^* \Rightarrow \eta).$$

It is obvious that :

$$<\gamma_1 \cup \gamma_2> \subset <\Phi> \cap <\xi^*> .$$

Now using the rule (m_2) we infer that $(\gamma_1 \cup \gamma_2)$ is an interpolant for the second partition.

Now let us assume that the last inference is (m_3), i.e. $\xi = \Gamma \Rightarrow ((\alpha \Rightarrow \neg\delta) \Rightarrow \eta)$. As before we omit case when $\Phi = \Gamma'$ and we assume that η is not of the form $\eta_1 \Rightarrow \eta_2$. If $\Phi \neq \Gamma'$, then we have only one possibility, namely $\Phi = \Gamma', (\alpha \Rightarrow \delta)$ and $\xi^* = \Gamma^* \Rightarrow \eta$. On account of the induction hypothesis we obtain that for the partitions $[\Gamma', \neg\alpha;\ \Gamma^* \Rightarrow \eta]$ $[\Gamma', \neg\delta;\ \Gamma^* \Rightarrow \eta]$ there are the interpolants γ_1 and γ_2, respectively. Thus $(\gamma_1 \cup \gamma_2)$ is an interpolant for $[\Gamma', (\alpha \Rightarrow \neg\delta);\ \Gamma^* \Rightarrow \eta]$ which completes our proof for case of the rule (m_3).

Let the last inference be (m_4), i.e. $\xi = \Gamma \Rightarrow \neg(\alpha \Rightarrow \delta)$. There is only one partition $[\Phi, \xi^*]$ such that $\Phi \neq \Gamma'$, namely $\Phi = \Gamma', \alpha \Rightarrow \delta$ and $\xi = \Gamma^* \Rightarrow \bot$. Consider this case. Let $\xi_1 = \Gamma \Rightarrow \neg\neg\alpha$ and $\xi_2 = \Gamma \Rightarrow \neg\delta$. By the induction assumption there are γ_1 and γ_2 such that

$$MS \vdash \Gamma' \Rightarrow (\neg\alpha \Rightarrow \gamma_1), \quad MS \vdash \gamma_1 \Rightarrow (\Gamma^* \Rightarrow \bot), \quad MS \vdash \Gamma' \Rightarrow (\delta \Rightarrow \gamma_2) \quad MS \vdash \gamma_2 \Rightarrow (\Gamma^* \Rightarrow \bot)$$

and

$$<\gamma_1> \subset <\Gamma', \neg\alpha> \cap <\Gamma^* \Rightarrow \bot>, \quad <\gamma_2> \subset <\Gamma' \Rightarrow \delta> \cap <\Gamma^* \Rightarrow \bot>.$$

Thus by $(\cup r)$, $(\cup l)$, $(\Rightarrow 1)$, (str_3) and (m_4) we infer that

$MS \vdash \Gamma \Rightarrow ((\alpha \Rightarrow \delta) \Rightarrow \neg\neg(\gamma_1 \cup \gamma_2))$ and $MS \vdash \neg\neg(\gamma_1 \cup \gamma_2) \Rightarrow (\Gamma^* \Rightarrow \bot)$. □

Note that

2.6 The system MS without the cut rule is separable. □

2.7 In a normal derivation of α only subformulas of α occur. □

It is not difficult to check that the following two rules

$$(\forall^\beta r) \qquad \frac{\Gamma \Rightarrow (\neg\beta \cup (\Delta \Rightarrow \alpha(a))}{\Gamma. \Rightarrow (\neg\beta \cup (\Delta \Rightarrow \forall x\alpha(x))}$$

$$(\exists^\beta l) \qquad \frac{\Gamma \Rightarrow \neg\beta \cup (\alpha(a) \Rightarrow \gamma)}{\Gamma \Rightarrow (\neg\beta \cup \exists x\alpha(x) \Rightarrow \gamma)}$$

where a does not occur as a free object variable in any formula in the conclusion of the mentioned rules, are valid in LM considered as predicate logic.

Let MS^Q be the system MS enlarged by the predicate part of the system BS and the rules $(\forall^\beta r)$, $(\exists^\beta l)$.

Conjuncture. The following conditions are equivalent

1) $\alpha \in LM$,
2) $MS^Q \vdash \alpha$,
3) α has a cut-free proof in MS^Q.

Notice that if the conjuncture is true then it is not difficult to prove Craig's interpolation lemma for predicate logic LM, the separation theorem and the subformula property for MS.

§3. LOGIC LMH

In [1] Gabbay proved that $LMH = LI + (\forall x \neg\neg\alpha(x) \Rightarrow \neg\neg \forall x\alpha(x))$ is complete for the class of all Kripke structures $\langle T, \leq \rangle$ with partially ordered worlds system T fulfilling

$$\wedge t \vee s(t \leq s \;\&\; \wedge u(s \leq u \rightarrow s = u)).$$

It follows from the above characterization of LMH that for any formula α

$$\alpha \in LK \quad \text{iff} \quad \neg\neg\alpha \in LMH.$$

The above theorem is often called the Glivenko type theorem for LMH.

Umezawa observed in [10] that the following schemas of formulas:

$$\neg\neg\forall x(\alpha(x) \cup \neg\alpha(x)),$$

$$\neg\forall x\alpha(x) \Rightarrow \neg\neg\exists x\alpha(x),$$

$$\neg\neg(\forall x\neg\neg\alpha(x) \Rightarrow \neg\neg\forall x\alpha(x)),$$

$$\neg\neg(\neg\forall x\alpha(x) \Rightarrow \neg\neg\exists x\neg\alpha(x)),$$

$$\neg(\neg\forall x\alpha(x) \cap \neg\exists x\neg\alpha(x)),$$

(*) $\neg\neg((\forall x\alpha(x) \Rightarrow \exists x\beta(x)) \Rightarrow \exists x(\alpha(x) \Rightarrow \beta(x))),$

$\neg\neg\exists x\forall y(\alpha(x) \Rightarrow \alpha(y)),$

(**) $((\forall x\alpha(x) \Rightarrow \beta) \Rightarrow \neg\neg\exists x(\alpha(x) \Rightarrow \beta))$

are equivalent to

(MH) $(\forall x\neg\neg\alpha(x) \Rightarrow \neg\neg\forall x\alpha(x))$.

To construct the formal system MHS for LMH let us take the system IS for intuitionistic logic and add the following rules:

$$(mh_1)\qquad \frac{\Gamma \Rightarrow (\neg\alpha \Rightarrow \neg\neg\eta) \qquad \Gamma \Rightarrow (\delta \Rightarrow \neg\neg\eta)}{\Gamma \Rightarrow ((\alpha\Rightarrow\delta) \Rightarrow \neg\neg\eta)}$$

$$(mh_2)\qquad \frac{\Gamma \Rightarrow (\alpha\Rightarrow \neg\neg\eta)}{\Gamma \Rightarrow \neg\neg(\alpha\Rightarrow\eta)}$$

$$(mh_3)\qquad \frac{\Gamma \Rightarrow \neg\neg\alpha(a)}{\Gamma \Rightarrow \neg\neg\forall x\alpha(x)}$$

where a does not appear as a free object variable in any formula in the conclusion of (mh_3) and $\neg\gamma$ is $\gamma \Rightarrow \bot$.

3.1 Let (r) be one of the rules (mh_1) - (mh_3). If the premise(s) of the rule (r) belong(s) to LMH, so does the conclusion.

PROOF. Suffices to observe that the rules (mh_1) and (mh_2) are intuitionistically valid. In the case of the rule (mh_3) the theorem follows from (MH).

3.2 Let π be a normal proof in the system MHS of a formula of the form $\neg\neg\alpha \Rightarrow \gamma$. Then there exists a subformula of γ, say γ_1, and a set of subformulas of γ, say Γ_1, such that either the formula of the form $\Gamma_1 \Rightarrow (\neg\neg\alpha \Rightarrow \gamma_1)$ is an axiom and it occurs in π or the formulas of the form $\Gamma_1 \Rightarrow \neg\alpha$ and $\bot \Rightarrow \gamma_1$ appear in π or γ_1 appears in π.

PROOF. Let us restore a normal MHS proof of a formula of the form $\neg\neg\alpha \Rightarrow \gamma$. Note that the subformula $(\alpha \Rightarrow \beta)$ of the formula of the form $(\alpha \Rightarrow \beta) \Rightarrow \gamma$ is decomposed only by the rules $(\Rightarrow 1)$ and (mh_1). In the latter γ must be of form $\neg\neg\gamma_1$ and in our case this rule does not play any role. So let π be a normal MHS-proof of a formula of the form $\neg\neg\alpha \Rightarrow \gamma$. Then π runs as follows:

$$
\begin{array}{c}
(str_1)\dfrac{\dfrac{\vdots}{\gamma_1} \quad \text{or} \quad \dfrac{\vdots}{\Gamma_1 \Rightarrow \neg\alpha} \quad \bot \Rightarrow \gamma_1}{\Gamma_1 \Rightarrow (\neg\neg\alpha \Rightarrow \gamma_1)} (\Rightarrow 1) \\
(r_1),\dots,(r_k) \left\{ \begin{array}{c} \overline{\overline{\quad\quad\quad}} \\ \ddots \\ \hline \neg\neg\alpha \Rightarrow \gamma \end{array} \right.
\end{array}
$$

where Γ_1 is a set of subformulas of γ (perhaps empty) and γ_1 is a subformula of γ. The rules $(r_1)\dots(r_n)$ operate only on the subformulas of γ. Note that either a formula of the form $\Gamma_1 \Rightarrow (\neg\neg\alpha \Rightarrow \gamma_1)$ is an axiom or it is derived from $\Gamma_1 \Rightarrow \neg\alpha$ and $\bot \Rightarrow \gamma_1$ by the rule $(\Rightarrow 1)$ or it is derived from γ_1 by (str_1). □

3.3 (Cut elimination for MHS). Any drivation in MHS reduces to a normal derivation.

PROOF. By 1.4 the only new cases arise when one premise (or both premises) of the cut rule is (are) the conclusion of the rule (mh_i), $i = 1,2,3$.

Assume that $R = 2$, i.e. $R_l = R_r = 1$, and that the theorem is proved for the grade n.

Case of (mh_1). Suppose that $cfl = (\alpha \Rightarrow \delta) \Rightarrow \neg\neg\eta$ and $g(cfl) = n+1$. Consider the following derivation π:

$$
\pi \left\{ \dfrac{(mh_1)\dfrac{\dfrac{\vdots}{\Gamma \Rightarrow (\neg\alpha \Rightarrow \neg\neg\eta)} \quad \dfrac{\vdots}{\Gamma \Rightarrow (\delta \Rightarrow \neg\neg\eta)}}{\Gamma \Rightarrow ((\alpha \Rightarrow \delta) \Rightarrow \neg\neg\eta)} \quad \dfrac{\vdots}{(\alpha \Rightarrow \delta) \Rightarrow \neg\neg\eta) \Rightarrow \gamma}}{\Gamma \Rightarrow \gamma} (cut) \right.
$$

Let π' be a normal proof of $((\alpha \Rightarrow \delta) \Rightarrow \neg\neg\eta) \Rightarrow \gamma$
The only interesting cases are when the end formula of π' is either the result of the rule $(\Rightarrow 1)$ or of (mh_1). In the latter γ must be of the form $\neg\neg\gamma_2$.

CASE 1. π' runs as follows:

$$\dfrac{\dfrac{\vdots}{\alpha \Rightarrow \delta} \qquad \dfrac{\vdots}{\neg\neg\eta \Rightarrow \gamma}}{((\alpha \Rightarrow \delta) \Rightarrow \neg\neg\eta) \Rightarrow \gamma}\,(\Rightarrow 1)$$

Then π can be transformed into:

$$\dfrac{\dfrac{\dfrac{\dfrac{\vdots}{\Gamma \Rightarrow (\delta \Rightarrow \neg\neg\eta)}}{\Gamma \Rightarrow (\neg\eta \Rightarrow \neg\delta)} \qquad \dfrac{(\Rightarrow 1)\,\dfrac{\dfrac{\vdots}{\alpha \Rightarrow \delta} \quad \bot \Rightarrow \bot}{\neg\delta \Rightarrow \neg\alpha} \qquad \dfrac{\dfrac{\vdots}{\Gamma \Rightarrow (\neg\alpha \Rightarrow \neg\neg\eta)}}{\neg\alpha \Rightarrow (\Gamma \Rightarrow \neg\neg\eta)}}{\neg\delta \Rightarrow (\Gamma \Rightarrow \neg\neg\eta)}\,(cut_1)}{\Gamma \Rightarrow \neg\neg\eta}\,(cut_2) \qquad \dfrac{\vdots}{\neg\neg\eta \Rightarrow \gamma}}{\Gamma \Rightarrow \gamma}\,(cut_3)$$

By the induction hypothesis on the grade (cut_1) - (cut_3) are eliminable.

CASE 2. π' runs as follows:

$$\dfrac{\dfrac{\vdots}{\neg(\alpha \Rightarrow \delta) \Rightarrow \neg\neg\eta} \qquad \dfrac{\vdots}{\delta \Rightarrow \neg\neg\gamma}}{((\alpha \Rightarrow \delta) \Rightarrow \neg\neg\eta) \Rightarrow \neg\neg\gamma}\,(mh_1)$$

Then π can be transformed into:

$$\dfrac{\dfrac{\dfrac{\dfrac{\dfrac{\vdots}{\Gamma \Rightarrow ((\alpha\Rightarrow\delta) \Rightarrow \neg\neg\eta)}}{\Gamma \Rightarrow (\neg\eta \Rightarrow \neg(\alpha\Rightarrow\delta))} \qquad \dfrac{\vdots}{\neg(\alpha\Rightarrow\delta)\Rightarrow \neg\neg\gamma}}{\Gamma \Rightarrow (\neg\gamma \Rightarrow \neg\neg\eta)}\,(\mathrm{cut}_1) \qquad \neg\neg\eta \Rightarrow \neg\neg\gamma}{\Gamma \Rightarrow (\neg\gamma \Rightarrow \neg\neg\gamma)}\,(\mathrm{cut}_2)}{\Gamma \Rightarrow \neg\neg\gamma}$$

On account of the induction hypothesis on the grade cut_1 and cut_2 are eliminable.

The case of (mh_2) is trivial so let us prove the theorem for case of (mh_3). Suppose that $\mathrm{cfl} = \neg\neg\forall x\alpha(x)$ and $g(\mathrm{cfl}) = n + 1$. Consider the following derivation:

$$\pi \left\{ \dfrac{\dfrac{\dfrac{\vdots}{\Gamma \Rightarrow \neg\neg\alpha(a)}}{\Gamma \Rightarrow \neg\neg\forall x\alpha(x)}\,(\mathrm{mh}_3) \qquad \dfrac{\vdots}{\neg\neg\forall x\alpha(x) \Rightarrow \gamma}}{\Gamma \Rightarrow \gamma}\,(\mathrm{cut}) \right.$$

Let π' be a normal proof of $\neg\neg\forall x\alpha(x) \Rightarrow \gamma$. If we omit trivial cases, then by 3.2 the only interesting case of π' is as follows:

$$\pi' \left\{ \dfrac{\dfrac{\vdots}{\dfrac{\dfrac{\dfrac{\dfrac{\dfrac{\vdots}{\Gamma_1 \Rightarrow (\alpha(b)\Rightarrow\bot)}}{\Gamma_1 \Rightarrow (\forall x\alpha(x)\Rightarrow\bot)}\,(\forall 1) \qquad \bot \Rightarrow \gamma_1}{\Gamma_1 \Rightarrow (\neg\neg\forall x\alpha(x) \Rightarrow \gamma_1)}\,(\Rightarrow 1)}{\neg\neg\forall x\alpha(x) \Rightarrow (\Gamma_1 \Rightarrow \gamma_1)}}{}}\,(r_1),\ldots,(r_k)}{\neg\neg\forall x\alpha(x) \Rightarrow \gamma} \right.$$

This is transformed into:

$$
\dfrac{\dfrac{\dfrac{\vdots}{\Gamma_1 \Rightarrow \neg\,\alpha(b)} \qquad \dfrac{\dfrac{\vdots}{\Gamma \Rightarrow \neg\neg\alpha(b)}}{\neg\,\alpha(b) \Rightarrow (\Gamma \Rightarrow \bot)}}{\Gamma_1 \Rightarrow (\Gamma \Rightarrow \bot)}\,(cut_1) \qquad \bot \Rightarrow \gamma_1}{\dfrac{\Gamma_1 \Rightarrow (\Gamma \Rightarrow \gamma_1)}{\dfrac{\Gamma \Rightarrow (\Gamma_1 \Rightarrow \gamma_1)}{\vdots \; (r_1),\ldots,(r_k)}}}\,(cut_2)
$$

$$\Gamma \Rightarrow \gamma$$

Above the formula $\Gamma_1 \Rightarrow \neg\neg\alpha(b)$ we write the same part of the derivation which previously occurred above $\Gamma \Rightarrow \neg\neg\alpha(a)$ replacing every occurrence of the free object variable a by b.

By the induction hypothesis on the grade (cut_1) and (cut_2) are eliminable.

Notice that if $\Gamma_1 \Rightarrow (\forall x\alpha(x) \Rightarrow \bot)$ or $\neg\neg\forall x\alpha(x) \Rightarrow (\Gamma_1 \Rightarrow \gamma_1)$ is an axiom then a reduction of the cut rule from π is trivial. Indeed, suppose that $\Gamma_1 \Rightarrow (\forall x\alpha(x) \Rightarrow \bot)$ is an axiom then Γ_1 is of the form $\forall x\alpha(x) \Rightarrow \bot$. Let us consider the following derivation:

$$
\dfrac{\dfrac{\Gamma_1 \Rightarrow \neg\,\forall\,\alpha(x) \qquad \dfrac{\dfrac{\dfrac{\vdots}{\Gamma \Rightarrow \neg\neg\,\alpha(a)}}{\Gamma \Rightarrow \neg\neg\,\forall x\alpha(x)}\,(mh_3)}{\neg\,\forall x\alpha(x) \Rightarrow (\Gamma \Rightarrow \bot)}}{\Gamma_1 \Rightarrow (\Gamma \Rightarrow \bot)}\,(cut_1) \qquad \bot \Rightarrow \gamma_2}{\dfrac{\Gamma \Rightarrow (\Gamma_1 \Rightarrow \gamma_1)}{\vdots \; (r_1),\ldots,(r_k)}}\,(cut_2)
$$

$$\Gamma \Rightarrow \gamma$$

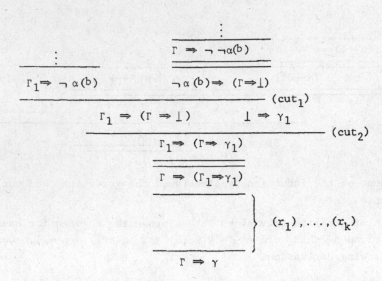

As usual, (cut_1) and (cut_2) are eliminable.

So the proof of cut-elimination for $R = 2$ and any grade of cut formula is over.

If the rank is >2, then using induction hypothesis on the rank and lemma 3.2 the cut-elimination theorem can be proved very easily. □

As a natural consequence of the above theorem we obtain.

COROLLARY. For any formula α the following conditions are equivalent:

1. $\alpha \in LMH$,
2. $MHS \vdash \alpha$,
3. α has a normal proof. □

3.4 The system MHS without the cut rule is separable. □

3.5 In a normal derivation of α only subformulas of α occur. □

3.6 (Craig's interpolation lemma). For any formula ξ, if $\xi \in LMH$, then ξ has an interpolant.

PROOF. Suppose that $\xi \in LMH$. By the corollary ξ has a normal proof in MHS. An account of 1.9 the only new cases arise when ξ is a conclusion of the rule (mh_i), $i = 1,2,3$.

Consider only the case of (mh_3). Let ξ be of the form $\Gamma \Rightarrow \neg\neg \forall x\alpha(x)$. There are two different partitions of ξ, namely:

1. $\Phi = \Gamma'$; $\xi^* = \Gamma^* \Rightarrow \neg\neg \forall x\alpha(x)$,
2. $\Phi = \Gamma', \neg \forall x\alpha(x)$; $\xi^* = \Gamma^* \Rightarrow \bot$.

The standard proof for (1) is omitted. Let us take (2).
By the induction hypothesis there is an interpolant for $\xi_1 = \Gamma \Rightarrow \neg\neg \alpha(a)$ (a does not appear as free object variable in Γ). Let $\phi_1 = \Gamma^*$, $\xi_1^* = \Gamma' \Rightarrow \neg\neg \alpha(a)$ be a partition of ξ_1 and let γ be a formula such that

$$MHS \vdash \Gamma^* \Rightarrow \gamma \quad \text{and} \quad MHS \vdash \gamma \Rightarrow (\Gamma' \Rightarrow \neg\neg \alpha(a)).$$

Moreover $\langle\gamma\rangle \subset \langle\Phi_1\rangle \cap \langle\xi_1^*\rangle$. Thus a does not occur free in γ. Applying the rules $(\Rightarrow 1)$, (mh_3) and (str_3) we infer that:

$$MHS \vdash \Gamma' \Rightarrow (\neg \forall x\alpha(x) \Rightarrow \neg\gamma) \quad \text{and} \quad MHS \vdash \neg\gamma \Rightarrow (\Gamma^* \Rightarrow \bot).$$

It is obvious that $\langle\neg\gamma\rangle \subset \langle\Phi\rangle \cap \langle\xi^*\rangle$ which proves that $\neg\gamma$ is the required interpolant. □

Problem. To modify the system MIS in such a way that the rules of inference specific for MHS are only LMH valid.

REFERENCES

[1] Gabbay, D. Applications of trees to intermediate logics, J.Symbolic Logic 37(1972), 135-138.

[2] ________ Semantic proof of Craig's theorem for intuitionistic logic and its extensions, I and II. Proc. 1969 Logic Colloquium, North-Holland Publ. Co., Amsterdam(1971), 391=410.

[3] ________ Craig interpolation theorem for intuitionistic logic and extensions, Part III. J.Symbolic Logic 42(1977), 269-271.

[4] López-Escobar, E.G.K. On the interpolation theorem for the logic of constant domains. J.Symbolic Logic 46(1981),87-88.

[5] ________ A second paper on the interpolation theorem for the logic of constant domains. J.Symbolic Logic 48 (1983), 595-599.

[6] ________ A natural deduction system for some intermediate logics. J.of Non-Classical Logic, 1(1982),

[7] Maksimowa,L. Craig's interpolation theorem and amalgamable varieties, Algebra i Logica 16(1977), 643-681.

[8] Schütte,K. Der Interpolationssatz der intuitionistischen Prädikatenlogik, Mathematische Annalen 148(1962), 192-200.

[9] ________ Proof theory, Springer-Verlag, Berlin-Heidelberg-New York (1977).

[10] Umezawa, T. On logics intermediate between intuitionistic and classical predicate logic, J.Symbolic Logic 24(1959) 141-153.

TYPES IN CLASS SET THEORIES

William N. Reinhardt
Dept. of Mathematics, Campus Box 426
University of Colorado
Boulder, CO 80309

Rolando Chuaqui
Universidad Catolica de Chile
Casilla 114-D
Santiago, Chile

This paper deals with the problem of defining order types for well ordered proper classes. This is a special case of the problem of defining equivalence types for an equivalence relation. We recall that the analogous problem for set theory with regularity is solved completely by Scott's method of introducing equivalence types as the equivalence class restricted to its members of minimal rank. It is curious that the situation changes so markedly when we pass to class set theory. As this problem may seem somewhat peripheral to most set theorists, we would like to make some remarks about our interest in it. The problem arises naturally when one tries to develop the theories of constructible classes within a class set theory (such as Kelley-Morse-Tarski); this is how Chuaqui originally became involved in the problem. Professor Chuaqui drew the problem to Reinhardt's attention in 1982, when he was preoccupied with problems involving the notion of intuitive provability. Because of this he was interested in some aspects of proof theory and in particular in ordinal notations. The problem of developing notations for larger and larger ordinals is analogous to the problem of getting types for ordinals. Since key ideas often show themselves more clearly and simply in classical settings than in constructive ones, one might hope these problems would illuminate ordinal notations; we were thus thinking largely of positive results. We have nothing to say from these investigations concerning ordinal notations, but it does still seem to (at least one of) us that the problems are related. The positive results we have obtained are modest but of interest. In particular, the method of defining order types appears to lead to a more conceptual development of constructible classes than any we know of in the literature. Also, in conjunction with ideas of Manuel Corrada it leads to a nice characterization of the set part of the impredicative theory of classes (i.e. Kelley Morse-Tarksi set theory). The negative results mentioned here were pointed out to us by several people at the symposium, especially Steve Simpson, Menachem Magidor, and W. Marek.

Suppose that we are given an equivalence relation $X \equiv Y$ on classes. We permit $\equiv$ to involve parameters Z. The question then is whether there is a term τ (which may depend on Z and on other parameters U as well) so that

$$1) \qquad \forall Z \exists U (\tau(X) = \tau(Y) \leftrightarrow X \equiv Y).$$

We can ask for more, that τ be a selector: $\tau(X) \equiv X$. We call the axiom which says that if $\equiv$ is an equivalence, then 1) holds, "τ is an $\equiv$-type" or Type $(\tau, \equiv)$ or even "τ" for short.

Surprisingly, it appears to be an open problem whether there is such a τ for the equivalence which holds between two ordered pairs when they correspond to the same unordered pair. That is, it appears to be unknown whether unordered pairs can be defined in class set theory.

On the other hand, it can be shown using methods of Lévy developed for investigating questions of definability, that there are models of set theory in which no such τ can be given for the relation of order isomorphism for well orderings (i.e. well order types cannot be defined). This happens in the model obtained by collapsing all cardinals below the second inaccessible to the first inaccessible. (This model was considered independently by Lévy and Rowbottom). We are indebted to Rich Laver for the following argument.

Let M be a model of ZF with two inaccessible cardinals $\kappa < \lambda$. Let N be obtained from M by collapsing all cardinals $\alpha < \lambda$ to κ, while adding no new elements of V_κ. The model of KM we consider is just $V^N_{\kappa+1}$. We observe that a definition of types here would give rise to a function $\Psi: \lambda \to V_{\kappa+1}$ in N with $\Psi(\alpha) = \Psi(\beta) \to \alpha = \beta$, namely $\Psi(\alpha) = \tau(x)$ where $x \simeq \alpha$. Such a function in N involves only an initial segment of the collapsing functions. We may think of N as $M_o[G]$, where $M_o = M[G_o]$ contains the relevant initial segment. We may represent Ψ by a relation $\Theta(\alpha, x)$ between objects in the ground model: $\Theta(\alpha, x)$ if $x \in \Psi(\alpha)$. By homogeneity Θ is determined by the empty condition over M_o, and hence Ψ is in M_o, giving in $M_o \Psi: \lambda \to V_{\kappa+1}$. This contradicts the inaccessibility of λ in M_o.

For example, for constructible classes, let τ pick the first class in question. (Here "first" means in a canonical ordering of the constructive hierarchy; to see that this makes sense requires knowing that the level at which a class first occurs is independent of the representation of the construction). Again, even in models M where no τ exists, the addition of the existence of τ to the theory of M may be conservative, so that in some sense it is as though there were a τ after all.

We first make the trivial observation that if we do not allow τ in the comprehension scheme, we may always add a τ satisfying the axiom we have indicated. Proof: If M is any model, consider the external classes $\{Y \epsilon M \mid M \text{ has } Y \equiv X$, and let τ select one member of each. Thus we are interested in whether τ can be added so that it both satisfies the equivalence type axiom, <u>and</u> other schemas of class theory. In the case of Kelly-Morse Tarski theory (which we shall call KM), this is the class comprehension schema.

Open Problem: Is KM together with τ for well order types a conservative extension of KM?

We <u>do</u> have

<u>THEOREM.</u> Let DC be the axiom of (transfinitely many) dependent choices for classes. That is, DC says that if $\forall\exists Y\ \theta(X,Y)$ then there is a sequence Z_α ($\alpha \epsilon OR$) of classes so that for all $\alpha, \theta(\langle Z_\beta \mid \beta<\alpha\rangle, Z_\alpha)$. (Here the sequence Z_α is of course to be construed as an appropriate class). Then

$$KM + DC + \text{"}\tau \text{ is an } \equiv\text{-type"}$$

is a conservative extension of KM + DC.

Note. DC is DC_{on} of Marek and Zbierski [1972].

<u>PROOF</u>: The proof is really the same as the proof that using DC one can add a well ordering of the reals without adding any new reals. However, we only need to use the simplest possible kind of forcing (that is, no complicated terms or corresponding boolean valued functions are needed). This was observed by Mostowski [1972] for the case of analysis; again, our proof is essentially the same. The result for set theory (without infinity) implies that for analysis, but the reverse does not seem to hold. We have in the forcing language one new atomic formula G(x,y) for the well order. (Of course it suffices to add a well order, as then the equivalence types are definable as "the G-first Y such that $Y \equiv X$"). We force with conditions which well order OR-many classes. A condition may be thought of as a well ordering of OR together with a (1-1) assignment of a class to each point of OR. Two distinct conditions can contain the same information. A condition p will have the form p = (W,Z); W,Z are proper classes, and so we take the ordered pair in an appropriate sense (e.g. $(W \times \cup \{0\}\ \ (Z \times \{1\}))$. We say that one condition extends (has more informa-

tion than) another in case the latter is an initial segment of the former; e.g.:

$$q \text{ extends } p \text{ iff } \exists f,X(W_p \simeq_f W_q \upharpoonright X \;\; \& \;\; \forall\alpha\; Z''_p(\{\alpha\} = Z''_q\{f(\alpha)\})).$$

(If necessary, we write $q \geq p$ for this).

We carry out forcing in one of the usual ways now, in the language with ϵ and G.

We shall adopt the point of view that we are giving an interpretation of the language of KM + G into that of KM. Thus to each formula of KM + G we associate a formula θ whose free variables are those of θ together with one new free variable. We shall use X,Y,... for the free variable of and p,q,... for the new free variable. To carry out the definition we first introduce the syntactic operation:

$$\neg_p\theta = \forall q \geq p \rightarrow \theta(p/q)$$

and the basic formula for interpreting G(X,Y),

$$p \Vdash^W G(X,Y) = \exists x,y\; (X = Z''_p\{x\} \;\&\; Y = Z''_q\{y\} \;\&\; (x,y)\;\; W_p)$$

which of course has free variables p,X,Y. In the definition of $\neg_p$, it is of course important to avoid collisions of variables with q, so q should not occur in θ.

Observe that $\neg_p\neg_p\theta(p)$ has the form

$$\neg_p(\forall q \geq p \neg\theta(q)) = \forall_q{}'\;\; p \neg\;\; (V_q \geq q' \neg\theta(q))$$

which is equivalent to

$$\forall q' \geq p\; \exists q \geq q'\theta(q)$$

We may now define $\Vdash\theta$ by induction on θ:

Atomic:

$$\Vdash (x \epsilon y) \;\; = \;\; x \epsilon y$$

$$\Vdash G(x,y) \;\; = \;\; \neg_p\neg_p\;\; (p \Vdash^W G(x,y))$$

Propositional:

$$\Vdash (\theta \;\&\; \phi) = (\Vdash \theta) \;\&\; (\Vdash \theta)$$

$$\Vdash (\neg\, \theta) = \neg_p \; (\Vdash \theta)$$

Quantifier:

$$\Vdash (\forall x \theta) = \forall x \; (\Vdash \theta)$$

To be definite, we may always choose p to be the first new variable not occurring in θ. Thus the new free variable of $\Vdash\theta$ will be p. We therefore shall write $p \Vdash \theta$.

Now we wish to show that the axiom of KM + DC, plus "G is a well ordering of the universe of classes", are forced. Since there are no provisions for changing the ϵ-relation, it is trivial that KM + DC remains true under this interpretation (it is the identity on the language of KM!). Thus we have only to see that the comprehension axiom of KM remains true when we allow G:

$$p \Vdash \; \exists Z \forall x (x \epsilon Z \leftrightarrow \theta \;\&\; x \epsilon V)$$

i.e.

$$\forall_q \geq p \; \exists r \geq q \; \exists Z \forall x (r \Vdash x \epsilon Z \leftrightarrow \theta \;\&\; x \epsilon V)$$

Given $q \geq p$, it will suffice to extend q to r so that for some Z

1) $r \Vdash x \epsilon Z \rightarrow \theta \;\&\; x \epsilon V$

& 2) $r \Vdash \theta \;\&\; x \epsilon V \rightarrow x \epsilon Z$

i.e.

1') $x \epsilon Z \rightarrow r \Vdash \theta \;\&\; x \epsilon V$

& 2') $\forall r' \geq r \; (x \epsilon V \;\&\; r' \Vdash \theta \rightarrow x \epsilon Z)$

We shall extend q to r so that for all $x \epsilon V$, $r \Vdash \theta$ or $r \Vdash \neg\, \theta$. We may then choose

$$Z = \{x \epsilon V \mid r \Vdash \theta\} \quad ,$$

in which case 1'), 2') will evidently be satisfied.

Of course, for each x we can extend r to r_x which settles θ. By DC we may thus form a sequence p_α ($\alpha \in OR$) such that if $\alpha<\beta$, $p_\alpha \le p_\beta$, and for each x some p_α settles θ for x: it is only necessary to enumerate V on OR, and observe that for any $\lambda \in OR$, we can amalgamate all the $p_\alpha (\alpha<\lambda)$ into one condition p_λ which extends each of the p_α. The formula $\theta(X,Y)$ to which we apply DC can say for example that if X is a relation with domain an ordinal λ, so that the $X''\{\alpha\} = p_\alpha$ settle the X_α and are increasing; then Y has domain $\lambda+1$ and continues X. To complete the proof it is only necessary to do the same for the full sequence p_α. This however is readily accomplished; to form W_r, we join all the $W_\alpha = W_{p_\alpha}$, using an indexing of ORxOR by OR. The Z_r is formed by assigning corresponding Z_α's under this correspondence. □

While this result is very nice from the point of view of someone who wishes to work in KM with DC, and also wishes to introduce equivalence types, it has the unappealing feature that it is surprisingly specific to KM. It does not necessarily carry over in the natural way to extensions of KM. For example, consider the strong set theory B of Bernays. This is an extension of KM which may be obtained by adding a reflection scheme asserting that any statement of KM true in the universe is reflected in some $V_{\kappa+1}$ $(V = \{V_\alpha | \alpha \in OR\})$: $x \in V \; \& \; \theta(x,X) \to \exists \kappa \in OR$ $\theta^{V_{\kappa+1}}(x, X \cap V_\kappa)$. In B, DC is provable. Thus, by the above theorem we can use types: but this provides no justification for allowing them in the reflection principle, only in the comprehension schema. The argument does not carry over. Thus the theorem shows that adding types does not strengthen KM very much, namely no more than DC does, but it does not say much about extensions of KM. Since DC is already present in B, B already does everything that can be done using types in KM. ($KM + \tau \vdash \sigma \implies KM + DC + \tau \vdash \sigma \implies KM + DC \vdash \sigma \implies B \vdash \sigma$). This fact, however, can be obtained from B more directly and we present the argument because it is suggestive. The idea is that equivalence types over a domain may always be added by the use of higher type objects (classes of classes, in the case of the domain of classes). But the reflection principles give the effect of being able to continue with such a process indefinitely. In particular, if $KM + \tau \vdash \sigma$, we may in B choose κ so that $V_{\kappa+1} \models KM$, and also so that $V_{\kappa+1} \models \sigma$ iff σ. But then there is $T : V_{\kappa+1} \to V_{\kappa+1}$ such that

$$(V_{\kappa+1}, \in, T) \models KM + \tau .$$

Hence $V_{\kappa+1} \models \sigma$, if σ only involves ϵ. This, however, shows that $B \vdash \sigma$. It does not answer the question whether $B + \tau$ is a conservative extension of B. We do however, have the following result.

THEOREM. Let B_μ be like Bernays strong set theory B, except that for each natural number n, we assume the reflection can always be made to a measureable cardinal of degree n. Then $B_\mu + \tau$ is a conservative extension of B_μ.

Note 1. We must explain in what sense τ is allowed in the reflection schema. The simplest way to formulate this is the following. We add not only τ but, for each κ<OR, τ_κ. (We can suppose τ is a well order of the classes; the strongest type existence statements are implied by this). The reflection principle is then:

$$\Theta(\tau,x,X) \rightarrow \exists\kappa \in MC^n \quad \Theta^{V_{\kappa+1}}(\tau_\kappa,x,X\cap V_\kappa).$$

Note 2. A measurable cardinal is of degree 1 if it has a normal measure which assigns the set of measurable cardinals measure 1; of degree n+1 if it admits a measure concentrating on measurable cardinals of degree n.

PROOF. We note that if κ is a MC, then it satisfies the Bernays reflection scheme (in fact it is universal 3rd order indescribable in the sense of Hanf and Scott [1961]). Let κ be of degree n+1. It will suffice to show that we can adjoin T, T_ν (ν<k) which satisfy the reflection to degree n scheme in $V_{\kappa+1}$. (Given a proof in $B_\mu + \tau$, it will involve only reflection to a certain degree, with ϵ-formulas of a fixed complexity; if n is a bound for each of these, and we choose a κ which reflects complexity n and is moreover of degree n+1, then in ($V_{\kappa+1}$, , T,T_ν) the proof is valid and hence the ϵ-formulas in question are true, inheriting this from $V_{\kappa+1}$). To see this, we take the T_ν to be arbitrary well orders of $V_{\nu+1}$, and consider the ultraproduct of the ($V_{\nu+1}$,ϵ,T_ν) by a measure which concentrates on degree n. This has the form ($V_{\kappa+1}$,ϵ,T); this is our choice of T. Now a statement Θ(T,x,X) is true in $V_{\kappa+1}$ iff it is true in almost all ν that $\Theta(T_\nu,x,X\cap V_\nu)$ holds in $V_{\nu+1}$. Since almost all ν are of degree n, this shows that $V_{\kappa+1}$ satisfies reflection to degree n when we adjoin T,T_ν. This completes the argument. □

It would be desirable to have a result of the above sort which is stable under extensions of the base theory. Of course, if we make restrictive assumptions, such as "all classes are constructible", then we can actually define equivalence types and this is perfectly stable under further extensions of the theory. There seems to be no reason why such a phenomenon could not occur with suitable nonrestrictive assumptions. However, we only have some results of the following form: if a definition of types is possible at all, then there is a definition of a specified kind.

We were led to the following idea by an analysis of Chuaqui's definition of well order types and levels in the ramified hierarchy in [1980]. Well order types seemed to be distinguished by finding some difference in the complete theory of the well order, or rather of the associated segment of the ramified hierarchy, allowing parameters for all the sets. When such differences ceased to appear, one obtained sufficiently large order types that a version of the elementary substructure argument established class comprehension. But of course, this is not surprising, as having the same complete theory entails comparatibility under elementary substructure.

DEFINITION. Let α be a well ordered class. By the structure $A(\alpha)$ we understand $(V\dot{\cup}\alpha, \epsilon_V, <_\alpha, x)_{x\in V}$. Here $\dot{\cup}$ is disjoint union. Note that the coding of α as a well order of V is unavailable in $A(\alpha)$. We put

$$\tau_n(\alpha) = \text{the } \Delta^1_n \text{ theory of } A(\alpha).$$

Note that if $\alpha \simeq \beta$, then $\tau(\alpha) = \tau(\beta)$ since the two structures $A(\alpha)$, $A(\beta)$ are isomorphic. Also note that truth for Δ^1_n sentences over V is definable in KM.

THEOREM. If there is a definition of types τ in KM, then for some n, $\tau_n(\alpha)$ is such a type.

PROOF: If τ is defined by a Δ^1_n formula, then the statements $x \in \tau(\alpha)$ for $x \in V$ are part of the Δ^1_n theory: to see this, note that $\exists W \subseteq V(W \simeq \alpha \,\&\, x \in \tau(W))$ is expressible in $A(\alpha)$. Hence if $\tau_n(\alpha) = \tau_n(\beta)$, $\tau(\alpha) = \tau(\beta)$, so $\alpha \simeq \beta$ by the assumption that τ is a type. □

Here is another result of the sort "if anything works, this will". This time, however, we need a theory stronger than KM.

THEOREM (of B): Let $F : On \to V$, and let $\mathcal{F}$ be a definable filter on On.

Define

$$\tau_F(X) = \{t\in V|\ \exists S\ \ \in F\{\kappa|t\in F_\kappa(X\cap R_\kappa)\}\}$$

$$= \{t\in V|(Fa.e.\kappa)\ (t\in F\ (X\cap R\ ))\}$$

Suppose that τ is a definition of types. Then there are F,F such that $\tau(X) = \tau_F(X)$.

PROOF: Suppose that τ is given by a Δ^1_n definition.

Let F be the filter generated by the S such that for some $X\subseteq V$,

$$S = \{\kappa(R_{\kappa+1},\epsilon,X\cap R_\kappa) \underset{n}{<} (\text{Univ},\epsilon,X)\}$$

Let $F_\kappa(X) = \tau^{R_{\kappa+1}}(X)$ for $X\subseteq R_{\kappa+1}$

Now $t\in\tau(X) \rightarrow$ on an F class, $t\in\tau^{R_{\kappa+1}}(X\cap R_\kappa)$, i.e. $t\in F_\kappa(X\cap R_\kappa)$ and $t\notin\tau(X) \rightarrow$ similarly $t\notin F_\kappa(X\cap R_\kappa)$. Consequently $\tau(X) = \tau_F(X)$ as claimed. □

Note: Argument works in KM if definition of σ is simple enough (Σ^1_1).

In conclusion we would like to indicate how to use the definition of types as complete theories to obtain the following conservative extension theorem. Note that while no DC is required, only sentences relative to sets are conserved. Here σ^V indicates the result of relativizing all quantifiers in σ to V.

THEOREM. $KM + \tau \vdash \sigma^V$ implies $KM \vdash \sigma^V$, i.e. adding types is conservative over the set part of KM.

Note: This follows from the development of constructible classes in KM. That such a development can be carried out was noticed by several people (including Solovay, Marek, Chuaqui), often in connection with extending Cohen's methods to class set theory. The matter is discussed in print in Mostowski and Marek [1974] and Chuaqui [1980]. See also Chuaqui and Marshall [1978].

PROOF. For this discussion, let us write $\tau_o(\alpha)$ for the first order theory of $(V \cup (V\times\alpha), \in_V, E_\alpha, x)_{x\in V}$, where E_α is the ϵ-relation for $L_\alpha(V)$ imposed on $V \times \alpha$. (The argument is relatively intensive to the particular version of the ramified hierarchy used). We can now define "small" well orders.

$$\text{Peq}(\alpha)\ \ \text{iff}\ \neg\ \exists\beta,\gamma(\beta<\gamma\leq\alpha\ \&\ L_\beta(V) < L_\gamma(V)\ )$$

and "large"

$Gd(\alpha)$ iff not $Peq(\alpha)$;

we then have the lemma:

LEMMA: $\forall\alpha,\beta(Peq(\alpha) \rightarrow (\alpha \approx \beta \leftrightarrow \tau_o(\alpha) = \tau_o(\beta))$

Thus: either all α are small, and τ_o is a type; or else some α is large, and we easily get comprehension for some $L_\beta(V)$, $\beta<\alpha$. To prove the lemma, it will suffice to suppose $\alpha<\beta$ and $\tau_o(\alpha) = \tau_o(\beta)$, and show $Gd(\beta)$. The argument assumes familiarity with constructible sets, but uses nothing which presents special problems for proper classes. We may suppose that α is minimal, and β also. Let D be the definable elements of $L_\alpha(V)$ (allowing parameters from V). Then $D<L_\alpha(V)$; but by a collapsing argument of the usual sort $D \approx L_{\alpha'}(V)$. Thus $L_\alpha \quad (V) \approx < L_\alpha(V)$, with say κ as critical point. Thus $L_\kappa(V) < L_{\kappa'}(V)$, with $\kappa<\kappa'$. But then $\tau_o(\kappa) = \tau_o(\kappa')$; by the minimality we must have had $L_\alpha(V) < L_\beta(V)$ all along. □

REFERENCES

Chuaqui, R. [1980] Internal and forcing models for the impredicative theory of classes, Dissertationes Mathematicae 176.

Chuaqui, R and Marshall, V. [1978] Constructibility in the impredicative theory of classes, in Mathematical Logic: First Brazilian Conference, ed. Arruda, da Costa and Chuaqui, Marcel Dekker, New York, pp. 179-202.

Hanf, W. and Scott, D. [1961] Classifying inaccessible cardinals, Notices A.M.S. 8, p. 445.

Lévy, A. [1965] Definability in axiomatic theory I, Proc. 1964 Internat.Congress Logic, Meth.and Phil. of Science, ed. Y. Bar-Hillel, North Holland, pp. 127-151.

________ [1963] Independence results in set theory by Cohen's method. IV, Notices Amer.Math.Soc. 10, p. 593.

Marek, W. and Zbierski, P. [1972] Axioms of choice in impredicative set theory, Bulletin d L'Académie Polonaise des Sciences, Série des sciences math., astr. et phys., 20.

Marek, W. and Mostowski, A. [1974] On extendability of ZF models to KM models, Keil conference 1974, Springer Lecture Notes in Mathematics, vol. 499.

Mostowski, A. [1972] Models of second order arithmetic with definable Skolem functions, Fundamenta Math. 75, pp. 223-234.

GENERIC EXTENSIONS WHICH DO NOT ADD RANDOM REALS

Jacques Stern
Université de Caen
Département de Mathématiques et de Mécanique
14032 Caen Cedex, France

We give a sufficient condition ensuring that a partially ordered set $\mathbb{P}$ has the property that no random real appears in the corresponding generic extensions. As an application we show that, provided ZF is consistent, the statement : "there is a (light face) Σ^1_2 non measurable set" is consistent with various other statements like:

i) Every $\underset{\sim}{\Sigma}^1_2$ set has the property of Baire.
ii) Every $\underset{\sim}{\Sigma}^1_2$ set is Ramsey.
iii) Every set has the property of Baire (in presence of the restricted axiom of choice DC).

1. NOTATIONS AND RESULTS.

1.1 In order to state our main technical result, we will use the framework of topological ordered sets. This framework was introduced by the author [7], in order to give an alternative proof of the following result of Shelah [5].

THEOREM 1. If the theory ZF is consistent, then the theory ZF+DC+ "every set of reals has the property of Baire" is consistent as well.

A topological ordered set is a partially ordered set endowed with a topology (which is not necessarily Hausdorff). A topological ordered set $\mathbb{P}$ is separable if it has a countable basis of open sets.

DEFINITION. An open subset V of a topological ordered set is coherent if any finite subset of V admits a common lower bound. A topological ordered set is locally coherent if it has a basis of coherent open sets. In this paper we shall be mainly interested in partially ordered sets which can be endowed with a topology which makes them separable and locally coherent. Actually these partially ordered sets have been considered in the literature. Recall that a partially ordered set $\mathbb{P}$ is σ-centered if it is a countable union of subsets (X_n) such that, for every n, finitely many members from X_n admit a common lower bound. Now, it quickly follows from this definition that the set $\mathbb{P}$, endowed with the topology generated by the sets (X_n), is separable and locally coherent. Conversely, any separable, locally coherent topological ordered-set is σ-centered and therefore the two notions coincide.

From the previous remark, it is clear that the whole paper could have been written using the notion of σ-centered partially ordered set, with some care (and some work) in order to handle iterations; the choice that was made reflects both the taste of the author and his conviction that the alternative approach is useful.

1.2 The following will be our main technical tool:

PROPOSITION. Let $\mathbb{P}$ be a separable, locally coherent, topological ordered set; then, in a generic extension corresponding to $\mathbb{P}$, no real is random over the ground model.

It should be said that our condition was inspired by several arguments previously used by A. Miller [2]. The proof of the proposition will appear below in section 2.

Using the remarks of section 1.1, the above proposition can also be read as follows:

PROPOSITION. If $\mathbb{P}$ is a σ-centered partially ordered set, then, in a generic extension corresponding to $\mathbb{P}$, no real is random over the ground model.

1.3 It is well known that if no real is random over L, then there is a (light face) Σ^1_2 set which is not Lebesgue measurable. In fact, any real α belongs to some Borel set of zero-measure with a code δ in L and we can pick the first such code $\delta(\alpha)$, with respect to the canonical well ordering of L. The set

$$\{(\alpha,\beta) \ : \ \delta(\alpha) \leq \delta(\beta)\}$$

is a Σ^1_2 non measurable subset of $\mathbb{R}^2$; from this set it is not difficult to build a Σ^1_2 non-measurable subset of $\mathbb{R}$.

From the previous remark, it is clear how proposition 1.2 can be used to build models, in which a non-measurable Σ^1_2 set exists. In this way, we prove:

THEOREM. Assume there is a standard model of ZF; then, there is a standard model of ZFC with the following properties

i) Every $\utilde{\Sigma}^1_2$ set has the property of Baire

ii) Every $\underset{\sim}{\Sigma}^1_2$ set is Ramsey

iii) There is a non measurable Σ^1_2 set.

This theorem should be compared with a result of Raisonnier and the author, which states that the measurability of all $\underset{\sim}{\Sigma}^1_2$ sets implies that all $\underset{\sim}{\Sigma}^1_2$ sets have the property of Baire as well [4]. The above theorem shows - inter alia - that there is no converse implication. This is not really new; the author had already observed that, in Shelah's model where all sets have the property of Baire [5], some $\underset{\sim}{\Sigma}^1_2$ set was not measurable; but this was through an indirect argument using some work of Raisonnier [3]; the proof that appears in the present paper is straightforward and gives a result which is more effective in character.

1.4 Combining proposition 1.2 with the construction of Shelah [5], we show that no hypothesis concerning the Baire property is strong enough to imply the measurability of Σ^1_2 sets.

THEOREM. Assume there is a standard model of ZF; then, there is a standard model of ZF+DC, such that:

i) all sets of reals have the property of Baire.

ii) there is a non-measurable Σ^1_2 set.

2. THE MAIN RESULT.

2.1 We start to give the proof of proposition 1.2; we let $\mathbb{P}$ be a separable, locally coherent, topological space. We want to show that any real which appears in some generic extension corresponding to $\mathbb{P}$ belongs to a Borel set of zero-measure coded in the ground model M. As usual, we don't work with actual real numbers but with elements of 2^ω; 2^ω is endowed with the product measure μ of the uniform probability measure on $\{0,1\}$. Any element of 2^ω in the generic extension is denoted by some term τ such that the boolean value

$$[\![\tau \text{ is an element of } 2^\omega]\!]$$

equals $\mathbb{1}$. We fix such a term τ.

LEMMA. For any coherent open subset V of $\mathbb{P}$ and any integer n, there exists an element s of 2^n, such that the following holds:

(1) $\forall p \in V \ \exists q \leqslant p \quad q \Vdash \tau$ extends s.

PROOF OF THE LEMMA. We fix an enumeration $s_1,\ldots,s_i,\ldots s_k$ of the elements of 2^n; if the conclusion of the lemma does not hold, then, for any index i, there is an element p_i of V such that

$\forall q \leq p_i \quad q \not\Vdash \tau$ extends s_i

which means

$p_i \Vdash \tau$ does not extend s_i;

now, any common lower bound to all the p_i's forces

τ does not extend any element of 2^n;

this gives a contradiction and finishes the proof of the lemma.

2.2 We now use the fact that $\mathbb{P}$ has a countable basis of coherent open sets; we fix an enumeration (V_n) of such a basis with the property that any element of the basis appears infinitely many times. Using the lemma, for such interger n, we can find an element s_n of 2^n such that

(2) $\forall p \in V_n \quad \exists q \leq p \quad q \Vdash \tau$ extends s_n.

We note that all this takes place in the ground model; still working in the ground model, we let

$$U_n = \{\alpha : \alpha \text{ extends } s_n\}$$

$$B = \limsup (U_n) = \bigcap_m \bigcup_{n \geq m} U_n.$$

Now, it is well known that if the series

$$\sum_n \mu(U_n)$$

converges, then, $\limsup (U_n)$ is of zero-measure; thus, B is a G_δ set of zero-measure coded in the ground model. Now, in any generic extension there is a G_δ set of zero-measure $\hat{B}$ which is built like B (in the sense that it has the same code). We claim that if β is the interpretation of the term τ, then β belongs to $\hat{B}$. If this is not the case, there exists a condition p of $\mathbb{P}$ and an integer m such that

$$p \Vdash \tau \notin \bigcap_{n \geq m} U_n.$$

This means

(3) $\forall n \geq m \quad b \Vdash \tau$ does not extend s_n.

Now, we pick an open neighborhood V of p in the basis that we have fixed; V appears infinitely many times in the sequence (V_n) so that we can pick an integer $n \geq m$ for which $V_n = V$. From property (2) above, we get, for some $q \leq p$:

$q \Vdash \tau$ extends s_n

but this contradicts (3). So we have shown that any real β, which appears in a generic extension, belongs to some G_δ set of zero measure coded in the ground model M, hence is not random over M.

3. LOCALLY COHERENT TOPOLOGICAL ORDERED SETS.

3.1 We now study the closure properties of the class of coherent topological ordered sets. This will be useful to give the proof of theorem 1.3. We first note the following:

PROPOSITION. Any separable, locally coherent, topological ordered set satisfies the c.c.c.

PROOF. If (V_k) is a sequence of coherent open sets which is a countable basis for $\mathbb{P}$, then, any set of cardinality $\aleph_1$ has two different members in some V_k, hence, it is not an antichain.

3.2 UD

UD is a set of conditions used to add a new element of ω^ω dominating all members of ω^ω lying in the ground model. UD consists of pairs (k,f) with $k\in\omega$ and $f\in\omega$; the ordering on UD is given by:

$(\ell,g) \leq (k,f)$ if ℓ exceeds k, $f\restriction k = g\restriction k$ and $\forall n\ f(n) \leq g(n)$

where $f\restriction k$ denotes the restriction of f to integers $<k$. UD is endowed with the topology induced by the usual topology on $\omega\times\omega^\omega$. Clearly, this topology has a countable basis; furthermore, if a condition (k,f) is given, then

$\{(k,g) \ : \ g\restriction k = f\restriction k\}$

is a coherent open set containing the given condition (k,f); thus, we have proved the following:

LEMMA. UD is a separable, locally coherent, topological ordered set.

3.3 Composition with UD.

Let $\mathbb{P}$ be a separable, locally coherent, topological ordered set; as in Stern [7], we will endow the one step iteration of $\mathbb{P}$ and UD with a topology. Actually, we will use a different topology but it turns out that this gives a simpler and more straightforward definition which is enough for our purposes.

Recall that $\mathbb{P} \tilde{\otimes}$ UD consists of triples (p,k,τ) where

i) $p \in \mathbb{P}$

ii) $k \in \omega$

iii) τ is a term of the forcing language and $p \Vdash \tau$ is a mapping from the integers to the integers

iv) for any $i<k$, $p \Vdash \tau(i) = j$ for some j.

We identify conditions (p,k,τ) and (p,k,σ) whenever $p \Vdash \tau = \sigma$.

The ordering on $\mathbb{P} \tilde{\otimes}$ UD is defined by

$(q,\ell,\sigma)\leq(p,k,\tau)$

if $q\leq p$, $\ell \geq k$ and

$q \Vdash \forall n \quad \sigma(n) \geq \tau(n)$
$q \Vdash \forall i<k \ \sigma(i) = \tau(i).$

We then define subsets of $\mathbb{P} \tilde{\otimes}$ UD by :

$$\Lambda\ (V,k,u,v) \ = \{(p,k,\tau) \ : p\in V \ \text{ and } \ p \Vdash \tau(u) = v\}$$
$$\Gamma\ (V) \ = \{(p,0,\tau) \ : p\in V \ \text{ and } \ p \Vdash \forall n\ \tau(n) = 0\}$$

where V is an open subset of $\mathbb{P}$, and k,u,v are integers; finally, we endow $\mathbb{P} \tilde{\otimes}$ UD with the topology generated by all these subsets. Actually, the same topology is obtained if we restrict the open set V to range over a countable basis of $\mathbb{P}$; from this, it follows that the topology defined on $\mathbb{P} \tilde{\otimes}$ UD is separable. Actually, we have

LEMMA: $\mathbb{P} \tilde{\otimes}$ UD is a separable, locally coherent, topological ordered set.

PROOF: given a condition (p,k,τ), we pick a coherent open subset V of $\mathbb{P}$ containing p; now for any integer $i<k$ there is an integer $j(i)$, such that:

$$p \Vdash \tau(i) = j(i)$$

we claim that the intersection U of the sets

$$\Lambda\ (V,k,i,j(i)) \quad i<k$$

is a coherent open neighborood of (p,k,τ); indeed if conditions $(p_1,k,\tau_1),\ldots,(p_r,k,\tau_r)$ are given in U, we can find a common lower bound q to $p_1,\ldots,p_r$; if σ is a term of the forcing language such that

$$q \Vdash \sigma \text{ is the supremum of } \tau_1,\ldots,\tau_r$$

then, it is easy to check that condition (q,k,σ) lies below all conditions $(p_1,k,\tau_1),\ldots,(p_r,k,\tau_r)$.

We now define an embedding k from $\mathbb{P}$ into $\mathbb{P} \tilde{\otimes}$ UD by:

$$k(p) = (p,0,\nu)$$

where ν is a term of the forcing language such that

$$[\![\forall n \quad \nu(n) = 0]\!] = \mathbb{1}.$$

It is not difficult to verify that k has the following properties:

i) k is a homeomorphism from $\mathbb{P}$ onto an open subset of $\mathbb{P} \tilde{\otimes}$ UD

ii) k is increasing

iii) if p,p' are incompatible, then, $k(p)$ and $k(p')$ are incompatible in $\mathbb{P} \tilde{\otimes}$ UD.

iv) if $\mathfrak{U}$ is a maximal antichain in $\mathbb{P}$, $k(\mathfrak{U})$ is a maximal antichain in $\mathbb{P} \tilde{\otimes}$ UD.

Embeddings satisfying properties i) to iv) are called sweet in [7] and we keep this terminology.

3.4 Composition with Mathias forcing.

As in the previous section, $\mathbb{P}$ is a separable, locally coherent, topological ordered set; we assume that some term $\mathcal{U}$ of the forcing language is such that

$$[\![\mathcal{U} \text{ is a Ramsey ultrafilter on } \omega]\!] = \mathbb{1}$$

We then consider the one step iteration of $\mathbb{P}$ and Mathias forcing (see Mathias [11]), which we denote by $\mathbb{P} \tilde{\otimes} M_{\mathcal{U}}$; $\mathbb{P} \tilde{\otimes} M_{\mathcal{U}}$ consists of triples (p,s,τ) such that

i) $p \in \mathbb{P}$

ii) s is a finite sequence of integers

iii) τ is a term of the forcing language and $p \Vdash \tau \in \mathcal{U}$.

We identify conditions (p,s,τ) and (p,s,σ) whenever $p \Vdash \tau = \sigma$. The ordering on $\mathbb{P} \tilde{\otimes} M_{\mathcal{U}}$ is defined by

$$(q,t,\sigma) \leqslant (p,s,\tau)$$

if $q \leqslant p$, t extends s, $q \Vdash t-s \subseteq \tau$ and $q \Vdash \sigma \subseteq \tau$.

In order to define a topology, we consider the sets

$$\Lambda(V,s) = \{(p,s,\tau) : p \in V\}$$
$$\Gamma(V) = \{(p,\emptyset,\tau) : p \in V \text{ and } p \Vdash \tau = \omega\}.$$

where V is an open set of $\mathbb{P}$ and s a sequence of integers; we endow $\mathbb{P} \tilde{\otimes} M_{\mathcal{U}}$ with the topology generated by all these subsets. As in the previous section, we can restrict the open sets V to range over a countable basis of $\mathbb{P}$, hence, the topology is separable.

LEMMA. $\mathbb{P} \tilde{\otimes} M_{\mathcal{U}}$ is a separable, locally coherent topological ordered set.

PROOF. Given a condition (p,s,τ), we pick a coherent open subset V of $\mathbb{P}$ containing p; we claim that $\Lambda(V,s)$ is a coherent open neighborhood of (p,s,τ); indeed, if conditions $(p_1,s,\tau_1),\ldots,(p_r,s,\tau_r)$ are given in $\Lambda(V,s)$ we can find a common lower bound q to $p_1,\ldots,p_r$; if σ is a term of the forcing language such that

$$q \Vdash \sigma = \bigcap_{i=1}^{r} \tau_i$$

then (q,s,σ) lies below all conditions $(p_1,s,\tau_1),\ldots,(p_r,s,\tau_r)$.

We finally define an embedding $k : \mathbb{P} \to \mathbb{P} \tilde{\otimes} M_{\mathcal{U}}$ by

$$k(p) = (p,\phi,\omega) ;$$

it is not difficult to check that it is a sweet embedding.

4. GENERIC MODELS WITHOUT RANDOM REALS.

4.1 In order to prove theorem 1.3, we start with a model of ZFC+V = L and, in this model, we define a transfinite sequence of Length $\aleph_1$ of separable, locally coherent topological ordered sets ($\mathbb{P}_\xi$), together with sweet embeddings $k_{\xi\zeta}: \mathbb{P}_\xi \to \mathbb{P}_\zeta$, $\xi<\xi<\aleph_1$ such that whenever $\xi<\zeta<\rho$ $k_{\xi\rho}= k_{\zeta\rho}k_{\xi\zeta}$. The definition is by transfinite induction:

i) $\mathbb{P}_o$ is UD

ii) if ξ is even $\mathbb{P}_{\xi+1}$ is $\mathbb{P}_\xi \tilde{\otimes}$ UD, where $\mathbb{P}_\xi \tilde{\otimes}$ UD is endowed with the topology defined in section 3.3. and $k_{\xi\xi+1}$ is the embedding defined in the same section; $k_{\zeta\xi+1}$ if defined by composition for $\zeta<\xi$.

iii) if ξ is odd, we note that since $\mathbb{P}_\xi$ is a separable locally coherent space, it is c.c.c and therefore the continuum hypothesis holds in any generic extension of the ground model via a generic subset of $\mathbb{P}_\xi$; as is well-known, the existence of a Ramsey ultrafilter follows from CH; using the so-called maximum principle we can find a term $\mathcal{U}$ in the forcing language corresponding to $\mathbb{P}_\xi$ such that:

$$[\![\mathcal{U} \text{ is a Ramsey ultrafilter}]\!] = \mathbf{1};$$

we then define $\mathbb{P}_{\xi+1}$ to be $\mathbb{P}_\xi \tilde{\otimes} M_{\mathcal{U}}$; the topology on $P_{\xi+1}$ and the embedding $k_{\xi\xi+1} : \mathbb{P}_{\xi+1} \to \mathbb{P}_{\xi+1}$ are defined according to section 3.4; $k_{\zeta\xi+1}$ is defined by composition for $\zeta<\xi$.

iv) if λ is a limit ordinal, $\mathbb{P}_\lambda$ is the direct limit of the system given by ($\mathbb{P}_\xi)_{\xi<\lambda}$ together with the maps $(k_{\xi\zeta})_{\xi<\zeta<\lambda}$; $(k_{\xi\lambda})$ are the limit maps. The topology on $\mathbb{P}_\lambda$ is the limit topology whose open sets are exactly the sets $X \subseteq \mathbb{P}_\lambda$ such that

$$\forall\xi<\lambda \quad k^{-1}_{\xi\lambda} (X) \text{ is open.}$$

<u>LEMMA</u>. $\mathbb{P}_\lambda$ is a separable, locally coherent topological ordered set; furthermore, the limit maps $k_{\xi\lambda}$ are sweet embeddings.

<u>PROOF.</u> We note that, for any $\xi<\lambda$, $k_{\xi\lambda}(\mathbb{P}_\xi$) is an open subset of $\mathbb{P}_\lambda$; this is because $k_{\xi\zeta}$ is an embedding from $\mathbb{P}_\xi$ onto an <u>open</u> subset of $\mathbb{P}_\zeta$; from this, it follows that $\mathbb{P}_\lambda$ has a countable basis; a basis for $\mathbb{P}_\lambda$ is obtained in the following way : take a countable basis for each $\mathbb{P}_\xi$; apply $k_{\xi\lambda}$; take the union of all these sets. Similarly, the local coherence of $\mathbb{P}_\lambda$ stems from the fact that $\mathbb{P}_\lambda$ is the union of the open sets $k_{\xi\lambda}$ ($\mathbb{P}_\xi$). The fact that $k_{\xi\lambda}$ is an homeomorphism of $\mathbb{P}_\xi$ onto an open subset of $\mathbb{P}_\lambda$ is proved by a similar argument. There

remains to check conditions ii) to iv) of the definition of sweet embeddings; but these are standard facts on direct limits.

4.2 We let $\mathbb{P}$ be the direct limit of the ordered sets ($\mathbb{P}_\xi$) and maps $k_{\xi\zeta}$ defined in the previous section. $\mathbb{P}$ is c.c.c so that in the corresponding generic extension M[G], cardinals are preserved and the continuum hypothesis holds; also, any real of M[G] appears in an intermediate generic extension obtained by forcing with some $\mathbb{P}_\xi$. Because $\mathbb{P}_\xi$ is a separable, locally coherent space, no random real can be found in such an intermediate extension; finally, no random real can exist in M[G], so that there is a Σ^1_2 non-measurable set in this model. We now prove the following.

<u>PROPOSITION.</u> In the model M[G], every $\underset{\sim}{\Sigma}^1_2$ set has the property of Baire.

<u>PROOF.</u> It is well known that $\underset{\sim}{\Sigma}^1_2$ sets have the property of Baire if and only if, for any real α, there exists a dense G_δ of cohen generic reals over $L[\alpha]$ (see eg [6]). Also, it is known that any model obtained by performing iterations of length $\aleph_1$ of c.c.c ordered sets has this property, as long as iterations with UD appear cofinally (for a proof see [7] section 1.3.2). From these observations, the proposition follows.

4.3 We now turn to the Ramsey property;

<u>PROPOSITION.</u> In the model M[G], every $\underset{\sim}{\Sigma}^1_2$ set is Ramsey.

<u>PROOF.</u> Let $\phi(x,\alpha)$ be a Σ^1_2 statement with a real parameter α; this parameter appears in an intermediate generic extension corresponding to some $\mathbb{P}_\xi$ and we may assume that ξ is odd so that $\mathbb{P}_{\xi+1} = \mathbb{P}_\xi \tilde{\otimes} M_{\mathcal{U}}$, where $\mathcal{U}$ is a term denoting a Ramsey ultrafilter. We let M_ξ (resp. $M_{\xi+1}$) be the generic extension corresponding to $\mathbb{P}_\xi$(resp. $\mathbb{P}_{\xi+1}$). We now work in M_ξ; $M_{\xi+1}$ is obtained from this model by forcing with Mathias forcing [1]; we let γ be a term of this forcing denoting the generic subset of ω wich is added; it is well known that for some condition of type (0,a),

(1) $(0,a) \Vdash \phi(\gamma,\hat{\alpha})$ or $(0,a) \Vdash \neg\phi(\gamma,\hat{\alpha})$

where $\hat{\alpha}$ is a term denoting α. Now, it is known also that, if g is a generic subset of ω added via Mathias forcing, any infinite subset of g is generic as well. From these observations it follows that

i) either any infinite subset g' of $g \cap a$ is such that

$$M_\xi(g') \models \phi(g',\alpha).$$

ii) or else any such infinite subset is such that

$$M_\xi(g') \models \neg\, \phi(g',\alpha).$$

By Shoenfield absoluteness lemma, we get, in the first case, that the following holds in M[G]

i) $P_\infty(g\cap a) \subseteq \{x : \phi(x,\alpha)\}$

where $P_\infty(g\cap a)$ is the set of infinite subsets of $g\cap a$; in the second case, we get similarly:

ii) $P_\infty(g\cap a) \subseteq \{x : \neg\phi(x,\alpha)\}$;

thus, the Ramsey property for $\underset{\sim}{\Sigma}^1_2$ sets is established.

4.4 We now turn to the proof of theorem 1.4.

If we look at Shelah's paper [5] we realize that this model of ZF+DC where all sets of reals have the property of Baire is built by a transfinite induction of length $\aleph_1$, similar to the one we have performed in section 4.3 but -of course- more elaborated; we also realize that the intermediate partially ordered set built after each countable step of the interaction is σ-centered; from these observations, theorem 1.4 follows.

For those who are interested in the approach via topological ordered sets, we will sketch an alternative proof based on our paper [7]. Contrary to what we have tried to do in the rest of the paper, it will not be possible to make this section self-contained; therefore, we assume that the reader is familiar with [7] and with the notion of sweet topological ordered set, which appears there.

DEFINITION. An ordered topological space is supersweet if it sweet and satisfies the following property:

(*) given conditions $p,q,q \leqslant p$, together with a neighborhood V of q, there exists a neighborhood W of p, such that any finite number of elements of W have a common lower bound in V.

Clearly, any supersweet space is locally coherent.

In order to prove theorem 1.4, one has to go through the constructions of [7] and check that they can be performed using supersweet ordered sets instead of sweet ones. This amounts to showing that supersweet sets are closed under amalgamation, composition with UD and direct limits (under sweet maps). Direct limits and amalgamation are easily handled so we restrict ourselves to composition. We refer to §3 of [7] and recall that the topology on $\mathbb{P} \overset{\sim}{\otimes}$ UD, which is defined there and which we consider now, is not the one which appears above in section 3.3. We claim that if $\mathbb{P}$ is supersweet, the same is true of $\mathbb{P} \tilde{\otimes}$ UD. We note that by section 1.1 of [7] (proposition 1), it is enough to verify the following weakened form of (*):

(**) For any condition p and any neighborhood V of p, there exists a neighborhood W of p such that any finite number of elements of W have a common lower bound in V.

Following the notations of [7], we consider a condition (p,k,τ), together with an open neighborhood V; we may assume that V is the intersection of some $\tilde{B}(i,k)$ with finitely many open sets $\Lambda(i_1,n_1),\ldots,\Lambda(i_s,n_s)$, plus possibly the set $k(\mathbb{P})$; we first choose extensions $q_1,\ldots,q_s$ of p lying in $B_{i_1},\ldots B_{i_s}$ respectively and such that $q_1 \Vdash \phi_{n_1}(\tau),\ldots,q_s \Vdash \phi_{n_s}(\tau)$; then, we pick open neighborhoods $B_{j_1},\ldots,B_{j_s}$ of $q_1,\ldots,q_s$ respectively such that:

any finite subset of B_{j_1} has a lower bound in B_{i_1} (similarly for $B_{j_2},\ldots$).

Next, we pick a neighborhood $B_{\bar{I}}$ of p included in B_i such that any element of $B_{\bar{I}}$ has extensions in $B_{j_1},B_{j_2},\ldots,B_{j_s}$. Finally, we choose a neighborhood B_j of p such that any finite subset of B_j has a lower bound in $B_{\bar{I}}$.

We let $W_o = \tilde{B}(j,k)\cap\Lambda(j_1,n_1)\cap\ldots\cap\Lambda(j_s,n_s)$. We let t be a finite set of integers such that $p \Vdash \tau\restriction k = t$; if conditions

$$(p_1,k,\tau_1),\ldots,(p_r,k,\tau_r)$$

are given in W_o, with

(1) $p_1 \Vdash \tau_1\restriction k = t,\ldots,p_r \Vdash \tau_r \restriction k = t$

then, we can build a common lower bound by considering an element q of $B_{\bar{I}}$ below $p,p_1,\ldots,p_r$ and a term σ of the forcing language such that

$$q \Vdash \sigma = \sup(\tau_1,\ldots,\tau_r).$$

<u>CLAIM 1:</u> Condition (q,k,σ) belongs to $\tilde{B}(i,k)$.

<u>PROOF OF CLAIM</u>: this is because $q \in B_{\bar{I}} \subseteq B_i$.

<u>CLAIM 2:</u> Condition (q,k,σ) belongs to $\Lambda(i_1,n_1),\ldots,\Lambda(i_s,n_s)$.

<u>PROOF OF CLAIM.</u> We focus on $\Lambda(i_1,n_1)$ and we write n for n_1 as no confusion can arise. We pick extensions $p'_1,\ldots,p'_r$ of $p_1,\ldots,p_r$ respectively, lying in B_{j_1} and forcing $\phi_n(\tau_1),\ldots,\phi_n(\tau_r)$; we also pick

an extension p' of q, lying in B_{j_1} : this is possible because q belongs to $B_{\bar{I}}$. Now, any common extension q to $p', p_1, \ldots, p_r$ lying in B_i forces

$$\phi_n(\tau_1)\wedge\ldots\wedge\phi_n(\tau_r)$$

as $\phi_n(\tau)$ is a statement like $\tau(u) \leqslant v$, $\tau(u) \geqslant v$, $\forall m\ \tau(m) = 0$, it is easy to see that

$$q \Vdash \phi_n(\sigma)$$

this shows why (q,k,σ) belongs to $\Lambda(i_1,n)$.

Now, we show that the statement

$$p_1 \Vdash \tau_1 \restriction k = t$$

can be ensured by taking (p_1,k,τ_1) in some open set W_1; for any integer $u<k$, there exists an integer v such that

$$p \Vdash \tau(u) = v;$$

we let W_1 be the intersection of all open sets $\Lambda(j,n)$, when ϕ_n is any of the corresponding statements $\tau(u) \leqslant v$, $\tau(u) \geqslant v$; clearly, if (p_1,k,τ_1) is in W_1 then p_1 can only decide $\tau_1 \restriction k$ to coincide with t.

Finally, we have shown that any finite subset of $W_o \cap W_1$ has a lower bound in V, except if $k(\mathbb{P})$ appears in the definition of V; in this case, we simply replace $W_o \cap W_1$ by $W_o \cap W_1 \cap k(\mathbb{P})$. This completes the proof of theorem 1.4.

REFERENCES

[1] Mathias, A.R.D. Happy families, Annals of Math.Logic, 12(1977)pp. 59-111.

[2] Miller, A.W. Some properties of measure and category. Transactions of the Amer.Math.Soc. 266(1981), pp.93-114.

[3] Raisonnier, J. A mathematical proof of S. Shelah's theorem on the measure problem and related results, Israel J.Math. (to appear).

[4] Raisonnier, J. Stern, J. The strength of measurability hypotheses. (to appear).

[5] Shelah, S. Can you take Solovay's inaccessible away? (to appear)

[6] Stern, J. Some measure theoretic results in effective descriptive set theory, Israel J.Math. 20(1975) pp. 97-110.

[7] Stern, J. Regularity properties of definable sets of reals. (to appear).